Textbook of Zoology

Textbook of Zoology

Edited by **Carlos Wyatt**

Callisto Reference
New York

Published by Callisto Reference,
106 Park Avenue, Suite 200,
New York, NY 10016, USA
www.callistoreference.com

Textbook of Zoology
Edited by Carlos Wyatt

International Standard Book Number: 978-1-63239-701-0 (Hardback)

Printed in China.

Contents

	Preface	**IX**
Chapter 1	**Selective foraging by non-native rainbow trout on invertebrates in Patagonian streams in Argentina** Cecilia Yanina Di Prinzio, Guillermo Omad, María Laura Miserendino and Ricardo Casaux	**1**
Chapter 2	**Description of a new genus and new species of freshwater crab (Brachyura: Gecarcinucidae) from the Western Ghats, Kerala, India** Sameer Kumar Pati and Arath Raghavan Sudha Devi	**15**
Chapter 3	***Cymothoa frontalis*, a cymothoid isopod parasitizing the belonid fish *Strongylura strongylura* from the Malabar Coast (Kerala, India): redescription, description, prevalence and life cycle** Aneesh Panakkool Thamban, Sudha Kappalli, Helna Ameri Kottarathil, Anilkumar Gopinathan and Trilles Jean Paul	**23**
Chapter 4	**Topographical distribution of blubber in finless porpoises (*Neophocaena asiaeorientalis sunameri*): a result from adapting to living in coastal waters** Xianyuan Zeng, Junhua Ji, Yujiang Hao and Ding Wang	**51**
Chapter 5	**Seasonality in anuran activity and calling season in a Brazilian subtemperate wetland** Simone da Silva Ximenez and Alexandro Marques Tozetti	**62**
Chapter 6	**Sexual dimorphism in echolocation pulse parameters of the CF-FM bat, *Hipposideros pratti*** Zi-Ying Fu, Xing-Yue Dai, Na Xu, Qing Shi, Gao-Jing Li, Bin Li, Juan Li, Jie Li, Jia Tang, Philip Hung-Sun Jen and Qi-Cai Chen	**71**
Chapter 7	**Differentiation of visual spectra and nuptial colorations of two *Paratanakia himantegus* subspecies (Cyprinoidea: Acheilognathidae) in response to the distinct photic conditions of their habitats** Chia-Hao Chang, Yi Ta Shao, Wen-Chung Fu, Kazuhiko Anraku, Yeong-Shin Lin and Hong Young Yan	**80**
Chapter 8	**Long-term monitoring of Gastropoda (Mollusca) fauna in planted mangroves in central Vietnam** Sofya Zvonareva, Yuri Kantor, Xinzheng Li and Temir Britayev	**89**
Chapter 9	**A new species of fish-eating rat, genus *Neusticomys* (Sigmodontinae), from Ecuador** J Delton Hanson, Guillermo D'Elía, Sheri B Ayers, Stephen B Cox, Santiago F Burneo and Thomas E Lee Jr	**105**

Chapter 10 **The research on the formation mechanism of extraordinary oxidative capacity of skeletal muscle in hibernating ground squirrels (*Spermophilus dauricus*)** 116
Shanfeng Jiang, Yunfang Gao, Yangmei Zhang, Kun Liu, Huiping Wang and Nandu Goswami

Chapter 11 **The poorly known genus *Ventrifurca* Roewer, 1913 revisited (Opiliones: Cranaidae)** 122
Osvaldo Villarreal M, Adriano B Kury and Ricardo Pinto-da-Rocha

Chapter 12 **Biological mosquito control is affected by alternative prey** 140
Ram Kumar, Priyanesh Muhid, Hans-Uwe Dahms, Jaigopal Sharma and Jiang-Shiou Hwang

Chapter 13 **Spatial heterogeneity of benthic copepods: a comparative aspect on composition, abundance, and correlation** 158
R Jayabarathi, G Padmavati and I Anandavelu

Chapter 14 **Stock assessment and management implications of an endemic fish, *Oxygymnocypris stewartii*, in the Yarlung Zangbo River in Tibet, China** 167
Bin Huo, Bao Shan Ma, Cong Xin Xie, You Jian Duan, Xue Feng Yang and Hai Ping Huang

Chapter 15 **Seasonal dynamics of the copepod community in a tropical monsoonal estuary and the role of sex ratio in their abundance pattern** 182
G. Vineetha, N. V. Madhu, K. K. Kusum and P. M. Sooria

Chapter 16 **The function of body coloration of the hai coral snake *Sinomicrurus japonicus boettgeri*** 201
Koji Mochida, Wan-Yu Zhang and Mamoru Toda

Chapter 17 **Abundance and diversity of amphipods (Crustacea: Peracarida) on shallow algae and seagrass in lagoonal ecosystem of the Mediterranean Tunisian coast** 207
Wahiba Zaabar, Rym Zakhama-Sraieb, Faouzia Charfi-Cheikhrouha and Mohamed Sghaïer Achouri

Chapter 18 **Habitat suitability models of mountain ungulates: identifying potential areas for conservation** 218
Prakash Kumar Paudel, Martin Hais and Pavel Kindlmann

Chapter 19 **Recolonizing wolves influence the realized niche of resident cougars** 234
L Mark Elbroch, Patrick E Lendrum, Jesse Newby, Howard Quigley and Daniel J Thompson

Chapter 20 **Diagnosability and description of a new subspecies of Indo-Pacific humpback dolphin, *Sousa chinensis* (Osbeck, 1765), from the Taiwan Strait** 245
John Y Wang, Shih Chu Yang and Samuel K Hung

Chapter 21 ***Alona kaingang* (Crustacea, Cladocera, Aloninae): a new species of the *pulchella*-group, with identification key to Neotropical species** 260
Francisco Diogo Rocha Sousa, Lourdes Maria Abdu Elmoor-Loureiro and Sandro Santos

Chapter 22 **Morphological variability of the cranium of *Lontra longicaudis* (Carnivora: Mustelidae): a morphometric and geographic analysis** **272**
Pablo César Hernández-Romero, José Antonio Guerrero and Carolina Valdespino

Permissions

List of Contributors

Preface

This book discusses the fundamentals as well as modern approaches of zoology. It delves into the embryology, evolution, habits and distribution of all animals of this planet. It classifies and analyses animals on the bases of physiology, ethology, biogeography, etc. Zoology has many branches such as comparative anatomy, animal physiology, behavioural ecology, etc. This book will provide comprehensive insights into the field of zoology. It will discuss in detail the new methodologies and evaluations related to this subject. Those in search of information to further their knowledge will be greatly assisted by this book. It is a collective contribution of a renowned group of international experts. It will serve as a valuable source of reference to graduate and post-graduate students as well as zoologists, researchers and professionals associated with zoology.

All of the data presented henceforth, was collaborated in the wake of recent advancements in the field. The aim of this book is to present the diversified developments from across the globe in a comprehensible manner. The opinions expressed in each chapter belong solely to the contributing authors. Their interpretations of the topics are the integral part of this book, which I have carefully compiled for a better understanding of the readers.

At the end, I would like to thank all those who dedicated their time and efforts for the successful completion of this book. I also wish to convey my gratitude towards my friends and family who supported me at every step.

Editor

1

Selective foraging by non-native rainbow trout on invertebrates in Patagonian streams in Argentina

Cecilia Yanina Di Prinzio[1,2,3*], Guillermo Omad[1,2,3], María Laura Miserendino[1,2,3] and Ricardo Casaux[1,2,3,4]

Abstract

Background: It is well known that fish predation alters ecosystem processes by top-down effects. Salmonids are described as aggressive, visually and size-selective predators. Thus, prey selection by the non-native rainbow trout was examined on a seasonal basis at two streams: Nant y Fall (NyF) and Cabeza de Vaca (CVA) at Patagonia, a region where this kind of information is lacking.

Results: The benthos density at NyF was higher than that at CVA, and at both streams, riffles supported higher macroinvertebrate densities than pools. The diet of trouts from both streams was dominated by aquatic macroinvertebrates, was diverse, and was varied seasonally. The individuals represented in the stomach contents were among the largest available at the streams. Diet diversity peaked during spring at NyF and during summer at CVA, whereas at both streams, the niche width peaked during spring.
Prey selectivity varied seasonally. The selected preys included both aquatic (Gasteropoda, Crustacea, Plecoptera, Trichoptera, Ephemeroptera, Coleoptera, Diptera, and Odonata) and terrestrial organisms (adult dipterans, Oligochaeta, Araneae, Homoptera, Hymenoptera, Orthoptera, and Hemiptera). Some infaunal invertebrates like oligochaetes and some small Coleoptera and Diptera larvae (mainly Chironomidae) were not selected by trouts.

Conclusions: Despite of the overall dominance of trichopteran species, the composition of the diet of the rainbow trout varied seasonally. This fish positively selected both aquatic and terrestrial organisms. We observed that in both streams, trouts consumed the larger individuals available in those environments.

Keywords: Selective predation; Non-native fish; Stream; Patagonia

Background

In freshwater environments, fish predation is considered an important selective force because it can shape the structure and composition of freshwater communities by top-down effects determining food web functioning and dynamics (Newman and Waters 1984; Nakano et al. 1999; Baxter et al. 2004; Wesner 2012).

It has been suggested that in order to increase fitness, predators actively choose prey that minimizes energy spent on capturing and handling while maximizing energy intake (Stephens and Krebs 1986). Consequently, understanding on the predator–prey relationships is one of the most important issues to comprehend the trophic interactions and their effects (Macchi et al. 1999). Worldwide, many studies have attempted to determine the mechanisms of prey selection by salmonids in streams, linking the composition of diets with available drifting prey (Allan 1978; Bisson 1978; Ringler 1979; Forrester et al. 1994; Nakano et al. 1999). Those studies demonstrated the selection of different prey taxa and size. Zaret (1980) and Peckarsky (1982) stated that selective predation on a particular prey size altered the structure of invertebrate communities. In different works, Sih (1987) and Wooster (1994) proposed that predators produce a strong selective force that is concentrated on vulnerable individuals, and this selection pressure can lead to changes in the behavior, growth, fecundity, and morphology of susceptible prey. To enhance recreational fishing, Patagonian streams have been affected by the introduction of salmonids since the early twentieth century. The rainbow trout (*Oncorhynchus mykiss*) has been the most widely distributed species in the region (Pascual et al. 2002; Baigun and

* Correspondence: cydiprinzio@yahoo.com.ar
[1]Consejo Nacional de Investigación en Ciencia y Tecnología (CONICET), Rivadavia 1917, 1033 Buenos Aires, Argentina
[2]Centro de Investigaciones Esquel de Montaña y Estepa Patagónica (CIEMEP), Sarmiento 890, 9200 Esquel, Argentina
Full list of author information is available at the end of the article

Ferriz 2003; Di Prinzio et al. 2009) generating a great economic resource input. Several studies focused on rainbow trouts' feeding strategies in Patagonian environments and analyzed the impact of this species on native fauna (Modenutti and Balseiro 1994; Pascual et al. 2002; Buria et al. 2007, 2009, 2010; Albariño and Buria 2011; Di Prinzio and Casaux 2012; Di Prinzio et al. 2013). However, the degree of prey selectivity displayed by this fish remains poorly understood.

Prey selectivity metrics are important measures of possible impacts on native communities and of changes to stream environments (Wooster 1994; Osenberg et al. 1996, 1999; Johnson et al. 2006). Therefore, we investigated in a seasonal basis prey selection by the rainbow trout at two streams in Patagonia.

Methods

Study area and site selection

The study area is located in the Northwest of the Chubut Province, Patagonia, Argentina, in the ecotone between the Subantarctic forest and the Patagonian steppe, and exhibits a marked altitudinal gradient (2,000 to 600 m above sea level (m.a.s.l.)). Perennial (*Austrocedrus chilensis*, *Nothofagus dombeyi*, and *Maitenus boaria*) and deciduous (*Nothofagus pumilio* and *Nothofagus anctarctica*) tree species constitute the Subantarctic forest. Rivers at the study area have a pluvionival regime; therefore, discharge pattern is bimodal, with one peak associated to winter rains (June to July) and a second one originated by snowmelt in spring (September to October) (Coronato and del Valle 1988). The watercourses selected for this study were Nant y Fall (NyF) and Cabeza de Vaca (CVA) streams (2nd order streams). The sampled site at NyF (43° 13′ 24″ S, 71° 25′ 17″ W) is located 3.7 km downstream of Rosario lake (19 km^2) in the Futaleufú-Yelcho basin, at an altitude of 690 m.a.s.l. NyF's sub-basin has 15% of the area dedicated to pastures, 24.8% correspond to lakes, and 60.2% to the timberland (H. Claveri unpublished information). Land-use adjacent to the lake and the stream is mainly pasture, with agricultural and extensive livestock grazing activity.

The sampled site selected at CVA (740 m.a.s.l.) (43° 30′ 02″ S, 71° 20′ 49″ W) is located 25.8 km from the headwater and belongs to the Carrenleufú basin. CVA's sub-basin is covered by pastures (29.4%) and timberland (65.4%), and 5.2% of the area corresponds to lakes (H. Claveri unpublished information). Land-use around the stream is primarily wood extraction.

Environmental characterization and sampling procedure

Sampling sites were visited in early autumn (May), late winter (September), and late spring (December) of 2005 and during late summer (March) of 2006, under stable environmental conditions. At each site, substrate size composition was estimated as percentages of boulders, cobbles, gravel, pebbles, and sand in the reach, using a 1-m^2 grid ($n = 3$). Digital pictures of the grid were obtained and processed in the laboratory. Water velocity (ms^{-1}) was measured in mid-channel (thalweg) on three occasions by timing a float (average of three trials) as it moved over a distance of 10 m (Gordon et al. 1994). Average depth (cm) was estimated from five measurements along one transversal profile across the channel with a calibrated stick. At each site (run-riffle areas), water temperature (°C) was measured with a mercury thermometer.

During each sampling, specific conductance (μS_{20} cm^{-1}), pH, and dissolved oxygen (DO, mg $O_2 l^{-1}$) were measured with a multiparameter probe (Hach sensION 156, Hach Instruments, Lovedale, CO, USA). For nutrient analyses, water samples were collected below the water surface and kept at 4°C prior to analysis and transported to the laboratory. Total suspended solids (TSS), nitrate + nitrite nitrogen ($NO_3 + NO_2$), ammonia (NH_4), and soluble reactive phosphate (SRP) were analyzed following standard methods (APHA 1994).

Biological data collection

Fish and macroinvertebrates were sampled seasonally at the four mentioned dates. Fish were sampled along reaches of 100 m long employing a portable backpack electrofishing gear (Coffelt Mark-10 CPS, output 350 V). The width of the sampling area was coincident with the stream width. *In situ*, the individuals caught were preserved in cooled containers. At laboratory, fish were counted, weighted (g), and measured in total length (cm) and the stomachs were separated and fixed with 90% alcohol for posterior diet analysis.

Macroinvertebrate samples were obtained using a Surber sampler (0.09 m^2; 250-mm mesh size). Three samples from riffles ($n = 3$) and three from pools ($n = 3$) were taken at each reach. Samples were fixed *in situ* with 4% formaldehyde for posterior analysis.

At the laboratory, all the invertebrates represented in the samples, both in fish stomach contents and in benthos samples, were sorted (using a binocular microscope at × 5 magnification), identified to the lowest taxonomic level possible using available keys (Domínguez and Fernández 2009) and counted. Individuals from stomach and benthos samples were identified to the same taxonomic level. Terrestrial invertebrates (including those adults of aquatic insects with aerial phases) in the stomach contents were grouped into a single category (terrestrial items). Algae and vegetal fragments were grouped into a single category (vegetal items). Inorganic material (i.e., little stones) was classified as inorganic items. Total body length of each individual, excluding antennae and terminal cerci, was measured to the nearest 0.1 mm using an ocular micrometer (Zeiss stereomicroscope Stemi DV4, Zeiss, Jena, Germany).

Data analysis

In order to calculate the contribution of a food item to the diet, the dietary coefficient (Q) (Hureau 1970) was employed. This method reduces the biases associated to the use of numeric or weight methods, because it is the product of the percentage by number (%F) and the percentage by mass (%M) of each prey type ($Q = \%F \times \%M$). According to this index, the prey items were separated into the following categories: main preys, $Q >$ 200; secondary preys, $200 > Q > 20$; and occasional preys, $Q < 20$.

To estimate diet width, the Levins (1968) index was calculated with 95% confidence limits as follows: $B = 1/\Sigma p^2_i$, $i = 1 \ldots n$, where p_i is the proportion of each prey type i in the diet and equals Ai expressed as fraction rather than percentage.

Diet diversity was assessed using the Shannon-Wiener index of diversity (Krebs 1989) according to:

$$H = -\sum_{i=l}^{s}(p_i \ln\ p_i)$$

where p_i is the fraction of items in benthos sample that are of category i.

Comparisons between proportions of preys in the rainbow trout stomach contents and in benthos samples were carried out using Ivlev's index. Ivlev (1961) electivity index (Ei) was selected because it handles the situation where a prey item is represented in the diet but absent from field samples or where an item occurs in the field but is absent from the stomach contents. The electivity index ranges between 1 (the prey is represented in the stomachs but absent from benthos samples) and −1 (the prey is absent from the stomach but present in benthos samples) as it is estimated according to the following equation:

$$\mathrm{I} = \frac{Ei - Bi}{Ei + Bi}$$

where Ei is the percentage by number of taxon i in the stomach contents and Bi is the percentage by number of taxon i in benthos samples. Additionally, to identify if trouts select preys according to their size, the Ivlev's electivity index was calculated for three size ranges of each prey species (small, medium, and large size).

Table 1 Physico-chemical parameters, annual mean standard deviation, and total range values from NyF and CVA streams (Patagonia, Argentina)

	NyF		CVA	
Physical parameters	Mean ± SD	Range	Mean ± SD	Range
Water temperature (°C)	10.4 ± 2.7	7.6 to 13.6	6.5 ± 2.4	4.0 to 9.4
Wet width (m)	21.1 ± 3.1	17.5 to 25.0	7.2 ± 2.9	4.6 to 11.0
Depth (m)	0.2 ± 0.1	0.19 to 0.32	0.2 ± 0.0	0.20 to 0.28
Water velocity (ms $^{-1}$)	0.8 ± 0.2	0.68 to 1.0	1.1 ± 0.2	0.93 to 1.15
Discharge (m^3 s^{-1})	4.6 ± 2.4	2.71 to 8.0	1.9 ± 0.9	0.92 to 3.08
Substrate composition (%)				
Boulder	20		15	
Cobble	25		30	
Pebble	10		15	
Gravel	30		25	
Sand	15		15	
Chemical parameters	Mean ± SD	Range	Mean ± SD	Range
Conductivity (μS cm^{-1})	114.0 ± 2.7	112.0 to 118.0	50.0 ± 14.0	38.0 to 63.0
Dissolved oxygen (mg l^{-1})	8.8 ± 1.6	7.7 to 11.1	11.1 ± 1.7	9.7 to 13.0
pH	7.6 ± 0.4	7.2 to 7.9	7.3 ± 0.1	7.1 to 7.4
Turbidity (TNU)	4.2 ± 3.9	2.0 to 10.0	10.0 ± 15.0	0 to 27.0
Ammonium (μg L^{-1})	1.05 ± 0.5	0.6 to 1.6	0.86 ± 0.7	0 to 1.7
Nitrate + nitrite nitrogen (μg L^{-1})	0.18 ± 0.2	0.1 to 0.3	0.17 ± 0.2	0 to 0.3
Soluble reactive phosphate (μg L^{-1})	0.37 ± 0.0	0.3 to 0.4	0.53 ± 0.2	0.2 to 0.7
Total suspended solids (mg L^{-1})	2.74 ± 0.8	1.7 to 3.6	2.38 ± 1.8	0.7 to 4.4

During the study period (early autumn, late winter, late spring of 2005, and late summer of 2006). SD, standard deviation.

To identify differences in prey availability and density between streams, habitats, and seasons, we employed one-way analysis of variance. Similarly, to identify seasonal changes in the size of macroinvertebrates from the benthos, we used ANOVA and the *post hoc* Newman-Keuls test (Sokal and Rohlf 1995). Prior to analysis data were tested for normality and homogeneity of variances using Kolmogorov-Smirnov and Levene's tests, respectively, and log $(x + 1)$-transformed when appropriate (Gotelli and Ellison 2005). Mann–Whitney test was used to verify whether the length distributions between individuals represented in stomach contents and in benthos samples were different (Sokal and Rohlf 1995).

To compare the size of rainbow trouts between streams, a Kruskal-Wallis test was applied (Sokal and Rohlf 1995).

In order to summarize the variation among seasons in the distribution of fish based on their diets, we performed a Non-Metric Multidimensional Scaling (N-MDS) ordination technique based on the Bray-Curtis similarity coefficient using the software Statistica (version 6.0) (Ludwing and Reynolds 1988; Marshall and Elliott 1997; Clarke and Warwick 1994). Pearson correlation matrices based on quantitative data of trout stomach contents were built for each stream (NyF and CVA). The proximity of points in a resulting 2-D plot graphic indicates a higher degree of similarity, whereas more dissimilar points are positioned further apart.

Results

Environmental features

At both sites, the substrate was dominated by gravel cobble (Table 1). The highest value of water temperature (late summer), wet width (late spring), depth (late spring), current speed (late spring), and discharge (late spring) was observed at NyF while the highest value of dissolved oxygen (late summer) and turbidity (late winter) was recorded at CVA (Table 1). Physicochemical conditions indicated that waters were circum-neutral and well oxygenated. Conductivity was significantly higher (Kruskal-Wallis, $p < 0.05$) at NyF than at CVA. Both streams presented similar (Kruskal-Wallis, $p > 0.05$) low nutrient levels as expected for low-order streams in mountain areas (Table 1).

Biological features

Macroinvertebrate density was higher at NyF than at CVA (ANOVA, $p < 0.05$); nevertheless, the seasonal trend in benthos density at both streams was similar (ANOVA, $p > 0.05$). At both streams, riffles supported higher macroinvertebrate densities than pools (ANOVA, $p < 0.05$), being Diptera at CVA and Trichoptera at NyF the most abundant groups (Figure 1).

Overall, at both streams, the benthos community was mainly composed of individuals belonging to small size classes (Figure 2). However, the size of the individuals changed seasonally (ANOVA, $p < 0.05$), and at both

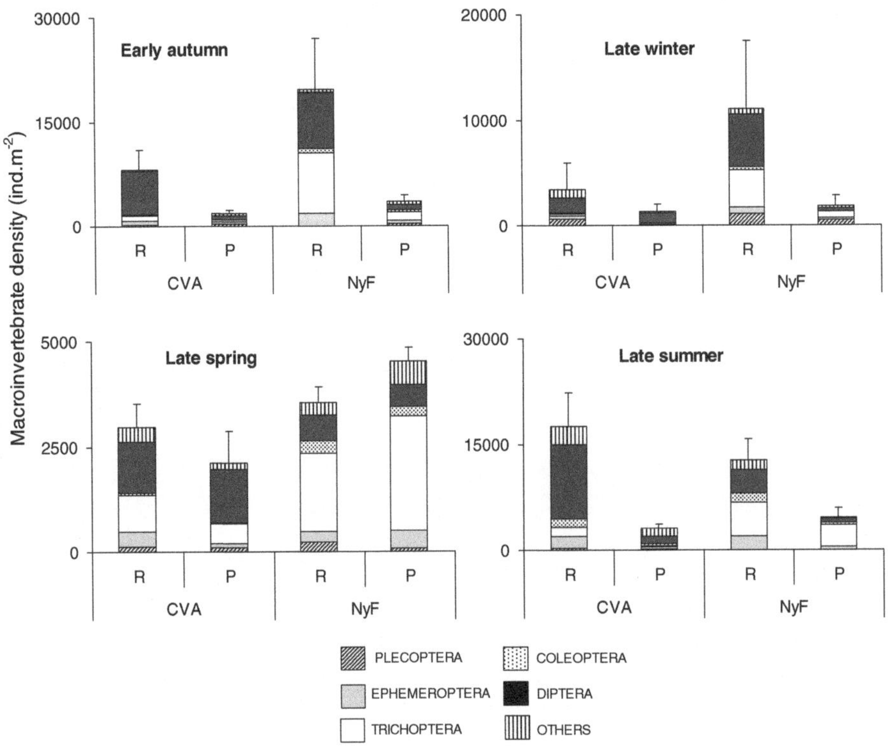

Figure 1 Mean invertebrate density (major groups) (±SD, ind m^2) by season. At riffles (R) and pools (P) ($n = 3$) of Cabeza de Vaca (CVA) and Nant y Fall (NyF) streams (Patagonia, Argentina) from May 2005 (early autumn) to March 2006 (late summer).

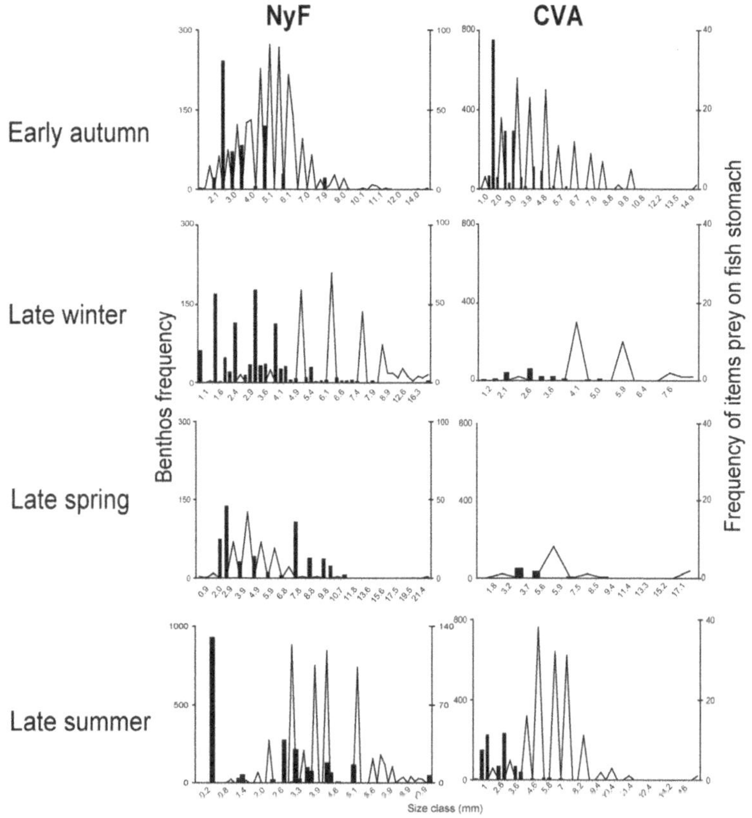

Figure 2 Frequency distribution of size classes. Macroinvertebrates in the environment (bars) and in rainbow trout *(Oncorhynchus mykiss)* stomach contents (line) at Nant y Fall (NyF) and Cabeza de Vaca (CVA) streams (Patagonia Argentina) during the study period (early autumn, late winter, late spring 2005, and late summer 2006).

streams, the larger individuals were observed during late spring (Newman-Keuls test, $p < 0.05$).

A total of 109 and 60 rainbow trouts were caught throughout the study period at NyF and CVA, respectively. Trouts from both streams were similar in total length and mass when comparing mean values (Kruskal-Wallis, $p > 0.05$) (Table 2). The high biomass value observed in NyF (988.7 g) corresponded just to one specimen. The diet of trouts from both streams was dominated by aquatic macroinvertebrates (Figure 3) and was diverse, and distinct preys were dominant in different seasons. In NyF crustaceans (*Hyalella araucana*) and trichopterans (*Parasericostoma ovale* and *Hudsonema flaminii*) conformed the diet almost the whole year; plecopteran preys appeared in the trout diet in spring at CVA (Figure 3). During early autumn and late summer, trouts from NyF foraged predominantly on *P. ovale* (Figure 2) which was the most abundant species in benthos samples (Additional file 1). Trouts from CVA preyed predominantly on the most abundant benthos species *Brachysetodes* sp., (see Additional file 1) in late spring and summer (Figure 2). During the remaining seasons, the main

Table 2 Morphometric measurements annual mean and total range values, seasonal niche broad, and seasonal prey diversity

	NyF (*n* = 109)		CVA (*n* = 60)	
Biological parameters	Mean ± SD	Range	Mean ± SD	Range
Fish length (cm)	11.1 ± 7.0	2.6 to 44.2	11.3 ± 4.2	5.2 to 9.6
Fish weight (g)	44.6 ± 157.0	0.1 to 988.7	20.3 ± 21.2	1.6 to 84.7
Diet broad	Levin's index			
Early autumn	0.032		0.263	
Late winter	0.247		0.169	
Late spring	0.331		0.588	
Late summer	0.037		0.332	
Prey diversity	Shannon index			
Early autumn	1.82		3.37	
Late winter	3.44		1.98	
Late spring	3.54		1.98	
Late summer	1.25		3.46	

Of *Oncorhynchus mykiss* (rainbow trout) from Nant y Fall (NyF) and Cabeza de Vaca (CVA) streams (Patagonia, Argentina) during the study period (early autumn, later winter, later spring of 2005, and later summer of 2006).

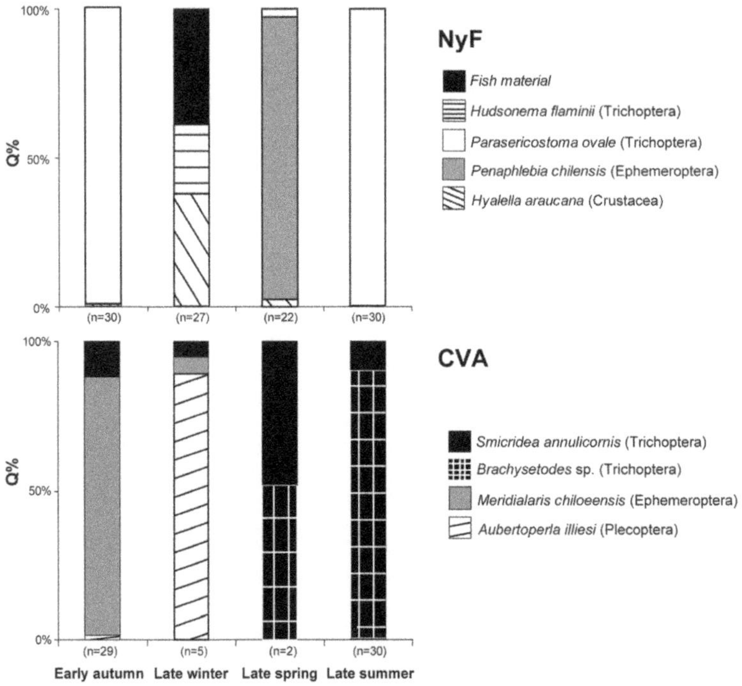

Figure 3 Diet composition (dietary coefficient *Q*%). *Oncorhynchus mykiss* at Nant y Fall (NyF) and Cabeza de Vaca (CVA) streams (Patagonia Argentina) during the study period (early autumn, late winter, late spring 2005, and late summer 2006); *n*, number of fish stomach analyzed.

species in the diet were invertebrates scarcely represented in benthos samples (i.e., *H. araucana* in late winter at NyF and *Aubertoperla illiesi* in late winter at CVA) (see Additional file 1).

The individuals represented in the stomach contents (NyF $\bar{x}$ = 4.9 ± 2.2 mm; CVA $\bar{x}$ = 5.0 ± 2.1 mm) were among the largest available at the streams (NyF $\bar{x}$ = 2.9 ± 2.1 mm; CVA $\bar{x}$ = 2.5 ± 1.4 mm). The prey size consumption at NyF did not show any evident pattern (Figure 4), whereas trouts at CVA consumed the larger individuals of all prey species along the year (Figure 5).

At NyF (H' = 3.54) and CVA (H' = 3.46), diet diversity peaked during late spring and late summer, respectively, whereas at both streams, the niche width peaked during late spring (Table 2).

The Ivlev index indicated that at NyF, the Ephemeroptera *Meridialaris chiloeensis* (early autumn and late spring), the Trichoptera *Oxyethira bidentata* (early autumn, late winter, and summer), the Diptera *Thienemanniella* sp. (early autumn), the Gasteropoda *Biomphalaria peregrina* (late summer), and the fish eggs were positively selected by rainbow trouts (Table 3). The species positively selected at CVA were the Plecoptera *A. illiesi* (early autumn), the Coleoptera *Stethelmis kaszabi* (early autumn and late summer), Muscidae larvae (early autumn, late winter, and late spring), and the Ephemeroptera *Nousia crena* (early autumn) (Table 3). Although present and abundant in the benthos throughout the year at both streams, some Oligochaeta (Lumbriculidae spp.), Trichoptera (*Smicridea frequens* and *Neoatopsyche brevispina*), Coleoptera (*Luchoelmis cekalovici* and *Austrolimnius* sp.), and Diptera (Empididae larvae, *Parapsectocladius* sp., *Pseudochironomus* sp., and *Ceratopogonidae* sp.) species were scarcely or not selected by trouts (Table 3). Terrestrial preys like Arachnidae, adult dipterans, Hymenoptera, Homoptera, Hemiptera, and Isopoda were positively selected by the rainbow trout at both streams (Table 3). At both streams, trouts selected prey items with body size ranging from medium to large (Table 4). In this sense, while at NyF, the larger prey items of Ephemeroptera: *N. crena* (early autumn) and *Andesiops torrens* (later summer); Plecoptera: *N. femina* (early autumn), *A. illiesi*, and *L. jaffuelli* (late winter and late spring); Trichoptera: *Brachysetodes* sp. (early autumn), *Smicridea annulicornis* (early autumn and late summer), and *P. ovale* (late winter and late spring); and Diptera: Simuliidae (early autumn and late winter) were selected, and at CVA, the larger prey items of Ephemeroptera: *N. crena* (late summer); Plecoptera: *Antarctoperla michaelseni* (early autumn); Trichoptera: *S. annulicornis* (early autumn and late summer), and Diptera: *Paratrichocladius* (late winter) were chosen.

The results of the MDS ordination supported data displayed in Figure 3 (Figure 6). At NyF, most of the trout captured during autumn and summer (stomachs dominated by *P. ovale*) were grouped to the negative

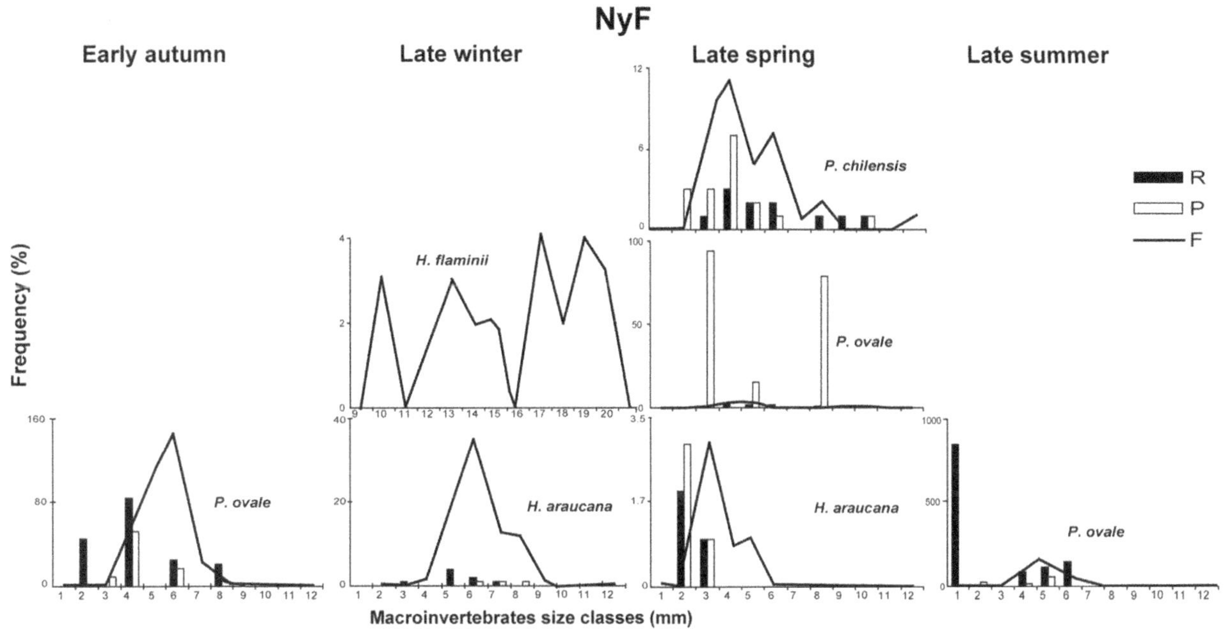

Figure 4 Annual distribution of size classes. The main preys (Q) in rainbow trout stomachs (F, solid line), species availability in benthic samples at riffles (R, black bars), and pools (P, white bars) at Nant y Fall (NyF) stream (Patagonia Argentina).

end of dimension 1, whereas those obtained during spring (stomachs dominated by *Penaphlebia chilensis*) were positioned on the lower right quadrant and those captured on winter (stomach content: fish material, *H. flaminii* and *H. araucana*) were mainly grouped to the positive side of dimension 2 (Figure 6). On the other hand, at CVA, the MDS ordination reflected that trout captured during summer was grouped towards the positive side of dimension 1. Dominant preys at this season were *Brachysetodes* sp. and *S. annulicornis* (Figure 6).

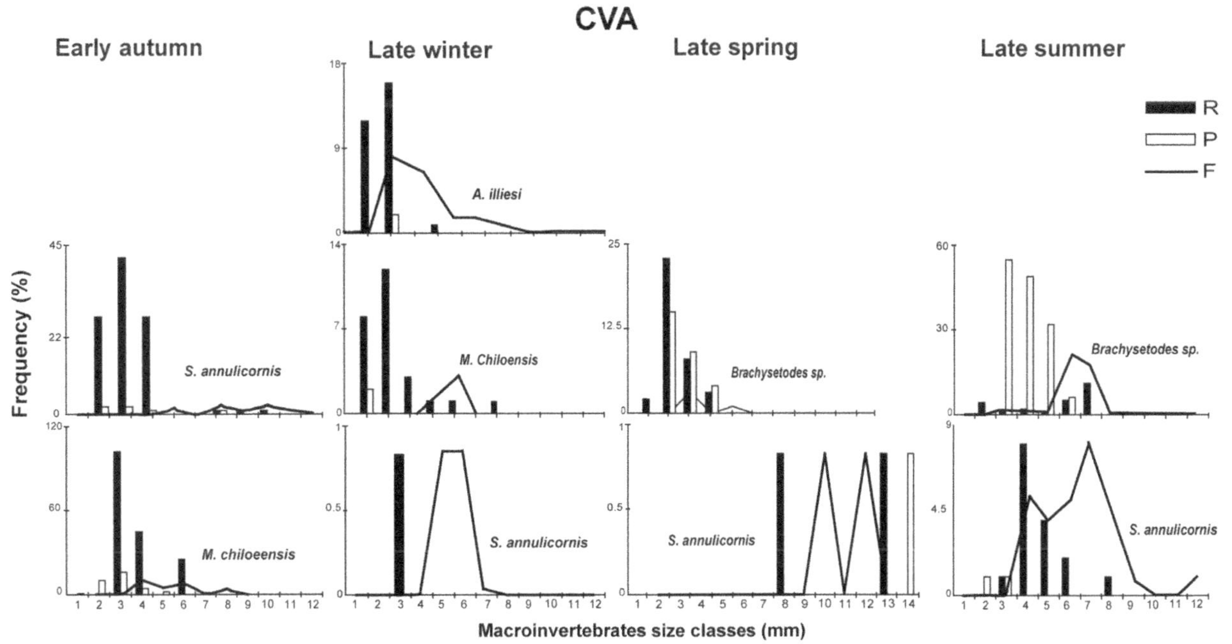

Figure 5 Annual distribution of size classes. The main preys (Q) in rainbow trout stomachs (F, solid line) and species availability in benthic samples at riffles (R, black bars) and pools (P, white bars) at Cabeza de Vaca (CVA) stream (Patagonia Argentina).

Table 3 Selectivity index (Ivlev) of prey in the diet of rainbow trout (*Oncorhynchus mykiss*)

		Early autumn		Late winter		Late spring		Late summer	
Order	**Species**	**NyF**	**CVA**	**NyF**	**CVA**	**NyF**	**CVA**	**NyF**	**CVA**
Oligochaeta	Lumbriculidae sp.*	−1	−1	−1	−1	−1	−1	−1	−1
Gasteropoda	*Chilina patagonica*	0.11	0.91	−1	---	−1	0.95	−1	0.41
	Biomphalaria peregrina	---	---	---	---	---	---	1	---
Crustacea	*Hyalella araucana*	0.75	---	0.98	---	0.31	---	1	---
	*Isopoda**	---	1	---	---		---	---	---
Araneae	Arachnidae*	---	---	1	---	1	---	---	---
Cladocera	*Daphnia daphnia*	---	---	1	---	---	---	---	---
Plecoptera	*Aubertoperla illiesi*	---	1	−0.30	0.71	---	−1	---	−1
	Limnoperla jaffuelli	−1	---	−0.20	−1	0.57	−1	---	---
	Antarctoperla michaelseni	−1	0.12	---	−1	−1	−1	−1	0.76
	Notoperlopsis femina	−0.70	1	−1	---	---	---	−1	−1
	Pelurgoperla personata	---	0.95	---	−1	---	−1	---	0.93
	Potamoperla myrmidon	---	---	---	---	---	---	---	1
	Udamocercia arumifera	---	---	---	---	---	−1	---	1
	Adult*	---	---	---	---	1	---	---	---
Ephemeroptera	*Nousia crena*	0.41	−1	---	−1	−1	−1	−1	0.85
	Meridialaris chiloeensis	1	0.64	−1	0.73	1	---	---	−1
	Meridialaris laminata	−1	---	1	---	−1	−1	---	---
	Penaphlebia chilensis	−1	−1	−0.30	---	0.74	−1	−1	1
	Penaphlebia flavidula	---	---	1	---	---	---	---	---
	Penaphlebia (A)*	---	---	---	---	1	---	---	---
	Andesiops torrens	0.35	0.78	−1	---	−1	−1	0.34	0.03
	Andesiops ardua	−0.90	−1	−1	−1	1	−1	−1	−1
	Ephemeroptera sp.	1	---	---	---	1	---	---	---
	Parasericostoma ovale	0.38	0.96	−0.80	−1	−0.70	1	0.48	1
Trichoptera	*Hudsonema flaminii*	0.50	0.72	1	−1	−1	−1	---	0.89
	Brachysetodes sp.	−0.30	0.46	−1	−1	−1	0.14	−1	0.94
	Triplectides sp.	---	−1	1	---	---	−1	---	−1
	Oxyethira bidentata	1	---	1	---	−1	---	1	---
	Neoatopsyche unispina	−1	0.97	−1	−1	1	−1	−1	−1
	Neoatopsyche brevispina	−1	−1	−1	---	---	---	−1	0.14
	Cailloma sp.	1	−1	−1	---	---	---	−1	0.93
	Smicridea annulicornis	0.45	0.04	−1	0.90	0.53	0.90	−0.60	0.64
	Smicridea frequens	−1	−1	−1	---	---	---	−1	---
	Limnephilidae	---	---	1	---	---	---	---	---
	Adult*	---	---	---	---	1	---	---	---
	Stethelmis kaszabi	1	1	−1	---	−1	---	−1	1
Coleoptera	*Luchoelmis* sp.	−1	1	−1	−1	−1	−1	−1	−1
	Elmidae (L)	−1	−1	---	1	−1	−1	−1	−1
	Luchoelmis cekalovici	−1	−1	−1	−1	−1	−1	−1	−1
	Austrolimnius sp.	−1	−1	−1	−1	---	−1	−1	−1
	Rhantus signatus	---	---	---	---	---	---	1	---

Table 3 Selectivity index (Ivlev) of prey in the diet of rainbow trout (*Oncorhynchus mykiss*)

	Tropisternus setiger	---	---	---	---	---	---	1	---
	Staphynilidae (A) *	1	---	---	---	---	---	---	---
	Nitidulidae (A) *	---	---	---	---	1	---	---	---
	Scirtidae (L)	---	---	---	---	1	---	---	---
	Coleoptera sp. *	---	1	0.32	1	---	---	---	---
	Athericidae (L)	−1	0.57	0.45	1	−1	−1	−1	---
Diptera	Muscidae (L)	---	1	---	1	1	---	---	---
	Simuliidae (L)	−0.70	−0.20	−0.80	−0.30	1	−1	−1	−0.50
	Empididae (L)	−1	−1	−1	−1	−1	−1	−1	1
	Paratrichocladius sp.	−0.30	1	−1	−0.70	0.68	−1	−1	−1
	Parapsectocladius sp.	−1	---	---	---	−1	−1	−1	−1
	Pseudochironomus sp.	---	---	---	−1	−1	−1	−1	---
	Thienemanniella sp.	1	−1	---	−1	---	−1	−1	−1
	Orthocladius sp.	1	---	−1	---	---	−1	---	---
	Lopescladius sp.	---	−1	---	−1	---	1	---	−1
	Ceratopogonidae sp.	---	---	−1	−1	−1	−1	---	−1
	Pupae sp.	1				1			
	Adult sp.*	1	1	1		1		1	
Odonata	*Cyanallagma interruptum*	---	---	1	---	---	---	---	---
Homoptera	Homoptera sp.*	1	1					1	
Hymenoptera	Hymenoptera sp.*	1						I	1
Orthoptera	Orthoptera sp.*							1	
Hemiptera	Hemiptera sp.*	---	---	---	---	---	---	1	---
Fish fragment	Eggs			1					

From Nant y Fall (NyF) and Cabeza de Vaca (CVA) streams (Patagonia, Argentina) during the study period (early autumn, late winter, late spring of 2005, and late summer of 2006). Taxa absence at both environment and trout stomach were represented by --- and terrestrial prey (*).

Discussion

Benthos density and composition varied seasonally and in a different way at both streams. Whereas at NyF, the highest benthos density occurred in early autumn, and at CVA, it occurred in late summer. This difference between streams is in line with observations by Miserendino and Pizzolón (2003, 2004), who documented at other Patagonian mountain rivers that macroinvertebrate diversity and density vary mainly due to environmental factors such as current speed and water temperature which differed between both streams. NyF was the most productive stream in terms of invertebrate density which results in a higher food availability for trouts. This agrees with the fact that lake-outlet streams, like NyF, are considered highly productive environments (Huryn and Wallace 2000; Brand and Miserendino 2012) and that the presence of patches of native forest upstream of the sampling site (Brand and Miserendino 2012) may help to buffer negative impacts on the structure of the invertebrate community related to the land-use activities there (Sponseller et al. 2001).

Overall, diet composition was dominated by Trichoptera species. These preys contributed strongly to the diet in late summer and early autumn in NyF and in late spring and summer in CVA, decreasing their importance as prey in the remaining seasons. Despite of the overall dominance of these species, the composition of the diet varied seasonally as observed in previous studies carried out in the region (Buria et al. 2009; Di Prinzio and Casaux 2012; Di Prinzio et al. 2013). Rainbow trouts from both streams tended to forage on the larger individuals of each prey species. Similarly, Buria et al. (2007) observed that at other three Patagonian streams, trouts foraged on the largest size classes of each prey species. According to Wissing and Hasler (1971), McCauley et al. (1974), and Rodgers and Qadri (1977), it seems that certain Trichoptera species contain more calories per gram than most of the macroinvertebrate taxa found in streams, and this could explain for the observed pattern in the diets of the analyzed specimens.

The diversity and width of the rainbow trout diet observed in this study is similar to that reported for this

Table 4 Selectivity index (Ivlev) of body size prey (mm) in the diet of rainbow trout (*Oncorhynchus mykiss*)

Early autumn						Late winter					
NyF			**CVA**			**NyF**			**CVA**		
	Size	**Ivlev**		**Size**	**Ivlev**		**Size**	**Ivlev**		**Size**	**Ivlev**
Chilina patagonica	S 4.87	1	*Chilina patagonica*	S 4.87	1	*Hyalella araucana*	S 2.85	−1	*Aubertoperla illiesi*	S 2.05	−1
	M 9.74	---		M 9.74	---		M 5.70	−0.89		M 4.10	−0.76
	L 14.61	−1		L 14.6	−1		L 8.55	−0.77		L 6.15	−0.09
Hyalella araucana	S 2.22	1	*Antactoperla michaelseni*	S 3.34	−0.44	*Aubertoperla illiesi*	S 1.78	−1	*Meridialaris chiloeense*	S 2.47	−1
	M 4.44	−0.37	*Antactoperla michaelseni*	M 6.68	0.59		M 3.56	−0.96		M 4.94	−0.95
	L 6.66	1		L 10.02	0.51		L 5.34	0.68		L 7.41	−0.65
Notoperlopsis femina	S 4.14	−0.75	*Pelurgoperla personata*	S 2.05	−1	*Limnoperla jaffueli*	S 2.6	−1	*Smicridea annulicornis*	S 3.65	---
	M 8.28	0.43		M 4.10	−0.95		M 4.72	−0.94		M 7.30	−0.82
	L 12.42	1		L 6.15	1		L 7.08	0.83		L 10.9	1
Nousia delicata	S 2.94	−1	*Meridialaris chiloeense*	S 2.67	−1	*Penaphlebia chilensis*	S 2.87	−1	Simuliidae	S 2.00	−1
	M 5.88	0.58		M 5.34	−0.96		M 5.74	−1		M 4.00	−0.91
	L 8.82	1		L 8.01	−0.72		L 8.61	−0.14		L 6.00	−0.56
Andesiops torrens	S 2.05	−0.22	*Andesiops torrens*	S 2.45	---	*Parasericostoma ovale*	S 3.42	−1	*Paratrichocladius*	S 1.78	---
	M 4.10	−0.03		M 4.90	---		M 6.84	0.05		M 3.56	---
	L 6.15	1		L 7.35	−0.68		L 10.2	0.71		L 5.34	0.91
Parasericostoma ovale	S 2.94	−0.95	*Parasericostoma ovale*	S 2.14	---	Athericidae	S 4.45	---			
	M 5.88	−0.14		M 4.28	---		M 8.90	−0.78			
	L 8.82	−0.21		L 6.42	−0.92		L 13.3	1			
Hudsonema flaminii	S 1.40	---	*Hudsonema flaminii*	S 2.22	−0.97	Simuliidae	S 2.31	---			
	M 2.80	−0.60		M 4.44	−0.85		M 4.62	0.83			
	L 4.20	−0.25		L 6.66	−0.35		L 6.93	1			
Brachysetodes sp.	S 1.29	---	*Brachysetodes* sp.	S 1.65	−0.61						
	M 2.58	0.21		M 3.30	−1						
	L 10.68	1		L 15.2	−1						
Simuliidae	S 1.84	−0.55	*Smicridea annulicornis*	S 3.65	−1						
	M 3.68	0.37		M 7.30	−0.88						
	L 5.52	0.04		L 10.9	0.21						
Paratrichocladius	S 1.78	---	Athericidae	S 4.20	−1						
	M 3.56	−0.39		M 8.40	−1						
	L 5.34	4		L 12.6	−0.87						
			Simuliidae	S2.09	−0.96						
				M 4.18	−0.82						
				L 6.27	−0.58						

Table 4 Selectivity index (Ivlev) of body size prey (mm) in the diet of rainbow trout (*Oncorhynchus mykiss*)

Late spring						Late summer					
NyF			CVA			NyF			CVA		
	Size	Ivlev		Size	Ivlev		Size	Ivlev		Size	Ivlev
Hyalella araucana	S 1.35	---	*Chilina patagonica*	S 1.34	---	*Andesiops torrens*	S 1.80	−1	*Chilina patagonica*	S 5.94	---
	M 2.70	−1		M 2.68	---		M 3.60	−0.61		M 11.8	1
	L 4.05	1		L 4.02	−0.96		L 5.40	0.63		L 17.8	−1
Limnoperla jaffueli	S 2.00	---	*Brachysetodes* sp.	S 1.91	−1	*Parasericostoma ovale*	S 2.77	−1	*Antactoperla michaelseni*	S 1.65	---
	M 4.00	0.57		M 3.82	−0.95		M 5.57	−0.55		M 3.30	−0.44
	L 6.00	0.57		L 5.73	−0.65		L 8.31	−0.55		L 4.95	1
Penaphlebia chilensis	S 4.31	−069	*Smicridea annulicornis*	S 4.47	---	*Smicridea annulicornis*	S 1.74	---	*Pelurgoperla personata*	S 1.78	---
	M 8.62	−0.67		M 8.94	−0.16		M 3.48	0.85		M 3.56	---
	L 12.9	−1		L 13.4	−0.16		L 5.22	0.85		L 5.34	−0.95
Parasericostoma ovale	S 3.74	−0.75							*Nousia crena*	S 2.27	−1
	M 7.48	0.38								M 4.57	−1
	L 11.2	−0.87								L 6.81	0.60
Smicridea annulicornis	S 3.67	1							*Andesiops torrens*	S 2.05	---
	M 7.34	---								M 4.10	−1
	L 11.0	−1								L 6.15	1
Paratrichocladius sp.	S 1.78	---							*Hudsonema flaminii*	S 2.14	−1
	M 3.56	−0.92								M 4.28	---
	L 5.34	−0.69								L 6.42	1
									Brachysetodes sp.	S 2.27	−1
										M 4.54	−1
										L 6.81	−0.91
									Neoatopsyche brevispina	S 2.89	−1
										M 5.78	−1
										L 11.0	−0.97
									Smicridea annulicornis	S 3.69	−1
										M 7.38	−0.17
										L 11.0	0.13
									Simuliidae	S 2.22	−1
										M 4.44	−1
										L 6.66	1

From Nant y Fall (NyF) and Cabeza de Vaca (CVA) streams (Patagonia, Argentina) during the study period. Body size taxa absence was represented by ---. S, small size; M, medium size; L, large size.

species in other Patagonian environments (Arismendi et al. 2012; Di Prinzio et al. 2013) but higher than the observed in the native catfish *Hatcheria macraei* (Ferriz 1994, 2012; Barriga et al. 2009; Di Prinzio and Casaux 2012). This fact supports the proposal by Di Prinzio et al. (2013) indicating that the expansion success of the rainbow trout in Patagonia could be explained, among other factors, by their high feeding plasticity, which

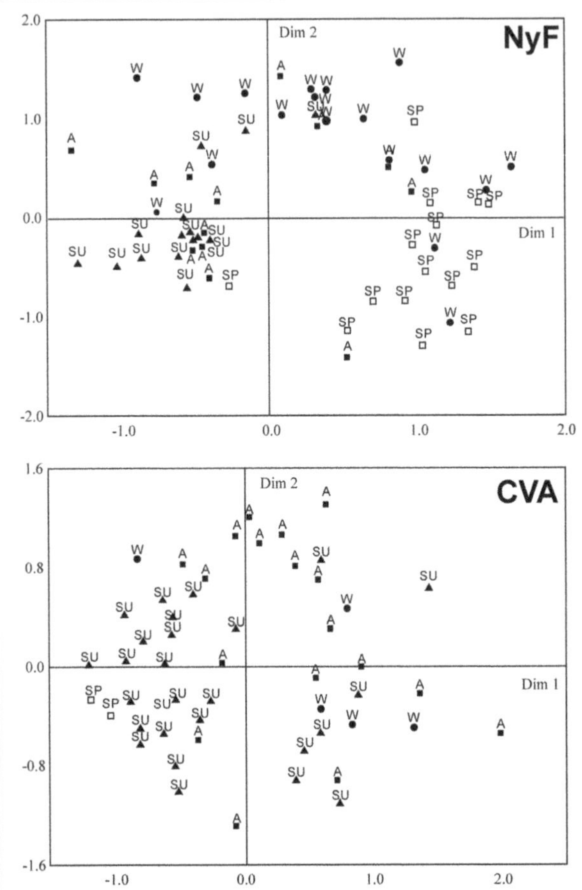

Figure 6 Results of the multidimensional ordination of *Oncorhynchus mykiss*. Based on stomach content data collected during the study period (early autumn, late winter, late spring 2005, and late summer 2006) at Nant y Fall (NyF) and at Cabeza de Vaca (CVA) streams (Patagonia Argentina). A, autumn; W, winter; SP, spring; SU, summer.

depends on different issues (prey-type characteristics, seasonal changes in resource supply, food-habitat utilization, etc.). This allows the trout to exploit temporarily abundant food resources and switch between specialized and generalized feeding strategies, buffering changes in food availability as observed in some streams in Patagonia (Di Prinzio et al. 2013).

Although prey selectivity varied seasonally, rainbow trout positively selected both aquatic (Gasteropoda, Crustacea, Plecoptera, Trichoptera, Ephemeroptera, Coleoptera, Diptera, and Odonata) and terrestrial organisms (adult dipteran, Oligochaeta, Araneae, Homoptera, Hymenoptera, Orthoptera, and Hemiptera), which is in line with other studies carried out in the area (Buria et al. 2009; Di Prinzio and Casaux 2012; Di Prinzio et al. 2013). Compared to CVA, at NyF, rainbow trouts positively selected a higher number of terrestrial species (Table 3). This difference between streams in the number of terrestrial prey selected could be explained, to some extent, by the higher riparian and littoral invertebrate richness registered at NyF (Miserendino et al. 2011). During late winter, trouts at NyF positively selected fish eggs. According to the egg size, these come from the rainbow trout spawning individuals who evidence that this trout auto-regulate, to some extent, their populations. On the other hand, some infaunal invertebrates like Oligochaeta and some small larvae of Coleoptera and Diptera (mainly Chironomidae) were not selected by trouts. This could be related, among other reasons, to the fact that trouts rely predominantly on visual cues to detect prey (Wilzbach et al. 1986; Angradi and Griffith 1990). For instance, compared drifting preys, McIntosh (2000) observed at New Zealand streams few oligochaetes in the diet of salmonid species.

The effects of the introduction of freshwater fish species are a matter of ongoing debate worldwide (Gozlan 2008, 2009; Leprieur et al. 2009; Gozlan et al. 2010). Unfortunately, the scarcity of biological data from the basins before and after the introduction prevents to clearly visualize which are the consequences and impacts of these actions. Frequently, introduced rainbow trouts and prey interaction results are unpredictable because the introduction is achieved in disturbed environments as those studied here. Moreover, fish can also have important effects on different levels of the ecosystems (Power 1992; Nyström et al. 2001; Townsend 2003). In this sense, it was observed at several countries, included Argentina, that fish introductions resulted in the elimination/reduction/alteration of native populations of both vertebrates and invertebrates (Flecker and Townsend 1994; Arismendi et al. 2009; McIntosh 2000; Di Prinzio and Casaux 2012; among others). Despite of the problems associated to the rainbow trout introduction, this fish is an emblematic species for sport fishing being this recreative activity an important source of incomes in the region. For being economically and ecologically sustainable, the administration of this activity requires an adequate knowledge on the species biology and on the prey–predator pattern, so this type of study helps to design mitigation and conservation guidelines because it encompasses not only the fish population but also the macroinvertebrate community and their biological interactions.

Conclusions

Our study reflects a dominance of trichopteran species in the rainbow trout diet. The fact that certain Trichoptera species contain more calories per gram than most of the macroinvertebrate taxa found in these streams could explain the observed pattern in the diets of the trout analyzed. The composition of the diet of the exotic trout varied seasonally. In both streams, rainbow trouts consumed the largest preys available in the environment and positively selected both aquatic and terrestrial preys.

Competing interests
The authors declare that they have no competing interests.

Authors' contributions
CYDP, GO, MLM, and RC participated in the field trips carried out environmental characterization. GO and MLM carried out the benthos sampled and benthos processed in laboratory. CYDP and RC carried out the fish sampled in the field. CYDP processed the fish material in laboratory and wrote the manuscript. RC made intellectual contributions and reviewed the manuscript. All authors read and approved the final manuscript.

Acknowledgements
We would like to thank Dra. Cecilia Brand for fieldtrip assistance and Dr. Miguel Archangelsky for fieldtrip assistance and for helpful comments on the English style. Thanks to anonymous reviewers for valuable comments that greatly improved the manuscript. This research was partially financed by the CONICET and PADI Foundation. This is Scientific Contribution n° 109 from LIESA.

Author details
[1]Consejo Nacional de Investigación en Ciencia y Tecnología (CONICET), Rivadavia 1917, 1033 Buenos Aires, Argentina. [2]Centro de Investigaciones Esquel de Montaña y Estepa Patagónica (CIEMEP), Sarmiento 890, 9200 Esquel, Argentina. [3]Laboratorio de Investigación en Ecología y Sistemática Animal (LIESA), Universidad Nacional de la Patagonia, Sarmiento 890, 9200 Esquel, Argentina. [4]Instituto Antártico Argentino, Cerrito 1248, 1010 Buenos Aires, Argentina.

References
Albariño RJ, Buria LM (2011) Altered mayfly distribution due to strong interactions with alien rainbow trout in Andean streams of Patagonia. Limnologica 41:220–227

Allan JD (1978) Trout predation and the size composition of the stream drift. Limnol Oceanog 23:1231–1237

Angradi TR, Griffith JS (1990) Diel feeding chronology and diet selection of rainbow trout (*Oncorhynchus mykiss*) in the Henry's Fork of the Snake River. Idaho Can J Fish Aquat Sci 47:199–209

APHA (1994) Standard methods for the examination of water and wastewater. American Public Health Association, Hanover MD, p 70

Arismendi I, Soto D, Penaluna B, Jara C, Leal C, León-Muñoz J (2009) Aquaculture, non-native salmonid invasions and associated declines of native fishes in Northern Patagonian lakes. Freshwater Biol 54(5):1135–1147, doi:10.1111/j.1365-2427.2008.02157.x

Arismendi I, Gonzalez J, Soto D, Penaluna B (2012) Piscivory and diet overlap between two non-native fishes in southern Chilean streams. Austral Ecol 37:346–354

Baigun C, Ferriz RA (2003) Distribution patterns of freshwater fishes in Patagonia (Argentina). Org Divers Evol 3:151–159

Barriga JP, Battini MA (2009) Ecological significances of ontogenetic shifts in the stream-dwelling catfish, *Hatcheria macraei* (Siluriformes, Trichomycteridae), in a Patagonian river. Ecol Freshw Fish 18(3):395–405, doi:10.1111/j.1600-0633.2009.00356.x

Baxter CV, Fausch KD, Murakami M, Chapman PL (2004) Fish invasion restructures stream and forest food webs by interrupting reciprocal prey subsidies. Ecology 85:2656–2663

Bisson JD (1978) Trout predation and the size-composition of stream drift. Limnol Oceanogr 23:1231–1237

Brand C, Miserendino ML (2012) Life cycle phenology, secondary production, and trophic guilds of caddisfly species in a lake-outlet stream of Patagonia. Limnologica 42:108–117

Buria L, Albariño R, Díaz Villanueva V, Modenutti B, Balseiro E (2007) Impact of exotic rainbow trout on the benthic macroinvertebrate community from Andean-Patagonian headwater streams. Fundam Appl Limnol 168:145–154

Buria LM, Albariño RJ, Modenutti BE, Balseiro EG (2009) Temporal variations in the diet of the exotic rainbow trout (*Oncorhynchus mykiss*) in an Andean-Patagonian canopied stream. Rev Chil Hist Nat 82:3–15

Buria L, Albariño R, Díaz Villanueva V, Modenutti B, Balseiro E (2010) Does predation by the introduced rainbow trout cascade down to detritus and algae in a forested small stream in Patagonia? Hydrobiologia 651:161–172

Clarke KR, Warwick RM (1994) Change in marine communities: an approach to statistical analysis and interpretation. Natural Environment Research Council, UK

Coronato FR, Del Valle HF (1988) Caracterización hídrica de las cuencas hidrograficas de la provincia del Chubut. Publicación Técnica, Cenpat-Conicet, Chubut, Puerto Madryn, p 184

Di Prinzio CY, Casaux RJ (2012) Dietary overlap among native and non-native fish in Patagonian low-order streams. Ann Limnol-Int J Lim 48:21–30

Di Prinzio CY, Casaux RJ, Miserendino ML (2009) Effects of land use on fish assemblages in Patagonian low order streams. Ann Limnol-Int J Lim 45:267–277

Di Prinzio CY, Miserendino ML, Casaux R (2013) Feeding strategy of the non-native rainbow trout, *Oncorhynchus mykiss*, in low-order Patagonian streams. Fisheries Manag Ecol 20:414–425

Domínguez E, Fernández HR (eds) (2009) Macroinvertebrados bentónicos sudamericanos. Sistemática y biología. Fundación Miguel Lillo, Tucumán, Argentina

Ferriz RA (1994) Diet of *Olivaichthys viedmensis* (Mac Donagh, 1931) and *Hatcheria macraei* (Girard, 1855) (Teleostei, Siluriformes) in the Limay River, Argentina. Nat Patag Cinc Biol 2:83–88

Ferriz RA (2012) Diet of Hatcheria macraei (Girard, 1855) (Teleostei, Siluriforms, Trichomycteridae) in the Chubut River, Argentina. Lat Am J Aquat Res 40(1):248–252

Flecker AS, Townsend CR (1994) Community-wide consequences of trout introductions into New Zealand streams. Ecol Appl 44:798–807

Forrester GE, Chase JG, McCarthy W (1994) Diel and density-related changes in food consumption and prey selection by brook charr in a New Hampshire stream. Environ Biol Fish 39:301–311

Gordon ND, McMahon TA, Finlayson BL (1994) Stream hydrology, an introduction for ecologists. Wiley and Sons, New York. pp, 444

Gotelli NJ, Ellison AM (2005) A primer of ecological statistics. Sinauer Associates Inc. Publishers, Sunderland, Massachusetts, U.S. A, p 510

Gozlan RE (2008) Introduction of non-native freshwater fish: is it all bad? Fish Fish 9:106–115

Gozlan RE (2009) Response by R Gozlan biodiversity crisis and the introduction of non-native fish: solutions, not scapegoats. Fish Fish 10:109–110

Gozlan RE, Britton JR, Cowx I, Copp GH (2010) Current knowledge on non-native freshwater fish introductions. J Fish Biol 76:751–786

Hureau J (1970) Biologie comparee de quelques poissons antarctiques (Nototheniidae). Bull Inst Oceanogr 68:1–244

Huryn AD, Wallace JB (2000) Life history and production of stream insects. Ann Rev Entomol 45:83–110

Ivlev VS (1961) Experimental ecology of the feeding of fishes. Yale University Press, New Haven, CT & London, p 302

Johnson RK, Hering D, Furse MT, Clarke RT (2006) Detection of ecological change using multiple organism groups: metrics and uncertainty. Hydrobiologia 566(1):115–137

Krebs CJ (1989) Ecological methodology. Harper Collins Publishers, New York, p 652

Leprieur F, Brosse S, García-Berthou E, Oberdorff T, Olden JD, Townsend CR (2009) Scientific uncertainty and the assessment of risks posed by non-native freshwater fishes. Fish Fish 10:88–97

Levins R (1968) Evolution in changing environments: some theoretical explorations. Princeton University Press, Princeton, NJ, p 132

Ludwing JA, Reynolds JF (1988) Statistical Ecology. Wily Interscience, New York

Macchi PJ, Cussac VE, Alonso MF, Denegri MA (1999) Predation relationships between introduced salmonids and the native fish fauna in lakes and reservoirs in northern Patagonia. Ecol Freshw Fish 8:227–236

Marshall S, Elliott M (1997) A comparison of univariate and multivariate numerical and graphical techniques for determining inter and intraspecific feeding relationships in estuarine fish. J Fish Biol 51:526–545

McCauley VJE, Tsumura K (1974) Calorific values of Chironomidae (Diptera). Can J Zool 52:581–586
McIntosh AR (2000) Aquatic predator–prey interactions. In: Collier KJ, Winterbourn MJ (eds) New Zealand stream invertebrates: ecology and implications for management. New Zealand Limnological Society, Hamilton, New Zealand
Miserendino ML, Pizzolón LA (2003) Distribution of macroinvertebrates assemblages in the Azul-Quemquemtreu river basin, Patagonia, Argentina. New Zeal J Mar Fesh 37:525–539
Miserendino ML, Pizzolón LA (2004) Interactive effects of basin features and land-use change on macroinvertebrate communities of headwater streams in the Patagonian Andes. River Res Appl 20:967–983
Miserendino ML, Casaux R, Archangelsky M, Di Prinzio CY, Brand C, Kutschker AM (2011) Assessing land-use effects on water quality, in-stream habitat, riparian ecosystems and biodiversity in Patagonian northwest streams. Sci Total Environ 409(3):612–624, doi:10.1016/j.scitotenv.2010.10.034
Modenutti BE, Balseiro EG (1994) Zooplankton size spectrum in four lakes of the Patagonian Plateau. Limnologica 24:51–56
Nakano S, Fausch KD, Kitano S (1999) Flexible niche partitioning via a foraging mode shift: a proposed mechanisms for coexistence in stream-dwelling charrs. J Anim Ecol 68:1079–1092
Newman RM, Waters TH (1984) Size-selective predation on *Gammarus pseudolimnaeus* by trout and sculpins. Ecology 65:1535–1545
Nyström P, Svensson O, Lardner B, Brönmark C, Granéli W (2001) The influence of multiple introduced predators on a littoral pond community. Ecology 82:1023–1039
Osenberg CW, Mittelbach GG (1996) The relative importance of resource limitation and predator limitation in food chains. In: Polis GA, Winemiller KO (eds) Food webs: integration of patterns and dynamics. Chapman & Hall, New York, New York, USA, pp 134–148
Osenberg CW, Sarnelle O, Cooper SD, Holt RD (1999) Resolving ecological questions through meta-analysis: goals, metrics, and models. Ecology 80(4):1105–1117
Pascual M, Macchi P, Urbanski J, Marcos F, Riva Rossi C, Novara M, Dell'Arciprete P (2002) Evaluating potential effects of exotic freshwater fish from incomplete species presence-absence data. Biol Inv 4:101–113
Peckarsky B (1982) Aquatic insect predatory-prey relations. Bioscience 32(4):261–265
Power ME (1992) Top-down and bottom-up forces in food webs: do plants have primacy? Ecology 73:733–746
Ringler NH (1979) Selective predation by drift feeding brown trout (*Salmo trutta*). J Fish Res Board Can 36:392–403
Rodgers DW, Qadri SU (1977) Seasonal variations in calorific values of some littoral benthic invertebrates of the Ottawa River, Ontario. Can J Zool 55:881–884
Sih A (1987) Predators and prey lifestyles: an evolutionary and ecological overview. In: Kerfoot WC, Sih A (eds) Predation: direct and indirect impacts on aquatic communities. University Press of New England, Hanover, pp 203–224
Sokal RR, Rohlf FJ (1995) Biometry, 3rd edn. W.H. Freeman and Company, New York, USA
Sponseller RA, Benfield EF, Valett HM (2001) Relationships between land use, spatial scale and stream macroinvertebrate communities. Freshwater Biol 46:1409–1424
Stephens DW, Krebs JR (1986) Foraging theory. Princeton University Press, Princeton
Townsend CR (2003) Individual, population, community, and ecosystem consequences of a fish invader in New Zealand streams. Conserv Biol 17:38–47
Wesner JS (2012) Predator diversity effects cascade across an ecosystem boundary. Oikos 121:53–60
Wilzbach ME, Cummins KW, Hall JD (1986) Influence of habitat manipulations on interactions between cutthroat trout and invertebrate drift. Ecology 67:898–911
Wissing TE, Hasler AD (1971) Intraseasonal change in caloric content of some freshwater invertebrates. Ecology 52:371–373
Wooster D (1994) Predator impacts on stream benthic prey. Oecologia 99:7–15
Zaret TM (1980) Predation and freshwater communities. Yale University Press, New Haven, CT

2

Description of a new genus and new species of freshwater crab (Brachyura: Gecarcinucidae) from the Western Ghats, Kerala, India

Sameer Kumar Pati[1*] and Arath Raghavan Sudha Devi[2]

Abstract

Background: Freshwater crabs of the Western Ghats have been severely underestimated in their diversity. Since freshwater crabs have a small spatial scale endemism, many remote and poorly studied localities in the Western Ghats need to be surveyed. In this study, a new genus and new species is described.

Results: *Arcithelphusa cochleariformis* gen. et sp. nov. is described from Ondayangadi, Wayanad district, Kerala, India. The new genus and new species is exceptional among Indian freshwater crabs due to its squarish, highly arched carapace, appearing box-like in frontal view, and long, stout, spoon-shaped male first gonopod.

Conclusions: With the current discovery, 14 genera and 36 species of gecarcinucid freshwater crabs are recognized in the Western Ghats of India. More new taxa can be expected from this exceptionally biodiversity-rich area if the necessary surveys are done.

Keywords: Crustacea; Decapoda; Gecarcinucidae; Freshwater crab; Taxonomy; Ecology; New genus; New species; Western Ghats; India

Background

About 35 species of freshwater crabs under 13 genera were so far reported from the Western Ghats of India (Alcock 1910; Roux 1931; Bott 1970; Bahir and Yeo 2007; Pati and Sharma 2013, 2014a, b; Klaus et al. 2014). In fact, the diversity of freshwater crabs is severely underestimated in the Western Ghats, a biodiversity 'hotspot' with exceptional freshwater diversity and endemism (Raghavan et al. 2014). This fact is evident from the discovery of a new genus and five new species during the last few years (Pati and Sharma 2013, 2014a; Klaus et al. 2014). Many areas of the Western Ghats were not well explored for freshwater crab diversity, and this is partly due to the difficulty in accessing remote localities. Since small spatial scale endemism is seen in freshwater crabs (Klaus et al. 2014), many remote and poorly studied localities in the Western Ghats need to be surveyed.

Recently, some freshwater crabs were collected from the Western Ghats of Kerala. After careful examination of the carapace morphology and male gonopod structures, these crabs were recognized as a new species in a new genus, *Arcithelphusa*. The specimens were compared with the species of morphologically related genera, *Ghatiana* Pati & Sharma, 2014 [type species: *Ghatiana aurantiaca* Pati & Sharma, 2014, by original designation; gender feminine], *Globitelphusa* Alcock, 1909 [type species: *Paratelphusa* (*Globitelphusa*) *bakeri* Alcock, 1909, by original designation; gender feminine], *Gubernatoriana* Bott, 1970 [type species: *Paratelphusa* (*Globitelphusa*) *gubernatoris* Alcock, 1909, by original designation; gender feminine], *Inglethelphusa* Bott, 1970 [type species: *Paratelphusa* (*Globitelphusa*) *fronto* Alcock, 1909, by original designation and monotypy; gender feminine], *Pilarta* Bahir & Yeo, 2007 [type species: *Pilarta anuka* Bahir and Yeo, 2007, by original designation and monotypy; gender feminine] and *Snaha* Bahir & Yeo, 2007 [type species: *Snaha aruna* Bahir and Yeo, 2007, by original designation; gender feminine].

Methods

Crab specimens were collected from shallow to deep burrows adjacent to water channels in betel nut plantations at Ondayangadi, about 5 km northeast of Mananthavady in

* Correspondence: sameer_pati@yahoo.co.in
[1]Zoological Survey of India, Western Regional Centre, Vidyanagar, Sector 29, P.C.N.T. Post, Akurdi, Pune 411 044, Maharashtra, India
Full list of author information is available at the end of the article

Wayanad district of Kerala (Figure 1). Specimens were preserved in 95% ethyl alcohol and deposited in the National Zoological Collections (NZC) of Zoological Survey of India (ZSI), Western Regional Centre (WRC), Pune. Measurement methods and terminology follow Ng and Tay (2001). Abbreviations used are as follows: cw, carapace width; cl, carapace length; ch, carapace height; fw, frontal width; S2 to S4, thoracic sternites 2 to 4, respectively; G1, male first pleopod; G2, male second pleopod; and P2 to P5, pereiopods 2 to 5, respectively.

Results

Taxonomy

Family Gecarcinucidae Rathbun, 1904

***Arcithelphusa* gen. nov.**

Type species. *Arcithelphusa cochleariformis* sp. nov., by present designation.

Diagnosis. Carapace smooth, squarish (cl/cw = 0.9), highly arched (ch/cw = 0.7), box-like in frontal view; anterolateral margin with short, oblique striations; epigastric cristae and postorbital cristae indistinct; external orbital angle poorly developed; epibranchial tooth indistinct; branchial regions inflated; cervical grooves shallow; H-shaped gastric groove visible; front narrow, 0.25 times cw, without frontal median triangle; epistomal median lobe lacking median tooth (Figures 2A,B and 3A,B). Suture between male thoracic sternites S2, S3 prominent, deep, not reaching lateral margins and between S3, S4 prominent, deep, reaching lateral margins (Figures 2C and 3C). Male abdomen narrowly triangular; sixth abdominal somite distinctly broader than long, shorter than telson, lateral margins slightly convex; telson longer than broad, narrowly triangular, with slightly concave outer margins, apex acutely triangular (Figures 2C and 3D). Third maxilliped exopod lacking flagellum (Figure 3E). G1 long, stout, appearing as lateral view of spoon in both dorsal and lateral views; terminal segment cylindrical, short, 0.30 times length of subterminal segment, curved more outward at mid-length towards dorsal side; subterminal segment basal half distinctly broader than distal half (Figures 2D,E,F and 3F,G,H). G2 with very short distal segment, 0.2 times length of basal segment (Figures 2G and 3I).

Etymology. The genus name, derived from *Arca* (Latin for 'box') in combination with the genus name *Thelphusa*, alludes to the highly inflated carapace of the crab, appearing box-like in frontal view. Gender is feminine.

Ecological notes. These crabs dwell in shallow to deep burrows along water channels of betel nut plantations (Figure 4), banana farms and rice field embankments (A.R. Sudha Devi, unpublished data).

Distribution. At present, the genus is known only from the Western Ghats in Wayanad district of Kerala.

***Arcithelphusa cochleariformis* sp. nov.**

Type material. India: adult male, holotype (cw 30.76 mm, cl 26.30 mm, ch 21.46 mm, fw 7.60 mm), Ondayangadi, 5 km northeast of Mananthavady, Wayanad district, Kerala (11.8231649° N and 76.0232338° E), altitude 767 m, 25 June 2011, coll. M.K. Smija (ZSI, WRC-C.1188); paratypes, three males (cw 27.28 to 32.20 mm, cl 22.56 to 26.94 mm, ch 19.36 to 21.38 mm, fw 7.20 to 7.68 mm), one female (cw 26.76 mm, cl 22.90 mm, ch 19.22 mm, fw 7.06 mm), same collection data as holotype (ZSI, WRC-C.1189); five males (cw 23.94 to 28.84 mm, cl 20.50 to 23.80 mm, ch 17.06 to 21.14 mm, fw 6.34 to 7.46 mm), one female (cw 32.26 mm, cl

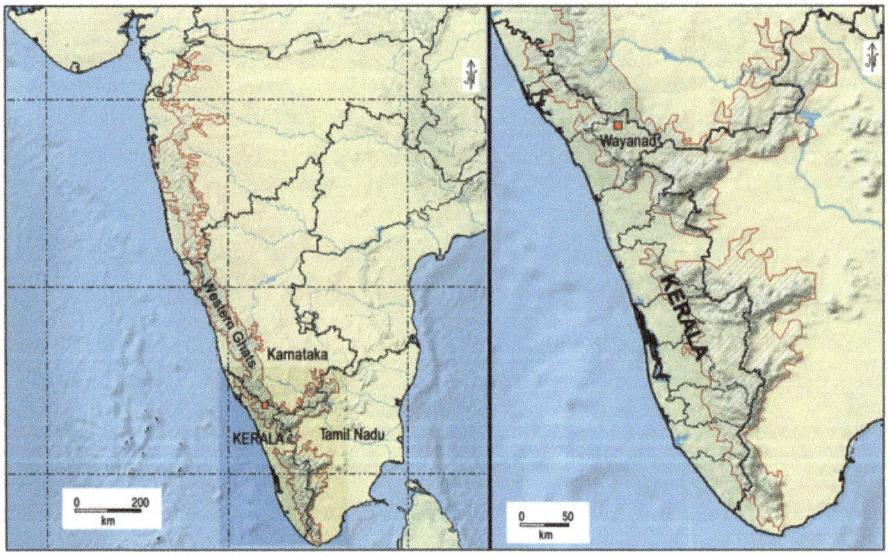

Figure 1 Map of collection site (red square mark) of *Arcithelphusa cochleariformis* sp. nov.

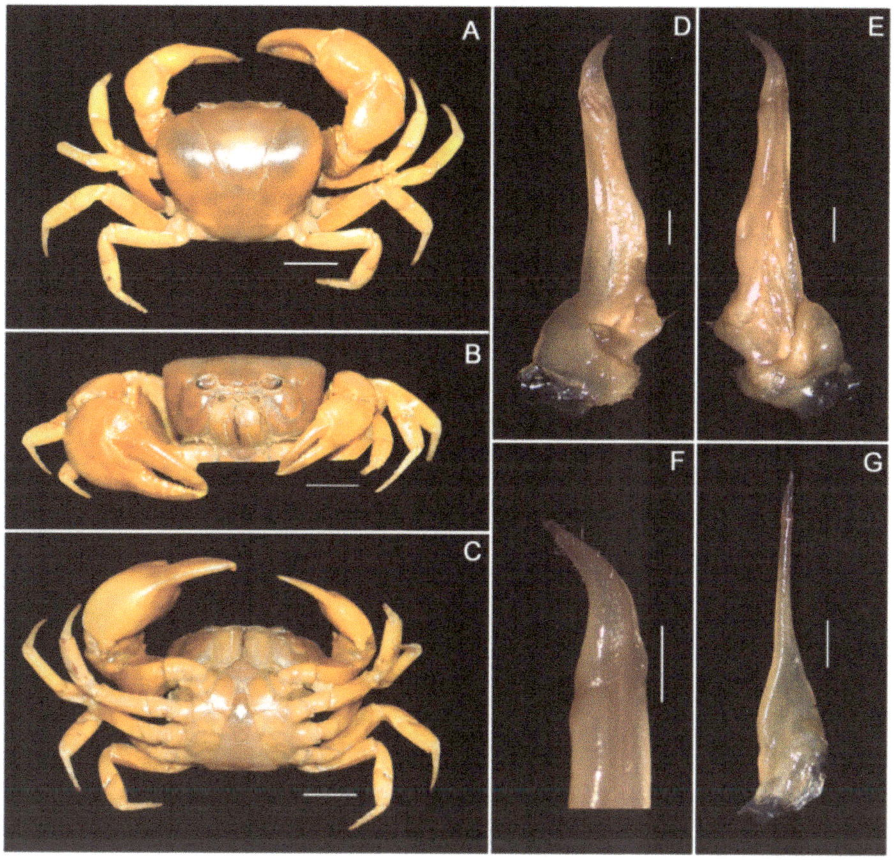

Figure 2 *Arcithelphusa cochleariformis* sp. nov., holotype male (ZSI, WRC-C.1188). **(A)** Carapace, dorsal view; **(B)** carapace, frontal view; **(C)** carapace, ventral view; **(D)** right G1, dorsal view; **(E)** right G1, lateral view; **(F)** right G1 terminal segment, lateral view; **(G)** right G2. Scale bars = 10 mm (A-C) and 1 mm (D-G).

25.76 mm, ch 22.24 mm, fw 7.40 mm), Ondayangadi, 5 km northeast of Mananthavady, Wayanad district, Kerala (11.82463552° N and 76.03024006° E), altitude 760 m, 16 November 2013, coll. Ammini (ZSI, WRC-C.1190).

Diagnosis. As for genus.

Description of holotype. Carapace smooth, squarish (cl/cw = 0.9), highly arched (ch/cw = 0.7), box-like in frontal view; anterolateral carapace highly inflated in frontal view; anterolateral margin with distinct, oblique, short striations; posterolateral margin with fine, oblique, short striations; epigastric cristae indistinct, represented by two lumps; postorbital cristae indistinct; external orbital angle small, poorly developed; epibranchial tooth indistinct; postorbital region flat; branchial regions inflated; subhepatic region slightly rugose; cervical grooves narrow, shallow, slightly broad anteriorly, almost reaching to level of postorbital cristae; mesogastric groove narrow, shallow, bifurcated posteriorly; H-shaped gastric groove visible; front smooth, flat, narrow, 0.25 times cw, without frontal median triangle; epistomal median lobe triangular, lacking median tooth (Figures 2A,B and 3A,B).

Chelipeds smooth, highly unequal, left chela larger; palm of larger chela almost equal in length to height and length of dactylus; teeth on fingers of both chelipeds blunt with four or five enlarged teeth on fingers of larger chela; distinct gap when fingertips of larger chela are in contact; tip of fingers blunt; carpal spine distinct, slightly blunt; outer surface of merus rugose (Figure 2A,B,C).

Ambulatory legs (P2 to P5) smooth, slender; dactylus equal in length to propodus; longest propodus (P3) twice as long as broad (Figure 2A,C).

Suture between male thoracic sternites S2, S3 prominent, deep, not reaching lateral margins and between S3, S4 prominent, deep, reaching lateral margins (Figures 2C and 3C).

Male abdomen narrowly triangular; sixth abdominal somite distinctly broader than long, shorter than telson, lateral margins slightly convex; telson longer than broad, narrowly triangular, with slightly concave outer margins, apex acutely triangular; male sternoabdominal cavity deep, long, reaching imaginary line joining bases of chelipeds (Figures 2C and 3D).

Mandibular palp with two joints, terminal joint bilobed. Exopod of first, second maxillipeds each with long flagellum. Exopod of third maxilliped lacking flagellum, longer than ischium; ventral sulcus of ischium deep, long, parallel to inner margins and towards inner sides; depression on merus towards inner sides (Figure 3E).

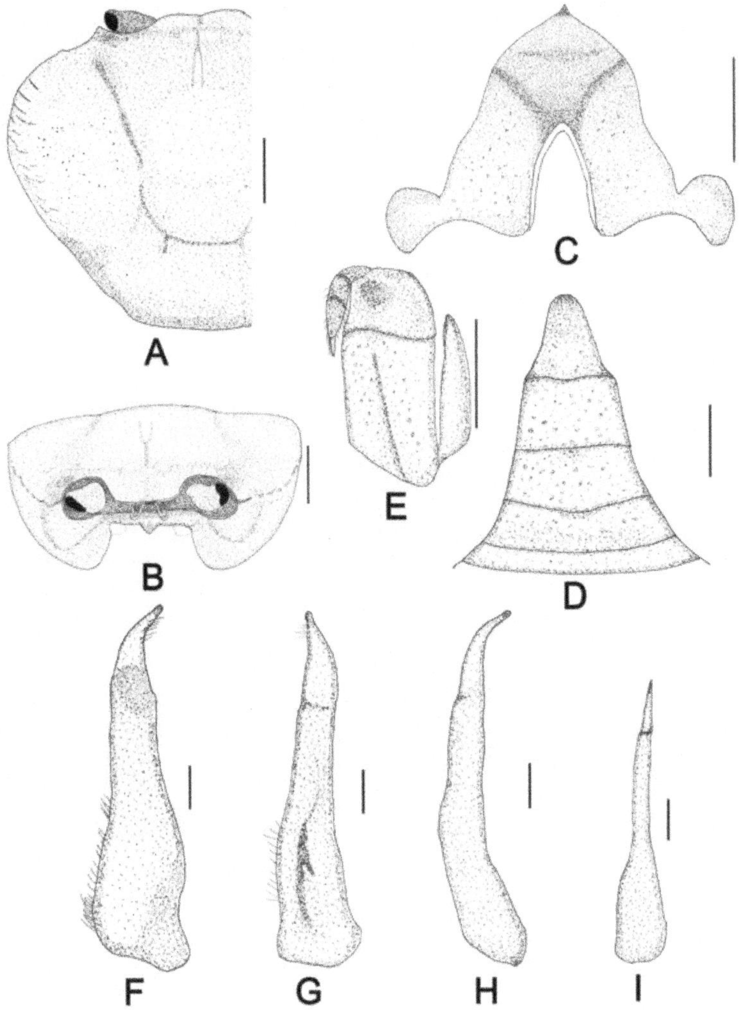

Figure 3 *Arcithelphusa cochleariformis* sp. nov., holotype male (ZSI, WRC-C.1188). **(A)** Carapace (left half), dorsal view; **(B)** carapace, frontal view; **(C)** anterior male thoracic sternites; **(D)** male abdomen; **(E)** left third maxilliped; **(F)** right G1, dorsal view; **(G)** right G1, ventral view; **(H)** right G1, lateral view; **(I)** right G2. Scale bars = 5 mm (A-E) and 1 mm (F-I).

G1 long, stout, appearing as lateral view of spoon in both dorsal and lateral views, with short, cylindrical terminal segment, 0.30 times length of subterminal segment; terminal segment curved more outward at mid-length towards dorsal side, tip narrow, blunt; subterminal segment basal half broader than distal half (Figures 2D,E,F and 3F,G,H). G2 short, with very short distal segment, 0.2 times length of basal segment (Figures 2G and 3I).

Colour. In life, carapace dark brown, chelipeds and ambulatory legs light brown.

Type locality. Ondayangadi, 5 km northeast of Mananthavady, Wayanad district, Kerala, India (11.8231649° N and 76.0232338° E) (altitude 767 m).

Paratypes. The paratypes do not differ much from the holotype in carapace morphology and G1 structure. However, minor differences were observed in carapace morphology of some specimens. In one paratype (ZSI, WRC-C.1190, largest male), the epistomal median lobe is broadly triangular in opposition to triangular epistomal median lobe in the holotype and other paratypes. The right chela is comparatively large in five male and two female paratypes ranging from smallest to largest size. In most of the paratypes (seven males and one female), the fingertips are blunt to sharp. Colour variations were noted among the paratypes (except male specimens of ZSI, WRC-C.1189): one female paratype (ZSI, WRC-C.1189) with dark brown carapace, and brown chelipeds and ambulatory legs; one paratype (ZSI, WRC-C.1190, largest male) has a yellowish brown dorsal surface medially, dark to reddish brown on the sides, and with mainly reddish brown chelipeds and ambulatory legs; the rest of the paratypes (ZSI, WRC-C.1190) have reddish brown carapaces, chelipeds and ambulatory legs.

Etymology. The species epithet is derived from *cochleariformis* (Latin for 'spoon-shaped'), referring to the shape of

Figure 4 Habitat of *Arcithelphusa cochleariformis* sp. nov. at type locality, Ondayangadi. **(A)** Betel nut plantations; **(B)** burrows along water channels (yellow arrow head marks); **(C)** live crab near a burrow.

the first male gonopod in dorsal and lateral views that resembles lateral view of a spoon.

Ecological notes. Specimens were found in shallow to deep burrows along the water channels of betel nut plantations (Figure 4), which were once rice fields (A.R. Sudha Devi, unpublished data). Forest or grasslands were the natural habitats for the new species before their conversion to rice fields. These semi-terrestrial crabs dwell in burrows adjacent to water channels or away from it in open or shaded areas of the banana and betel nut plantations. These crabs are also found in marshy lands and along the paddy field embankments. Normally, their burrow openings are surrounded by wet clods (Figure 4B,C). The burrows are waterlogged during rainy season whereas the burrow openings are sealed with mud during dry seasons in order to maintain humidity inside.

Arcithelphusa cochleariformis is an omnivore species, which mainly feeds on a wide range of plant material (grasses, aquatic vegetation, leaves of colocasia, green algae, creepers and areca nut shells), earthworms and insects (grasshoppers, brown planthoppers and praying mantes) (A.R. Sudha Devi, unpublished data). They also survive on decaying leaves and detritus. Decayed leaves and remnants of food material are often seen accumulated near their burrows. Juvenile crabs prefer aquatic plants, grasses, green algae, moss plants, insect larvae, decayed leaves and detritus as their food. These crabs are mainly nocturnal and are found active during early morning and night. They very rarely forage during the day time. Adult crabs prefer to stay inside the burrows while juveniles mostly spend their time in water. Their predators include birds (herons and egrets) and mammals (wild boars, mongooses and cats). The new species was sympatric with *Vela carli* (Roux, 1931).

Distribution. The species is presently known only from the type locality, Ondayangadi and other localities of Wayanad district in Kerala (Table 1).

Discussion

Arcithelphusa is a monotypic genus and is immediately distinguished from all other Indian gecarcinucid genera except *Ghatiana*, *Globitelphusa*, *Gubernatoriana*, *Inglethelphusa*, *Pilarta* and *Snaha* by the absence of a flagellum on the third maxilliped exopod (Figure 3E). The flagellum on the exopod of the third maxilliped is also absent or vestigial in *Ghatiana*, *Globitelphusa*, *Gubernatoriana*, *Inglethelphusa*, *Pilarta* and *Snaha* (see Pati and Sharma 2014a: Figures three (D),

Table 1 Distribution of *Arcithelphusa cochleariformis* sp. nov. in Wayanad district, Kerala

Locality	Latitude (N)	Longitude (E)	Altitude (m)	Habitat
Bavali	11.8520258°	76.1138744°	709	Paddy fields
Chettapalam	11.7981135°	76.191415°	752	Paddy fields
Ellumannam	11.7867502°	75.9622899°	755	Paddy fields
Kambalakkad	11.6756376°	76.0754501°	789	Betel nut plantations
Muthanga	11.6690787°	76.3695167°	864	Paddy fields
Ondayangadi (type locality)	11.8231649°	76.0232338°	767	Betel nut plantations and paddy fields
Padinjarathara	11.685401°	75.974814°	753	Betel nut plantations
Pakkom	11.8020239°	76.1015508°	724	Paddy fields
Panamaram	11.74057°	76.07579°	724	Betel nut plantations and paddy fields
Thirunelly	11.9081185°	75.9971183°	809	Paddy fields

five (D) for *Ghatiana*, Figure seven (D) for *Gubernatoriana*; Alcock 1910: Plate 8, Figure thirty-one (b) for *Globitelphusa*, Plate 9, Figure thirty-four (b) for *Inglethelphusa*; Bahir and Yeo 2007: Figure twenty-five (B) for *Pilarta*, Figure twenty-seven (A) for *Snaha*). However, the highly arched carapace with a narrow front and the prominently curved G1 terminal segment easily distinguish the new genus from *Ghatiana*, *Globitelphusa*, *Gubernatoriana*, *Inglethelphusa*, *Pilarta* and *Snaha* (Figures 2B,D,E,F and 3B,F,H; see Pati and Sharma 2014a: Figures two (B), three (E), four (B), five (E) for *Ghatiana*; Bott 1970: Plate 33, Figures thirteen, fifteen for *Globitelphusa*; Pati and Sharma 2014b: Plate 7, Figure seven (B,D), Plate 8, Figure eight (B,D) for *Gubernatoriana*, Plate 9, Figure nine (B,D) for *Inglethelphusa*; Bahir and Yeo 2007: Figures twenty-five (F), twenty-six (B) for *Pilarta*, Figures twenty-seven (C), twenty-eight (B) for *Snaha*). Further, the prominent and deep suture between thoracic sternites S2, S3 and S3, S4 separates the new genus from the above genera except *Snaha* (Figures 2C and 3C; see Bahir and Yeo 2007: Figures twenty-eight (C), twenty-nine (A)). As in the genera *Ghatiana*, *Gubernatoriana* (except *G. basalticola* Klaus, Fernandez & Yeo, 2014) and *Inglethelphusa*, the G2 distal segment of the new genus is short (Figures 2G and 3I; see Pati and Sharma 2014a: Figures three (H), five (H) for *Ghatiana*; Pati and Sharma 2014b: Plate 7, Figure seven (F), Plate 8, Figure eight (F) for *Gubernatoriana*, Plate 9, Figure nine (F) for *Inglethelphusa*) whereas *Globitelphusa*, *Pilarta* and *Snaha* have a long G2 distal segment (see Bahir and Yeo 2007: Figure twenty-five (H) for *Pilarta*, Figures twenty-seven (G), twenty-nine (E) for *Snaha*). The recently described species, *G. basalticola*, has an elongated G2 distal segment and more wide frontal margin, indicating that the species should be placed in a new genus (Klaus et al. 2014). Other than its short to vestigial G2 distal segment, *Arcithelphusa* also resembles *Ghatiana*, *Gubernatoriana* and *Inglethelphusa* in its squarish appearance of the carapace (Figure 2A; see Pati and Sharma 2014a: Figures two (A), four (A) for *Ghatiana*; Pati and Sharma 2014b: Plate 7, Figure seven (A), Plate 8, Figure eight (A) for *Gubernatoriana*, Plate 9, Figure nine (A) for *Inglethelphusa*). Nonetheless, the new genus is distinguished from *Ghatiana*, *Gubernatoriana* and *Inglethelphusa* by the narrowly triangular male abdomen and distinctly curved, comparatively short G1 terminal segment, 0.3 times the length of the subterminal segment (Figures 2C,D and 3D,F) (vs. the T-shaped male abdomen and slightly curved to straight, comparatively long G1 terminal segment, 0.35 to 0.80 times the length of the subterminal segment) (see Pati and Sharma 2014a: Figures two (C), three (B,E), four (C), five (B,E) for *Ghatiana*; Pati and Sharma 2014b: Plate 7, Figure seven (C,D), Plate 8, Figure eight (C,D) for *Gubernatoriana*, Plate 9, Figure nine (C,D) for *Inglethelphusa*).

The new genus is also similar to the genus *Cylindrotelphusa* Alcock, 1909 in some carapace morphology and male gonopod structures, notably in the highly convex carapace, narrow front, small external orbital angle, narrowly triangular male abdomen, narrowly triangular telson with concave outer margins and an acutely triangular apex, highly curved G1 terminal segment, and short G2 distal segment (Figures 2A,B,C,D,E,F,G and 3B,D,F,H,I; see Bahir and Yeo 2007: Figures four (A,D), five (A,B,C)). However, the new genus, in addition to lacking a flagellum on the third maxilliped exopod, can be differentiated from *Cylindrotelphusa* by the relatively squarish (cl/cw = 0.9) and more convex carapace (ch/cw = 0.7), striated anterolateral margin, indistinct postorbital cristae, indistinct epibranchial tooth, prominent and deep suture between thoracic sternites S2, S3 and S3, S4, stout G1 with a distinctly curved, short terminal segment, 0.3 times the length of the subterminal segment (Figures 2A,C,D and 3A,C,F) [vs. the relatively transverse (cl/cw = 0.8) and less convex carapace (ch/cw = 0.5), serrated anterolateral margin, distinct postorbital cristae, small, blunt epibranchial tooth, less prominent and shallow suture between thoracic sternites S2, S3 and S3, S4, slender G1 with a sickle-shaped, long terminal segment, 0.55 times the length of the subterminal segment] (see Bahir and Yeo 2007: Figures four (A), five (A,C)). *Arcithelphusa cochleariformis*, new species, is highly distinct among the other Indian freshwater crabs due to its squarish, highly arched carapace, appearing box-like in frontal view, and long, stout, spoon-shaped G1. The new species is morphologically similar to *Cylindrotelphusa steniops*, particularly due to its deep carapace, narrow front, poorly developed external orbital angle, prominently curved G1 terminal segment and short G2 distal segment. However, both the species are markedly different in many other morphological features as discussed above.

Altogether, 14 genera and 36 species of gecarcinucid freshwater crabs are now found in the Western Ghats of India: *Arcithelphusa* gen. nov. [*A. cochleariformis* sp. nov.]; *Baratha* Bahir & Yeo, 2007 [*B. peena* Bahir & Yeo, 2007, *B. pushta* Bahir & Yeo, 2007]; *Barytelphusa* Alcock, 1909 [*B. cunicularis* (Westwood, 1836), *B. guerini* (H. Milne Edwards, 1853), *B. mccanni* (Chopra & Das, 1935)]; *Cylindrotelphusa* [*C. steniops*]; *Gecarcinucus* H. Milne Edwards, 1844 [*G. edwardsi* Alcock, 1909, *G. jacquemontii* H. Milne Edwards, 1844]; *Ghatiana* [*G. aurantiaca*, *G. hyacintha* Pati & Sharma 2014]; *Gubernatoriana* [*G. basalticola*, *G. gubernatoris*, *G. pilosipes* (Alcock, 1909), *G. triangulus* Pati & Sharma 2014]; *Inglethelphusa* [*I. fronto*]; *Lamella* Bahir & Yeo, 2007 [*L. lamellifrons* (Alcock, 1909)]; *Pilarta* [*P. anuka*]; *Snaha* [*S. aruna*, *S. escheri* (Roux, 1931)]; *Travancoriana* Bott, 1969 [*T. charu* Bahir & Yeo, 2007, *T. convexa* (Roux, 1931), *T. granulata* Pati & Sharma, 2014, *T. kuleera* Bahir & Yeo, 2007, *T. pollicaris* (Alcock, 1909), *T. schirnerae* Bott, 1969]; *Vanni* Bahir & Yeo, 2007 [*V. ashini* Bahir & Yeo, 2007, *V. deepta* Bahir & Yeo, 2007, *V. giri* Bahir & Yeo, 2007, *V. malabarica* (Henderson, 1912), *V. nilgiriensis* (Roux, 1931), *V. pusilla* (Roux, 1931), *V. travancorica* (Henderson, 1912)]; *Vela* Bahir & Yeo, 2007 [*V. carli*, *V. pulvinata* (Alcock, 1909), *V. virupa* Bahir & Yeo, 2007].

Ecological requirements (habitat, food and feeding habits) are the same for *A. cochleariformis* and the sympatric species, *V. carli* (A.R. Sudha Devi, unpublished data). Both the species live in burrows adjoining to stagnant or running water and also in burrows away from water channels in open or shaded areas. *Vela carli* is primarily an omnivore or scavenger, but sometimes it shows opportunistic carnivore behaviour by preying upon small rats and water snakes. Suitable habitats containing optimum water levels (marshy land, paddy fields, betel nut or banana plantations) and availability of rich food sources, particularly from plant materials (including decaying leaves) and plant-associated animals (earthworms, insects), are the key factors for survival of these crabs in the Western Ghats of Wayanad district. In these circumstances, possibilities of occurrence of *A. cochleariformis* in other areas of Kerala and neighbouring states with similar conditions of habitat cannot be ignored.

Arcithelphusa cochleariformis is exposed to several natural and man-made threats. Predation of these crabs by birds and mammals is a noteworthy natural disturbance. Local tribal communities also consume large quantities of these crabs for their food. Applications of various pesticides including herbicides and insecticides may pose a serious threat to this species. Paddy fields and banana plantations in the region have already been contaminated by these chemicals whereas betel nut plantations are yet free from pesticide applications. Competition of the new species with the sympatric *V. carli* for habitat and food is also a threat. However, availability of adequate natural resources might be a reason for the successful colonization of these crabs. A severe threat to the new species is currently not recognized, which is evident from their abundant population and wide distribution in Wayanad district (A.R. Sudha Devi, unpublished data).

Conclusions

New taxa of freshwater crabs are described and compared with the related genera and species. In total, 36 species in 14 genera of gecarcinucid crabs are now recognized from the Western Ghats of India. More new taxa can be expected from this exceptionally biodiversity-rich area if the necessary surveys are done. The present discovery of the new species is useful for taxonomists as well as conservation managers.

Competing interests

The authors declare that they have no competing interests.

Authors' contributions

ARS collected the material and data. SKP conceived the study, prepared the illustrations and drafted the manuscript. Both SKP and ARS participated in the final identification, described the new taxa and approved the final manuscript

Authors' information

SKP, Ph.D., is working as a carcinologist in the Zoological Survey of India, Western Regional Centre, Pune, Maharashtra. ARS, Ph.D., is an associate professor in the Department of Zoology and Research Centre, Mary Matha Arts and Science College, Mananthavady, Wayanad, Kerala.

Acknowledgements

We are thankful to the Director Dr. K. Venkataraman, Zoological Survey of India, Kolkata and the Officer-in-Charge Dr. P.S. Bhatnagar, Zoological Survey of India, Western Regional Centre, Pune for the constant encouragement and facilities.

Author details

[1]Zoological Survey of India, Western Regional Centre, Vidyanagar, Sector 29, P.C.N.T. Post, Akurdi, Pune 411 044, Maharashtra, India. [2]Department of Zoology and Research Centre, Mary Matha Arts and Science College, Vemom P.O., Mananthavady 670 645, Kerala, India.

References

Alcock A (1910) Catalogue of the Indian decapod Crustacea in the collection of the Indian Museum. Part I. Brachyura. Fasciculus II. The Indian fresh-water crabs - Potamonidae. Indian Museum, Calcutta

Bahir MM, Yeo DCJ (2007) The gecarcinucid freshwater crabs of southern India (Crustacea: Decapoda: Brachyura). Raffles B Zool 16:309–354

Bott R (1970) Die Süßwasserkrabben von Europa, Asien, Australien und ihre Stammesgeschichte. Eine Revision der Potamoidea und Parathelphusoidea (Crustacea, Decapoda), vol 526. Abh Senckenb Naturforsch Ges, Frankfurt, pp 1–338

Klaus S, Fernandez K, Yeo DCJ (2014) Phylogeny of the freshwater crabs of the Western Ghats (Brachyura, Gecarcinucidae). Zool Scr 43(6):651–660, doi:10.1111/zsc.12078

Ng PKL, Tay FWM (2001) The freshwater crabs of Sri Lanka (Decapoda: Brachyura: Parathelphusidae). Zeylanica 6(2):113–199

Pati SK, Sharma RM (2013) A new species of freshwater crab, *Travancoriana granulata* n. sp. (Brachyura: Gecarcinucidae) from the southern Western Ghats of India. Zoosyst Evol 89(2):275–281, doi:10.1002/zoos.201300013

Pati SK, Sharma RM (2014a) Description of *Ghatiana*, a new genus of freshwater crab, with two new species and a new species of *Gubernatoriana* (Crustacea: Decapoda: Brachyura: Gecarcinucidae) from the Western Ghat Mountains, India. J Nat Hist 48(21–22):1279–1298, doi:10.1080/00222933.2013.859315

Pati SK, Sharma RM (2014b) Freshwater crabs (Crustacea: Decapoda: Brachyura: Gecarcinucidae) in the collection of the Western Regional Centre, Pune. Rec Zool Surv India 363:1–44

Raghavan R, Dahanukar N, Philip S, Iyer P, Kumar B, Daniel B, Molur S (2014) The conservation status of decapod crustaceans in the Western Ghats of India: an exceptional region of freshwater biodiversity. Aquatic Conserv: Mar Freshw Ecosyst. doi:10.1002/aqc.2490

Roux J (1931) Crustacés Décapodes d'eau douce de l'Inde méridionale. Rev Suisse Zool 38(1):31–62

3

Cymothoa frontalis, a cymothoid isopod parasitizing the belonid fish *Strongylura strongylura* from the Malabar Coast (Kerala, India): redescription, description, prevalence and life cycle

Aneesh Panakkool Thamban[1], Sudha Kappalli[1*], Helna Ameri Kottarathil[1], Anilkumar Gopinathan[2] and Trilles Jean Paul[3]

Abstract

Background: *Cymothoa frontalis* Milne Edward, 1840 is a very poorly described cymothoid, notwithstanding the previous redescription of the female. Pertinently, to date, the host of *C. frontalis* has not been identified with adequate precision. Most of the descriptions of cymothoids carried out hitherto were based primarily on females, but practically ignoring other life cycle stages. The present paper redescribes the female and describes other life cycle stages of the species *C. frontalis* to get better precision in their identification.

Results: The female phase of *C. frontalis* is redescribed according to type specimens extant in the NMNH, Paris, and also by the data obtained from live specimens collected during the present study. The general morphology and appendages of various life cycle stages of the species are described. Among 80 fish species from 35 families examined, *C. frontalis* was recovered only from *Strongylura strongylura* signifying its oligoxenous host specificity, the prevalence and intensity being 68.65% and 1.9, respectively. Each host fish in more than 85% of the population was infested with a pair of *C. frontalis*, in three combinations, predominantly with male-female pair (70.9%). *C. frontalis* exhibited strict site specificity attaching to the buccal cavity of the host fish. The study has also identified three major phases (marsupial, free living and infective) in the life cycle of *C. frontalis*. The zygotic-staged marsupiumites were developed through five sequential ontogenetic stages. The manca released from the marsupium become infective after a brief period of free swimming life. During the infective phase, *C. frontalis* completes remaining life cycle stages with successive moulting. Further, six successive stages of the ovigerous females have also been identified.

Conclusions: The present redescription of the female and the description of transitional, male, juvenile and larvae of *C. frontalis* facilitate precise identification of the species at any stage of the life cycle. Further, the strict host and site specificities of the parasite, as borne out from the present study, and its high degree of prevalence in the host make *C. frontalis* as an ideal model organism to study the strategies to be adopted for the management of parasites infesting edible fishes.

Keywords: *Cymothoa frontalis*; Cymothoidae; Redescription; Description; Prevalence; Life cycle; *Strongylura strongylura*; India

* Correspondence: ksudha50@rediffmail.com
[1]Post Graduate Department of Zoology and Research Centre, Sree Narayana College, Kannur 670 007, India
Full list of author information is available at the end of the article

Background

Cymothoids are ectoparasitic isopods parasitizing teleost fishes inhabiting diverse ecosystems (Trilles 1968, 1994; Brusca 1981; Seng and Seng 1990; Horton and Okamura 2001; Trilles et al. 2011; Hadfield et al. 2013; Aneesh et al. 2014) and have been shown to cause deleterious effects on their hosts (Trilles and Hipea-Jacquotte 2012). For instance, the anemone fish *Amphiprion akallopisos, Stolephorus commersonii* and *Hemiramphus far* were severely damaged due to the parasitic infestation by *Renocila heterozota, Nerocila phaeopleura* and *Mothocya plagulophora,* respectively (Bowman and Mariscal 1968; Rajkumar et al. 2006; Gopalakrishnan et al. 2010). Five species of *Nerocila* (*N. acuminata, N. phaiopleura, N. loveni, N. depressa* and *N. longispina*) were reported to have induced severe tissue damage in seven fish species such as *Gymnothorax eurostus, Thryssa malabarica, Thryssa mystax, Thryssa setirostris, Coilia dussumieri, Opisthopterus tardoore, Escualosa thoracata* and *Ambassis ambassis* (Bowman 1960; Aneesh et al. 2013a). Further, seasonal fluctuations in the population of four *Nerocila* species were reported from the Malabar Coast of India (Aneesh et al. 2013a). Recent studies conducted in our laboratory have shown profuse infestation of the Indian mackerel *Rastrelliger kanagurta* and the slender needle fish *Strongylura leiura* by *Norileca indica* and *Mothocya renardi,* respectively (Aneesh et al. 2011, 2013b). Most of the taxonomic descriptions of the cymothoids are based primarily on ovigerous females; the various life cycle stages (excepting perhaps the premanca/manca of few *Cymothoa* sp.) have not received much attention thus far (Richardson 1905; Sartor and Pires 1988; Thatcher et al. 2003, 2007; Trilles and Bariche 2006; Hadfield et al. 2011). Keeping these in view, the present paper addresses the question of precise identification of the cymothoid *C. frontalis* infesting the belonid fish inhabiting the Malabar Coast.

The genus *Cymothoa* Fabricius, 1793 comprises about 50 species which usually infest the tongue of the host fish (Trilles 1994; Schotte et al. 2008); some of them are reported to induce atrophy of the host's tongue (Brusca 1981; Romestand and Trilles 1977; Hadfield et al. 2011, 2013).

Cymothoa frontalis (Milne Edwards, 1840) is a poorly described one, and though the female stage of this species was redescribed by Schioedte and Meinert (1884), their Latin redescription seems inadequate in terms of modern standards. In spite of being a protandric hermaphrodite, only the female stage was subjected to detailed examination thus far; further, its host was not accurately identified even though a single host '*Cyclopterus*' was doubtfully reported by White (1847). The present study depicts a redescription of the female stage of *C. frontalis.* It also describes all the other life cycle stages (transitional stage, male, juvenile manca and premanca) of this cymothoid using the live samples collected afresh from the Indian spot tail needle fish *Strongylura strongylura* (Beloniformes, Belonidae). The study also brings out the results of our thorough observation on the host specificity, site specificity, prevalence, intensity and life cycle of the species.

Methods

Live marine fishes were collected freshly from the Ayyikkara fish-landing centre (Lat. 11°51′N, Long. 75°22′E; Malabar Coast, India). The various body parts of the fish were examined (immediately after collection) for the presence of cymothoids using a hand lens. The recovered parasites with their host were brought to the laboratory and subjected to detailed examination with the aid of a stereomicroscope (Leica-S6D, Leica Microsystems Ltd., Milton Keynes, UK). The cymothoids thus obtained were preserved in 75% ethanol (after Ramakrishna 1980) and subjected to further taxonomical identification. The survey was conducted twice a week round the year from November 2009 to November 2012. The prevalence (*P*) and intensity (*I*) of the recovered *C. frontalis* were calculated as per the methods described in Margolis et al. (1982) and Bush et al. (1997). The total length of the fish host as well as the parasite (*C. frontalis*) was measured to the nearest 1 mm. Mouthparts and appendages of the parasite were carefully dissected out using a dissecting needle and a fine pair forceps and observed under the microscope (Leica DM-750). The drawings of the observed mouthparts and appendages were performed using a camera lucida attached to the microscope. The moult stages and the moult-related changes of the parasite were determined through the macroscopical and microscopical observations on the characteristic changes occurring in the carapace and the appendages (after Sudha et al. 2012). The reproductive activity was assessed through microscopic observations of the ovary (after dissecting out) and marsupiumites (after Sudha et al. 2012; Aneesh 2014). Classification of the marsupial stages was carried out taking cues from Bakenhaster et al. (2006). Fish taxonomy and host nomenclature were performed according to FishBase (Froese and Pauly 2012). Voucher specimens were deposited in the collections of the Parasitic Crustacean Museum, Crustacean Biology Research Laboratory, Sree Narayana College, Kannur, Kerala, India.

Results

Systematics

Suborder: Cymothoida Wagele, 1989
Superfamily: Cymothooidea Leach, 1814
Family: Cymothoidae Leach, 1814
Genus: *Cymothoa* Fabricius, 1793
Restricted synonymy: Fabricius 1793: 503; Milne Edwards 1840: 264-269; Schioedte and Meinert 1884: 223-225;

Kussakin 1979: 289; Brusca 1981: 185; Trilles 1994: 137; Hadfield et al. 2013: 153.

Type species: *Oniscus oestrum* Fabricius, 1793, by subsequent designation (Kussakin 1979).

Cymothoa frontalis Milne Edward, 1840

Cymothoa frontale (*Cymothoé frontal*) Milne Edwards 1840: 271.

Cymothoa frontalis - White 1847: 110-?; Dana 1853: 750-751, pl. 49, Figure twelve (a-b); Schioedte and Meinert 1884: 226-228, tab. VI (Cym. XXIV), Figures one and two; Gerstaecker 1901: 261; Nierstrasz 1931: 136; Trilles 1975: 980-981, pl. 1; Avdeev 1978: 282; Avdeev 1979: 53; Avdeev 1990: 32-42; Trilles 1994: 143.

Material examined:

Type material: Ovigerous female (LT. 24 mm) and non-ovigerous female (LT. 18 mm), Mer des Indes, Dussumier, *Cymothoa frontalis* M. Edw., type (MNHN N° 202).

Non-type material: From *S. strongylura*, Ayyikkara fish landing centre (Lat. 11°51′N, Long. 75°22′E), Malabar Coast, India; collected by Aneesh, P. T., Helna A. K. and Sudha, K. and deposited in the PCM. Ovigerous female (LT 26 mm) with manca in the brood pouch (PCM N° CF-09), 12 July 2011; non-ovigerous female without brood pouch (LT 28 mm) (PCM N° CF-06), 16 August 2010; ovigerous female with pre-manca in the brood pouch (LT 23 mm) (PCM N° CF-04), 19 May 2010; ovigerous female with eggs in the brood pouch (LT 23 mm) (PCM N° CF-01), 8 December 2009; ovigerous female with eggs in the brood pouch (LT 21 mm) (PCM N° CF-03), 16 January 2010; male (LT 13 mm) (PCM N° CF-02), 8 December 2009; transitional stage (LT 20 mm) (PCM N° CF-08), 25 April 2011; juvenile (LT 8 mm) (PCM N° CF-07), 13 March 2011; 30 manca released by the specimen PCM N° CF-09 (LT 3.2 to 3.4 mm), (PCM N° CF-10), 12 July 2011; 20 pre-manca collected from the brood pouch of the specimen PCM N° CF-04 (LT 2.2 to 2.4 mm) (PCM N° CF-05), 19 May 2010.

Redescription of the ovigerous female

Body about 2.6 times as long as wide, widest at pereonites 5 to 6 (Figures 1D, 2, 3 and 4; Table 1). Cephalon partially immersed in poorly developed amphicephalic process of pereonite 1, as long as wide, anterior margin truncate. Eyes relatively smaller than male, not much developed but moderately distinct. Coxae not visible in dorsal view. Pereonite 1 longest, pereonites 2 to 4 subequal in length; pereonites 5 to 7 decreasing in length progressively, pereonite 7 shortest. The pereonites increase in width from 1 to 6; pereonite 6 widest and slightly wider than pereonite 7. All the pleonites visible, increasing progressively in width posteriorly from 1 to 5; pleonite 5 widest and longest. Pleotelson 1.6 times as wide as long, wider than pleonite 5, posterior margin broadly rounded.

Antennule stouter than antenna, subequal in length to antenna, composed of 8 articles, extending to anterior margin of pereonite 1; articles 1 to 3 slightly wider than others. Antenna with 9 articles, decreasing gradually in width, distal margin of articles 6, 7 and 8 with 3 to 6 setae. Article 9 with 3 to 6 terminal aesthetascs. Mandible palp 3 segmented, without setae in ovigerous female. Maxillule with 4 apical spines slightly recurved. Bi-lobed maxilla with 2 spines on median lobe and 1 spine on lateral lobe. Maxilliped with oostegial lobe with many plumose setae and a palp with 3 small apical and 2 lateral recurved spines on article 3.

Pereopods gradually increase in size from 1 to 7, without spines; basis increasing in width progressively from pereopods 1 to 7. Pleopods not distinctly visible in dorsal view; pleopod 2 without appendix masculina. Endopods of pleopods 4 and 5 moderately and strongly folded, respectively. Uropod rami short, not reaching the distal margin of pleotelson; rami subequal in length, curved and apically rounded.

Brood pouch with 4 pairs of overlapping oostegites arising from the bases of pereopods 2, 3, 4 and 6. Number of eggs or larvae per brood pouch ranges from 320 to 520 according to the size of the female.

Description of the male

Relatively smaller than ovigerous female and transitional stage (Figures 1B, 5 and 6; Table 1). Body 2.67 times as long as wide. Cephalon not immersed in pereonite 1. Eyes small but more prominent than in the female.

Pereonite 1 longest, pereonites 5 and 6 subequal in width and widest. Coxae not visible in dorsal view. All pleonites visible increasing progressively in width posteriorly; pleonite 5 widest and longest. Pleotelson 2 times as wide as long, slightly wider than pleonite 5, posterior margin broadly rounded.

Eight articled antennule stouter and subequal in length to antenna. First 3 articles slightly wider than others. Antennae 9 articled, decreasing gradually in width, distal margin of articles 6, 7 and 8 with 3 to 6 setae; article 9 with 3 terminal aesthetascs. Mandible palp 3 segmented, articles 2 and 3 with several setae on distolateral margin. Maxillule with 4 apical spines slightly recurved; maxilla bi-lobed, with 2 spines on median and lateral lobes. Maxilliped without oostegial lobe and a palp with 3 small apical and 2 lateral recurved spines on article 3.

Pereopods 1 to 7 gradually increasing in size, without spines; basis increasing in width progressively. Penes visible on sternite 7. Pleopods not distinctly visible in dorsal view. Pleopod 2 with appendix masculina. Endopodite of pleopods 4 moderately and 5 strongly folded. Uropod rami, nearly reaching the distal margin of pleotelson; rami subequal in length, curved and apically rounded.

Figure 1 *Cymothoa frontalis*. **(A)** Host fish *Strongylura strongylura*. **(B)** Male. **(C)** Transitional. **(D)** Female. **(E)** Pre-manca. **(F)** Manca. **(G)** Juveniles on lower beak of the host fish *S. strongylura*. **(H, I)** *C. frontalis* on *S. strongylura* (*Cf* (M), *Cymothoa frontalis* male; *Cf* (F), *Cymothoa frontalis* female).

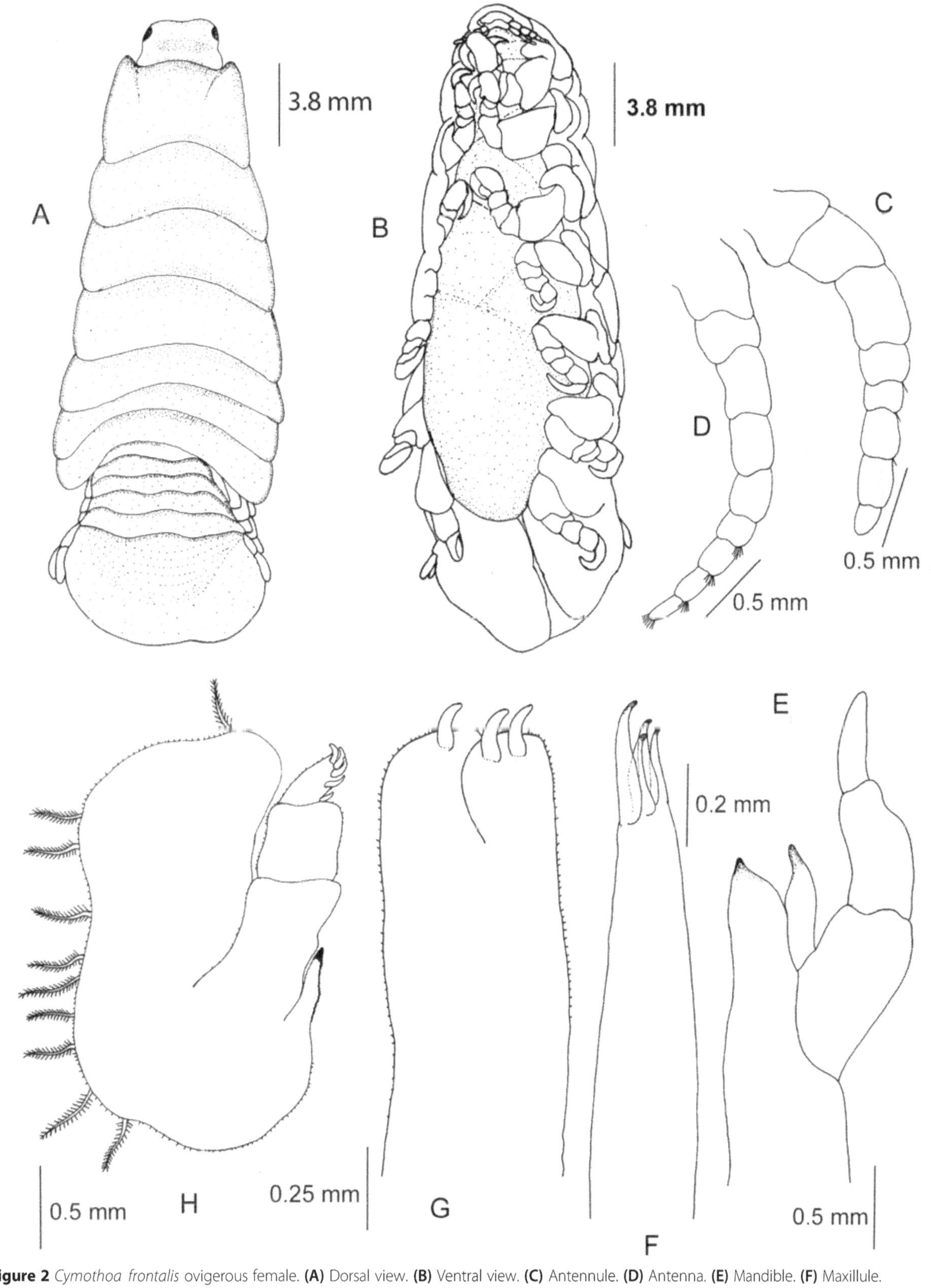

Figure 2 *Cymothoa frontalis* ovigerous female. **(A)** Dorsal view. **(B)** Ventral view. **(C)** Antennule. **(D)** Antenna. **(E)** Mandible. **(F)** Maxillule. **(G)** Maxilla. **(H)** Maxilliped.

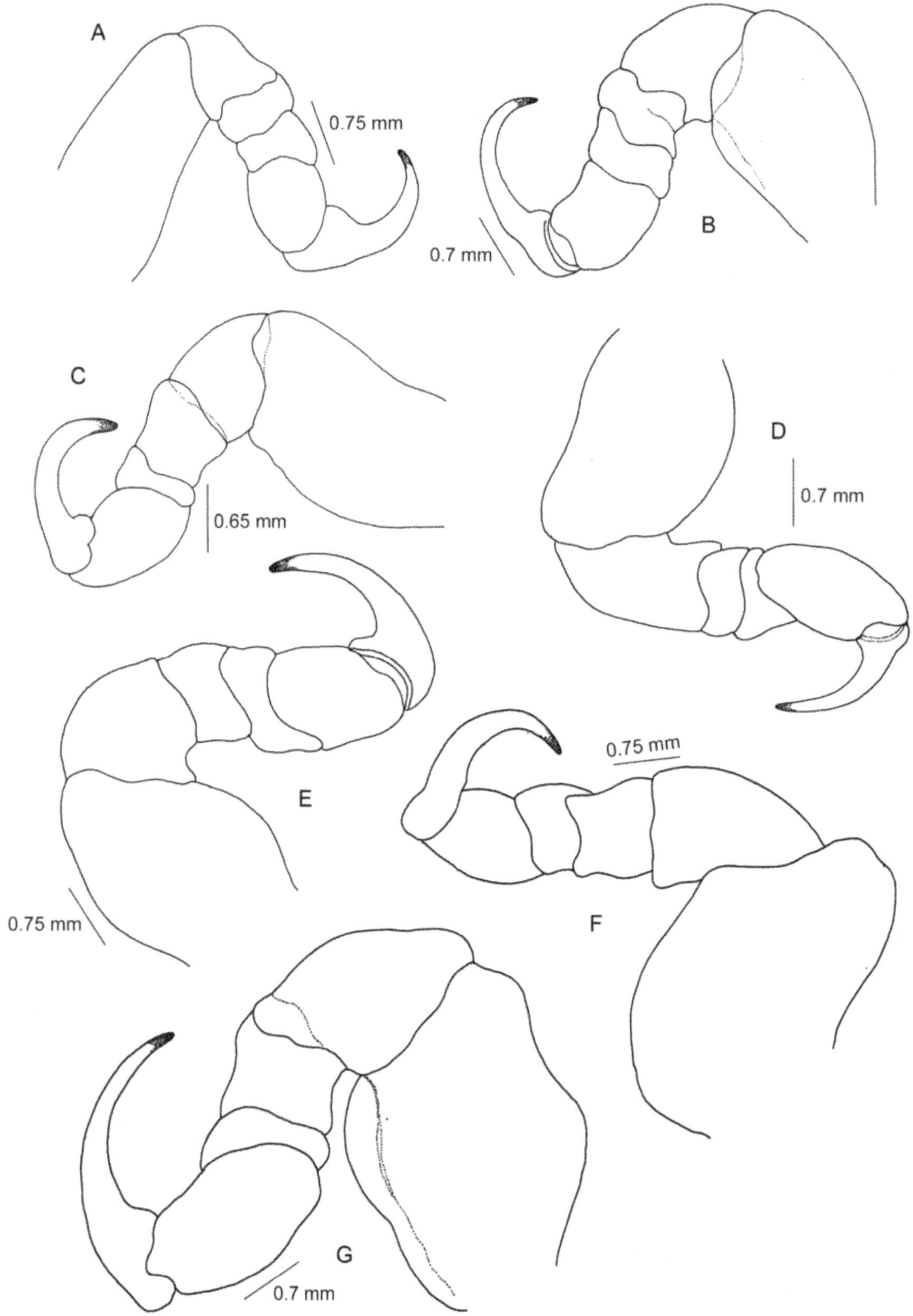

Figure 3 *Cymothoa frontalis* ovigerous female. **(A-G)** Pereopods 1 to 7.

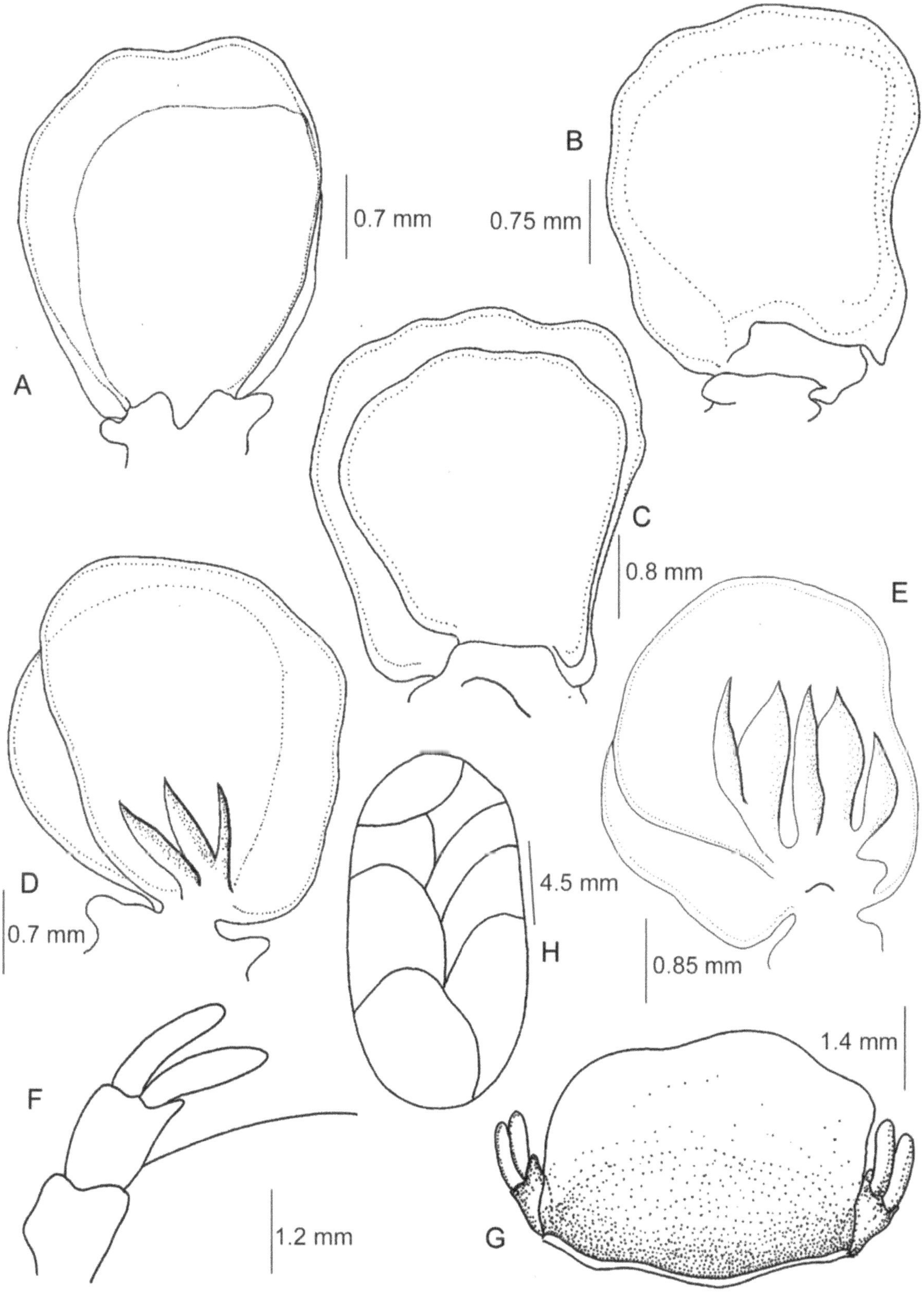

Figure 4 *Cymothoa frontalis* ovigerous female. **(A-E)** Pleopods 1 to 5. **(F)** Uropod with rami. **(G)** Uropod and pleotelson. **(H)** Brood pouch.

Table 1 Morphological characteristics of the different stages of the life cycle of *Cymothoa frontalis*

Characteristics	Ovigerous female	Male	Transitional stage	Juvenile	Manca	Pre-manca
Colour	Light grey with highly pigmented dorsal surface	Creamy white with a pigmented lateral and posterior margin of pereonites	Similar to that of ovigerous female in late stage	Transparent with scattered chromatophores	Clear with scattered chromatophores particularly on uropods and pleotelson	Similar to that of manca
Site of infestation	On the floor of the buccal cavity, the cephalon facing the opening of the host mouth	On the floor of the buccal cavity, on gill arch just behind the female	Similar to that of female	Usually in pairs, attached on the lower mucous membrane of the beak of the host	Found in the brood, then released in to water at inter-moult stage and lead a short free swimming life, and finally the manca either attaches to the body surface of the specific host or may die	Seen with in the brood pouch
Size	2.6 times as long as wide. Widest at pereonites 5 to 6	Body 2.67 times as long as wide	Body 2.7 times as long as wide	Body approximately 3 times as long as wide	Body 3.5 times as long as wide	Body 4 times as long as wide
Cephalon	Partially immersed in poorly developed amphicephalic processes of pereonite 1, as long as wide	Cephalon not immersed in pereonite 1	Cephalon partially immersed in poorly developed amphicephalic processes of pereonite 1, as long as wide, anterior margin truncate	Anterior margin of cephalon rounded, not immersed in pereonite 1	Cephalon wider than long, 1.6 times as wide as long and golden yellow in colour	Cephalon 1.4 times as wider than long and not immersed in pereonite 1
Pereon	With 7 pereonites. Coxae not visible in dorsal view	With 7 pereonites Coxae not visible in dorsal view	With 7 pereonites. Coxae not visible in dorsal view	With 7 pereonites. Coxae not visible in dorsal view	With 7 pereonites. Coxae not visible in dorsal view	With 7 pereonites. Coxae not visible in dorsal view
Pleon	With 5 pleonites	With 5 pleonites	With 5 pleonites	With 5 pleonites	With 5 pleonites	With 5 pleonites
Pleotelson	Pleotelson 1.6 times as wide as long, wider than pleonite 5, posterior margin broadly rounded	Pleotelson 2 times as wide as long, wider than pleonite 5, posterior margin broadly rounded	Pleotelson 1.8 to 2 times as wide as long	Pleotelson as long as wide, apical margin rounded with many setae	Pleotelson 1.2 times as wide as long and the apical margin with 18 to 20 plumose setae	Apical margin without setae, dark pigmentation
Eye	Eyes small, not very developed but moderately distinct	Eyes small, much prominent than that of female	Eyes small and moderately distinct	Eyes prominent, ovate and conspicuous in dorsal view	Eyes black, prominent, ovate and conspicuous in dorsal view	Eyes black, prominent, ovate and conspicuous in dorsal view
Antennule	Antennule, stouter than antenna, subequal in length to antenna, composed of 8 articles; articles 1 to 3 slightly wider than others	Antennule, stouter than antenna, subequal in length to antenna, 8 articled; first three articles slightly wider than other	Similar to that of ovigerous female	Antennule 8 articled, extending to anterior margin of pereonite 1; articles 5 to 8 with 3 to 5 spinules and article 8 with few terminal aesthetascs	Antennule, 8 articled, extending to anterior margin of pereonite 2; article 5, 7 and 8 with 2 setae and article 6 with one seta. All the articles with 3 to 4 spinules and article 8 with four terminal aesthetascs	8 articled antennules smaller than antennae. Without setae and spinnules
Antenna	Antenna 9 articled, decreasing gradually in width, distal margin of articles 6, 7 and 8 with 3 to	Antenna 9 articled, decreasing gradually in width, distal margin of articles 6, 7 and 8 with 3 to 6 setae; article 9 with 3 to 6 terminal aesthetascs	Similar to that of ovigerous female	Antenna longer than antennule, 9 articled, extending to anterior margin of pereonite 2; articles 6 to 9 with few	Antenna longer than antennules, 11 articled, extending to anterior margin of pereonite 4; all the articles with	Antenna is larger than antennule and 11 articled. Without setae and spinnules

Table 1 Morphological characteristics of the different stages of the life cycle of *Cymothoa frontalis* *(Continued)*

	6 setae; article 9 with 3 terminal aesthetascs			spinnules, inner margin of article 3 with 4, article 4 with 5 and terminal of article and 9 with 3 setae; distal margin of article 4 and 5 with 1 plumose setae	spinnules and article 11 with 4 terminal aesthetascs. Inner margin of article 3 with 1 plumose seta	
Mandible	Mandible palp 3 segmented, without setae	Mandible palp 3 segmented, articles 2 and 3 with several setae on distolateral margin	Similar to that of male	Mandible palp article 2 with 4 to 6 marginal seta; article 3 with 5 to 8 marginal setae and 1 apical recurved seta	Mandible palp article 2 with 1 seta arising from the distal margin; article 3 with 3 marginal setae and 1 apical recurved seta	Not well developed. Palp without setae
Maxilla	With 2 spines on median lobe and 1 spine on lateral lobe	Bi-lobed, with 2 spines on median and lateral lobes	Similar to that of male	Similar to male stage	Similar to male stage	Not well developed
Maxillule	With 4 apical spines, slightly recurved	With 4 apical spines, slightly recurved	With 4 apical spines, slightly recurved	Similar to male stage	Similar to male	Not well developed
Maxilliped	Maxilliped with oostegial lobes with many plumose setae and a palp with 3 small apical and 2 lateral recurved spines on article 3	Without oostegial lobes and 5 apical recurved spines on article 3 of palp	Similar to that of male	Similar to male stage	Similar to that of male	Not well developed
Pereopods	Gradually increased in size from 1 to 7, without spines; broad basis increasing in width progressively from p1-p7	Pereopods 1 to 7 gradually increased in size, without spines; basis increasing in width progressively	Similar to that of ovigerous female	Pereopod 1 without spines, pereopods 2 to 5 with 1 to 2 spines on posterior margin of propodus; pereopod 3 and 5 with 1 spine on the posterior margin of merus and the vestigial pereopod 7 with 3 to 4 spines on carpus and 4 to 6 spines on propodus. Pereopods 1 to 3, dactylus with vestigial teeth; dactylus of pereopods 2 and 3, about 2 times longer than propodus, pereopod 7 with short and blunt dactylus	6 pereopods; pereopod 1, with 1 spine on merus and carpus. Pereopod 2 with 1 spine on posterior margin of merus; pereopod 3 with 1 spine on the posterior margin of merus and propodus. One spine on the posterior margin of propodus of pereopod 4; pereopod 5 and 6 with 1 spine on carpus and 2 to 3 spines on the posterior margin of propodus. Dactyls of pereopods 1 to 3 toothed, 4 to 6 not toothed	Only 6 pereopods. Pereopod 7 is missing. Without spines
Penes	Absent	Penes visibles on sternite 7	Rudimentary penes visible on sternite 7	Poorly developed penes on sternite 7	Absent	Absent
Brood pouch	Brood pouch made up of 4 pairs of overlapping oostegites arising from the bases of pereopods 2, 3, 4 and 6	-	-	-	-	

Table 1 Morphological characteristics of the different stages of the life cycle of *Cymothoa frontalis* *(Continued)*

Pleopods	Pleopods not distinctly visible in dorsal view. Pleopod 2 without appendix masculine. Endopods of pleopods 4 and 5 moderately or strongly folded, respectively	Pleopods not distinctly visible in dorsal view. Pleopod 2 with appendix masculine. Endopodite of pleopods 4 and 5 moderately or strongly folded	Pleopod 2 with appendix masculina like that of male	The lateral margin of exopodite of all pleopods with few setae and the median margin of protopod with 4/5 retinaculae. Pleopod 2 with thick and vestigial appendix masculina, as long as endopodite. Pleopods 1 and 3 subequal. Endopods of pleopods 4 and 5 moderately folded	Pleopods not distinctly visible in dorsal view. Pleopod 1 with 4 to 6 plumose setae on endopodite and 14 to 18 plumose setae on exopodite. Pleopods 2 to 5, apical margin of exopodite with 14 to 18 plumose setae	Exopodite without plumose setae. appendix masculine not developed
Uropods	Uropod rami short, not reaching the distal margin of pleotelson; rami subequal in length, curved and apically rounded	Uropod rami short, not reaching the distal margin of pleotelson; rami subequal in length, curved and apically rounded	Pleotelson 1.8 to 2 times as wide as long and uropods similar to those of ovigerous female	Pleotelson as long as wide, apical margin rounded with many setae. Uropods extending beyond the distal margin of pleotelson, inner margin of basis with 3 to 4 plumose setae; rami subequal in length but exopodite slightly shortest, curved and apically round; exopodite with 40 to 44, and the endopodite with 60 to 64 setae	Pleotelson 1.2 times as wide as long and the apical margin with 18 to 20 plumose setae. Uropod basis extending beyond the distal margin of pleotelson; rami subequal extending strongly beyond the distal margin of pleotelson, apically rounded with 8 to 12 plumose setae. Uropods and pleotelson with many dark chromatophores	Uropods extending beyond the pleotelson, rami subequal; dark pigmentation only on the uropods and pleotelson

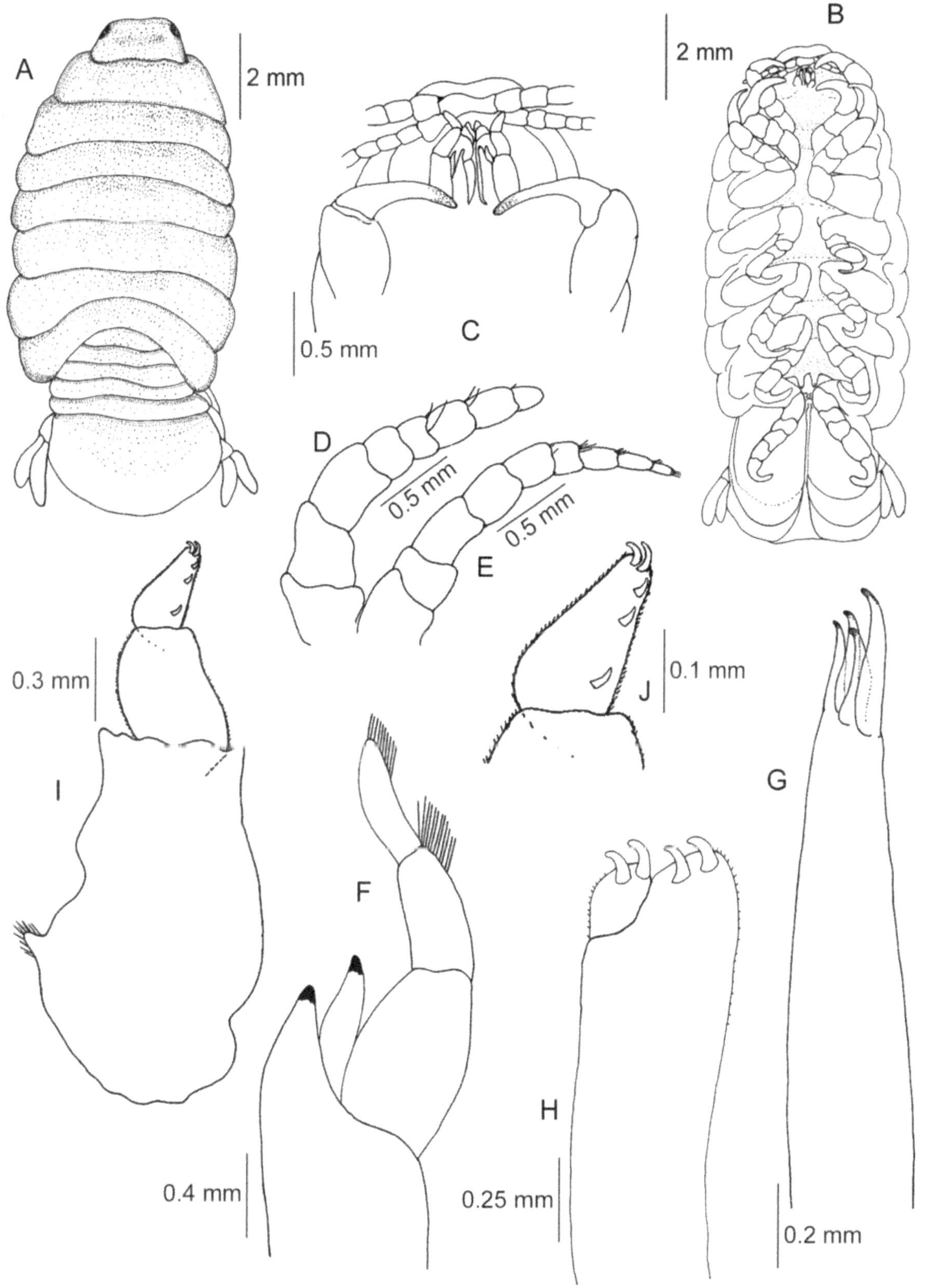

Figure 5 *Cymothoa frontalis* male. **(A)** Dorsal view. **(B)** Ventral view. **(C)** Cephalon ventral view. **(D)** Antennule. **(E)** Antenna. **(F)** Mandible. **(G)** Maxillule. **(H)** Maxilla. **(I)** Maxilliped. **(J)** Maxilliped apex.

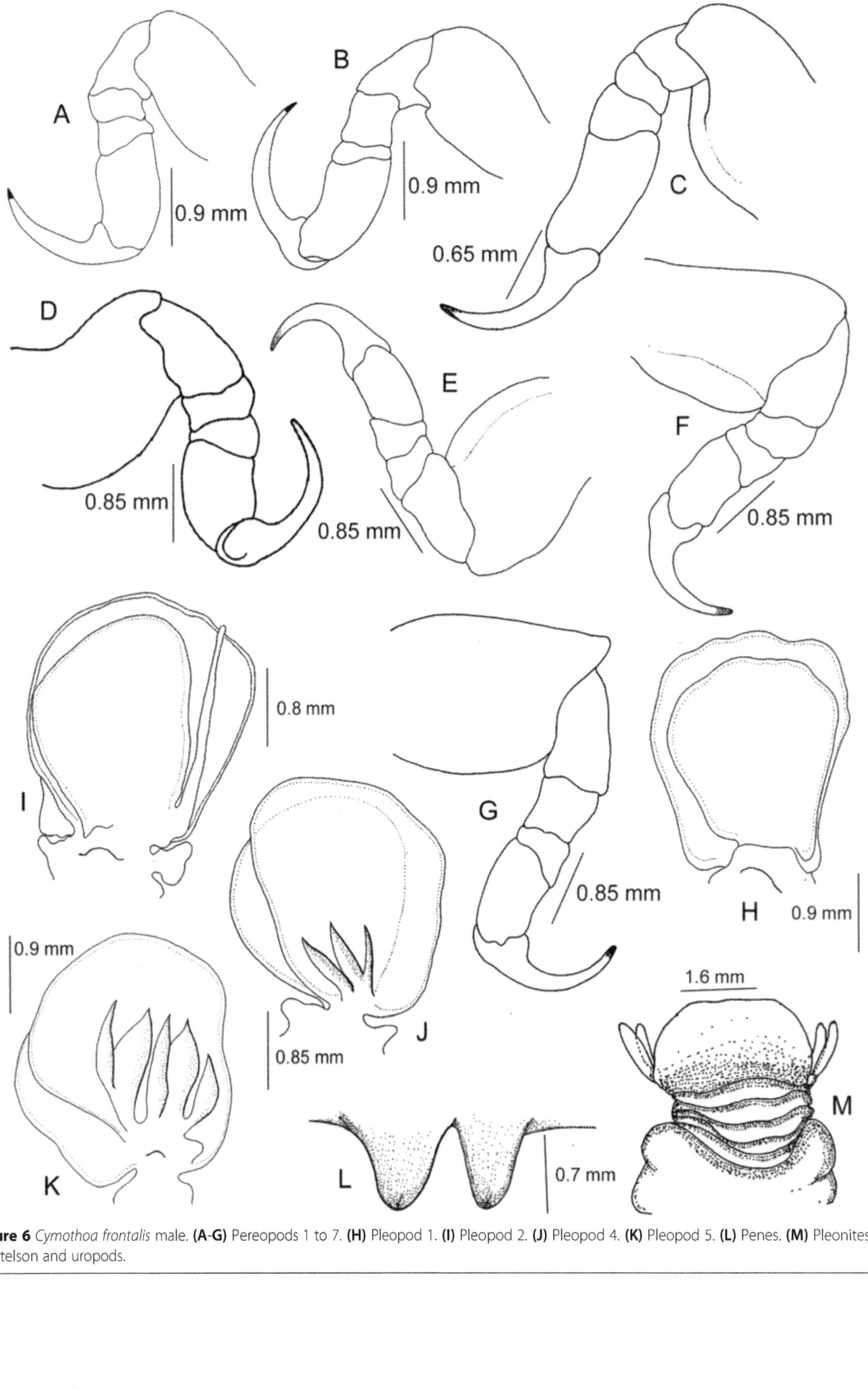

Figure 6 *Cymothoa frontalis* male. **(A-G)** Pereopods 1 to 7. **(H)** Pleopod 1. **(I)** Pleopod 2. **(J)** Pleopod 4. **(K)** Pleopod 5. **(L)** Penes. **(M)** Pleonites, pleotelson and uropods.

Description of the transitional stage

Body 2.7 times as long as wide (Figures 1C and 7; Table 1); Eyes small and moderately distinct. Cephalon partially immersed in poorly developed amphicephalic process of pereonite 1, as long as wide, anterior margin truncate.

Pereonites, pleonites, antennules, antenna and mandible palp are similar to those of the ovigerous female and maxilla, maxilliped similar to those of the male. Rudimentary penes are visible on sternite 7. Pleotelson 1.8 to 2 times as wide as long, slightly wider than pleonite 5 and uropods similar to those of the ovigerous female. Pereopods and pleopods similar to those of female and male.

Description of the juvenile

Body approximately 3 times as long as wide, relatively narrower and smaller than those of the female, transitional stage and male (Figures 1G, 8, 9 and 10; Table 1). Eyes prominent, ovate and conspicuous in dorsal view; anterior margin of cephalon rounded, not immersed in pereonite 1. Pereonite 1 longest, pereonites 2, 3 and 4 subequal; pereonites 4 and 5 widest and subequal. Pereonites 5 to 7 gradually decreasing in length. Pleonites gradually decreasing in width towards posterior but subequal in length. Pleotelson as long as wide, apical margin rounded with many setae.

Antennule 8 articled, reaching the anterior margin of pereonite 1; articles 5 to 8 with 3 to 5 spinules and article 8 with few terminal aesthetascs. Antenna longer than antennule, 9 articled, extending to anterior margin of pereonite 2; articles 6 to 9 with few spinules, inner margin of articles 3 and 4 with 4 and 5 setae, respectively; distal margin of article 9 with 3 setae; articles 4 and 5 with 1 plumose setae. Article 2 of mandible palp with 4 to 6 marginal setae; third article with 5 to 8 marginal setae and one apical recurved seta. Maxillule, maxilla and maxilliped similar to those of the male stage.

Pereopod 1 without spines, pereopods 2 to 5 with 1 or 2 spines on posterior margin of propodus; pereopods 3 and 5 with 1 spine on the posterior margin of merus and the vestigial pereopod 7 with 3 to 4 spines on carpus

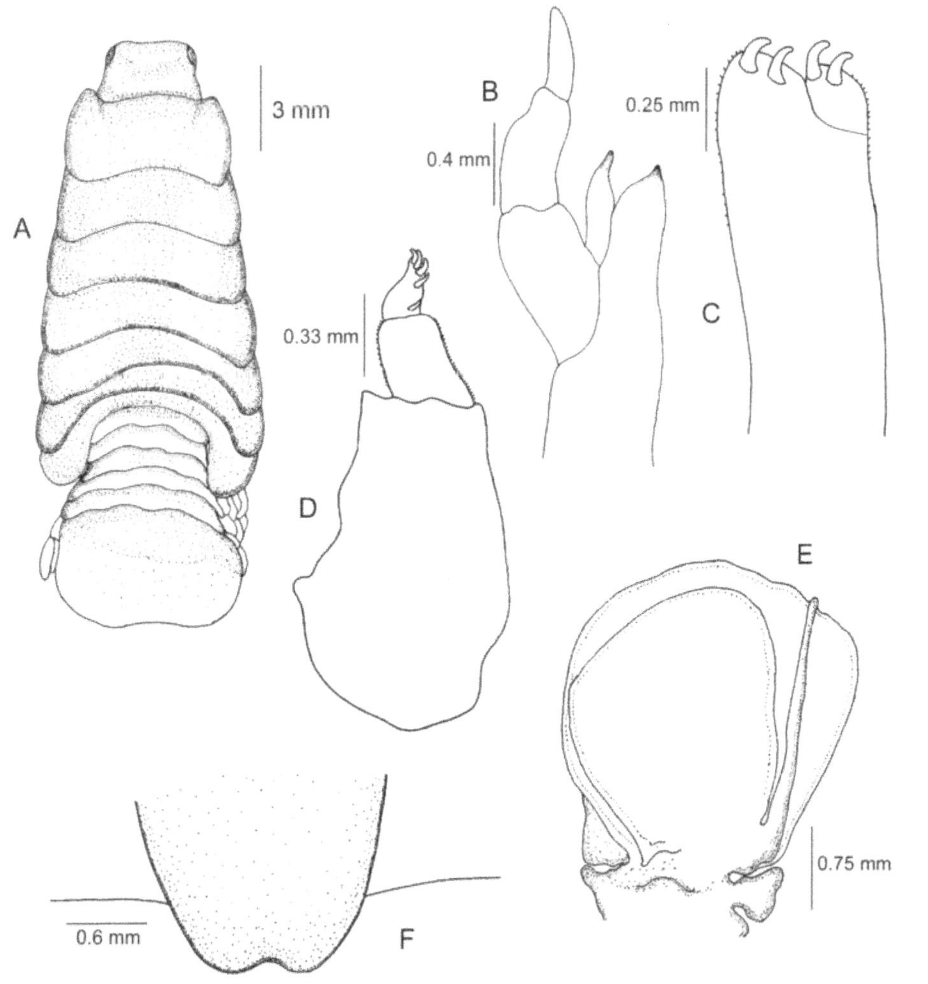

Figure 7 *Cymothoa frontalis* transitional stage. **(A)** Dorsal view. **(B)** Mandible. **(C)** Maxilla. **(D)** Maxilliped. **(E)** Pleopod 2. **(F)** Penes.

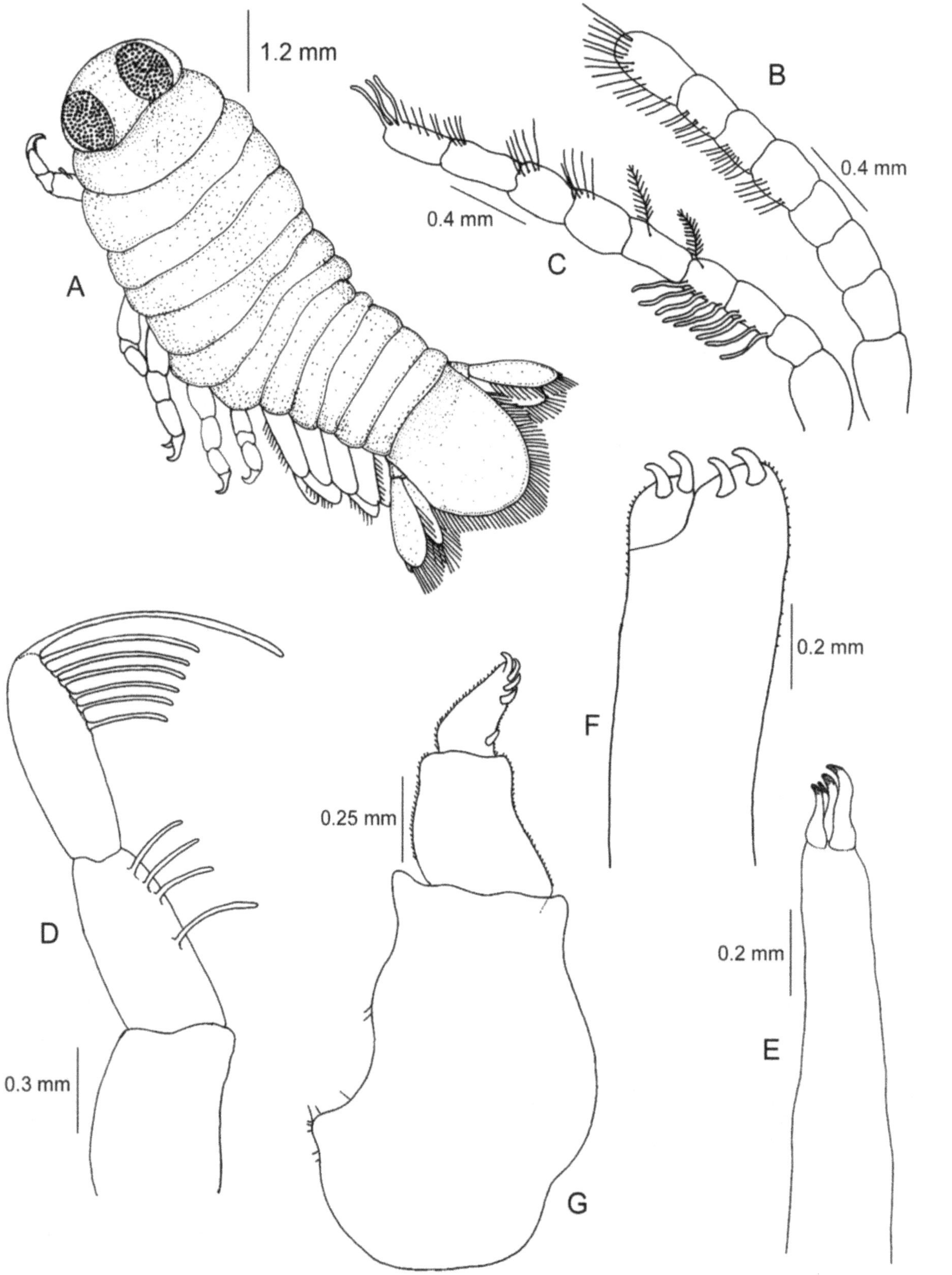

Figure 8 *Cymothoa frontalis* juvenile. **(A)** Dorsal view. **(B)** Antennule. **(C)** Antenna. **(D)** Mandible palp. **(E)** Maxillule. **(F)** Maxilla. **(G)** Maxilliped.

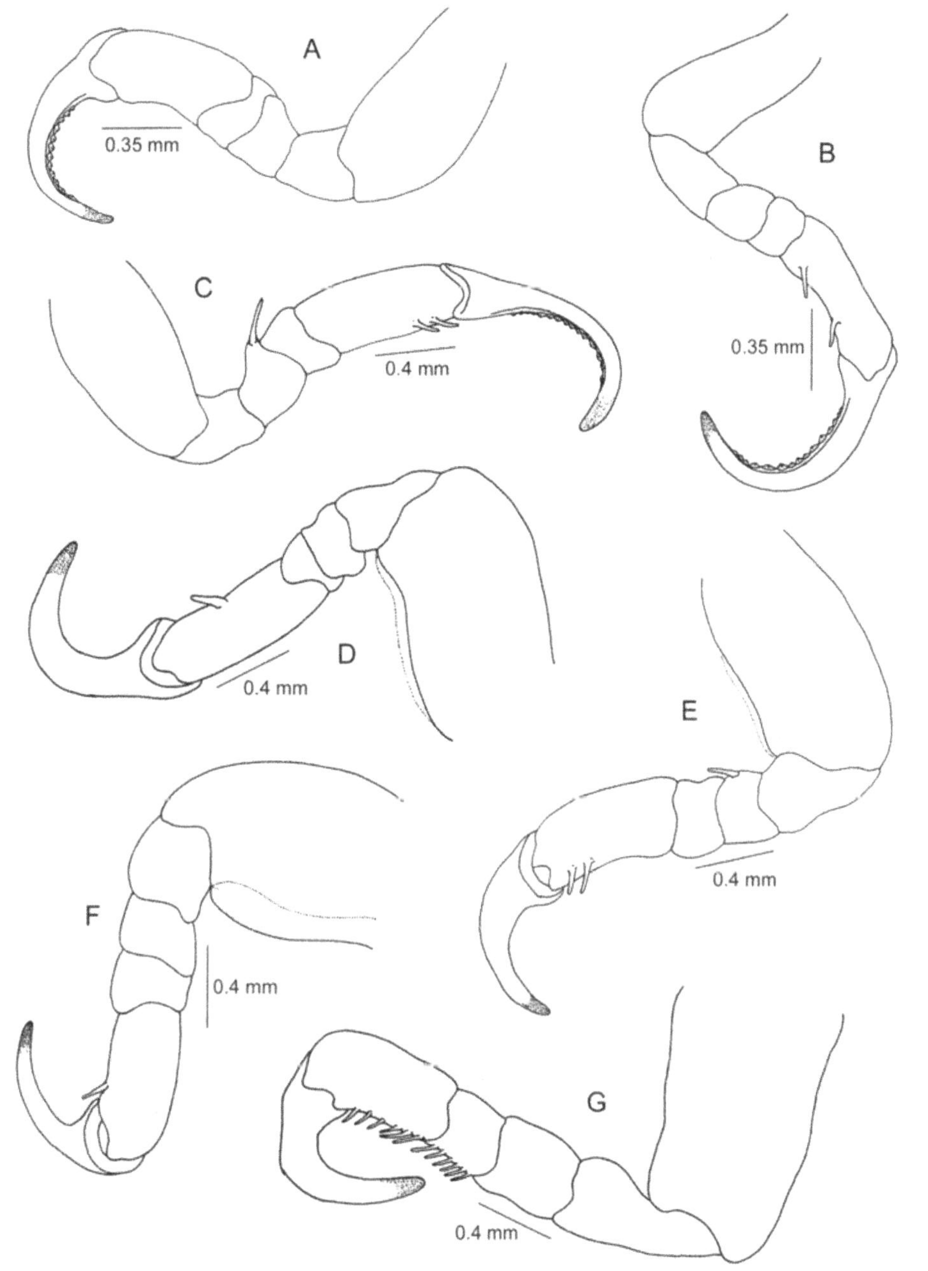

Figure 9 *Cymothoa frontalis* juvenile. **(A-F)** Pereopods 1 to 6. **(G)** Pereopod 7 with short and blunt dactylus.

and 4 to 6 spines on propodus. Pereopods 1 to 3 dactylus with vestigial teeth; dactylus of pereopods 2 and 3, about 2 times longer than propodus, pereopod 7 with short and blunt dactylus. Penes not developed. The lateral margin of exopodite of all pleopods with small setae, but not on the apical margin. Median margin of protopod with 4/5 retinaculae. Pleopod 2 with thick and vestigial appendix masculina, as long as endopodite. Pleopods 1 and 3 subequal. Endopod of pleopods 4 and 5 moderately folded. Uropod extending beyond the distal margin of pleotelson, inner margin of basis with 3 to 4 plumose setae; rami slightly curved, endopodite relatively large, exopodite with an apical recurved spine. Endopodite margin with 60 to 66 setae. Exopodite with 40 to 44 setae, restricted to the inner margin.

Description of the manca

The manca developed from the pre-manca through moulting shows distinct morphological characters (Figures 1F, 11 and 12; Table 1).

Body elongated and transparent, 3.5 times as long as wide, eyes black, prominent, ovate and conspicuous in

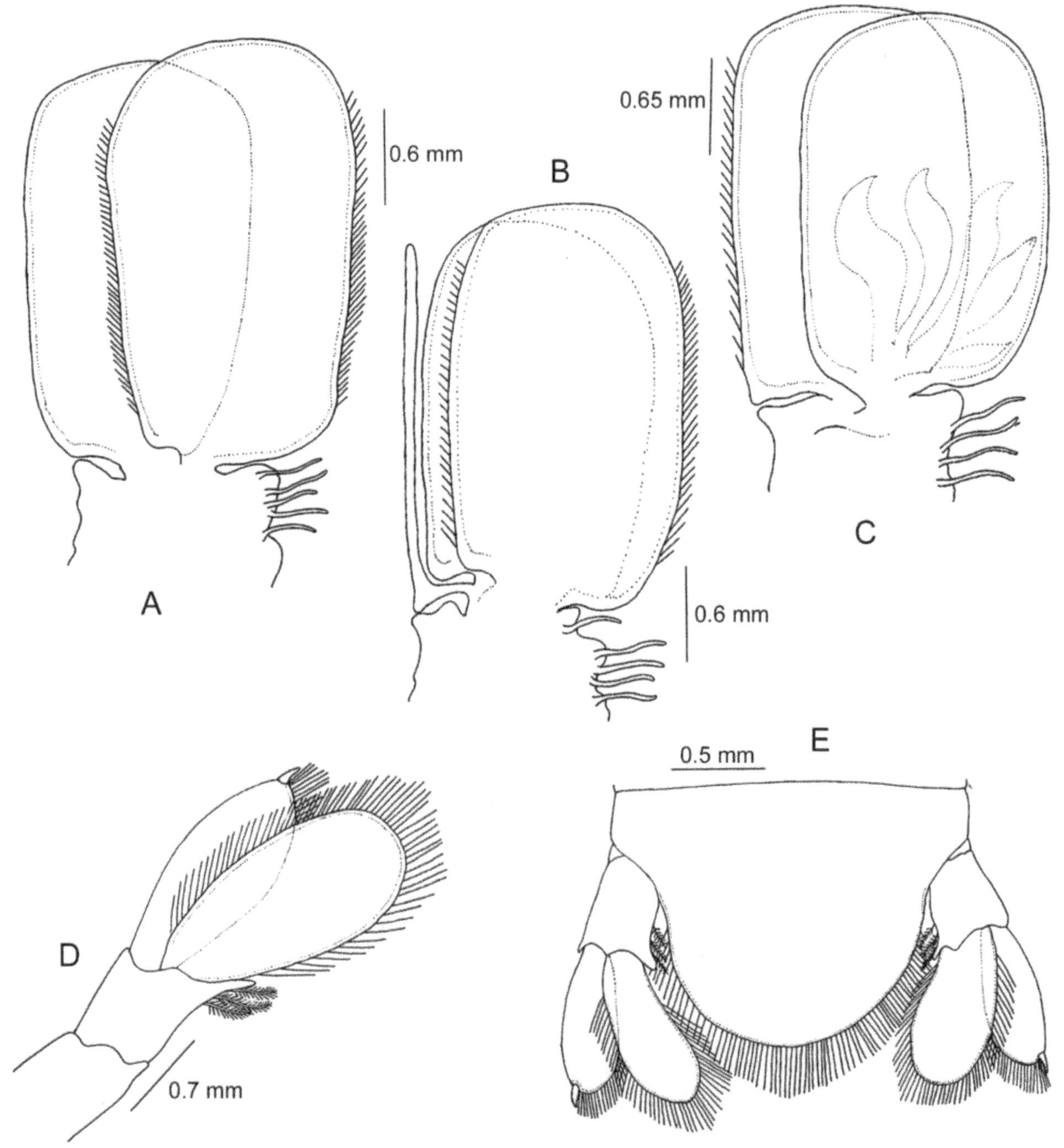

Figure 10 *Cymothoa frontalis* juvenile. **(A)** Pleopod 1. **(B)** Pleopod 2 with vestigial appendix masculina. **(C)** Pleopod 5. **(D)** Uropod with rami. **(E)** Pleotelson and uropods.

dorsal view. Cephalon 1.6 times as wide as long and golden yellow in colour.

Pereonite 1 relatively long; pereonites 2, 3 and 4 subequal in length; longer than pereonites 5 and 6; pereonites 5 and 6 subequal in length; pereonite 7 short and narrow; pereonites 4 and 5 widest. All pleonites visible and subequal in width and length; 4.8 times as wide as long. Pleotelson 1.2 times as wide as long and the apical margin with 18 to 20 plumose setae.

Eight articled antennule extends to anterior margin of pereonite 2; articles 5, 7 and 8 each with 2 setae and article 6 with 1 seta; all the articles with 3 to 4 spinules and article 8 with 4 terminal aesthetascs. Antenna longer than antennule, 11 articled, extending to anterior margin

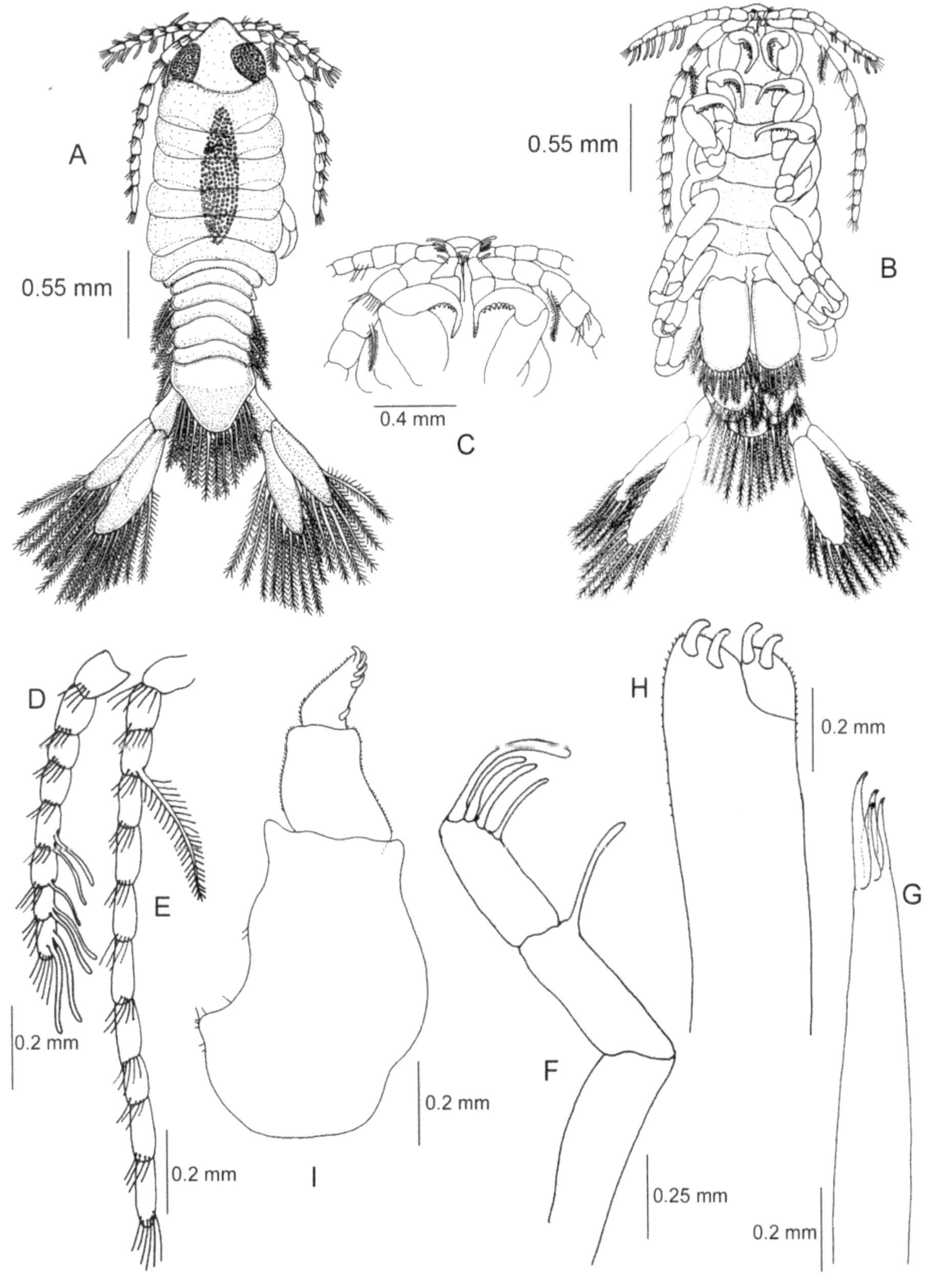

Figure 11 *Cymothoa frontalis* manca. **(A)** Dorsal view. **(B)** Ventral view. **(C)** Cephalon ventral view. **(D)** Antennules. **(E)** Antenna. **(F)** Mandible palp. **(G)** Maxillule. **(H)** Maxilla. **(I)** Maxilliped.

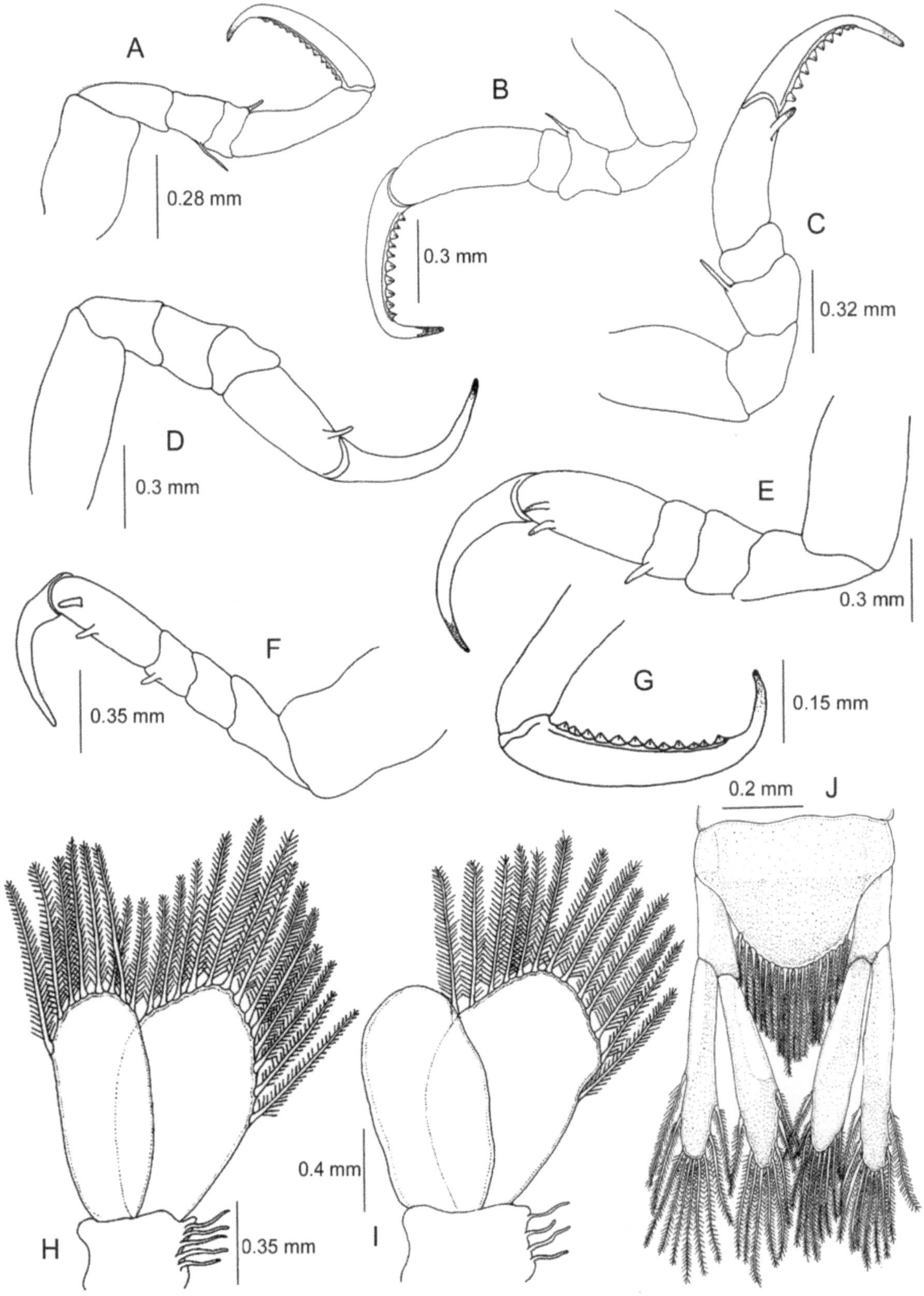

Figure 12 *Cymothoa frontalis* manca. **(A-F)** Pereopods 1 to 6. **(G)** Dactylus of pereopod 3. **(H)** Pleopod 1. **(I)** Pleopod 3. **(J)** Pleotelson and uropods.

of pereonite 4; all the articles with spinules and article 11 with 4 terminal aesthetascs; inner margin of article 3 with 1 plumose seta. Article 2 of the mandible palp with 1 seta arising from the disto-lateral margin; article 3 with 3 marginal setae and 1 apical recurved seta. Maxillule, maxilla and maxilliped similar to those of the male stage.

Six pereopods; pereopod 1, with 1 spine on merus and carpus. Pereopod 2 with 1 spine on posterior margin of merus; pereopod 3 with 1 spine on the posterior margin of merus and propodus; 1 spine on the posterior margin of propodus of pereopod 4; pereopods 5 and 6 with 1 spine on carpus and 2 to 3 spines on the posterior margin of propodus. Dactyls of pereopods 1 to 3 toothed, 4 to 6 not toothed. Pleopods not distinctly visible in dorsal view. Pleopod 1 with 6 to 8 plumose setae on endopodite and 14 to 18 plumose setae on exopodite. Pleopods 2 to 5, apical margin of exopodite with 14 to 18 plumose setae. Uropod rami subequal, extending strongly beyond the distal margin of pleotelson, apically rounded with 8 to 12 plumose setae. Uropods and pleotelson with many dark chromatophores.

Description of the pre-manca

Body elongated and transparent, 4 times as long as wide, eyes black, prominent, ovate and conspicuous in dorsal view (Figures 1E and 13; Table 1). Cephalon 1.4 times as wide as long, not immersed in pereonite 1. Yolk globules are visible in the pereon, between pereonites 2 and 6. Pereonite 1 slightly long, pereonite 7 short and narrow. Pereonites 4 and 5 widest. All pleonites visible and subequal in width and length; 3.6 times as wide as long. Pleotelson as wide as long, without plumose setae.

Antenna larger than antennule. Eight articled antennule extends to anterior margin of pereonite 2; antenna longer than antennule, 11 articled, extending to anterior margin of pereonite 4; all articles without setae and spinules. Mouthparts are not well developed; mandible palp articles without setae and spines; maxillule, maxilla and maxilliped with poorly developed apical non-recurved spines.

Six pereopods; pereopods 1 to 6 without spines. Distal margin of dactylus of pereopod 1 to 6 slightly indented and then forming a narrow terminal part. Dactylus of pereopod not toothed. Pleopods not distinctly visible in dorsal view. Pleopods 1 to 5 without plumose setae. Uropod rami subequal, extending strongly beyond the distal margin of pleotelson, apically rounded without plumose setae. Uropods and pleotelson with dark chromatophores.

Size: Ovigerous females, 21 to 28 mm; transitional stage, 15 to 23 mm; male, 11 to 13 mm; juvenile, 7 to 9 mm; manca, 3 to 3.4 mm; pre-manca, 2.2 to 2.4 mm.

Host: Present Indian material is from *S. strongylura* (Van Hasselt 1823). Until now, a single specimen was doubtfully reported by White (1847), from a host belonging to the genus *Cyclopterus.*

Distribution: Asian Sea (Milne Edwards 1840) and Indian Ocean (White 1847); Singapore (Heller 1868) and Bangkok (Schioedte and Meinert 1884); Malabar Coast, Kerala, India (present study).

Colour: Female and transitional stage - light grey with highly pigmented dorsal surface; male - creamy white with a pigmented lateral and posterior margin of pereonites and pleonites; juvenile - transparent with scattered pigmentation throughout the body. Posterior margin and lateral side of pereonites and pleonites and anterior margin of pleotelson with lining chromatophores, Manca and pre-manca - clear with scattered chromatophores particularly on uropods and pleotelson; cephalon golden yellow in colour.

Occurrence and prevalence of infestation by *C. frontalis* on *S. strongylura*

Eighty species of fishes belonging to 35 families were examined from November 2009 to November 2012 for the presence of *C. frontalis*; however, this cymothoid was recovered only from the belonid fish *S. strongylura*, signifying its oligoxenous host specificity (Table 2).

Among 134 members of host fish (*S. strongylura*) examined, 92 were infested by *C. frontalis*, the prevalence being 68.65%. A total of 171 *C. frontalis* including 67 ovigerous females, 14 transitional stages, 68 males and 22 juveniles were recovered; the mean intensity being equal to 1.9 (Table 3) (Figure 1B,C,D,E,F,G). Out of 92 infested members of *S. strongylura*, 79 were found with two members of *C. frontalis* appearing like a pair, but in three combinations such as male-female, male-transitional and juvenile-juvenile, with the male-female pair being predominant (56 out of 79); the large ovigerous female was attached on the floor of the buccal cavity, their cephalon facing towards the opening of the host mouth; male settled just behind the female on the gill arch (Figure 1A,G,H,I). In the case of the male-transitional pair (12 out of 79), both were found attached on the tongue of the host. The juvenile pair (11 out of 79) prefers the mucous membrane of the lower beak of the fish (Figure 1G). Unpaired females or transitional stages were also observed in 11 and 2 instances, respectively.

Life cycle of *C. frontalis*

The present study identified three major phases (such as marsupial, free living and infective) in the life cycle of *C. frontalis.* The details of each phase are represented in Table 4 and Figures 14 and 15. The marsupial phase encompasses one zygotic, three embryonic (ES I, ES II and ES III) and two larval (pre-manca and manca) stages, and each stage is found to be distinct from one another by their precise morphological characteristics (Figure 14). Round zygotic-staged eggs appear in light florescent

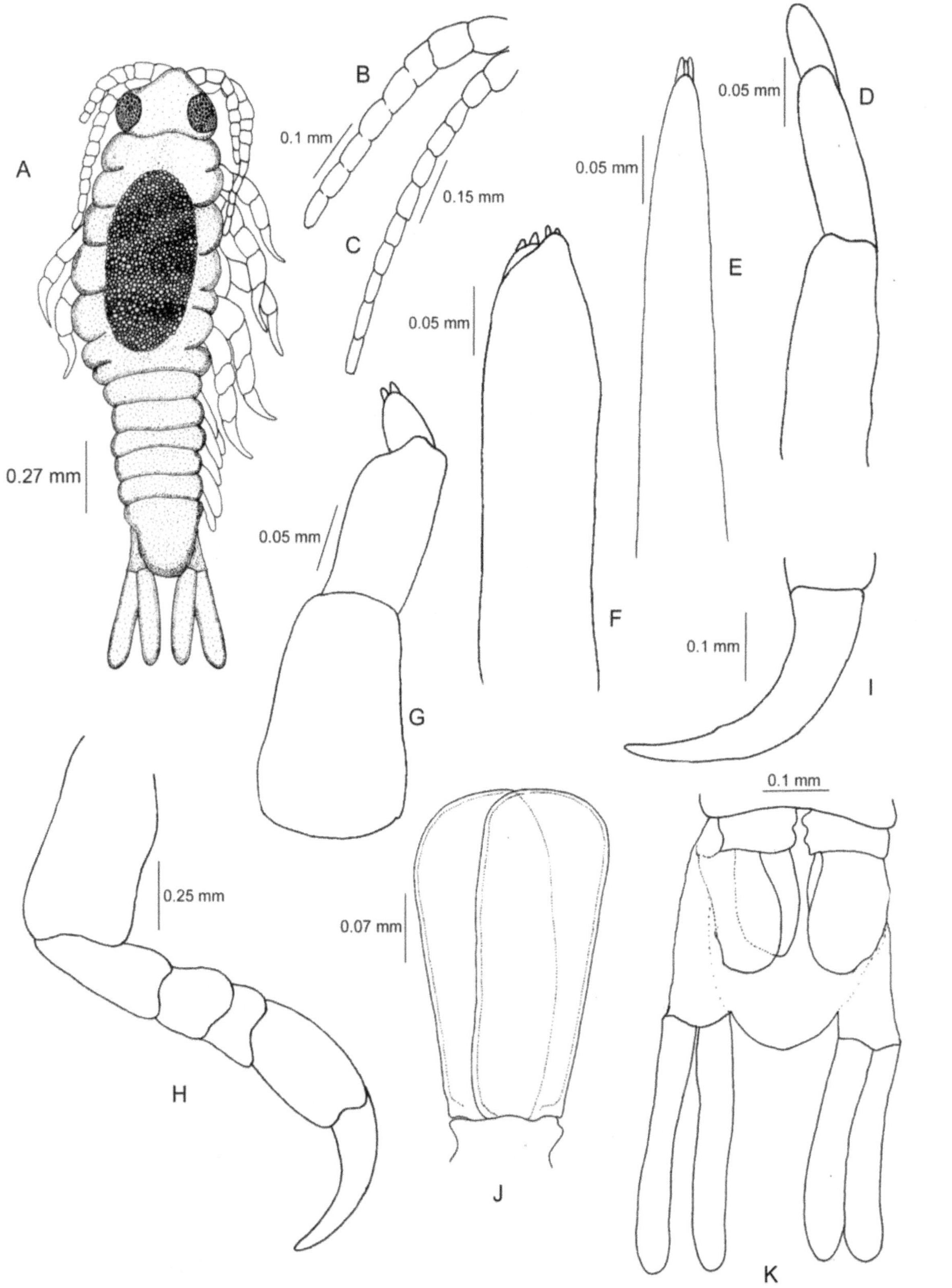

Figure 13 *Cymothoa frontalis* pre-manca. **(A)** Dorsal view. **(B)** Antennule. **(C)** Antenna. **(D)** Mandible palp. **(E)** Maxillule. **(F)** Maxilla. **(G)** Maxilliped. **(H)** Pereopod 3. **(I)** Dactylus of pereopod 3. **(J)** Pleopod 2. **(K)** Ventral view of pleotelson and uropods.

Table 2 List of marine fishes examined from Malabar Coast, India, during the period from November 2009 to November 2012

Serial number	Family	Fish species	Number of fishes examined	Infestation by *Cymothoa frontalis*
1	Scombridae	*Rastrelliger kanagurta* (Cuvier, 1817)	1,329	-
		Thunnus obesus (Lowe, 1839)	69	-
		Scomberomorus guttatus (Bloch and Schneider, 1801)	323	-
		Scomberomorus commerson (Lacepède, 1800)	425	-
		Auxis thazard thazard (Lacepede, 1800)	37	-
2	Carangidae	*Parastromateus niger* (Bloch, 1795)	136	-
		Decapterus russelli (Ruppell, 1830)	287	-
		Megalaspis cordyla (Linnaeus, 1758)	97	-
		Carangoides malabaricus (Bloch and Schneider, 1801)	64	-
		Chorinemus orientalis (Temminck and Schleyel, 1844)	22	-
		Scomberoides tol (Cuvier, 1832)	14	-
3	Terapontidae	*Terapon jarbua* (Forsskal, 1775)	312	-
		Terapon puta Cuvier, 1829	89	-
4	Dussumieriidae	*Dussumieria acuta* Valenciennes, 1847	422	-
		Dussumieria elopsoides Bleeker, 1849	122	-
5	Pristigasteridae	*Ilisha melastoma* (Bloch and Schneider, 1801)	798	-
		Pellona ditchela Valenciennes, 1847	928	-
		Ilisha megaloptera (Swainson, 1839)	222	-
6	Clupeidae	*Escualosa thoracata* (Valenciennes, 1847)	803	-
		Opisthopterus tardoore (Cuvier, 1829)	398	-
		Tenualosa ilisha (Hamilton, 1822)	318	-
		Anodontostoma chacunda (Hamilton, 1822)	144	-
		Sardinella albella (Valenciennes, 1847)	313	-
		Sardinella fimbriata (Valenciennes, 1847)	89	-
		Amblygaster clupeoides Bleeker, 1849	49	-
		Sardinella gibbosa (Bleeker, 1849)	375	-
		Amblygaster sirm (Walbaum, 1792)	42	-
		Sardinella melanura (Cuvier, 1829)	395	-
		Hilsa kelee (Cuvier, 1829)	79	-
		Tenualosa toli (Valenciennes, 1847)	284	-
7	Stromateidae	*Pampus argenteus* (Euphrasen, 1788)	131	-
		Pampus chinensis (Euphrasen, 1788)	87	-
8	Ambassidae	*Ambassis ambassis* (Lacepède, 1802)	596	-
9	Dasyatidae	*Himantura bleekeri* (Blyth, 1860)	27	-
10	Megalopidae	*Megalops cyprinoides* (Broussonet, 1782)	187	-
11	Engraulidae	*Coilia dussumieri* Valenciennes, 1848	280	-
		Stolephorus commersonii Lacepede, 1803	859	-
		Thryssa dussumieri (Valenciennes, 1848)	366	-
		Thryssa malabarica (Bloch, 1795)	502	-
		Thryssa mystax (Bloch and Schneider, 1801)	381	-
		Thryssa setirostris (Broussonet, 1782)	376	-
12	Ariidae	*Plicofollis tenuispinis* (Day, 1877)	123	-
		Hexanematichthys sagor (Hamilton, 1822)	96	-

Table 2 List of marine fishes examined from Malabar Coast, India, during the period from November 2009 to November 2012 *(Continued)*

		Arius acutirostris Day, 1877	44	-
13	Hemiramphidae	*Hemiramphus far* (Forsskal, 1775)	45	-
		Hemiramphus lutkei (Valencinnes, 1847)	85	-
		Rhynchorhamphus malabaricus Collette, 1976	48	-
14	Belonidae	*Strongylura strongylura* (van Hasselt, 1823)	134	+
		Strongylura leiura (Bleeker, 1850)	408	-
		Tylosurus acus melanotus Bleeker, 1850	19	-
		Tylosurus crocodilus crocodilus (Peron and Lesuer, 1821)	94	-
15	Leiognathidae	*Gazza minuta* (Bloch, 1795)	93	-
		Eubleekeria splendens (Cuvier, 1829)	108	-
		Photopectoralis bindus (Valenciennes, 1835)	35	-
		Leiognathus sp	29	-
16	Cynoglossidae	*Cynoglossus dubius* Day, 1873	306	-
		Cynoglossus arel (Bloch and Schneider, 1801)	293	-
17	Trichiuridae	*Lepturacanthus savala* (Cuvier, 1829).	34	-
18	Mugilidae	*Mugil cephalus* Linnaeus, 1758	96	-
		Mugil curema Valenciennes, 1836	39	-
19	Sphyraenidae	*Sphyraena forsteri* Cuvier, 1829	21	-
20	Lutjanidae	*Lutjanus johni* (Bloch, 1792)	29	-
		Lutjanus malabaricus (Bloch and Schneider, 1801).	32	-
21	Nemipteridae	*Nemipterus japonicus* (Bloch, 1791)	22	-
22	Exocoetidae	*Cypselurus oligolepis* (Bleeker, 1865)	34	-
		Exocoetus volitans Linnaeus, 1758	23	-
		Parexocoetus mento (Valenciennes, 1847)	15	-
23	Chanidae	*Chanos chanos* (Forsskal, 1775)	26	-
24	Carcharhinidae	*Rhizoprionodon acutus* (Rüppell, 1837)	39	-
25	Hemiscyllidae	*Chiloscyllium punctatum* Müller and Henle, 1838	62	-
26	Chirocentridae	*Chirocentrus dorab* (Forsskal, 1775)	25	-
27	Muraenidae	*Strophidon sathete* (Hamilton, 1822)	9	-
28	Muraenesocidae	*Congresox talabonoides* (Bleeker, 1853)	7	-
29	Sillaginidae	*Sillago sihama* (Forsskal, 1775)	22	-
30	Menidae	*Mene maculata* (Bloch and Schneider, 1801)	29	-
31	Polynemidae	*Eleutheronema tetradactylum* (Shaw, 1804)	32	-
32	Cichlidae	*Etroplus suratensis* (Bloch, 1790)	76	-
33	Labridae	*Parajulis poecilepterus* (Temminck and Schlegel, 1845)	37	-
34	Bagridae	*Mystus gulio* (Hamilton, 1822)	27	-
35	Rachycentridae	*Rachycentron canadum* (Linnaeus, 1766)	12	-

yellow colour, the size being 1.2 to 1.3 mm. Early-staged tissue differentiation seems to occur during ES I, and the embryo at this stage attains subspherical or ovoid shape without changing the colour from the preceding stage. The ES II-staged marsupiumite is characterized by the development of cephalic lobe, early limb bud and pigmented eye spot, and at this stage, the embryo becomes elongated. Marsupiumite at ES III appears as curved and segmented with thoracic and abdominal limb buds. By this stage (ES III), the embryonic development seems to be completed, and the egg gets ready to hatch out the larva called pre-manca; within the marsupium, it moults and transforms into the manca. Significantly, within the marsupium, the development of marsupiumites of *C.*

Table 3 Parasitological index of the parasitic cymothoid *Cymothoa frontalis*, along the Malabar Coast, India

Parasite and host fish	NFE	NFI	*P*	TNP (*I*)			
				F	T	M	J
Cymothoa frontalis Milne Edward,1840	134	92	68.65	67	14	68	22
Strongylura strongylura (Belonidae)				171 (1.9)			

P, prevalence; NFE, number of fishes examined; NFI, number of fishes infested; TNP, total number of collected parasites; *I* intensity of infestation; F, female of *C. frontalis*; T, transitional of *C. frontalis*; M, male of *C. frontalis*; J, juvenile stage of *C. frontalis*.

frontalis is synchronous evidenced by the presence of only one of the aforesaid developmental stages at a time. After a brief period of residence in the marsupium, the manca is seen released into the surrounding water where it leads a free swimming life. (During the present study, the free swimming manca was successfully reared for a maximum of 8 days in our laboratory.) Subsequently, manca enters into the infectious stage, quite likely as a single pair by parasitizing its host fish (*S. strongylura*). Through moulting, the paired manca becomes the juveniles which are usually found attached to the mucous membrane of the lower beak of the host (Figure 1G). Both members of the juvenile pair would undergo moulting (sequential?) which in turn induces remarkable morphological changes, leading to their transformation into functional paired males which settle within the host's buccal cavity. Within the buccal cavity, both males undergo moulting. However, by this process one of the (male) members becomes transitional but retains some of the male characters (Figure 7C,D,E). The transitional further undergoes two sequential moults; interestingly, by the culmination of the second moult (also referred to

Table 4 Life cycle of *Cymothoa frontalis*

Stages of life cycle			Remarks
Marsupial stage	Zygotic stage		Round or oval, light florescent yellow (size: 1.2 to 1.3 mm)
	Embryonic stages (ES)	ES I	Subspherical to ovoid embryo, structural differentiation not evident, light florescent yellow in colour (size: 1.3 to 1.5 mm length and 1.25 to 1.3 mm width)
		ES II	Elongate/oblong embryo with a cephalic end and early limb bud developed. Pigmentation in the eyespot started. (size: 1.6 to 1.85 mm length and 1.3 to 1.35 mm width)
		ES III	Curved and segmented embryos with pigmented eyespots, thoracic and abdominal limb buds appeared. Dorsal surface shows chromatophores. (size: 1.9 to 2.0 mm length and 1.3 to 1.400 mm width)
	Larval stages	Pre-manca	The first hatched larvae remain in the marsupium and undergo moulting to transform into manca. (Description is given in the text)
		Manca	Final marsupial stage released in the surrounding medium. (Description is given in the text)
Free living stage	Manca		Free swimming stage. Swim and search for specific host (*S. strongylura*)
Infective stage	Manca		Infectious stage attaches to the host (*S. strongylura*) and then undergoes moulting to transform into juvenile
	Juvenile		It undergoes moulting to transform into male. (Description is given in the text)
	Male		It undergoes a moulting process and enters into the transitional stage. (Description is given in the text)
	Transitional stage		It undergoes a moult to transform into the female stage 1 (Fs 1). (Description is given in the text)
	Female	Fs 1	Ovigerous female with fully grown ovary and with newly formed brood plates, but no marsupiumites. Undergo oviposition to form Fs 2
		Fs 2	Ovigerous female with growing ovary and carry first set of brood and marsupiumites show synchrony in their development
		Fs 3	Ovigerous female with growing ovary and carrying empty old brood pouch, after releasing manca. It undergoes a moult to transform into Fs 4
		Fs 4	Ovigerous female with growing ovary and no brood pouch/brood plates. It undergoes a moult to transform into Fs 5
		Fs 5	Ovigerous female with newly formed brood plates but no marsupiumites. It undergoes oviposition to form Fs 6
		Fs 6	Ovigerous female carrying second brood and marsupiumites show synchrony in their development

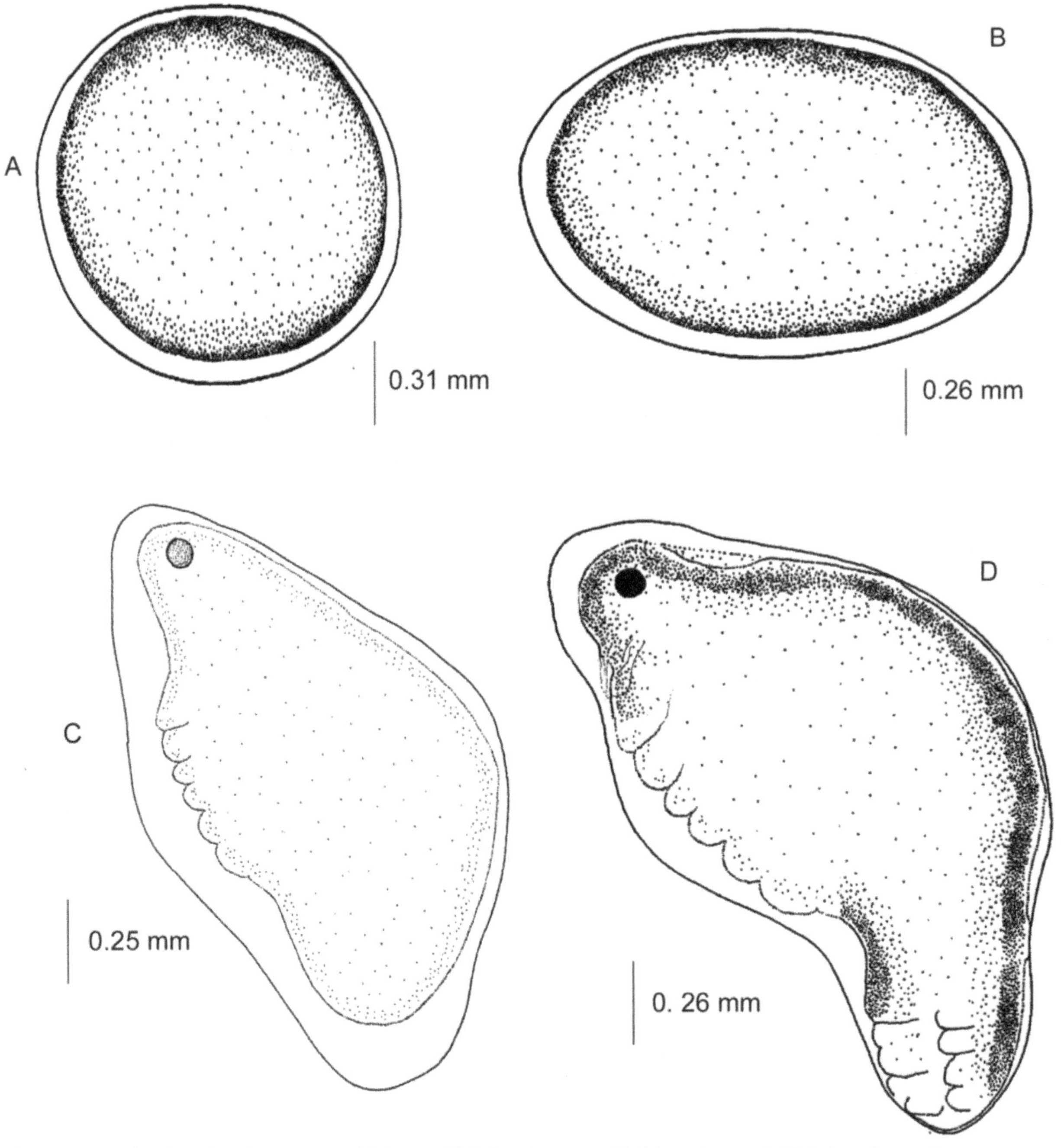

Figure 14 *Cymothoa frontalis* embryonic stages. **(A)** Zygote. **(B)** Embryonic stage 1. **(C)** Embryonic stage 2. **(D)** Embryonic stage 3.

as 'oostegition moult'), it transforms into an ovigerous female with brood pouch or marsupium (a cavity formed by the overlapped, plate-like oostegites developed from the endopodites of the second, third, fourth and sixth pereopods). The female life is continued further by engaging itself in reproduction and also in parental care, by keeping the eggs and larvae in their brood pouch, until they are released. The brood pouch is formed and removed in an alternative manner through oostegition and de-oostegition moults, respectively; oostegition moult occurs before the oviposition (spawning), followed by marsupial life and larval (manca) release, after which de-oostegition moult ensues. Based on the presence or absence of the brood pouch and/or marsupiumites, six consecutive stages of the ovigerous female (Fs 1 through Fs 6) were identified (Figure 15 and Table 4). The Fs 1-staged female that developed right from the transitional through the first oostegition moult is characterized by the presence of a growing ovary and a well-developed marsupium, but without marsupiumites; oviposition ensues and the eggs

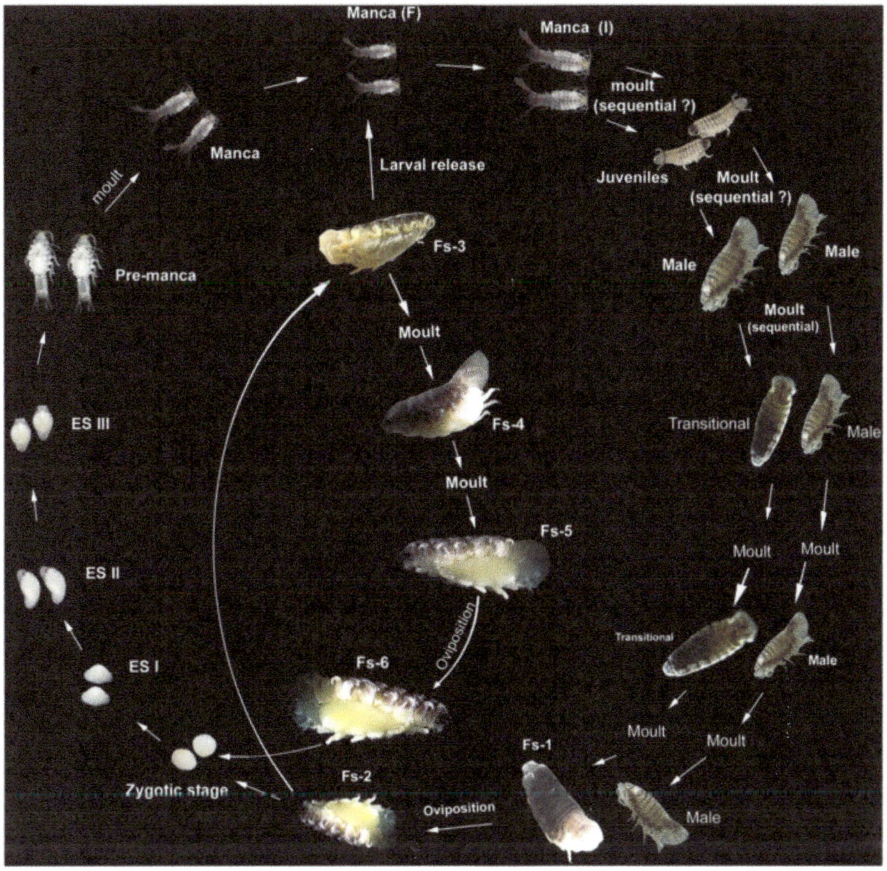

Figure 15 Schematic representation of the life cycle of *Cymothoa frontalis*. Manca (F), manca at free living stage; Manca (I), manca at infestive stage; ES, embryonic stage; Fs, female stage.

are deposited in the marsupium and this brooded ovigerous female is designated to be at Fs 2. Upon completion of the marsupial life, manca are released and the female with the old empty marsupium is referred to be at Fs 3 stage. The females at Fs 3 stage undergo de-oostegition moult (to remove the old marsupium) and enter Fs 4 stage, during which time it undergoes oostegition moult and transforms into Fs 5 stage with newly formed second marsupium and growing ovary. Excepting the meagre size difference and for the presence of a well-developed oostegial lobe of the maxilliped, the females at Fs 5 stage morphologically resemble the females at Fs 1 stage. Subsequent to oviposition, the females at Fs 5 would be transformed into Fs 6 stage, characterized by the presence of a second set of marsupiumites. Except the size differences, the females at Fs 6 stage would resemble those at Fs 2 stage.

Discussion

With the redescription of the females of *C. frontalis* and the description (of the transitional stage, the male, the juvenile and the larvae) as depicted in the present paper, we update the diagnosis of the species (*C. frontalis*) carried out by Milne Edwards (1840) and Schioedte and Meinert (1884); to our knowledge, this report appears to be the first of its kind, wherein the cymothoid crustacean is being described by collating the data from all the stages in its life cycle.

The presence of transitional stage, as observed in the life cycle of *C. frontalis* (present study), is an indication of the protandrous sex reversal, akin to what has been reported in other cymothoids (Bowman 1960; Williams and Williams 1985; Tsai and Dai 1999). Significantly, at this (transitional) stage, *C. frontalis* is found to retain some of the male characteristics including the appendix masculina and two spines on the median and lateral lobes of the maxilla; the penes, however, become rudimentary at this stage. After its transformation into the ovigerous female, the spines on the lateral lobe of maxilla are reduced to one and the appendix masculina and penes are seen to have completely disappeared.

Apart from the intense body tanning, the presence of plumose setae on the oostegial lobe of the maxilliped (Figure 2H) during the female phase makes *C. frontalis* distinct from those of other cymothoids described so far. On the other hand, *C. frontalis* resembles another cymothoid such as *Cymothoa spinipalpa* (described by

Thatcher et al. 2007) in the possession of folded endopodite of the fourth and fifth pleopod.

The pre-manca and manca of *C. frontalis* possess appendages in all pereonites except the seventh one. In pre-manca, pereopods lack spines and there are no setae on pleopods, uropod and pleotelson. Manca, on the other hand, possesses six pairs of spiny pereopods, and their pleopods, uropod rami, pleotelson are with plumose setae facilitating the free swimming life. The toothed dactylus on the first three pairs of pereopods would help the larva (manca) to cling on to the host fish during the infective stage. Contrary to this, in the manca of *Agarna malayi* and *N. indica*, the dactyli of all pereopods are devoid of teeth; teeth are, however, present on the propodus of the first three pereopods (Aneesh et al., unpublished). In pre-manca and manca of *C. frontalis*, the uropod basis and subequal rami are extremely extended beyond the distal margin of the pleotelson. Such an extension is not prominent in the case of the cymothoid *M. renardi* (Aneesh et al., unpublished). In the juvenile stage of *C. frontalis*, pleopodal setae and pareopodal spines and teeth are found to be more or less in a reduced state. By the onset of the adult male stage, all these structures completely disappeared and the limbs become short; the carapace gets gradually calcified and arched with intense tanning. In the female stage, the telson is kept in an upright position enabling the pleopods under constant vibration to aerate the marsupiumites.

Notwithstanding the fact that a single host '*Cyclopterus*' was doubtfully reported by White (1847), until now, the host fish of *C. frontalis* was not accurately identified. Out of the 80 fish species examined during the present study, *C. frontalis* was recovered only from the belonid fish *S. strongylura*, showing its oligoxenous host specificity. Rarely though, oligoxenous host specificity has also been reported in other cymothoid species such as *N. indica*, *M. renardi* and *N. longispina* parasitizing *R. kanagurta*, *S. leiura* and *A. ambassis*, respectively (Aneesh et al. 2011, 2013a, b).

In *C. frontalis*, broods are held in the female marsupium. This observation agrees with the previous reports on other cymothoids *Elthusa vulgaris* (Brusca 1978), *Anilocra pomacentri* (Adlard and Lester 1995) and *Glossobius hemiramphi* (Bakenhaster et al. 2006). The life cycle of *C. frontalis* passes through marsupial, free living and infective phases. During the marsupial phase, six sequential and morphologically distinct stages including one zygotic, three embryonic (ES I-III) and two larval stages (pre-manca and manca) are completed. Marsupiumites of cymothoids *E. vulgaris*, *A. pomacentri* and *G. hemiramphi* have also been reported to have five distinct ontogenic stages (Brusca 1978; Adlard and Lester 1995; Bakenhaster et al. 2006). The synchronized pattern of marsupial development as observed in *C. frontalis* (present study) looks comparable with what was reported in *G. hemiramphi* (Bakenhaster et al. 2006). However, in *C. frontalis*, the exact duration of the successive development of each marsupial stage from their preceding stage is not clear. The final marsupial stage (manca) of *C. frontalis* showed close resemblance to that of *A. pomacentri* (Adlard and Lester 1995) and *C. spinipalpa* (Thatcher et al. 2007). In *A. pomacentri*, the manca is released within 2 days after they appeared in the marsupium (Adlard and Lester 1995). In the present study, it is observed that after a brief period of marsupial residence, the manca of *C. frontalis* is released into the water and lead a free swimming life until it infests the host fish (*S. strongylura*). Deviating from this pattern, the manca of *G. hemiramphi* have been shown to infest other species of fish temporarily, before settling on its specific host *Hemiramphus brasiliensis* (Bakenhaster et al. 2006). The exact mechanism that helps manca to find its specific host is unclear. The physical characteristics of this larval form are suggested to facilitate precise attachment to the specific host (Bakenhaster et al. 2006). Evidences are not available to suggest if the environmental cues or host-derived factors are likely to influence the host-specific attachment. The success rate of the larvae for host attachment is also not available.

From the manca stage of *C. frontalis*, the transformation to successive stages (juveniles, male, transitional and female stages) is accomplished by the intervening moult. The present report on the existence of six successive stages (Fs 1 to Fs 6) of the ovigerous females of *C. frontalis* could be considered first of its kind due to a paucity of any such attempt on other cymothoid species.

Admittedly, the duration of the life cycle of *C. frontalis* could not be assessed during the present study. Previous reports suggest that the average duration to complete the life cycle of cymothoids is nearly 1 year; the entire life cycle of *G. hemiramphi* on its fish host (*H. brasiliensis*) is completed within 1 year (Bakenhaster et al. 2006). In *A. pomacentri*, the female without dislodging from the host or host death was shown to have an average life span of 13.5 months (Adlard and Lester 1995).

Conclusions

Not only would the present redescription of the female and description of the remaining stages in the life cycle of *C. frontalis* provide us with added information on its biology, but the study would also help us identify the cymothoid parasite very precisely. Further, the observations made through the present study could give us valuable cues for identification of other (parasitic) cymothoids as well. The occurrence of ovigerous females of *C. frontalis* at different reproductive phases throughout the study period indicates its high fecundity rate which could be a parasitic adaptation. Further, very pertinently, the features such as strict host specificity, site specificity, high rate of prevalence, characteristic life cycle stages, protandric hermaphroditism, the existence of different classes of ovigerous females and occurrence of fecund females all the way through different moult stages make *C. frontalis* an ideal model organism to study the physiological basis of parasitism and hermaphroditism which in turn would be helpful to study the strategies to be adopted for the management of parasites infesting edible fishes.

Abbreviations

LT: Total length; MNHN: National Museum of Natural History of Paris, France; PCM: Parasitic Crustacean Museum, Crustacean Biology Research Laboratory, Sree Narayana College, Kannur, Kerala, India.

Competing interests

The authors declare that they have no competing interests.

Authors' contributions

APT is the main worker on the topic and prepared the draft of the manuscript. SK is the mentor who critically followed the work with timely discussion, interpreted the results and corrected the manuscript. HAK helped the main worker for observation and identification, helped in the drawing and is involved in the interpretation of results. AG critically reviewed the manuscript and made significant comments for the improvement of the paper. TJP helped to describe and redescribe the specimen and compared with the type material, reviewed the article and made necessary corrections/modifications. All authors read and approved the final manuscript.

Acknowledgements

The authors gratefully acknowledge University Grant Commission, New Delhi (F.No: 38-218/2009 (SR); dated: 24/12/2009) and Kerala State Council for Science Technology and Environment, Government of Kerala (No. (T) 093/SRS/2011/CSTE; dated: 25/06/2011) for financial support to carry out this work.

Author details

[1]Post Graduate Department of Zoology and Research Centre, Sree Narayana College, Kannur 670 007, India. [2]School of Biosciences and Technology, VIT University, Vellore 632014, Tamil Nadu, India. [3]UMR 5119 (CNRS-UM2-IRD-UM1-IFREMER), Equipe Adaptation Ecophysiologique et Ontogenèse, University of Montpellier 2, CC. 092, 34095 Montpellier, Cedex 05, France.

References

Adlard RD, Lester RJG (1995) The life-cycle and biology of *Anilocra pomacentri* (Isopoda, Cymothoidae), an ectoparasitic isopod of the coral-reef fish, *Chromis nitida* (Perciformes, Pomacentridae). Aust J Zool 43(3):271–281. doi:10.1071/ZO9950271

Aneesh PT (2014) Studies on parasitic crustaceans infesting the fishes of Malabar Coast, Ph. D. thesis, Kannur University

Aneesh PT, Sudha K, Arshad K, Anilkumar G, Trilles JP (2011) Studies on the occurrence of *Norileca indica* (Milne Edwards), a parasitic cymothoid, parasitizing the Indian mackerel, *Rastrelliger kanagurta* (Scomberidae) from the Malabar coast of Kerala. In: Abstract volume of national conference on emerging trends in biological research, Madras University, Chennai, 21–22 February, 2011, 51

Aneesh PT, Sudha K, Arshad K, Anilkumar G, Trilles JP (2013a) Seasonal fluctuation of the prevalence of cymothoids representing the genus *Nerocila* (Crustacea, Isopoda), parasitizing commercially exploited marine fishes from the Malabar Coast, India. Acta Parasitol 58(1):80–90. doi:10.2478/s11686-013-0112-3

Aneesh PT, Sudha K, Helna AK, Arshad K, Anilkumar G, Trilles JP (2013b) Simultaneous multiple parasitic crustacean infestation on banded needlefish, *Strongylura leiura* (Belonidae) from the Malabar Coast, India. Int J Scientific Res Publications 3(7):367–375

Aneesh PT, Sudha K, Helna AK, Anilkumar G, Trilles J-P (2014) Multiple parasitic crustacean infestation on belonid fish *Strongylura strongylura*. In: Wehrtmann IS, Bauer RT (eds) Proceedings of the summer meeting of the Crustacean Society and the Latin American Association of Carcinology, Costa Rica, July 2013. ZooKeys 457:339–353

Avdeev VV (1978) Notes on the distribution of the marine Cymothoidae (Isopoda, Crustacea) in the Australian-New Zealand region. Folia Parasitol 25:281–283

Avdeev VV (1979) New species of the genus Cymothoa (Isopoda, Cymothoidae) from the coastal regions of Northern and North-Western Australia. Parazitologiya 13:50–55

Avdeev VV (1990) Morpho-physiological adaptation in ecto and mesoparasitic Isopoda of the suborder Flabellifera. Zoologicheski Zhurnal SSSR (en Russse) 69:33–42

Bakenhaster MD, McBride RS, Price WW (2006) Life history of *Glossobius hemiramphi* (Isopods: Cymothoidae): development, reproduction, and symbiosis with its host *Hemiramphus brasiliensis* (Pises: Hemiramphidae). J Crustacean Biol 26(3):283–294

Bowman TE (1960) Description and notes on biology of *Lironeca puhi*, n.sp. (Isopoda, Cymothoidae), parasite of the Hawaiian Moray Eel, *Gymnothorax eurostus* (Abbott). Crustaceana 1:82–91

Bowman TE, Mariscal RN (1968) *Renocila heterozoata*, a new cymothoid isopod, with notes on its host, the anemone fish *Amphiprion akallopisos*, in the Seychelles. Crustaceana 14:97–104

Brusca RC (1978) Studies on the cymothoid fish symbionts of the eastern Pacific (Crustacea: Isopoda: Cymothoidae). II. Systematics and biology of *Lironeca vulgaris* Stimpson 1857. Allan Hancock Occasional Papers New Series 2:1–19

Brusca RC (1981) A monograph on the Isopoda Cymothoidae (Crustacea) of the eastern Pacific. Zool J Linnean Soc 73:117–199

Bush AO, Lafferty KD, Lotz JM, Shostak AW (1997) Parasitology meets ecology on its own terms: Margolis et al. revised. J Parasitol 83:575–583

Dana JD (1853) Crustacea, Part II. Fam. Cymothoidae. In: United States Exploring Expedition during the Years 1838, 1839, 1840, 1841, 1842, under the command of Charles Wilkes, U.S.N. Vol. 14, C. Sherman, Philadelphia, 746–65.

Fabricius JC (1793) Entomologia Systematica emendata et acuta, secundum classes, ordines, genera, species adjectis synonimis, locis, observationibus, descriptionibus. Christ. Gottl. Proft., Hafniae, 519

Froese R, Pauly D (2012) FishBase. World Wide Web electronic publication. Available from: http:// www. Fishbase org, Version (March 2013).

Gerstaecker A (1901) Isopoda. In Dieklassen and ordnungen der. Arthropoden, wissenschaftlich dargestelt in wort and Bild., Funther Band. II Abtheilung. Crustacea (ed. H. B. Bronn), 8–278. C. F. Wintersche verlagshandlung

Gopalakrishnan A, Rajkumar M, Jun S, Trilles J-P (2010) Occurrence of double parasitism on black-barred halfbeak fish from the southeast coast of India. Chinese J Oceanol Limnol 28:832–835

Hadfield KA, Bruce NL, Smit NJ (2011) *Cymothoa hermani* sp. nov. (Isopoda, Cymothoidae, Crustacea), a parasitic isopod, collected off the Zanzibar coast, Tanzania from the mouth of a parrotfish (Scaridae). Zootaxa 2876:57–68

Hadfield KA, Bruce NL, Smit NJ (2013) Review of the fish-parasitic genus *Cymothoa* Fabricius, 1793 (Isopoda, Cymothoidae, Crustacea), from the southwestern Indian Ocean, including a new species from South Africa. Zootaxa 3640:152–176

Heller C (1868) Crustaceen, Ordo Isopoda. In: Reise der Österreichischen Fregatte Novara um die Erde in den Jahren 1857, 1858, 1859 unter den Befehlen des Commodore B. von Wüllerstorf- Urbair. Zoologischer Theil, Zweiter Band, III Abtheilung. Aus der Kaiserlich-Königlichen Hof- und Staatsdruckerei, Wien, 130–147.

Horton T, Okamura B (2001) Cymothoid isopod parasites in aquaculture: a review and case study of a Turkish sea bass (*Dicentrarchus labrax*) and sea bream (*Sparus auritus*) farm. Dis Aquat Organ 46:181–188

Kussakin OG (1979) Marine and brackish water isopod Crustacea. Suborder Flabellifera. Akademy of Science, U.S.S.R

Margolis L, Esch GW, Holmes JC, Kuris AM, Schad GA (1982) The use of ecological terms in parasitology (report of an ad hoc Committee of the American Society of Parasitologists). J Parasitol 68:131–133

Milne Edwards H (1840) Histoire Naturelle des Crustacés comprenant l.anatomie, la physiologie et la classification de ces animaux. Vol. III. Librairie Encyclopédique de Roret, Paris

Nierstrasz HF (1931) Die Isopoden der Siboga Expedition. 3. Isopoda Genuina. 2. Flabellifera. Siboga-Expeditie 32:123–233

Rajkumar M, Perumal P, Trilles JP (2006) A note on the double parasitism (Copepod and Isopod) in Commerson's Anchovy fish (India). J Environ Biol 27:613–614
Ramakrishna G (1980) Techniques of collection and preservation of parasitic crustaceans. In: Proceedings of workshop - techniques in parasitology. Zoological Survey of India 109–113
Richardson H (1905) A monograph on the isopods of North America. Bulletin United States National Museum 54:1–727
Romestand B, Trilles J-P (1977) Dégénérescence de la langue des Bogues (*Boops boops* L., 1758) (Téléostéens, Sparidae) parasités par *Meinertia oestroides* (Risso, 1826) (Isopoda, Flabellifera, Cymothoidae). Zeitschrift fur Parasitenkunde 54:47–53
Sartor SM, Pires AM (1988) The occurrence of *Cymothoa liannae*, a new species of cymothoid isopod from Brazil, with a comparative study of its post-marsupial development. Crustaceana 55:147–156
Schioedte JC, Meinert F (1884) Symbolae ad monographiam Cymothoarum Crustaceorum Isopodum Familiae. IV. Cymothoidae. TRIB. II. Cymothoinae. TRIB. III. Livonecinae. Naturhistorisk Tidsskrift, Ser. III, XIV: 221.454, pls VI. XVIII (Cym. XXIV. XXXVI).
Schotte M, Boykon CB, Bruce NL, Poore GCB, Taiti S, Wilson GDF (Eds) (2008 onwards). World list of marine freshwater and terrestrial isopod crustaceans. Available online at http://www.marinespecies.org/isopoda. [Accessed 19 Oct 2010]
Seng LK, Seng LT (1990) The Genus *Nerocila* (Isopoda: Flabellifera) from the marine fish *Triacanthus breverostris*, Penang, Malaysia. J Biol Sci 1:87–100
Sudha K, Supriya NT, Krishnakumar V, Anilkumar G, Chang ES (2012) Hemolymph ecdysteroid titres in a brachyuran crab (*Uca triangularis*) that concomitantly undergoes molting and reproduction. Zool Stud 51(7):966–976
Thatcher VE, de Loyola J, Silva E, Jost GF, Souza-Conceicao JM (2003) Comparative morphology of *Cymothoa* spp. (Isopoda, Cymothoidae) from Brazillian fishes, with the description of *Cymothoa catarinensis* sp. Nov. and redescription of *C. excise* Perty and *C. oestrum* (Linnaeus). Rev Bras Zool 2:541–552
Thatcher VE, de Araujo GS, TAX de lima J, Chellappa S (2007) *Cymothoa spinipalpa* sp. nov. (Isopoda, Cymothoidae) a buccal cavity parasite of the marine fish, *Oligoplites saurus* (Bloch & Schneider) (Osteichthyes, Carangidae) of Rio Grande do North State, Brazil. Rev Bras Zool 24:238–245
Trilles J-P (1968) Recherches sur les Isopodes "Cymothoidae" des côtes francaises. Vol. 1 : bionomie et parasitisme. Vol. 2. Biologie generale et. Sexualite. Deuxieme these, Systematique et faunistique. University of Montpellier, France, No. Enregistrement CNRS AO2305
Trilles J-P (1975) Les Cymothoidae (Isopoda, Flabellifera) des collections du Museum national d'Histoire naturelle de Paris. III. Les Cymothoinae Schioedte et Meinert, 1884. Genre *Cymothoa* Fabricius, 1787. Bulletin du Muséum National d' Histoire Naturelle, Paris 3ème série, n° 318, Zoologie 225:977–993
Trilles J-P (1994) Les Cymothoidae (Crustacea, Isopoda) du monde. Prodrome pour une faune. Studia Marina 21/22 (1–2) (1991): 5–288.
Trilles J-P, Bariche M (2006) First record of the Indo-Pacific *Cymothoa indica* (Crustacea, Isopoda, Cymothoidae), a Lessepsian species in the Mediterranean Sea. Acta Parasitol 51:223–230
Trilles J-P, Hipea-Jacquotte R (2012) Symbiosis and parasitisme in the Crustacea. In traité de zoologie by P.P. Grassé, J. Forest and c. von Vaupel Klein (Eds). The Crustacea 3:239–319
Trilles J-P, Ravichandran S, Rameshkumar G (2011) A checklist of the cymothoidae (Crustacea, Isopoda) recorded from Indian fishes. Acta Parasitol 56(4):445–459
Tsai ML, Dai CF (1999) *Ichthyoxenus fushanensis*, new species (Isopoda: Cymothoidae) parasite of the freshwater fish, *Varicorhinus bacbatulus*, from northern Taiwan. J Crustacean Biol 19:917–923
White A (1847) List of the specimens of Crustacea in the collection of the British Museum. Printed by order of the Trustees, London, London
Williams LB, Williams EH Jr (1985) Brood pouch release of *Anilocra chromis* Williams and Williams (Isopoda: Cymothoidae) a parasite of brown chromis, *Chromis multilineatus* (Guichenot), in the Caribbean. Crustaceana: Int J Crustacean Res 49:92–95

4

Topographical distribution of blubber in finless porpoises (*Neophocaena asiaeorientalis sunameri*): a result from adapting to living in coastal waters

Xianyuan Zeng[1,2], Junhua Ji[1,2], Yujiang Hao[1*] and Ding Wang[1*]

Abstract

Background: Blubber has many functions, among which energy storage, thermoregulation, buoyancy, and hydrodynamic streamlining are the most frequently cited. Within and between taxa, variations in its structure and distribution likely reflect different adaptations of a species to its life history requirements, environment, health, and function. Here, we use ultrasound to describe the distribution of blubber in the finless porpoise (*Neophocaena asiaeorientalis sunameri*) based on examinations of 34 fresh cadavers recovered as accidental fisheries bycatch.

Results: Measurements of blubber depth determined by ultrasound positively correlated with conventional measurements using a scalpel and calipers. Whereas conventional surgical incision and visual examination revealed two layers of blubber, ultrasound revealed up to three layers; thus, ultrasound reveals additional structural detail in blubber while crude necropsy techniques do not. Across life history categories, ultrasound revealed the distribution of inner blubber to be topographically consistent with that of full-depth blubber. Blubber in the dorsal region was stratified into three layers and was significantly thicker than that in the lateral and ventral regions, in which a middle layer was normally absent.

Conclusions: Ultrasound provides a fast, effective, and accurate means to determine blubber thickness and structure, and thus, assessment of the health of fresh finless porpoise carcasses. Blubber depth is determined largely by the thickness of the inner and middle layers, wherein lipids are concentrated. The thickening of blubber in the dorsal thoracic-abdominal region suggests multiple roles of thermal insulation, lipid storage, and, we speculate, to facilitate vertical stability in the complex shallow and estuarine waters in which this animal absent of a dorsal fin occurs.

Keywords: Blubber distribution; Finless porpoise; Cetacean; Postmortem; Ultrasound imaging

Background

Blubber performs a variety of functions in aquatic mammals, such as acting as an energy reserve, for thermal insulation, providing buoyancy, and streamlining the body (Ryg et al. 1988). Lipids are almost entirely accumulated within this tissue, which can constitute a large proportion of an animal's gross mass (Read 1990). It has become common to assess the condition of marine mammals on the basis of blubber measurements, sometimes in conjunction with other morphological parameters (Gomez-Campos et al. 2011; Miller et al. 2011).

Blubber not only performs multiple functions, but these functions may also vary in different regions of an animal. For instance, in ringed seals (*Pusa hispida*), blubber from the posterior region has been deemed over-insulated and to function primarily as a fat reserve; accordingly, it is from this region that blubber depth should, preferably, be measured to assess body condition (Ryg et al. 1988). Koopman (1998) determined sites of energy storage in harbor porpoises (*Phocoena phocoena*) to occur predominately in the thoracic-abdominal region, whereas that blubber in the tailstock region performed a hydrodynamic role, streamlining the peduncle. Biochemical and histological findings in harbor porpoises and bottlenose dolphins (*Tursiops truncatus*) have further demonstrated fat deposition and mobilization to

* Correspondence: hao.yj@ihb.ac.cn; wangd@ihb.ac.cn
[1]Institute of Hydrobiology, Chinese Academy of Sciences, 7 South Donghu Road, Wuchang District, Wuhan 430072, China
Full list of author information is available at the end of the article

occur primarily within the inner and middle layers of blubber in the thoracic-abdominal region, whereas the outer comparably metabolically inert layer functions primarily in thermal insulation (Koopman et al. 2002; Samuel and Worthy 2004; Struntz et al. 2004). Given this recognized variation in the roles of blubber within different layers or from different regions of a body in different species, we deem it necessary to have a basic appreciation of the normal topographical distribution of blubber in a given species before otherwise sampling from it - a practice that might lead to obtaining erroneous measures or metrics of body condition.

East Asian finless porpoises (*Neophocaena asiaeorientalis sunameri*) are one of the smallest odontocetes to occur in coastal waters from the Taiwan Strait north to the Bohai/ Yellow Sea in China, and the waters of Korea and Japan (Jefferson and Wang 2011). Investigations on blubber distribution in this porpoise are required to lay the groundwork for body condition assessment based on metrics of blubber thickness. To obtain baseline data of this nature, two techniques have been commonly used: ultrasound imaging on live animals and necropsy of carcasses (Beck and Smith 1995; Mellish et al. 2007; Montie et al. 2008). However, the number of live finless porpoises available for ultrasound imaging is limited, since captive animals in commercial aquaria are scattered and behavioral training must be undertaken before ultrasound examinations can be performed (Brook 2001). The alternative and more traditional, often unpleasant, and laborious method involves necropsy of carcasses (that are prone to rapid decomposition, especially when laboratories are far away from the field) (Ryg et al. 1990; Heng et al. 2009). Postmortem abdominal ultrasonography is generally considered to be of limited or no value, particularly for highly decomposed cadavers because gas rapidly accumulates in the gastrointestinal tract and blood vessels (Charlier et al. 2013; Heng et al. 2009). Nevertheless, we explore the possibility that ultrasound images of blubber might provide meaningful data, given that abdominal gases might not have the same effect on these tissues in fresh carcasses.

In this present study, we perform ultrasound imaging on 34 finless porpoise carcasses to: 1) validate postmortem ultrasound virtopsy on the blubber layer, 2) describe the distribution of blubber unique to this subspecies, and 3) speculate on the possible adaptive significance of blubber distribution in finless porpoises.

Methods

Animals

Finless porpoises are routinely, accidentally caught in gill nets in the Bohai/Yellow Sea. They are usually discovered within a few hours of death. During May of 2014, 34 such animals were recovered from fishing operations in Penglai, Shangdong. All cadavers exhibited no or little scavenger damage and had died within 12 h of reception; all also appeared to be in good condition, with the exception of neonates, which, according to their external appearance, were very thin. Upon receipt of individuals, ultrasound and physical examinations were performed immediately at a nearby fishing port; animals were weighed and their standard body and girth lengths were measured (American Society of Mammalogists 1961).

Animals were categorized into one of the following life history groups: 1) neonates retaining rostral hairs, unhealed umbilicus, and fetal folds (Dearolf et al. 2000); 2)

Table 1 Individual information of postmortem finless porpoises used in the present study

Animal	Total body length (cm)	Life history category	Sex
PL51221	77.2	Neonate	F
PL51622	77.2	Neonate	F
PL51925	79.8	Neonate	F
PL51722	85.4	Neonate	F
PL50811	72.2	Neonate	M
PL51913	82.1	Neonate	M
PL51911	82.3	Neonate	M
PL51211	83.8	Neonate	M
PL51621	107.6	Juvenile	F
PL51724	111.6	Juvenile	F
PL50821	114.1	Juvenile	F
PL50621	114.5	Juvenile	F
PL51721	116.1	Juvenile	F
PL51522	117.5	Juvenile	F
PL51021	119.5	Juvenile	F
PL51923	120.1	Juvenile	F
PL51723	122.6	Juvenile	F
PL51521	123.8	Juvenile	F
PL52212	104.4	Juvenile	M
PL51011	111.2	Juvenile	M
PL51711	112.9	Juvenile	M
PL52211	119.8	Juvenile	M
PL52011	120.2	Juvenile	M
PL50812	126.8	Juvenile	M
PL50911	128.2	Juvenile	M
PL51511	140.3	Adult	M
PL51311	140.4	Adult	M
PL50711	183.2	Adult	M
PL51922	140.4	Adult	F
PL51421	147.3	Adult	F
PL51321	160.9	Lactating	F
PL51921	172.7	Lactating	F
PL52021	175.1	Lactating	F
PL51924	186.7	Lactating	F

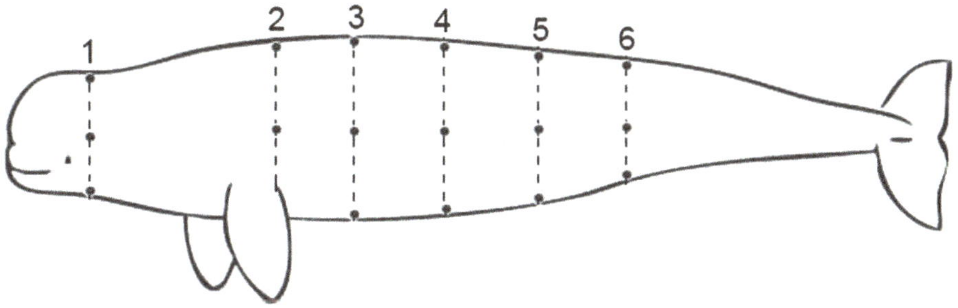

Figure 1 Sites of ultrasound imaging on finless porpoises. Numbers represent girths of the body: (1) head, (2) axillary, (3) mid-thorax, (4) umbilicus, (5) genital slit, and (6) anus. Black spots indicate imaging sites.

juveniles, including both females and males, of body length < 136.7 cm (calculated on basis of sexual maturation at 5 years of age and the age-length curve established for this population [Gao and Zhou 1993]); 3) adult males of body length ≥ 136.7 cm; 4) lactating females with milk in mammary glands; and 5) other females (neither pregnant nor lactating). Detailed information for these animals is provided in Table 1.

Ultrasound equipment

A LOGIQ Book XP ultrasound unit (General Electric Co., Schenectady, NY, USA), in combination with a broadband (3 to 5 MHz) curvilinear array transducer, was used to conduct blubber imaging and measurements. All images were stored in a flash disk. Acoustic gel was not required as there is no air layer on the surface of the skin (Brook 2001).

Protocol of ultrasound imaging and validation of blubber imaging

To determine the distribution of blubber, the body wall (including skin) was imaged at 18 sites (Figure 1). The body was geographically divided into six regions: head (just behind the blowhole), axillary (just caudal insertion of the flipper), mid-thorax (midway between the axillary

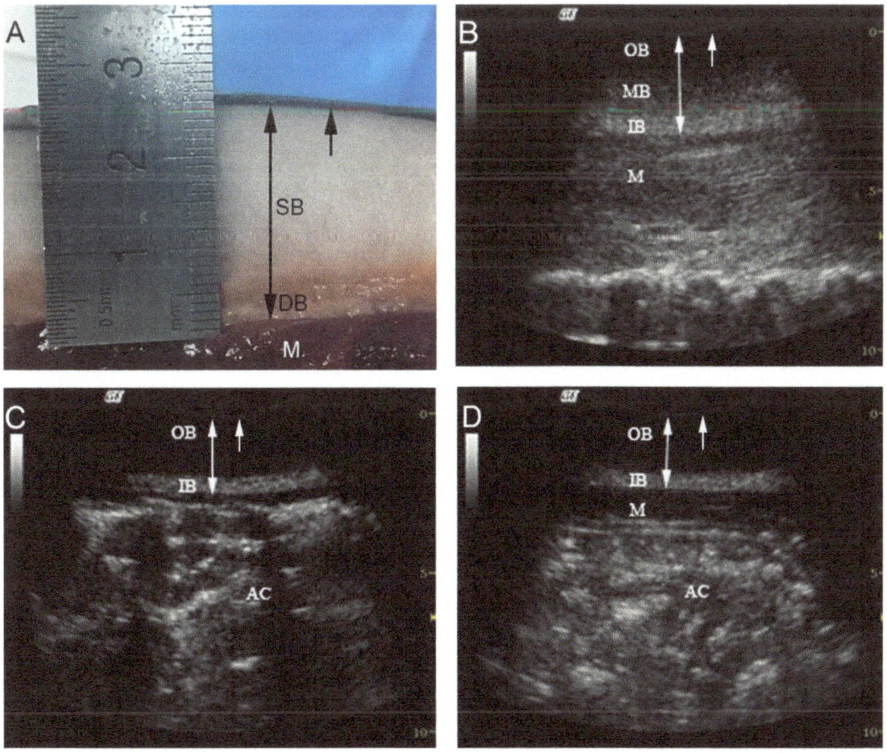

Figure 2 Blubber structure of finless porpoises. (A) A longitudinal section through the blubber layer at the ventral aspect of the umbilicus region. Longitudinal blubber images at the **(B)** dorsal, **(C)** lateral, and **(D)** ventral aspects of the umbilicus region, respectively. Lines with double-headed arrows indicate blubber depth; the arrows indicate the skin layer. AC, abdominal cavity; DB, deep blubber layer; IB, inner blubber layer; M, muscle layer; MB, middle blubber layer; OB, outer blubber layer; SB, superficial blubber layer. Scale marks are in centimeters.

and umbilicus), umbilicus, genital slit (midway between the umbilicus and anus), and anus regions. Given their bilateral symmetry, imaging was only performed on the left side; carcasses lay laterally, with the transducer perpendicular to the body surface during examination. Blubber was scanned at dorsal, lateral, and ventral aspects of each region. Because of the dorsal midline groove and ridge in these animals, dorsal sites were sampled immediately below the dorsal midline, whereas others were in the lateral and ventral midline of the body (Jefferson and Wang 2011). Blubber measurements were taken from longitudinal images with an ultrasound inbuilt electronic caliper.

To validate blubber measurements of postmortem ultrasound images, a necropsy was carried out following imaging on one randomly selected carcass, on which blubber depth measured by ultrasound was compared with that determined by a steel caliper at 18 sites (Figure 1). This research was licensed by the Ministry of Science and Technology of the People's Republic of China, with all animals procured and examination procedures strictly adhering to Chinese ethical guidelines and legal requirements.

Statistical analysis

All data analyses were performed with software SPSS 19.0 for Windows (SPSS Inc., Armonk, NY, USA). Normality of data was verified by Shapiro-Wilk test. Kendall's tau-b test was used to compare blubber depth measured by ultrasound unit with that determined by caliper. Intra- or inter-life history groups were compared by one-way ANOVA with Student-Newman-Keuls test for *post hoc* multiple comparisons. The other females (life history category 5) consisted of two individuals only and were excluded from analysis, and the adult males (life history category 3) were excluded from comparison among life history groups because this group comprised three individuals only, for which the body length of two was also close to the critical value.

Results

Blubber structure and measurements in actual and ultrasound images

Dissection revealed two discrete layers of blubber throughout the entire body, with distinct boundaries between the muscle, the deep and superficial blubber layers, and the skin (Figure 2A); the deep blubber layer close to muscle was structurally heterogeneous and pink colored, whereas that of the superficial blubber was more homogeneous in texture and much lighter in color. Ultrasound similarly revealed multilayered blubber. Blubber in the dorsal region posterior to the head was stratified into three layers (inner, middle, and outer layers) (Figure 2B), in contrast to that blubber in lateral and ventral regions, which was normally absent of the middle layer and consisted of two layers only (Figure 2C,D). The inner layer was significantly hyper-echoic and consistently fascicular, with the boundaries between it and the middle blubber and muscle layers being very clearly defined; the middle layer was relatively hypo-echoic and apparently punctiform, with its echo intensity gradually decreasing towards the outer layer; and the outer layer was anechoic, with its border far from clear from the middle layer. In the head region, only two layers of blubber were apparent, with the inner layer being significantly more hypo-echoic than that of the other regions. Blubber depth determined by ultrasound significantly correlated with actual caliper measurements (Kendall's tau-b test: $\tau = 0.634$, $N = 18$, $P < 0.01$) (Table 2).

Full-depth blubber

The distribution of full-depth blubber (combined inner, middle, and out layers) was similar across life history

Table 2 Blubber depths and corresponding ranks through entire body of actual and ultrasound measurements, respectively

Body region and aspect	Actual measurements		Ultrasound measurements	
	Blubber depth (cm)	Rank	Blubber depth (cm)	Rank
Head				
Dorsal	1.17	1	1.23	3
Lateral	1.20	2	1.16	1
Ventral	1.35	3	1.20	2
Axillary				
Dorsal	1.85	12	1.69	10
Lateral	1.79	10	1.53	5
Ventral	1.68	5	1.71	11.5
Mid-thorax				
Dorsal	2.38	17	1.87	15
Lateral	1.80	11	1.73	13
Ventral	1.78	9	1.82	14
Umbilicus				
Dorsal	2.49	18	2.17	17
Lateral	1.75	7.5	1.67	9
Ventral	1.43	4	1.59	7
Genital slit				
Dorsal	2.12	16	2.18	18
Lateral	1.95	13.5	1.58	6
Ventral	2.05	15	1.90	16
Anus				
Dorsal	1.95	13.5	1.71	11.5
Lateral	1.70	6	1.48	4
Ventral	1.75	7.5	1.61	8

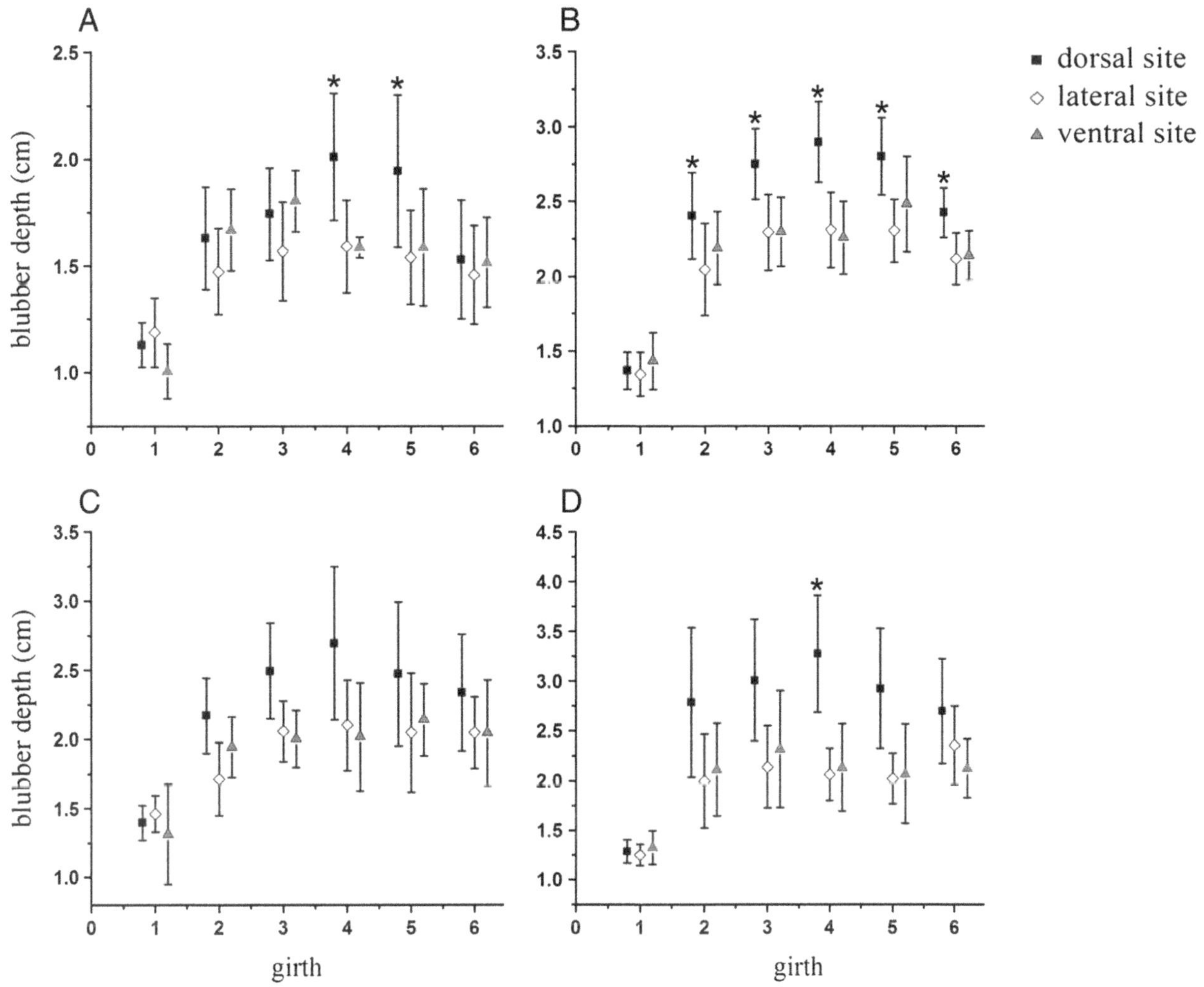

Figure 3 Blubber depth map throughout entire body. Blubber depth (mean ± SD) at the dorsal, lateral, and ventral aspects of six body regions in **(A)** neonates (*n* = 8), **(B)** juveniles (*n* = 17), **(C)** adult males (*n* = 3), and **(D)** lactating females (*n* = 4). Multiple comparisons of blubber depths at the dorsal, lateral, and ventral sites around the girth by Student-Newman-Keuls test: asterisk (*) indicates sites where the depth was significantly different ($P < 0.05$) from corresponding sites.

categories (Figure 3). Blubber depth in the head region was consistent and significantly thinner than it was in the posterior regions at corresponding sites (Figure 4). Dorsally, blubber thickness increased gradually from the head, peaked around the umbilical region, and decreased towards the anal area, whereas it was relatively uniform in thickness in both the lateral and ventral regions of the thorax and abdomen. Analysis of variance for a randomized block design revealed that posterior to the head, mean blubber depths of dorsal, lateral, and ventral regions differed significantly across life history groups (neonates: $F_{2,134} = 8.639$, $P < 0.01$; juvenile: $F_{2,289} = 66.165$, $P < 0.01$; adult males: $F_{2,46} = 7.360$, $P < 0.05$; lactating females: $F_{2,64} = 17.093$, $P < 0.01$). Furthermore, mean blubber depth was significantly thicker from the dorsal region (neonates: 1.66 cm, juveniles: 2.46 cm, adult males: 2.26 cm, and lactating females: 2.66 cm) than that from both lateral (neonates: 1.47 cm, juveniles: 2.11 cm, adult males: 1.90 cm and lactating females: 1.97 cm) and ventral regions (neonates: 1.53 cm, juveniles: 2.14 cm, adult males: 1.91 cm and lactating females: 2.01 cm) (Student-Newman-Keuls test, $P < 0.05$); no significant difference was found between the lateral and ventral regions. Dorsal blubber was thicker around the girth than at the corresponding lateral and ventral sites, but the significance in differences of thickness varied throughout the thoracic-abdominal region, and no difference was found between the lateral and ventral sites (Figure 3). Among life history categories, blubber depth of neonates was the thinnest at almost all sites (Table 3). Blubber of lactating females was thicker at the dorsal thoracic and abdominal sites than that of the juveniles, although a

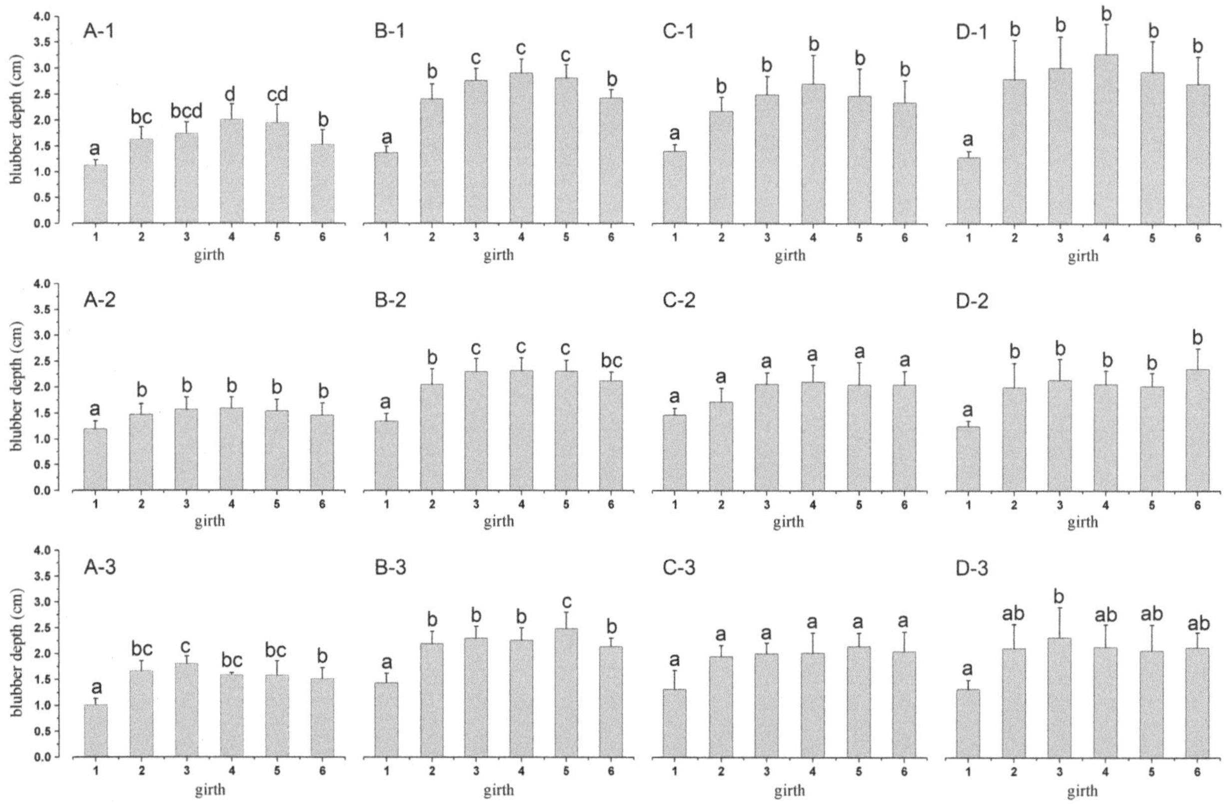

Figure 4 Comparisons of blubber depths among body regions. Blubber depths (mean + SD) were dorsally, laterally, and ventrally compared among body regions of neonates ($n = 8$) **(A-1, A-2, and A-3)**, juveniles ($n = 17$) **(B-1, B-2, and B-3)**, adult males ($n = 3$) **(C-1, C-2, and C-3)**, and lactating females ($n = 4$) **(D-1, D-2, and D-3)**. Blubber depths of six body regions followed by the same letter were not significantly different at $P = 0.05$ (Student-Newman-Keuls test).

significant difference was apparent only in the umbilicus region. In contrast, juvenile blubber was thicker at lateral and ventral abdominal sites than that of the lactating females, though a significant difference was apparent only in the genital slit region.

The inner blubber layer

Posterior to the head, across life history categories, the pattern in distribution of the inner blubber layer was topographically consistent with that of full-depth blubber (Figures 3 and 5). Thickness of the inner layer in the dorsal region gradually increased from the axillary, peaked in the umbilical, and declined to the anal region (Figure 6). However, thickness of the inner layer in the thoracic-abdominal region was uniform both laterally and ventrally. Posterior to the head, mean inner blubber thicknesses of the dorsal, lateral, and ventral regions differed significantly across life history groups (neonates: $F_{2,96} = 14.618$, $P < 0.01$; juvenile: $F_{2,227} = 97.113$, $P < 0.01$; adult males: $F_{2,38} = 43.562$, $P < 0.01$; lactating females: $F_{2,53} = 38.010$, $P < 0.01$). Additionally, mean inner blubber thickness was significantly thicker in the dorsal region (neonates: 0.39 cm, juveniles: 0.70 cm, adult males: 0.54 cm and lactating females: 0.81 cm) than that in the lateral (neonates: 0.32 cm, juveniles: 0.44 cm, adult males: 0.35 cm and lactating females: 0.44 cm) and ventral regions (neonates: 0.32 cm, juveniles: 0.43 cm, adult males: 0.37 cm, and lactating females: 0.43 cm) (Student-Newman-Keuls test, $P < 0.05$), but no significant difference was apparent between the lateral and ventral regions. The dorsal inner layer was thicker around the girth than that at the corresponding lateral and ventral sites in the thoracic-abdominal region (Figure 5). In addition, significant differences were found in most girths posterior to the head of juveniles, adult males, and lactating females, but only in the umbilical region of neonates. Among life history categories, the inner layer of neonates was significantly thinner at dorsal sites than that it was for lactating females and juveniles, but the inner blubber layer in juveniles was thinner at the dorsal sites posterior to the head than that in lactating females, though the significance in difference was not found (Table 4). No significant differences in the inner blubber thickness were found at the lateral and ventral sites among life history groups.

Table 3 Comparisons of blubber depths among life history categories

Body region and aspect	Neonates ($n = 8$)	Juveniles ($n = 17$)	Lactating females ($n = 4$)
Head			
Dorsal	1.13 ± 0.10 A	1.37 ± 0.12 B	1.29 ± 0.12 B
Lateral	1.19 ± 0.16	1.34 ± 0.15	1.25 ± 0.11
Ventral	1.01 ± 0.13 A	1.43 ± 0.19 B	1.33 ± 0.17 B
Axillary			
Dorsal	1.63 ± 0.24 A	2.40 ± 0.29 B	2.79 ± 0.75 B
Lateral	1.47 ± 0.20 A	2.04 ± 0.31 B	2.00 ± 0.47 B
Ventral	1.67 ± 0.19 A	2.19 ± 0.24 B	2.11 ± 0.47 B
Mid-thorax			
Dorsal	1.74 ± 0.22 A	2.75 ± 0.24 B	3.01 ± 0.61 B
Lateral	1.57 ± 0.23 A	2.29 ± 0.25 B	2.14 ± 0.41 B
Ventral	1.80 ± 0.15 A	2.30 ± 0.23 B	2.32 ± 0.59 B
Umbilicus			
Dorsal	2.01 ± 0.30 A	2.90 ± 0.27 B	3.28 ± 0.59 C
Lateral	1.59 ± 0.22 A	2.31 ± 0.25 B	2.06 ± 0.26 B
Ventral	1.59 ± 0.05 A	2.26 ± 0.24 B	2.13 ± 0.44 B
Genital slit			
Dorsal	1.94 ± 0.36 A	2.80 ± 0.26 B	2.93 ± 0.60 B
Lateral	1.54 ± 0.22 A	2.30 ± 0.21 C	2.02 ± 0.25 B
Ventral	1.59 ± 0.27 A	2.48 ± 0.32 C	2.07 ± 0.50 B
Anus			
Dorsal	1.53 ± 0.28 A	2.42 ± 0.17 B	2.70 ± 0.53 B
Lateral	1.46 ± 0.23 A	2.12 ± 0.17 B	2.35 ± 0.39 B
Ventral	1.52 ± 0.21 A	2.14 ± 0.16 B	2.12 ± 0.30 B

At each site, blubber depths (mean ± SD) (cm) of life history groups followed by the same letter were not significantly different at $P = 0.05$ (Student-Newman-Keuls test). There were no significant differences among life history categories at sites without letters.

Discussion

Blubber structure on ultrasound images

When surgically incised (by scalpel), two layers of blubber were visually apparent in the dorsal region of finless porpoises, but up to three layers were apparent in this same region using ultrasound. Blubber of both harbor porpoises and bottlenose dolphins has been similarly described with two visually discernable layers, which have subsequently proven to comprise three layers by combination of histological and biochemical techniques (Koopman et al. 2002; Samuel and Worthy 2004). In bottlenose dolphins, the lipid content was significantly greater and adipocytes were enlarged in the middle blubber layer, compared to that in the outer layer (McClelland et al. 2012; Struntz et al. 2004). Accordingly, greater lipid reserves in the finless porpoise middle blubber layer likely caused the distinction between it and the outer layer in echo density of ultrasound images. The hyper-echo intensity of inner layer resulted from its heterogeneous structure, which was similar to the report in fin whales (*Balaenoptera physalus*), wherein the inner layer is structurally complex and consists of large, loose fiber knots, and bundles (Lockyer et al. 1984).

Topographic distribution of full-depth and inner blubber

Across life history groups, finless porpoise full-depth blubber was distributed in a similar manner to that of the inner blubber layer, and full-depth blubber was significantly thicker in the dorsal region than that in the lateral and ventral regions where an obvious middle layer was absent. In bottlenose dolphins, blubber depth correlated positively with lipid content and adipocyte cell size within inner and middle layers across life history stages (Struntz et al. 2004). It is also likely that lipid in finless porpoises is stored primarily in the middle and inner layers, the combined thickness of which determines blubber depth. Finless porpoise head region blubber depth was also significantly thinner than it was in the thoracic-abdominal region, similar to accounts in harbor porpoises (Koopman 1998) and belugas (*Delphinapterus leucas*) (Doidge 1990). Such distribution might indicate

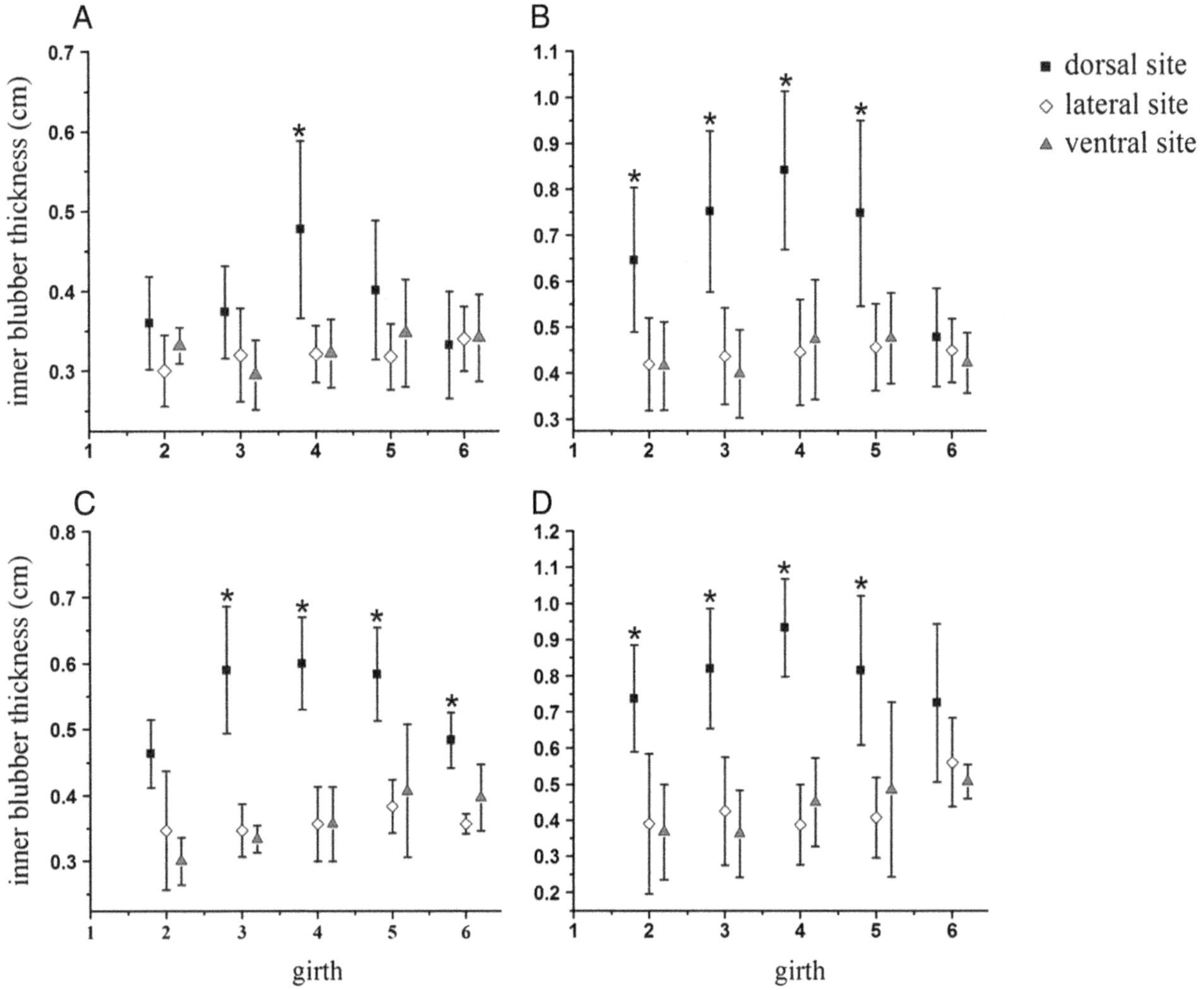

Figure 5 Inner blubber thickness map throughout entire body. Thickness of the inner layer (mean ± SD) at the dorsal, lateral, and ventral aspects of five body regions in **(A)** neonates ($n = 8$), **(B)** juveniles ($n = 17$), **(C)** adult males ($n = 3$), and **(D)** lactating females ($n = 4$). Multiple comparisons of inner blubber thicknesses at the dorsal, lateral, and ventral sites around the girth by Student-Newman-Keuls test: asterisk (*) indicates sites where the thickness was significantly different ($P < 0.05$) from corresponding sites.

blubber in the dorsal region of the thorax, and abdomen serves as an energy reserve in finless porpoises. Furthermore, measurements of blubber depth taken from the dorsal thoracic-abdominal regions are potential indicators to appraise an individual's nutritional state.

Given their extended genital openings, and the initiation of breeding in May (Lee et al. 2013), it is possible that all lactating females in this present study had given birth within a few days of capture. Accordingly, in terms of their body condition, these females could also be regarded as later-stage pregnant individuals. Blubber of lactating females was thicker in the dorsal regions of the thorax and abdomen than it was in juveniles and neonates, which is consistent with previous findings that pregnant females possess the greatest energy reserves among life history groups (Gomez-Campos et al. 2011; Miller et al. 2011). In contrast, blubber of lactating females was thinner in the lateral and ventral regions of the abdomen than it was in juveniles - an observation similar to that reported for harbor seals (*Phoca vitulina*), in which pregnant females had significantly thinner blubber in the ventral abdominal region than non-pregnant individuals, possibly to reduce resistance for swimming (Beck and Smith 1995). However, it is also possible that the situation is comparable to that described for rabbits and sheep, wherein the maternal abdominal wall consists of thin muscles and skin layer, which acts as a thermal window (Gilbert and Power 1986; Hart and Faber 1965), and that during finless porpoise gestation, the relatively thin blubber in the lateral-ventral region of the abdomen might facilitate dissipation of heat produced from fetal metabolism, thus maintaining a more stable and suitable temperature for fetal development. Amongst all finless porpoise life history categories, blubber of neonates was consistently the thinnest. Similarly, neonatal bottlenose dolphins possess the

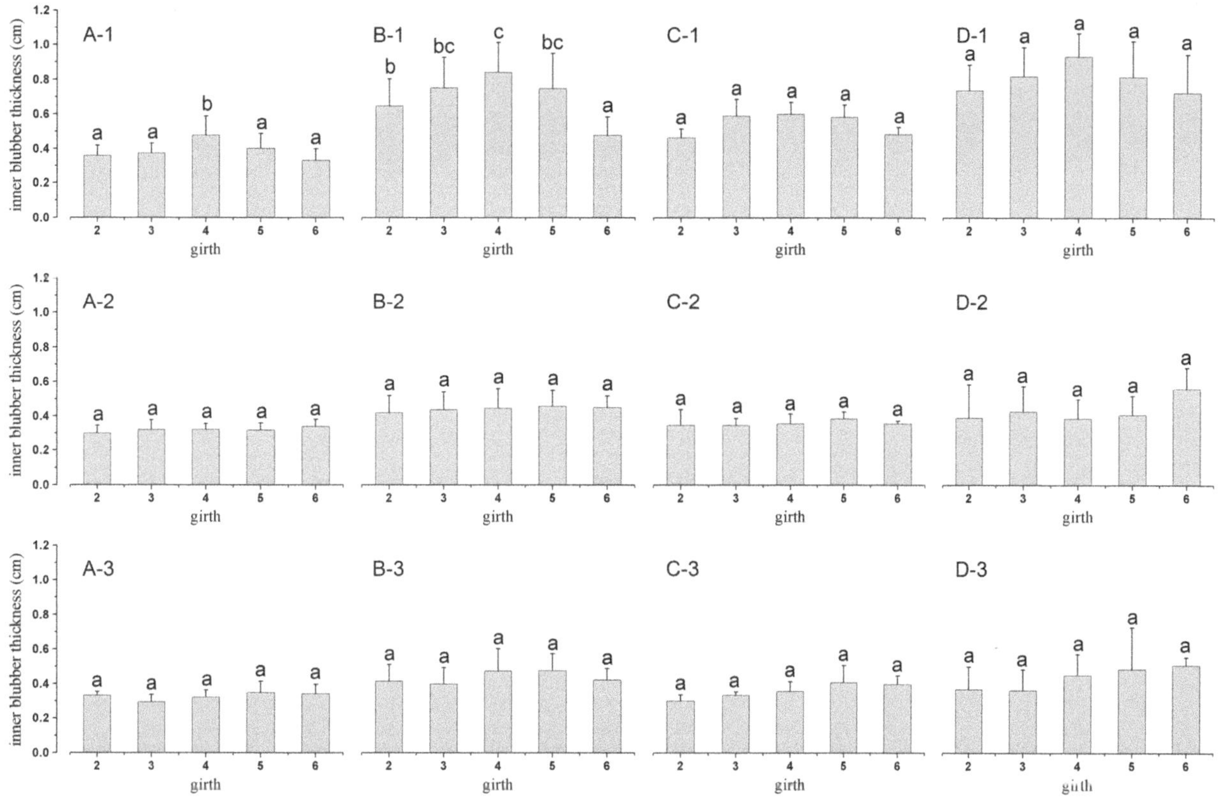

Figure 6 Comparisons of inner blubber thicknesses among body regions. Inner blubber thicknesses (mean + SD) were dorsally, laterally, and ventrally compared among body regions of neonates ($n = 8$) **(A-1, A-2, and A-3)**, juveniles ($n = 17$) **(B-1, B-2, and B-3)**, adult males ($n = 3$) **(C-1, C-2, and C-3)**, and lactating females ($n = 4$) **(D-1, D-2, and D-3)**. Inner blubber thicknesses of five body regions followed by the same letter were not significantly different at $P = 0.05$ (Student-Newman-Keuls test).

least lipid content, smallest adipocyte size, and thinnest blubber depth across ontogeny except fetuses (Struntz et al. 2004). The thin blubber in neonates may facilitate successful parturition.

Adaptive significance of specific blubber distribution in finless porpoises

During early stages of food deprivation in harbor porpoises and other toothed whales, blubber in the 'neck' region, and area lateral to the dorsal fin, is the first to be metabolized, whereas that blubber in the thoracic-abdominal region tends to be more stable (Kastelein and Battum 1990). This may indicate that blubber in the thoracic-abdominal region has other important, less metabolic functions. It is well established that the dorsal fin in cetaceans vertically stabilizes the body position to prevent sideslip (Fish 2002). This also, however, limits turning performance at small radii. For instance, although various cetaceans can bank towards the inside of a turn at small degrees, only finless porpoises and belugas that lack the dorsal fin are capable of banks forming a 90° angle (Fish 2002; Xiao and Wang 2005). Such ability might be advantageous in complex environments, such as the shallow, coastal, and estuarine waters inhabited by these species (Laidre et al. 2008; Jefferson and Wang 2011). In the absence of a dorsal fin, any loss in vertical stability might be compensated for by unique patterns of distribution of blubber throughout the animal. We report the dorsal blubber around the girth to be thicker than at the corresponding lateral and ventral sites in the thoracic-abdominal region - a pattern of distribution inconsistent with that reported for harbor porpoises, which are otherwise similar in body size but possess a dorsal fin, in which blubber depth varies little around the girth (Koopman 1998). This pattern in blubber distribution is, however, consistent with those reported for belugas and seals (Doidge 1990; Ryg et al. 1988). Given lipids are less dense than water (Kipps et al. 2002), additional lipids stored in the dorsal region could render an animal's center of gravity deeper in the body than otherwise might be the case. This in turn might facilitate normal swimming stabilization and even rapid and agile maneuvering at small radii. Therefore, it is possible that the unusual distribution of blubber we describe for finless porpoises might compensate for the loss of the dorsal fin and be an adaptation for nimble swimming in coastal and estuarine waters.

Table 4 Comparisons of inner blubber thicknesses among life history categories

Body region and aspect	Neonates ($n = 8$)	Juveniles ($n = 17$)	Lactating females ($n = 4$)
Axillary			
Dorsal	0.36 ± 0.06 A	0.65 ± 0.16 B	0.74 ± 0.15 B
Lateral	0.30 ± 0.04	0.42 ± 0.10	0.30 ± 0.04
Ventral	0.33 ± 0.02	0.42 ± 0.10	0.37 ± 0.13
Mid-thorax			
Dorsal	0.37 ± 0.06 A	0.75 ± 0.18 B	0.82 ± 0.17 B
Lateral	0.32 ± 0.06	0.44 ± 0.11	0.43 ± 0.15
Ventral	0.30 ± 0.04	0.40 ± 0.10	0.36 ± 0.12
Umbilicus			
Dorsal	0.48 ± 0.11 A	0.84 ± 0.17 B	0.93 ± 0.14 B
Lateral	0.32 ± 0.04	0.45 ± 0.12	0.39 ± 0.11
Ventral	0.32 ± 0.04	0.47 ± 0.13	0.45 ± 0.12
Genital slit			
Dorsal	0.37 ± 0.04 A	0.75 ± 0.20 B	0.82 ± 0.21 B
Lateral	0.32 ± 0.04 A	0.46 ± 0.09 BC	0.41 ± 0.11 AB
Ventral	0.35 ± 0.07	0.48 ± 0.10	0.49 ± 0.24
Anus			
Dorsal	0.33 ± 0.07 A	0.48 ± 0.11 A	0.73 ± 0.22 B
Lateral	0.34 ± 0.04 A	0.45 ± 0.07 B	0.56 ± 0.12 C
Ventral	0.34 ± 0.05 A	0.42 ± 0.07 B	0.51 ± 0.05 C

At each site, inner blubber thicknesses (mean ± SD) (cm) of life history groups followed by the same letter were not significantly different at $P = 0.05$ (Student-Newman-Keuls test). There were no significant differences among life history categories at sites without letters.

Conclusions

Ultrasonography provided rapid and valuable insights into blubber structure, measurements, and distribution that have enabled a faster and more pleasant, in addition to potentially more reliable appraisal of the condition of a cetacean than would otherwise have proven possible using conventional approaches, such as cadaver dissection and visual examination. Our findings suggest blubber in the dorsal region of the thorax and abdomen functions both as an energy reserve and for thermal insulation. Furthermore, measurements of blubber depth taken from this region likely provide an indication of body condition in this subspecies. However, further efforts are required to understand structure and functions of blubber in the 'neck' region of odontocetes, which was ignored generally in previous and present studies. Finally, the distribution of blubber with the absence of a dorsal fin in finless porpoises might facilitate stabilization and maneuvering in coastal and estuarine waters, although additional research is required to investigate this matter further.

Competing interests

The authors declare that they have no competing interests.

Authors' contributions

YH and DW designed and conceived the project. XZ and JJ conducted all the experimental procedures and data collection and analysis. XZ and YH wrote the paper. All authors read and approved the final manuscript.

Acknowledgements

We thank Shenglun Zhang, Mingen Yin, and other staff in the Ocean World of Penglai for the assistance in the carcass collection. We are especially grateful to Xiong Li and Jianrong Zhang for the advice on ultrasound and physiological examinations of this project. This study was financially supported by the National Natural Science Foundation of China (No. 31430080).

Author details

[1]Institute of Hydrobiology, Chinese Academy of Sciences, 7 South Donghu Road, Wuchang District, Wuhan 430072, China. [2]University of Chinese Academy of Sciences, 19 Yuquan Road, Shijingshan District, Beijing 100049, China.

References

American Society of Mammalogists (1961) Standardized methods for measuring and recording data on the smaller cetaceans. J Mammal 42:471–6

Beck GG, Smith TG (1995) Distribution of blubber in the northwest Atlantic harp seal, *Phoca groenlandica*. Can J Zool 73(11):1991–8

Brook F (2001) Ultrasonographic imaging of the reproductive organs of the female bottlenose dolphin, *Tursiops truncatus aduncas*. Reproduction 121(3):419–28

Charlier P, Chaillot P-F, Watier L, Ménétrier M, Carlier R, Cavard S, Hervé C, de la Grandmaison GL, Huynh-Charlier I (2013) Is postmortem ultrasonography a useful tool for forensic purposes? Med Sci Law 53(4):227–34

Dearolf JL, McLellan WA, Dillaman RM, Frierson D, Pabst DA (2000) Precocial development of axial locomotor muscle in bottlenose dolphins (*Tursiops truncatus*). J Morphol 244(3):203–15

Doidge DW (1990) Integumentary heat loss and blubber distribution in the beluga, *Delphinapterus leucas*, with comparisons to the narwhal, *Monodon monoceros*. Can Bull Fish Aquat Sci 224:129–40

Fish FE (2002) Balancing requirements for stability and maneuverability in cetaceans. Integr Comp Biol 42(1):85–93

Gao A, Zhou K (1993) Growth and reproduction of three populations of finless porpoise, *Neophocaena phocaenoides*, in Chinese waters. Aquat Mamm 19(1):3–12

Gilbert RD, Power GG (1986) Fetal and uteroplacental heat production in sheep. J Appl Physiol 61(6):2018–22

Gomez-Campos E, Borrell A, Aguilar A (2011) Assessment of nutritional condition indices across reproductive states in the striped dolphin (*Stenella coeruleoalba*). J Exp Mar Biol Ecol 405(1):18–24

Hart FM, Faber JJ (1965) Fetal and maternal temperatures in rabbits. J Appl Physiol 20(4):737

Heng HG, Selvarajah GT, Lim HT, Ong JS, Lim J, Ooi JT (2009) Serial postmortem abdominal radiographic findings in canine cadavers. Forensic Sci Int 192(1–3):43–7

Jefferson TA, Wang JY (2011) Revision of the taxonomy of finless porpoises (genus *Neophocaena*): the existence of two species. J Mar Anim Ecol 4(1):3–16

Kastelein R, Van Battum R (1990) The relationship between body weight and morphological measurements in harbour porpoises (*Phocoena phocoena*) from the North Sea. Aquat Mamm 16:48–52

Kipps EK, McLellan WA, Rommel SA, Pabst DA (2002) Skin density and its influence on buoyancy in the manatee (*Trichechus manatus latirostris*), harbor porpoise (*Phocoena phocoena*), and bottlenose dolphin (*Tursiops truncatus*). Mar Mammal Sci 18(3):765–78

Koopman HN (1998) Topographical distribution of the blubber of harbor porpoises (*Phocoena phocoena*). J Mammal 79(1):260–70

Koopman H, Pabst D, McLellan W, Dillaman R, Read A (2002) Changes in blubber distribution and morphology associated with starvation in the harbor porpoise (*Phocoena phocoena*): evidence for regional differences in blubber structure and function. Physiol Biochem Zool 75(5):498–512

Laidre KL, Stirling I, Lowry LF, Wiig O, Heide-Jørgensen MP, Ferguson SH (2008) Quantifying the sensitivity of arctic marine mammals to climate-induced habitat change. Ecol Appl 18(sp2):S97–125

Lee YR, An YR, Park KJ, Sohn H, An DH, Kim SA (2013) Age and reproduction of the finless porpoises, *Neophocaena asiaeorientalis*, in the Yellow Sea, Korea. Anim Cells Syst 17(5):366–73

Lockyer CH, McConnell LC, Waters TD (1984) The biochemical composition of fin whale blubber. Can J Zool 62(12):2553–62

McClelland SJ, Gay M, Pabst DA, Dillaman R, Westgate AJ, Koopman HN (2012) Microvascular patterns in the blubber of shallow and deep diving odontocetes. J Morphol 273(8):932–42

Mellish J-AE, Horning M, York AE (2007) Seasonal and spatial blubber depth changes in captive harbor seals (*Phoca vitulina*) and Steller's sea lions (*Eumetopias jubatus*). J Mammal 88(2):408–14

Miller CA, Reeb D, Best PB, Knowlton AR, Brown MW, Moore MJ (2011) Blubber thickness in right whales *Eubalaena glacialis* and *Eubalaena australis* related with reproduction, life history status and prey abundance. Mar Ecol-Prog Ser 438:267–83

Montie EW, Garvin SR, Fair PA, Bossart GD, Mitchum GB, McFee WE, Speakman T, Starczak VR, Hahn ME (2008) Blubber morphology in wild bottlenose dolphins (*Tursiops truncatus*) from the Southeastern United States: influence of geographic location, age class, and reproductive state. J Morphol 269(4):496–511

Read AJ (1990) Estimation of body condition in harbor porpoises, *Phocoena phocoena*. Can J Zool 68(9):1962–6

Ryg M, Smith TG, Øritsland NA (1988) Thermal significance of the topographical distribution of blubber in ringed seals (*Phoca hispida*). Can Bull Fish Aquat Sci 45(6):985–92

Ryg M, Lydersen C, Markussen NH, Smith TG, Øritsland NA (1990) Estimating the blubber content of phocid seals. Can Bull Fish Aquat Sci 47(6):1223–7

Samuel AM, Worthy GA (2004) Variability in fatty acid composition of bottlenose dolphin (*Tursiops truncatus*) blubber as a function of body site, season, and reproductive state. Can J Zool 82(12):1933–42

Struntz D, McLellan W, Dillaman R, Blum J, Kucklick J, Pabst D (2004) Blubber development in bottlenose dolphins (*Tursiops truncatus*). J Morphol 259(1):7–20

Xiao J-Q, Wang D (2005) Construction of ethogram of the captive Yangtze finless porpoise, *Neophocaena phocaenoides asiaeorientalis*. Acta Hydrobiol Sin 29(3):253–8 (in Chinese with an English abstract)

5

Seasonality in anuran activity and calling season in a Brazilian subtemperate wetland

Simone da Silva Ximenez[1] and Alexandro Marques Tozetti[2*]

Abstract

Background: Most tropical regions have a climate characterized by marked seasonal rainfall patterns, and these seasonal patterns of rainfall directly affect anuran activity. However, in regions with temperate climates, the main aspect of climate related to anuran activity is the thermal regime. Thus, transitional climate regions represent good opportunities to study the effect of abiotic factors on anuran activity. In this study, we present new data on the activity pattern and calling season of Neotropical anurans in a subtemperate climate. Anuran activity was assessed based on the rate of capture of specimens in pitfall traps and calling surveys. The field study was conducted between May 2010 and April 2011 in wetlands in southernmost Brazil.

Results: An analysis based on directional (circular) statistics showed that general activity in the studied anurans was significantly seasonal. In addition, the general activity pattern of most species was regulated by temperature, not by rainfall, and approached that observed in regions with a temperate climate. However, we did not record a well-defined peak in the number of species displaying calling activity. This parameter did not exhibit any influence of temperature variations or rainfall.

Conclusions: The observed general activity pattern is different from that expected for the majority of anurans inhabiting eastern South America. The present study showed that the general activity patterns of most anurans in the wetland region in southernmost Brazil are regulated by temperature and not rainfall, similar to the pattern of anuran assemblies from temperate climates. However, the nonexistence of a significant relationship between calling activity and any of the environmental variables tested suggests the presence of a different environmental factor (e.g., photoperiod or the length of the hydroperiod of the relevant water bodies) as a trigger for the levels of general calling activity.

Keywords: Anuran; Behavior; Calling; Temperature; Wetlands

Background

Many environmental factors act on such anuran activity patterns as reproduction, movements, and foraging. For example, reproductive activity depends on the acoustic quality of the habitat, availability of females, and presence and density of males at the breeding site (Both and Grant 2012). Nevertheless, reproduction is primarily associated with climate-specific abiotic factors (Bertoluci 1998; Oseen and Wassersug 2002; Xavier and Napoli 2011). Although some Neotropical anuran species breed throughout the year, general activity and reproduction typically exhibit remarkable seasonality (Jorgensen 1992; Bertoluci 1998). Other activities performed by anurans, such as dispersion, migration, or foraging, also appear to respond primarily to climatic variables. A wide range of methods are currently available for evaluating anuran activity, and the capture rate is the most frequently used indicator of general activity (Oliveira et al. 2013; Martins et al. 2014). Indeed, capture rate is an indirect measure of any activity performed by animals on the surface of a substrate (e.g., foraging, habitat exploration, dispersal, and migrations), as such activities increase their capture probability.

Associations between peaks of general activity and a particular behavior driven by physiological or metabolic events are poorly studied in the wild, especially in Neotropical anurans. An exception is reproductive activity,

* Correspondence: alexandro.tozetti@gmail.com
[2]Laboratório de Ecologia de Vertebrados Terrestres, Universidade do Vale do Rio dos Sinos—UNISINOS, São Leopoldo Av. Unisinos, 950, Cristo Rei, São Leopoldo RS 93022-000, Brazil
Full list of author information is available at the end of the article

which can be evaluated through calling surveys. Environmental factors such as light intensity, relative humidity, and air temperature are relatively well-documented habitat cues that affect the calling activity of anurans (Fukuyama and Kusano 1992; Hatano et al. 2002; Almeida-Gomes et al. 2007). In contrast, photoperiod is a less well-studied climatic factor relative to seasonal changes in anuran behavior (Bradshaw and Holzapfel 2007). There is evidence that amphibians in subtemperate wetlands could be tracking photoperiod (as previously reported for temperature and rainfall) to determine their seasonal variations in activity (Canavero and Arim 2009). In addition, there has recently been significant improvement in the number of ecological studies of amphibians in the Neotropics, though most of these studies are concentrated in forested or tropical open habitats. Less attention has been dedicated to subtemperate open habitats, such as wetlands. Southern Brazilian wetlands possess unique climatic characteristics, such as colder winters, in comparison with other Brazilian regions (Huckembeck et al. 2012; Oliveira et al. 2013). Given the divergent opinions regarding the effects of climatic factors on the activity patterns of anurans (Oseen and Wassersug 2002; Saenz et al. 2006), transitional climatic regions represent an important natural laboratory for testing ecological hypotheses concerning the evolutionary process in ectotherms. In most Brazilian habitats (and other tropical areas), there is a small annual thermal amplitude (Maluf 2000), and precipitation appears to be the main determining factor of anuran activity patterns (Heyer 1973; Boquimpani-Freitas et al. 2007; Oliveira et al. 2013). However, in temperate climate regions, annual temperature variation appears to be the main factor that determines ectothermic activity (Gilbert et al. 1994; Navas and Bevier 2001; Oseen and Wassersug 2002). However, it is important to emphasize that studies performed in transitional regions between tropical and temperate climates have yielded conflicting results (Bernarde and Anjos 1999; Bernarde and Machado 2001; Conte and Rossa-Feres 2006).

The southernmost region of Brazil is in a transition zone between subtropical and temperate climates (Maluf 2000), thus constituting an important study area for understanding the activity patterns (reproductive or not) of Neotropical anurans. The relatively low temperatures, by tropical standards, and unpredictable rainfall can act as abiotic filters to limit activity and the establishment of species in this region. Studies on anuran general activity are important for species conservation purposes (Eterovick et al. 2005) and typically yield information about the species' ecology and natural history (Santos et al. 2008). In Brazil, especially in Rio Grande do Sul State, the science of anuran general activity has not yet been pursued (Oliveira et al. 2013); thus, there is a lack of knowledge of this subject.

The present study aimed to evaluate the role of abiotic factors in the general activity pattern during the vocalization season in species associated with subtemperate wetlands in southernmost Brazil.

Methods

Study site

The study was conducted between May 2010 and April 2011 in an area formed by a mosaic of temporary wetlands (32° 08′ S and 52° 11′ E) located in Rio Grande, Rio Grande do Sul, southernmost Brazil. The samples were concentrated in an area of approximately 200 ha, which included a portion of the most extensive water bodies with the longest hydroperiods found at the study site. The predominant vegetation consisted of native field grass, which was relatively well preserved (Oliveira et al. 2013). The local climate was classified as humid subtemperate with an annual average temperature of 18.1 °C and an average temperature of 12.7 °C during the coldest month. The climatic seasonality in the study area was characterized by warm and cold seasons, which differed from much of the Brazilian territory climatic seasonality marked by dry and wet seasons (Maluf 2000; Tozetti et al. 2010).

During sampling, the mean minimum temperature of the air during the warm season ranged from 13.08 to 21.25 °C. During the cold season, the mean minimum temperature of the air ranged from 6.45 to 16.43 °C. The rainfall accumulation was 347.7 mm. Because of the small monthly variation in rainfall, the wettest period was defined by plotting the rainfall data (16.7 mm). June, July, October, and December of 2010 and January and April of 2011 were defined as dry months. The humid months were May, August, September, and November of 2010 and February and March of 2011. The climatological data were obtained from Meteorological Station 83,995 of Rio Grande, located 5 km from the study area.

Evaluation of anuran activity

Anuran activity was divided into two types: (1) general activity and (2) calling activity. These forms of activity were evaluated using the following two methods: general activity was evaluated in terms of the capture rate in pitfall traps with drift fences (Greenberg et al. 1994), and calling activity was evaluated through calling surveys. The capture rate in this study was considered an indicator of general activity. General activity corresponded to any activity by the animals on the surface of the substrate (e.g., foraging, habitat exploration, searching for partners, and migration) and increased the probability of capture. Because these methods are common in herpetology, they will be described only briefly in this paper.

We used two pitfall sets, each with two 40-m-long lines, distributed at selected locations around the water bodies (preferential sampling; Krebs 1999), with a distance of 3 km between them. These sites were defined in a manner so that flooding of the buckets was avoided. Each line had four 100-L buckets united by drift fences, for a total of 16 buckets (for more details, see Ximenez et al. 2014). The traps remained open for four consecutive days per month between May 2010 and April 2011. Each captured animal was identified and measured. Whenever possible, the sex of the specimen was determined. The animals were then marked by toe clipping and released 5 m away from one of the ends of the capture trap. The markings used aimed to differentiate between captures and recaptures. Recaptures represented less than 0.001 % of the captures.

A calling survey was performed twice per month between 1800 and 2300 h from May 2010 to April 2011 for a sample total of 120 h. To standardize the sampling effort, all samples were collected in a predefined area of approximately 1 ha located in the vicinity of the pitfall traps, for a total of two sampling sites. An estimate of the number of calling males was made during each sampling night for each site. The data were grouped into one of the following abundance classes (adapted from Bertoluci and Rodrigues 2002): a) 1–4 individuals in calling activity, b) 5–9 individuals, c) 10–20 individuals, or d) >20 individuals. In terms of the duration of the vocalization season, the species were classified as follows: 1) annuals—species that vocalized more than 70 % of the year or were recorded for at least 2 months in each season (warm and cold); 2) warm season—species that vocalized only during the warm months (November to April); and 3) cold season—species that vocalized only during the cold months (May to October). We calculated the frequency of calling occurrence, which corresponded to the number of samples (= months) in each vocalizing species divided by the total number of samples (12) of each species.

Data analysis

For all statistical analyses, we used data on anuran abundance to evaluate general activity and richness indices to evaluate calling activity. To evaluate general activity, only species with ten or more captured individuals were considered, thus making the analysis more robust. Additionally, arboreal species and/or species with "digital discs" were excluded from the analysis of general activity because they were able to escape from the traps (Enge 2001; Oliveira et al. 2013).

A circular statistical analysis (Zar 1999) was applied to evaluate the existence of a seasonal pattern in both general (estimated by the capture rate in traps) and calling activities (estimated by the number of calling species). The months were represented numerically by angles (intervals of 30 °C), and both the number of captures in the traps and number of calling species each month were used as frequencies for each month observed (Prado et al. 2005; Both et al. 2008). The following parameters were considered in the analysis: 1) the median angle (μ), which corresponded to the average period of the year (sampled days) in which most of the species were active and 2) the length of the mean vector (r), which indicated the clustering level of observations around μ. This value varied from 0 (least clustering) to 1 (greatest clustering), and 3) the circular standard deviation (SD). The seasonality tendencies for both activities were assessed using a nonparametric Rayleigh test (z). A significant Rayleigh test indicated the seasonal activity of the anuran assemblage studied. The niche breath of the calling season of each species was calculated using the inverse of the Simpson diversity index (Simpson 1949):

$$B = 1/\sum_{i=1}^{n} p_i^2$$

In the equation, p is the proportion of use (calling registers) of temporal category i (month survey), and n is the total number of categories. The values obtained for the diversity index as applied in our study range from 1 to 10, in which 1 corresponds to exclusive calling in only 1 month (specialists) and 10 corresponds to the use of all months in the same proportion (generalists; see Protázio et al. 2014). For a temporal niche analysis, we used the minimum number of calling males previously defined in each abundance class category in the sample (adapted from Bertoluci and Rodrigues 2002). Niche overlap was calculated using Pianka's overlap index (Pianka 1973). The overlap index between species 1 and 2 (O_{12}) with temporal calling activities p_{1i} and p_{2i}, respectively, was calculated as follows:

$$O_{12} = O_{21} = \sum_{i=1}^{n} p_{2i}p_{1i} / \sqrt{\sum_{i=1}^{n} (p_{2i}{}^2)(P_{1i}{}^2)}$$

In the equation, p represents the proportion of calling males in each month category i, n is the number of categories (months), and j and k represent compared species. The possible overlap scores range from 0 (no overlap) to 1 (complete overlap). We used a null model to investigate the presence of nonrandom patterns in the temporal distribution of calling. This null model was based on the niche overlap module of EcoSim (Gotelli and Entsminger 2003). For these null models, a data matrix with species, months, and the proportion of the number of calling males in each month was created. We performed 1000 randomizations of the data set to simulate random patterns that would be expected in the absence of seasonality in calling behavior (Gotelli and Graves 1996).

Comparisons of the capture rates between the months were analyzed using a Kruskal-Wallis analysis of variance followed by a Kruskal-Wallis post hoc test when necessary (Zar 1999). We applied a Mann–Whitney test to evaluate the variation in general activity between the warm/cold and dry/humid months (i.e., a Mann–Whitney *U* test: Zar 1999). A multiple regression was performed to detect possible associations between the climatic variables and the general and calling activities of the anurans (Zar 1999). The following climatic variables were considered for all regressions: maximum air temperature, minimum air temperature, relative air humidity, and accumulated rainfall. All statistical tests were performed with a significance level of 0.05 using the program Statistica 8.0.

Results

General activity

The circular statistical analysis revealed that anuran general activity was significantly seasonal ($P < 0.001$; Table 1). We captured 535 individuals belonging to nine species (Table 2). General activity was greater during warm months ($N = 392$) than during cold ones ($N = 143$, $U = 160.50$, $P = 0.008$). Multiple regression revealed the existence of a significant and positive relationship between the general activity, relative humidity, maximum and minimum temperatures, and accumulated rainfall ($R^2 = 0.22$, $P = 0.05$; Table 2). However, general activity did not vary significantly between dry and humid months (Table 2). When the species were evaluated separately, the relationship between general activity and abiotic variables was only significant for *Physalaemus biligonigerus* ($R^2 = 0.40$, $P < 0.001$), *Pseudopaludicola falcipes* ($R^2 = 0.25$, $P = 0.02$), *Leptodactylus gracilis* ($R^2 = 0.23$, $P = 0.03$), and *Elachistocleis bicolor* ($R^2 = 0.51$, $P < 0.001$; Table 1).

General activity was significantly higher during warm months for *E. bicolor* ($N = 103$), *L. gracilis* ($N = 27$), and *P. biligonigerus* ($N = 53$) than during cold months ($N = 1$, $N = 4$, $N = 13$; $U = 77.50$, $P < 0.001$; $U = 130.00$, $P < 0.001$; $U = 130.00$, $P < 0.001$, respectively; Table 2). Only *P. falcipes* presented greater general activity during cold months ($N = 16$) than during warm months ($N = 1$; $U = 190.00$, $P < 0.01$; Table 2). The general activity of *Leptodactylus* cf. *latrans*, *Physalaemus gracilis*, and *Pseudis minuta* did not vary significantly between warm and cold months (Table 2). There was no significant variation between dry and humid months for any of the species (Table 2). The multiple regression model showed a positive and significant relationship between the maximum air temperature, minimum air temperature, relative air humidity, accumulated rainfall, and activities of *P. biligonigerus* ($R^2 = 0.40$, $P = 0.0004$), *P. falcipes* ($R^2 = 0.25$, $P = 0.02$), *L. gracilis* ($R^2 = 0.23$, $P = 0.03$), and *E. bicolor* ($R^2 = 0.51$, $P = 0.000007$). The activity of the other species did not respond significantly to any of the climatic variables (Table 2).

Table 1 Anuran activity patterns in subtemperate wetlands of Brazil. The seasonality evaluation of general (estimated by the number of captures) and calling activities of anurans in subtemperate wetlands in southernmost Brazil

	Annual cycle 2010–2011	
Variables	General activity	Calling activity
Number of observations (*N*)	535	55
Mean vector (*μ*)	335.172°	315.000°
Mean vector length	0.449	0.159
Circular deviation pattern (SD)	72.523°	109.952°
Rayleigh test (*P*)	<0.001	0.251

Calling activity

The circular statistical analysis revealed that anuran calling activity was not significantly seasonal (Table 1). Calling records were obtained for 13 species distributed in six families (Fig. 1). Two periods with a higher number of calling species were detected: between August and November and between January and March. A peak in the number of calling species occurred during October ($N = 7$ species) and September, February, and March ($N = 6$) (Fig. 1).

Regarding calling activity, five species were classified as "annual" (*Dendropsophus sanborni*, *Hypsiboas pulchellus*, *Physalaemus* sp., *P. minuta*, and *Scinax squalirostris*), six as "warm season" (*Dendropsophus minutus*, *E. bicolor*, *L. gracilis*, *Odontophrynus maisuma*, *P. falcipes*, and *Rhinella dorbignyi*) and two as "cold season" (*L.* cf. *latrans* and *Scinax granulatus*). The species with a higher frequency of calling were *H. pulchellus* (83.33 %) and *S. squalirostris* (75 %), whereas *O. maisuma* and *L.* cf. *latrans* only called during 1 month (Fig. 1). The number of species exhibiting calling activity did not show a significant relationship to any of the environmental variables ($R^2 = 0.22$; $P = 0.53$; sample number = 12). Because of the difficulty in differentiating the vocalizations of *P. biligonigerus* and *P. gracilis*, these species were not differentiated and were recorded as *Physalaemus* sp.

For all species, a higher calling activity occurred during warm months, which coincided with a peak in the capture rate in the traps. However, when examining the species separately, the calling and general activity periods only overlapped for *E. bicolor*, *L. gracilis*, and *Physalaemus* sp. Males of *L.* cf. *latrans* had a calling activity peak (October) that occurred before the general activity peak (November to January). An identical result occurred for *P. minuta*, which had calling peaks in September, October, March, and April, and general activity peaks in November, December, June, and July. However, *P. falcipes* males concentrated their vocalizations during warm months (November and January) and presented the greatest general activity during cold months (August and September). It should be emphasized that this comparison between the

Table 2 Associations between anuran activity and different abiotic variables

Family	Captures		Annual activity variation		Activity variation between seasons				Activity and climatic factors	
Species	N(warm/cold)	N(dry/humid)	Kruskal-Wallis results		Mann–Whitney results for				Multiple regression results	
					Warm/cold		Dry/humid			
			H	*P*	*U*	*P*	*U*	*P*	R^2	*P*
Hylidae										
Pseudis minuta (Günther, 1858)	29/16	25/20	10.59	0.48	287.50	0.99	279.50	0.86	0.09	0.48
Leiuperidae										
Physalaemus biligonigerus (Cope, 1861)	53/13	27/39	24.03	0.01*	130.00	<0.001*	232.50	0.25	0.40	0.0004*
Physalaemus gracilis (Cope, 1861)	34/18	30/22	19.66	0.05*	252.50	0.43	283.50	0.92	0.13	0.26
Pseudopaludicola falcipes (Hensel, 1867)	1/16	4/13	15.75	0.15	190.00	<0.01*	256.00	0.51	0.25	0.02*
Leptodactylidae										
Leptodactylus gracilis (Duméril and Bibron, 1840)	27/4	14/17	20.54	0.03*	130.00	<0.001*	248.00	0.41	0.23	0.03*
L cf. *latrans*	141/63	133/71	35.47	0.0002*	285.50	0.96	242.50	0.35	0.10	0.43
Microhylidae										
Elachistocleis bicolor (Guérin-Méneville, 1838)	103/1	50/54	40.07	0.0001*	77.50	<0.001*	259.59	0.56	0.51	0.000007*
Bufonidae										
Rhinella dorbignyi (Duméril and Bibron, 1841)	2/8	3/7	–	–	–	–	–	–	–	–
Cycloramphidae										
Odontophrynus maisuma (Rosset, 2008)	4/2	3/3	–	–	–	–	–	–	–	–
All species	535	535	31.59	<0.01*	160.50	0.008*	265.00	0.63	0.22	0.05*

A comparison and correlation between anuran activity (evaluated by the number of captures) and different abiotic variables between May 2010 and April 2011 in subtemperate wetlands in southernmost Brazil. The number of samples (*n* samples) used for all statistical tests was 48. See "Methods" section for a detailed explanation

N the number of captures in pitfalls, *NA* not applicable

*Significant variation

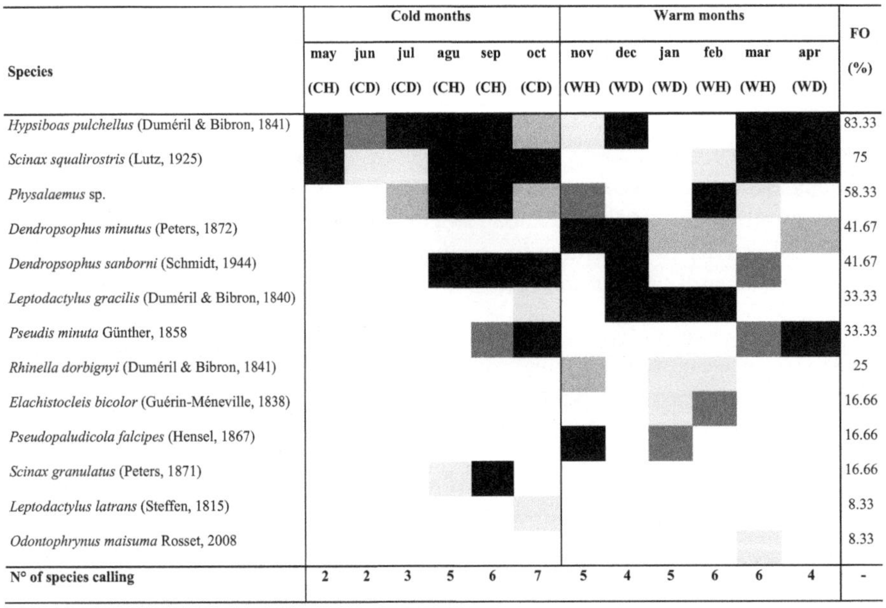

Fig. 1 Anuran calling activity in southernmost Brazil. Calling activity sampling between May 2010 and April 2011 in subtemperate wetlands in southernmost Brazil. The abundance classes of individuals' calling activities: ▢ 1–4 individuals; ▢ 5–9; ▢ 10–20; and ■ > 20. *FO%* occurrence frequency, *CD* cold and dry month, *CH* cold and humid month, *WD* warm and dry month, *WH* warm and humid month

calling and general activity periods was not possible for *R. dorbignyi* and *O. maisuma* because of the low number of captures (<10), similar to the hylids because of their ability to climb and escape from the traps (with the exception of *P. minuta*).

Greater values of temporal niche breath were observed for *H. pulchellus* (B = 6.72), followed by *D. sanborni* (B = 4.81), and *Physalaemus* sp. (B = 4.13) (Table 3). Lower values were observed for *E. bicolor*, *L.* cf. *latrans*, and *O. maisuma* (B = 1). A niche overlap analysis revealed that the patterns varied from complete overlap to no overlap between species (Table 3). The mean overlap was 0.227 ± 0.267, and only 10.4 % of all species pairs showed a temporal overlap greater than 0.6. Greater overlap values were observed for *R. dorbignyi* and *P. falcipes* (0.9845988), *Physalaemus* sp. and *S. fuscovarius* (0.731399), and *S. squalirostris* and *H. pulchellus* (0.9035631). The null model showed that the observed temporal niche overlap index (0.227) did not vary significantly more than expected by chance (0.218) ($P > 0.05$). This suggests that the assemblage calling activity is randomly distributed.

Discussion

Our data suggest that the general activity of the studied anuran populations exhibits a seasonal pattern predominantly related to air temperature but not rainfall. This pattern was different from that expected for the majority of eastern South America (Bertoluci 1998; Vasconcelos et al. 2010). The relatively harsh winter in southernmost Brazil possibly imposes limitations on anuran activity due to a reduction in metabolic rate (Lillywhite 1974). During the coldest months, there would be a decreased in mobility and, consequently, in the probability of capture in traps (Toledo and Haddad 2005; Saenz et al. 2006).

The potential for cold as a factor limiting general activity was reinforced by the fact that captures during cold months were higher on days with higher minimum temperatures. In addition, species are expected to exhibit different ranges of optimal operative body temperature (Huey 1991). Although this hypothesis is speculative, it could explain the peak of activity during warm months in *E. bicolor*, *L. gracilis*, and *P. biligonigerus*, whereas *P. falcipes* presented a higher general activity during colder months. Another aspect of life history that affects general activity is the differences in the level of dependency on aquatic habitats. After an intense rain, the habitat could remain flooded for many weeks, promoting a long period of intense activity in some species and with no relationship to the precipitation rate. This effect appears to occur in *L.* cf. *latrans*, *P. gracilis*, and *P. minuta*, which exhibited no seasonal variation in activity pattern.

Although the peaks in calling activities correspond to warmer and moister months (Navas and Bevier 2001), many studies noted the importance of evaluating the effects of environmental variables independently for each species (Steelman et al. 2010). This pattern is reinforced by our data, which showed particular behaviors in certain species, such as *P. falcipes*, whose general activity was higher during colder months. However, this pattern may have been a secondary response because some of the coldest months had isolated rainfall peaks, which may have increased the species' reproduction (Carvalho-e-Silva et al. 2000; Loebmann and Haddad 2010; Barreto et al. 2012).

Despite the occurrence of isolated peaks of rainfall during the coldest months, the region displayed a relatively homogeneous distribution pattern of rain throughout the year (Santos et al. 2008; Lingnau 2009; Oliveira et al. 2013). This pattern could further reinforce the effects of other factors, affording humidity only a secondary role as a moderator of anuran activity in southernmost Brazil. This scenario reduces the accuracy of predictions on activity patterns based on air temperature and humidity (Steelman et al. 2010). There was no significant relationship between calling activity and any of the environmental variables tested. We did not record a well-defined peak in the number of species showing calling activity. Such a peak has been demonstrated for other anuran communities in other locations in Brazil (Bertoluci 1998; Eterovick and Sazima 2000) and Uruguay (Canavero et al. 2008), including environments climatically similar to our study area (Bernarde and Anjos 1999; Conte and Machado 2005; Santos et al. 2008). Note that different species could respond differently to abiotic variables that influence reproductive activity (Hartel et al. 2007; Wells 2007). In addition to climatic variables, these triggers include other environmental parameters, such as photoperiod (Both et al. 2008; Canavero et al. 2009), which can make the detection of a general pattern of answers to the climatic variables difficult. In an attempt to predict mathematically the activity patterns of an anuran community in subtemperate wetlands in Uruguay, Canavero et al. (2008) observed that the annual pattern of calling activity is determined by a response to a set of seasonal variables or a variable that synthesizes the entire seasonal variation, such as photoperiod. This principle reinforces the fact that not all species respond to specific variables such as temperature or rainfall. Photoperiod appears to be a good predictor of activity in *H. pulchellus*, *P. minuta*, and *P. gracilis*. As also observed by Canavero et al. (2008), these species also presented a broad period of activity almost throughout the year in the present study.

The relatively low values of niche breath suggest that many species presents a short season of calling activity. Consequently, species would have a smaller overlap during reproductive periods, as occurred, e.g., between *L. latrans*, *D. minutus*, *E. bicolor*, and *O. maiusma* and between *O. maiusma*, *E. bicolor*, and *L. gracilis*. Finally, this behavior would reduce the overlap between species during the

Table 3 Values of niche overlap (*O*) and niche breath (*B*) in calling activities for 13 anuran species in subtemperate wetlands in southernmost Brazil

	Dmin	*Dsan*	*Ebic*	*Hpul*	*Lgra*	*Llat*	*Omai*	*Pfal*	*Phys*	*Pmin*	*Rdor*	*Sfus*	*Ssqu*
Dmin		0.3367491	0.1905002	0.09861692	0.6654356	0	0	0.6622442	0.1925226	0.1391423	0.666469	0	0.0695677
Dsan			0	0.5644163	0.3296402	0.4821027	0.2651565	0	0.6667762	0.4935692	0	0.4978963	0.6974541
Ebic					0.4459272	0	0	0	0.3566882	0	0,09901475	0	0.01404555
Hpul					0.0008878645	0.05973157	0.4778526	0.01158963	0.5650231	0.6384265	0.01182861	0.3745081	0.9035631
Lgra						0.01486424	0	0.1442043	0.1606475	0.007599841	0.1030245	0	0.0106476
Llat							0	0	0.1070065	0.5112835	0	0	0.2949566
Omai								0	0.01783441	0.2678165	0	0	0.5618221
Pfal									0,1903211	0	0.9845988	0	0
Phys										0.3200138	0.2295631	0.731399	0.5588524
Pmin											0	0.3649998	0.660339
Rdor												0	0.001390717
Sfus													0.2994721
Ssqu													
O(mean)	0.34	0.10	0.33	0.36	0.11	0.12	0.12	0.28	0.26	0.21	0.18	0.34	0.34
B(mean)	3.63	4.81	1.00	6.72	2.97	1.00	1.00	1.47	4.13	3.51	1.41	1.07	5.63

Niche overlap was estimated by Pianka's overlap index and niche breath by the inverse of the Simpson diversity index (*B*). *Dmin, Dendropsophus minutus; Dsan, Dendropsophus sanborni; Ebic, Elachistocleis bicolor; Hpul, Hypsiboas pulchellus; Lgra, Leptodactylus gracilis; Llat, Leptodactylus* cf. *latrans; Omai, Odontophrynus maisuma; Phys, Physalaemus sp; Pmin, Pseudis minuta; Pfal, Pseudopaludicola falcipes; Rdor, Rhinella dorbignyi; Sfus, Scinax fuscovarius; Ssqu, Scinax squalirostris* breeding seasons, thus reducing the possibility of competition for physical and acoustic space and diluting the effect of predators on nests (Kopp et al. 2010; Sinsch et al. 2012).

The majority of the species showed calling activity that was restricted to the warm season (46.15 %) or extended throughout the entire year (38.46 %). Even species that presented calling activity restricted to the cold season (*L.* cf. *latrans* and *S. granulatus)* were active during the transitional months between the cold and warm seasons. This result reinforces the finding that temperature was the predominant climatic factor for determining anuran calling activity. Note, however, that the calling season of *S. granulatus* may have been underestimated because of the low abundance of individuals in the studied area (personal observations).

The species with the longest calling periods were *H. pulchellus* and *S. squalirostris*. This pattern was most likely enhanced by their relatively high tolerance to low temperatures (Both et al. 2008). Species with higher tolerance to low temperatures may have greater reproductive success (Martins 2009), which could explain why those species were two of the most abundant in the study region. The prolonged calling pattern of *H. pulchellus* has also been recorded in other populations in southern South America (Achaval and Olmos 1997; Trindade et al. 2010) and in Uruguay (Canavero et al. 2008). However, the calling period of *S. squalirostris* in other subtemperate regions was concentrated during the warmer months between September and April (Achaval and Olmos 1997; Kwet and Di-Bernardo 1999). This result suggests a certain plasticity of the species regarding seasonality and reproductive activity. Such plasticity also appears to occur in *P. minuta*, whose reproductive activity was previously identified as brief (Santos et al. 2008) or long (Langone 1994; Melchiors et al. 2004). The low values of niche overlap reinforce the potential of the entire community to keep calling through all the year. Another important aspect of our work was that our results showed a temporal disjunction between the general and calling activities for some species (*L.* cf. *latrans*, *P. minuta*, and *P. falcipes*). The limitation of an ecological interpretation of activity data often occurs because of the low availability of basic information on the biology of the observed species, particularly regarding reproductive biology.

Conclusions

The present study is one of the first to describe anuran general activity in southern Brazil. The study showed that the general activity patterns of most anurans in the studied wetland habitats appear to be regulated by temperature and not rainfall, similar to the pattern of anuran assemblages in temperate climates (Oseen and Wassersug 2002; Saenz et al. 2006). The nonexistence of a significant relationship between calling activity and any of the environmental variables tested suggested the presence of a different environmental factor (e.g., photoperiod or the length of the hydroperiod of water bodies) or some untested biotic elements as a trigger for the levels of general calling activity. Nevertheless, the observed general activity pattern is different from that expected for most of eastern South America.

Competing interests

The authors declare that they have no competing interests.

Authors' contributions

SSX was responsible for the field samples, performed most of the statistical analyses, and drafted the first version of the manuscript. AMT conceived the study, determined the sampling design, and reviewed the manuscript. All authors read and approved the final manuscript.

Acknowledgements

We thank the Rio Grande do Sul Research Foundation (FAPERGS) and the National Council for Scientific and Technological Development (CNPq) for the financial support and the Coordination for the Improvement of Higher Education Personnel (CAPES) for the Master's Degree fellowship.

Author details

[1]Instituto de Ciências Biológicas, Universidade Federal do Rio Grande—FURG, Av. Itália km 8, Rio Grande RS 96203-900, Brazil. [2]Laboratório de Ecologia de Vertebrados Terrestres, Universidade do Vale do Rio dos Sinos—UNISINOS, São Leopoldo Av. Unisinos, 950, Cristo Rei, São Leopoldo RS 93022-000, Brazil

References

Achaval F, Olmos A (1997) Anfíbios y reptiles del Uruguay. Barreiro y Ramos S. A, Montevideo

Almeida-Gomes M, Van Sluys M, Rocha CFD (2007) Calling activity of *Crossodactylus gaudichaudii* (Anura: Hylodidae) in an Atlantic Rainforest area at Ilha Grande, Rio de Janeiro, Brasil. Belg J Zool 137:203–207

Barreto GS, Tinoco MS, Couto-Ferreira D, Browne-Ribeiro HC (2012) Distribuição de *Pseudopaludicola* aff. *falcipes* (Anura; Leiuperidae) na restinga do litoral norte da Bahia, Brasil. Revista Latino-Americana de Conservação 2(2)-3(1):27–36

Bernarde PA, Anjos L (1999) Distribuição espacial e temporal da anurofauna no Parque Estadual Mata dos Godoy, Londrina, Paraná, Brasil (Amphibia: Anura). Comunicações do Museu de Ciência e Tecnologia 12:127–140

Bernarde PS, Machado RA (2001) Riqueza de espécies, ambientes de reprodução e temporada de vocalização da anurofauna em Três Barras do Paraná, Brasil (Amphibia: Anura). Cuadernos de Herpetologia 14(2):93–104

Bertoluci J (1998) Annual patterns of breeding activity in Atlantic Rainforest anurans. J Herpetol 32(4):607–611

Bertoluci J, Rodrigues MT (2002) Seasonal patterns of breeding activity of Atlantic Rainforest anurans at Boracéia, Southeastern Brazil. Amphibia-Reptilia 23:161–167

Boquimpani-Freitas L, Marra RV, Van Sluys M, Rocha CFD (2007) Temporal niche of acoustic activity in anurans: interspecific and seasonal variation in a neotropical assemblage from south-eastern Brazil. Amphibia-Reptilia 28:269–276

Both C, Grant T (2012) Biological invasions and the acoustic niche: the effect of bullfrog calls on the acoustic signals of white-banded tree frogs. Biol Lett 8:714–716

Both C, Kaefer IL, Santos TG, Zechin STZ (2008) An austral anuran assemblage in the Neotropics: seasonal occurrence correlated with photoperiod. J Nat Hist 42(3):205–222

Canavero A, Arim M (2009) Clues supporting photoperiod as the main determinant of seasonal variation in amphibian activity. J Nat Hist 43(48):2975–2984

Canavero A, Arim M, Naya DE, Camargo A, Rosa I, Maneyro R (2008) Calling activity patterns in anuran assemblage: the role of seasonal trends and weather determinants. Northwest J Zool 4(1):29–41

Carvalho-e-Silva SP, Izecksohn E, Carvalho-e-Silva AMPT (2000) Diversidade e ecologia de anfíbios em restingas do sudeste brasileiro. In: Esteves FA, Lacerda LD (eds) Ecologia de Restingas e Lagoas Costeiras. NUPEM/UFRJ, Rio de Janeiro

Conte CE, Machado RA (2005) Riqueza de espécies e distribuição espacial e temporal em comunidade de anuros (Amphibia, Anura) em uma localidade de Tijucas do Sul, Paraná, Brasil. Rev Bras Zoo 22(4):940–948

Conte CE, Rossa-Feres DC (2006) Diversidade e ocorrência temporal da anurofauna (Amphibia, Anura) em São José dos Pinhais, Paraná, Brasil. Rev Bras Zoo 23(1):162–175

Enge KM (2001) The pitfalls of pitfall traps. J Herpetol 35:467–478

Eterovick PC, Sazima I (2000) Structure of an anuran community in a montane meadow in southeastern Brazil: effects of seasonality, habitat, and predation. Amphibia-Reptilia 21:439–461

Eterovick PC, Carnaval ACOQ, Borges-Nojosa DM, Silvano DL, Segalla MV, Sazima I (2005) Amphibian declines in Brazil: an overview. Biotropica 37(2):166–179

Fukuyama K, Kusano T (1992) Factors affecting breeding activity in a stream-breeding frog, *Buergeria buergeri*. J Herpetol 26:88–91

Gilbert M, Leclair R Jr, Fortin R (1994) Reproduction of the northern leopard frog (*Rana pipiens*) in floodplain habitat in the Richelieu River, P. Quebec, Canada. J Herpetol 28:465–470

Gotelli NJ, Graves GR (1996) Null models in ecology. Smithsonian Institution Press, Washington, DC

Greenberg CH, Neary DG, Harris LD (1994) A comparison of herpetofaunal sampling effectiveness of pitfall single-ended and double-ended funnel traps used with drift fences. J Herpetol 28:319–324

Hartel T, Sas I, Pernetta AP, Geltsh IC (2007) The reproductive dynamics of temperate amphibians: a review. North-West J Zool 3:127–145

Hatano FH, Rocha CFD, Van Sluys M (2002) Environmental factors affecting calling activity of a tropical diurnal frog (*Hylodes phyllodes*: Leptodactylidae). J Herpetol 36:314–318

Heyer WR (1973) Ecological interactions of frog larvae at a seasonal tropical location in Thailand. J Herpetol 7:337–361

Huckembeck S, Claudino M, Correa F, Bastos RF, Loebmann D, Tozetti AM, Garcia AM (2012) Pattern activity and microhabitat use of *Pseudis minuta* Günther, 1858 (Anura, Hylidae) in a subtropical Biosphere Reserve in Brazil. Braz J Biol 72:331–336

Huey RB (1991) Physiological consequences of habitat selection. Am Nat 137:91–115

Jorgensen CB (1992) Growth and reproduction. In: Feder ME, Burggren WW (eds) Environmental physiology of the amphibians. The University of Chicago Press, Chicago

Kopp K, Signorelli L, Bastos RP (2010) Distribuição temporal e diversidade de modos reprodutivos de anfíbios anuros no Parque Nacional das Emas e entorno, estado de Goiás, Brasil. Iheringia, Série Zoológica 100(3):192–200

Krebs CJ (1999) Ecological methodology. Addison Wesley Educational Publishers, Menlo Park, California

Kwet A, Di-Bernardo M (1999) Pró-Mata: Anfíbios-Amphibien-Amphibians. EDIPUCRS, Porto Alegre

Langone JA (1994) Ranas y Sapos del Uruguay (Reconocimiento y Aspectos Biológicos). Museo Dámaso Antonio Larranaga, Serie de Divulgación 5:1–123

Lillywhite HB (1974) How frogs regulate their body temperature. Environ Southwest 465:3–6

Lingnau R (2009) Distribuição temporal, atividade reprodutiva e vocalizações em uma assembleia de anfíbios anuros de uma floresta ombrófila mista em Santa Catarina, sul do Brasil. Pontifícia Universidade Católica do Rio Grande do Sul, Dissertation

Loebmann D, Haddad CFB (2010) Amphibians and reptiles from a highly diverse area of the Caatinga domain: composition and conservation implications. Biota Neotrop 10(3):227–256

Maluf JRT (2000) Nova classificação climática do Estado do Rio Grande do Sul. Revista Brasileira de Agrometeorologia 8(1):141–150

Martins LA (2009) Comportamento reprodutivo e social de *Scinax squalirostris* (Lutz, 1925) (Anura, Hylidae) sob influência de fatores ambientais. Pontifícia Universidade Católica do Rio Grande do Sul, Dissertation

Martins LS, Verrastro L, Tozetti AM (2014) The influences of habitat on body temperature control in a southern population of *Liolaemus occipitalis* (Boulenger, 1885) in Brazil. South American Journal of Herpetology 9(1):9–13

Melchiors J, Di-Bernardo M, Pontes GMF, Oliveira RB, Solé M, Kwet A (2004) Reprodução de *Pseudis minuta* (Anura, Hylidae) no sul do Brasil. Phyllomedusa 3(1):61–68

Navas CA, Bevier CR (2001) Thermal dependency of calling performance in the eurythermic frog *Colostethus subpunctatus*. Herpetologica 57:384–395

Oliveira MCLM, Santos MB, Loebmann D, Tozetti AM (2013) Diversity and associations between coastal habitats and anurans in southernmost Brazil. Anais da An Acad Bras Cienc 85:575–582

Oseen KL, Wassersug RJ (2002) Environmental factors influencing calling in sympatric anurans. Oecologia 133:616–625

Pianka ER (1973) The structure of lizard communities. Annu Rev Ecol Evol Syst 4:53–74

Prado CPA, Uetanabaro M, Haddad CFB (2005) Breeding activity patterns, reproductive modes, and habitat use by anurans (Amphibia) in a seasonal environment in the Pantanal, Brazil. Amphibia-Reptilia 26(2):211–221

Protázio AS, Albuquerquea RL, Falkenberga LM, Mesquita DO (2014) Acoustic ecology of an anuran assemblage in the arid Caatinga of northeastern Brazil. J Nat Hist 48:1–20

Saenz D, Fitzgerald LA, Kristen AB, Richard NC (2006) Abiotic correlates of anuran calling phenology: the importance of rain, temperature, and season. Herpetol Monogr 20(1):64–82

Santos TG, Kopp K, Spies MR, Trevisan R, Cechin SZ (2008) Distribuição temporal e espacial de anuros em área de Pampa, Santa Maria, RS. Iheringia, Série Zoologia 98(2):244–253

Sinsch U, Lümkemann K, Rosar K, Schwarz C, Dehling JK (2012) Acoustic niche partitioning in an anuran community inhabiting an Afromontane wetland (Butare, Rwanda). Afr Zool 47(1):60–73

Steelman CK, Dorcas ME (2010) Anuran calling survey optimization: developing and testing predictive models of anuran calling activity. J Herp 44(1):61–68

Toledo LF, Haddad CFB (2005) Acoustic repertoire and calling behavior of *Scinax fuscomarginatus* (Anura, Hylidae). J Herp 39(3):455–464

Tozetti AM, Pontes GMF, Borges-Martins M, Oliveira RB (2010) Temperature preferences of *Xenodon dorbignyi*: field and experimental observations. Herpetol J 20:277–280

Trindade AO, Oliveira SV, Cappellari LH (2010) Anfíbios anuros de uma área da serra do sudeste, Rio Grande do Sul (Caçapava do Sul). Biodiversidade Pampeana 8(1):19–24

Vasconcelos TS, Santos TG, Haddad CFB, Rossa-Feres DC (2010) Climatic variables and altitude as predictors of anuran species richnesss and number of reproductive modes in Brazil. J Trop Ecol 26(4):423–432

Wells KD (2007) The ecology and behavior of amphibians. The University of Chicago Press, Chicago

Xavier AL, Napoli MF (2011) Contribution of environmental variables to anuran community structure in the Caatinga Domain of Brazil. Phyllomedusa 10(1):45–64

Ximenez SS, Oliveira MCLM, Santos MB, Tozetti AM (2014) The influence of habitat configuration on anuran species composition in subtemperate wetlands of southernmost Brazil. South American Journal of Herpetology 9(3):235–243

Zar JH (1999) Biostatistical analysis. Prentice-Hall, New Jersey

6

Sexual dimorphism in echolocation pulse parameters of the CF-FM bat, *Hipposideros pratti*

Zi-Ying Fu[1], Xing-Yue Dai[1], Na Xu[1], Qing Shi[1], Gao-Jing Li[1], Bin Li[1], Juan Li[1], Jie Li[1], Jia Tang[1], Philip Hung-Sun Jen[2,3,4*] and Qi-Cai Chen[1*]

Abstract

Background: Previous studies of sexual dimorphism in the echolocation pulses of the constant frequency-frequency modulating (CF-FM) bat have been mainly concentrated on the difference in the frequency of the CF component of the predominant second harmonic while neglected other pulse parameters. However, recent studies have shown that other pulse parameters of the predominant second harmonic are also biologically significant to the bat hunting. To complement and advance these studies, we have examined sexual dimorphism of multiple parameters (e.g., duration, frequency, bandwidth of the FM component, and repetition rate of emitted pulses) of the echolocation pulses of the CF-FM bat, *Hipposideros pratti*.

Results: Our studies of the predominant second harmonic show that on average the male bat has higher frequency of the CF component, wider FM bandwidth, and higher pulse repetition rate while the female bat has longer duration of the CF and FM components.

Conclusions: These observations suggest that bats may potentially use this sexual dimorphism in echolocation pulse parameters for social communication and species and sex identification.

Keywords: CF-FM bat; Echolocation pulse; Sexual dimorphism; *Hipposideros pratti*

Background

During hunting, insectivorous bats mainly emit either multiple frequency modulating (FM) or constant frequency-frequency modulating (CF-FM) echolocation pulses and analyze the returning echoes to extract information about the target features (e.g., size, shape, texture, velocity, range etc.) (Griffin 1958). Based on the frequency content of these echolocation pulses, these bats have been conveniently called FM or CF-FM bats (Griffin 1958; Neuweiler 2003; Ulanovsky and Moss 2008). The echolocation pulses of these bats are typically multiple harmonics with the first one being the fundamental frequency. Also, the echolocation pulse energy is not evenly distributed among different harmonics. For example, the CF-FM bat emits a long CF component followed by a short FM component, and the pulse energy is predominantly distributed in the second harmonic (H_2) (i.e., predominant harmonic) (Griffin 1958; Jones and Teeling 2006; Neuweiler 2003; Suga 1990; Vanderelst et al. 2013). It has been proposed that the CF component is used for echo detection and evaluation of Doppler-shifted echo frequency while the FM component is effective for echo ranging and fine target feature analysis (Fitzpatrick et al. 1991; Jakobsen et al. 2013; Kober and Schnitzler 1990; O'Neill and Suga 1982; Schnitzler and Denzinger 2011; Simmons and Vernon 1971).

In addition to using the echolocation pulses for hunting, the bats also use these pulses as social signals for communication, individual recognition, threatening, and guidance (Möhres 1967; Knörnschild et al. 2012; Kazial et al. 2008; Yovel et al. 2009; Voigt-Heucke et al. 2010). Other studies also showed that the echolocation pulses could be used for identification of species (Russo et al. 2007; Schuchmann and Siemers 2010), age (Jones and Ransome 1993; Masters et al. 1995), and group affiliation (Jameson and Hare 2009; Masters et al. 1995). Presumably, these social interactions are based on the fact that

* Correspondence: jenp@missouri.edu; chenqc@mail.ccnu.edu.cn
[2]College of Special Education, Binzhou Medical College, Yantai, Shandong, China
[1]School of Life Sciences and Hubei Key Lab of Genetic Regulation and Integrative Biology, Central China Normal University, Wuhan 430079Hubei, China
Full list of author information is available at the end of the article

there are variations on pulse parameters among individual bats and between female and male bats (sexual dimorphism).

Sexual dimorphism in the echolocation pulses has been reported for both FM and CF-FM bats (Feng et al. 2003; Grilliot et al. 2009; Jones et al. 1992; Jones et al. 1993; Kazial and Masters 2004; Knörnschild et al. 2012; Russo et al. 2001; Siemers et al. 2005; Suga et al. 1987). Interestingly, most of these studies have been devoted to examining sexual dimorphism of echolocation pulses in the CF-FM bat (Feng et al. 2003; Jones et al. 1992; Knörnschild et al. 2012; Russo et al. 2001; Siemers et al. 2005). This is perhaps due to the fact that this bat performs unique Doppler-shifted compensation behavior during which the bat appropriately adjusts the frequency of the emitted pulse to compensate for the Doppler-shifted frequency of the returning echo. Furthermore, most of these studies mainly examined the sexual dimorphism of the frequency and duration of the H_2 of the echolocation pulses (conveniently referred as CF hereafter) while neglected other pulse parameters. However, recent studies have shown that the FM component of the CF-FM pulse not only plays an essential role in shaping the response properties of central auditory neurons (the inferior colliculus) (Fu et al. 2010) but also facilitates the selectivity of these neurons in duration (Luo et al. 2008; Macías et al. 2013), amplitude (Fu et al. 2011), and the recovery cycle (Tang et al. 2011). These studies suggest that the CF-FM bat may utilize the information provided by the variation in the CF and FM components of the echolocation pulses throughout the entire course of hunting for successful prey capture. These studies also implicate that bats may utilize the variation in the FM components and other pulse parameters if any for social interactions.

In this study, we studied the sexual dimorphism of the echolocation pulses by analyzing multiple pulse parameters (e.g., duration, frequency, bandwidth of the FM component, and repetition rate of echolocation pulses) of the predominant H_2 which has been shown to be essential for successful echolocation in behavioral and neurophysiological studies (Neuweiler et al. 1987; O'Neill and Suga 1982; Schnitzler and Henson 1980; Schnitzler and Denzinger 2011; Suga et al. 1987; Suga 1990). We report here that the male bat has higher CF, wider FM bandwidth, and higher pulse repetition rate (PRR) while the female bat has longer duration of both CF and FM components (conveniently referred as CF duration and FM duration hereafter).

Methods

Animals and morphological studies

In this study, we used 34 adult *Hipposideros pratti* (*H. Pratti*) (13 males and 21 females, 34.5–48.1 g body weight (b. w.)) caught from a cave near Xianning city of Hubei providence of People's Republic of China in May of 2013 and 2014. They were caught from the same colony by means of mist net with the permission of Central China Normal University. These bats were housed in groups (three to five bats each group) in a cage within a temperature- and humidity-controlled animal room and were given larvae of *Tenebrio molitor* plus vitamin-mineral-enriched water ad libitum. The bats were trained to eat larvae and drink water for about 1 week. The age category of the bat was determined on the basis of whether phalangeal epiphyses were fused with the diaphysis, and its sex was assessed by inspecting the genitalia (Racey 1988). Each bat was weighed to the nearest 1 g with a Pesola balance, and the length of its right forearms was measured to the nearest 0.1 mm with dial calipers. This study was conducted with the approval of the Institutional Animal Care and Use Committee of Central China Normal University, Wuhan, Hubei, People's Republic of China.

Recordings of echolocation pulses

The recording of the echolocation pulses was made a few days after each bat was acclimated to the laboratory condition. The recording of the echolocation pulses was made at about 7 p.m. in May of 2013 and 2014. During recording, each bat was hanging on the ceiling of the experimental anechoic room and the recording microphone was placed 1 m below. The echolocation pulses were recorded only from each sedentary bat in real time with a handheld ultrasound detector (Petterson D1000X; Pettersson Elektronik AB, Uppsala, Sweden) before being fed into a notebook computer. Each bat was recorded 1–3 times and each time lasted 3–5 s.

Pulse analysis

The echolocation pulses were analyzed using the software BatSound pro 3.31b (Pettersson Elektronik AB, Uppsala, Sweden), with a fast Fourier transformation (FFT) size of 1024 points and a Hanning window using a cursor and visual determination on a screen. We only analyzed high-quality pulses with good signal-to-noise ratio. To characterize each echolocation pulse, we analyzed the CF, FM bandwidth, and durations of both CF and FM components of the H_2 as well as the PRR. The time parameters were taken from amplitude waveforms, and frequency parameters were taken from spectrograms. In the present study, when we calculated the CF difference between the recorded pulses from male and female bats, we firstly calculated the CF average of all pulses recorded from a same male or female bat as a male or female individual, i.e., as a sampling unit. Then, we put each CF average from each male or female bat together and calculated the total mean CF value from all male and female individuals.

Statistical analysis

All data obtained were processed and plotted using Sigmaplot 2000 (Systat Software, San Jose, CA, USA). They were then quantitatively examined and statistically compared using SPSS 13.0 (Student's t test at $p < 0.05$) (SPSS, Chicago, IL, USA).

Results

The forearm and body weight of *Hipposideros pratti*

The measurement of the forearm and body weight of these 34 bats is shown in Table 1. It is clear that there are differences between male and female bats and the former have longer average forearm length but slightly lighter body weight than the latter. However, these differences are not statistically significant ($p > 0.05$).

Echolocation pulses of *Hipposideros pratti*

In this study, 1574 individual echolocation pulses were recorded from these 34 bats ranging from 24 to 76 pulses (average 46 pulses) from each bat. Fig. 1 shows the sonogram and power spectrum of a representative echolocation pulse. The echolocation pulse consists of 4 harmonics with the first harmonic being the fundamental frequency. Each of these harmonics starts with a long CF component followed by a short FM component. The analysis of the power spectrum of the pulse clearly shows that the second harmonic (H_2) is the predominant one that contains the highest energy. This is evident by the fact that the H_2 has the darkest band in Fig. 1a and the highest peak in Fig. 1b. The measurements of CF, FM bandwidth, CF duration, FM duration, and PRR of the H_2 of echolocation pulses from all these 34 bats are shown in Table 2. Although there are differences among these multiple pulse parameters, the H_2 of this bat has an average of about 6.3-ms CF component at 60 kHz followed by 2.5-ms FM component with 10-kHz bandwidth. Each hanging bat emitted its echolocation pulses at a rate about 14 pulses per second.

Table 1 Measurements of the forearm and body weight of 13 male and 21 female *Hipposideros pratti*

	Sex (sample size)		Student's t test p
	Male ($n = 13$)	Female ($n = 21$)	
Forearm length (mm)			
Range	82.00–89.06	81.00–87.54	>0.05
Mean ± sd	86.41 ± 2.36	85.21 ± 2.05	
Body weight (g)			
Range	34.6–48.1	34.5–47.9	>0.05
Mean ± sd	41.2 ± 4.6	39.7 ± 4.1	

n number of bats

Sexual dimorphism of echolocation pulse parameters in *Hipposideros pratti*

We examined the sexual dimorphism of echolocation pulses in this bat species by separately plotting the distribution of average value of these multiple parameters of echolocation pulses obtained from each male and female bat. As shown in Fig. 2, the predominant CF of the male bat was between 58.8 and 60.5 kHz with most above 59.5 kHz (61.5 %, Fig. 2(A-1)) while that of the female bat was between 58.0 and 59.9 kHz with most below 59.5 kHz (57.1 %, Fig. 2(A-2)). This sexual dimorphic difference in CF (0.63 kHz) was statistically significant (59.91 ± 0.43 vs. 59.28 ± 0.56 kHz, $n = 13$ for male, $n = 21$ for female, $T = 3.509$, $p < 0.01$, t test, Fig. 2(A-3)). The FM bandwidth of the male bat ranged from 9.1 to 12.7 kHz with most above 9.5 kHz (69.28 %, Fig. 2(B-1)) while that of the female bat ranged from 8.1 to 10.9 kHz with most below 9.5 kHz (61.9 %, Fig. 2(B-2)). This sexual dimorphic difference in FM bandwidth was also statistically significant (10.08 ± 1.01 vs. 9.38 ± 0.81 kHz, $n = 13$ for male, $n = 21$ for female, $T = 2.239$, $p < 0.05$, t test, Fig. 2(B-3)). The PRR of the male bat ranged between 4.0 and 31.0 Hz with most (61.5 %) higher than 15.0 Hz (Fig. 2(C-1)). Different from this observation, the PRR of the female bat ranged between 2.7 and 27.0 Hz and most (85.7 %) of them were below 15 Hz (Fig. 2(C-2)). A statistical analysis revealed a significant sexual dimorphic difference in the average PPR of echolocation pulse of this bat species (16.3 ± 7.8 vs. 10.9 ± 6.2 Hz, $n = 13$ for male, $n = 21$ for female, $T = 2.288$, $p < 0.05$, t test, Fig. 2(C-3)).

The distribution of the averaged CF and FM duration of the predominant H_2 component of the echolocation pulses of each male and female *H. pratti* is shown in Fig. 3. The CF duration of the predominant H_2 of the male bat was between 4.2 and 8.8 ms with more than half (69.2 %) shorter than 6 ms (Fig. 3(A-1)). In contrast, the CF duration of the predominant H_2 of the female bat was between 5.4 and 10.9 ms with most (81.0 %) longer than 6.0 ms (Fig. 3(A-2)). This sexual dimorphic difference in the CF duration of the predominant H_2 was statistically significant (5.7 ± 1.1 vs. 7.3 ± 1.6 ms, $n = 13$ for male, $n = 21$ for female, $T = -3.066$, $p < 0.05$, t test, Fig. 3(A-3)). The FM duration of the predominant H_2 of the male bat was between 1.3 and 2.9 ms with 61.5 % shorter 2.5 ms while that of the female bat ranged from 2.0 to 3.2 ms with most (71.4 %) longer than 2.5 ms (Fig. 3(B-1) vs. (B-2)). The sexual dimorphic difference in this measurement was also statistically significant (2.4 ± 0.3 vs. 2.7 ± 0.4 ms, $n = 13$ for male, $n = 21$ for female, $T = -3.058$, p < 0.05, t test, Fig. 3(B-3)).

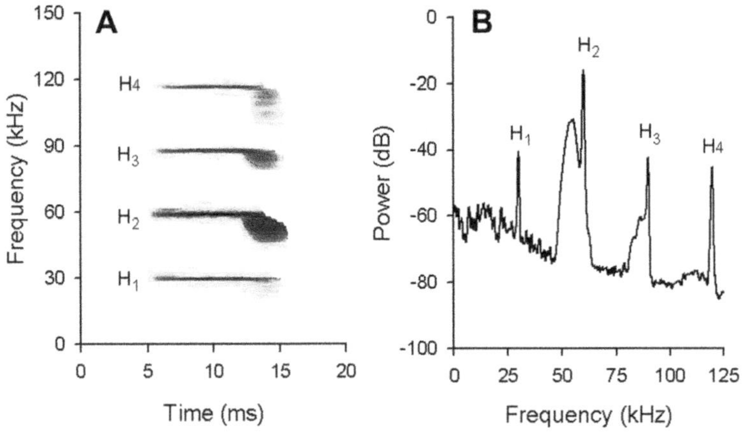

Fig. 1 The sonogram (**a**) and power spectrum (**b**) of a representative echolocation pulse emitted by a hanging *Hipposideros pratti*. The CF (kHz), FM bandwidth (kHz), CF duration (ms), and FM duration (ms) of this echolocation pulse were 60.0, 12.8, 7.5, and 2.7. Clearly, the second harmonic is the predominant one that contains the highest energy

Discussion

The forearm length, body weight, and echolocation pulses of *Hipposideros pratti*

The forearm length, body weight, and the predominant CF of *H. pratti* measured in the present study (Fig. 1, Tables 1 and 2, bats caught in Hubei province) were similar to the measurements of the same bat species reported in a previous study (bats caught in Fujian, Guangxi, and Henan provinces, Zhang et al. 2009) but different from that reported in another study (bats caught in Guizhou province, Chen et al. 2002). The difference in these measurements of the same bat species may be due to geographic difference. Regardless of these different reports, our present study did not find any sexual dimorphism in the forearm length and body weight of *H. pratti* (Table 1).

Sexual dimorphism of multiple parameters of the predominant H_2

In this study, we observed the sexual dimorphism of multiple parameters in the predominant H_2 of the echolocation pulses of *H. pratti* (Figs. 2 and 3, Table 2). On average, the male bat has higher predominant CF, wider FM bandwidth, and higher PRR while the female bat has longer duration of the CF and FM components. This observed sexual dimorphism in the multiple pulse parameters in this bat species is not likely related to the difference in the forearm length and body weight since these two latter physical parameters were not significantly different between the male and female bats (Table 1).

Table 2 Parameters of the predominant second harmonic (H_2) of the echolocation pulses emitted by *Hipposideros pratti*

Parameters	Mean ± sd	Median	Min-max
CF frequency (kHz)	59.5 ± 0.6	59.7	58.0–60.5
FM bandwidth (kHz)	9.6 ± 0.9	9.5	8.1–12.7
CF duration (ms)	6.7 ± 1.6	5.9	4.2–10.9
FM duration (ms)	2.6 ± 0.4	2.6	1.6–3.2
Pulse repetition rate (Hz)	13.0 ± 7.2	11.5	2.7–31.0

CF constant frequency, *FM* frequency modulation

Previous studies on sexual dimorphism of echolocation pulses of the CF-FM bats were mainly focused on the predominant CF (Jones et al. 1992, 1993; Knörnschild et al. 2012; Russo et al. 2001; Feng et al. 2003; Siemers et al. 2005; Suga et al. 1987). For comparison, data on the sexual dimorphism of the predominant CF in different CF-FM bats examined in these previous and the present studies are assembled in Table 3. It is clear that significant sexual dimorphism in the predominant CF is observed in all but two bat species (e.g., *Rhinolophinae euryale* and *Rhinolophinae mehelyi*, Russo et al. 2001). It is also clear that the predominant CF of the same bat species (*Rhinolophinae euryale)* varies with geographic location (Russo et al. 2001; Siemers et al. 2005). This finding is similar to a recent report (Jiang et al. 2010).

As shown in Table 3, sexual dimorphism in the predominant CF varies with bat species. There is no consistent pattern in the sexual dimorphism in the predominant CF difference in these bat species. For example, in Rhinolophidae, the CF was higher in the female than in the male bat but the opposite was observed in other studies. This is also true in the predominant CF in all Hipposiderinae bat species studied. On the one hand, these different observations on the sexual dimorphism of the predominant CF might be due to species-specific difference, different methods of analysis of the echolocation pulses, or due to the different ways that the echolocation pulses were recorded (Luo et al. 2007). On the other hand, it has been suggested that these different observations might be due to geographic location, different dimensions of vocal cord, and sex hormone level (Neuweiler et al. 1987; Russo et al. 2001). While the largest sexual dimorphic CF difference is

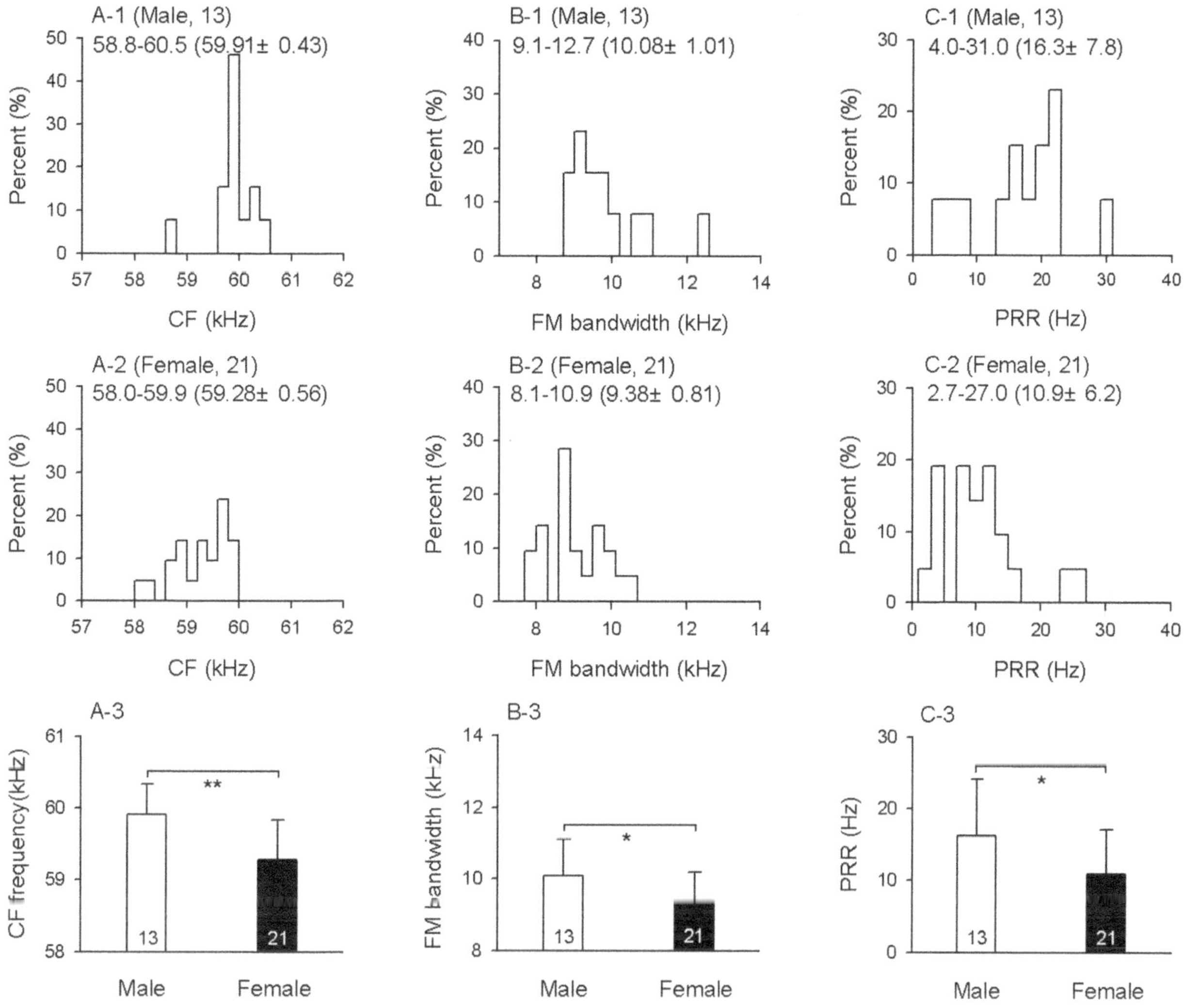

Fig. 2 Sexual dimorphism of the CF, FM bandwidth, and PRR of echolocation pulses emitted by hanging *Hipposideros pratti*. *A-1*, *B-1*, *C-1* vs. *A-2*, *B-2*, *C-2*: the distribution of CF, FM bandwidth, and PRR of echolocation pulses recorded from male and female bats. *A-3*, *B-3*, *C-3*: statistical comparison of these three pulse parameters recorded from male and female bats. The number within each parenthesis indicates the sample size. Student's *t* test: $*p < 0.05$; $**p < 0.01$

10.49 kHz in *Hipposideros caffer* (Jones et al. 1993), the smallest sexual dimorphic CF difference is only 0.8 kHz in *Rhinolophinae euryale* (Siemers et al. 2005). The CF difference of 0.63 kHz between male and female bats (*H. pratti*) in the present study is comparable and significant (Fig. 2(A-1) vs. (A-2)). Because the bats were from a same colony of the same cave, there were the possible relativity among the individuals and the possible similarity among the recorded pulses.

In addition to examining sexual dimorphism in the predominant CF, we also observed that the FM bandwidth was significantly wider and the PRR was significantly higher in male than in female *H. pratti* (Fig. 2). On the contrary, the duration of both CF and FM was significantly longer for female than for male bat (Fig. 3). Whether these multi-parametric dimorphic differences are due to the length and width of vocal cord between the male and female bats remains to be studied. If they were different significantly, combined with the fact that the prey detection may be a primary selective force influencing the echolocation features of bats, it might help us to realize our data. In the case of female bats, they have lower CF than male bats due to their laryngeal structure; to increase pulse intensity, they may shorten the bandwidth (Lazure and Fenton 2011); and to get adequate information, they may increase the duration, simultaneously leading to the PRR reduction considering that low-duty-cycle bats separate pulse and echo in time (Lazure and Fenton 2011; Ho et al. 2013). And the echolocation features of male bats are opposite to the female bats, which may also be beneficial for them to make best use of the predation space and distinct each other and resist jamming.

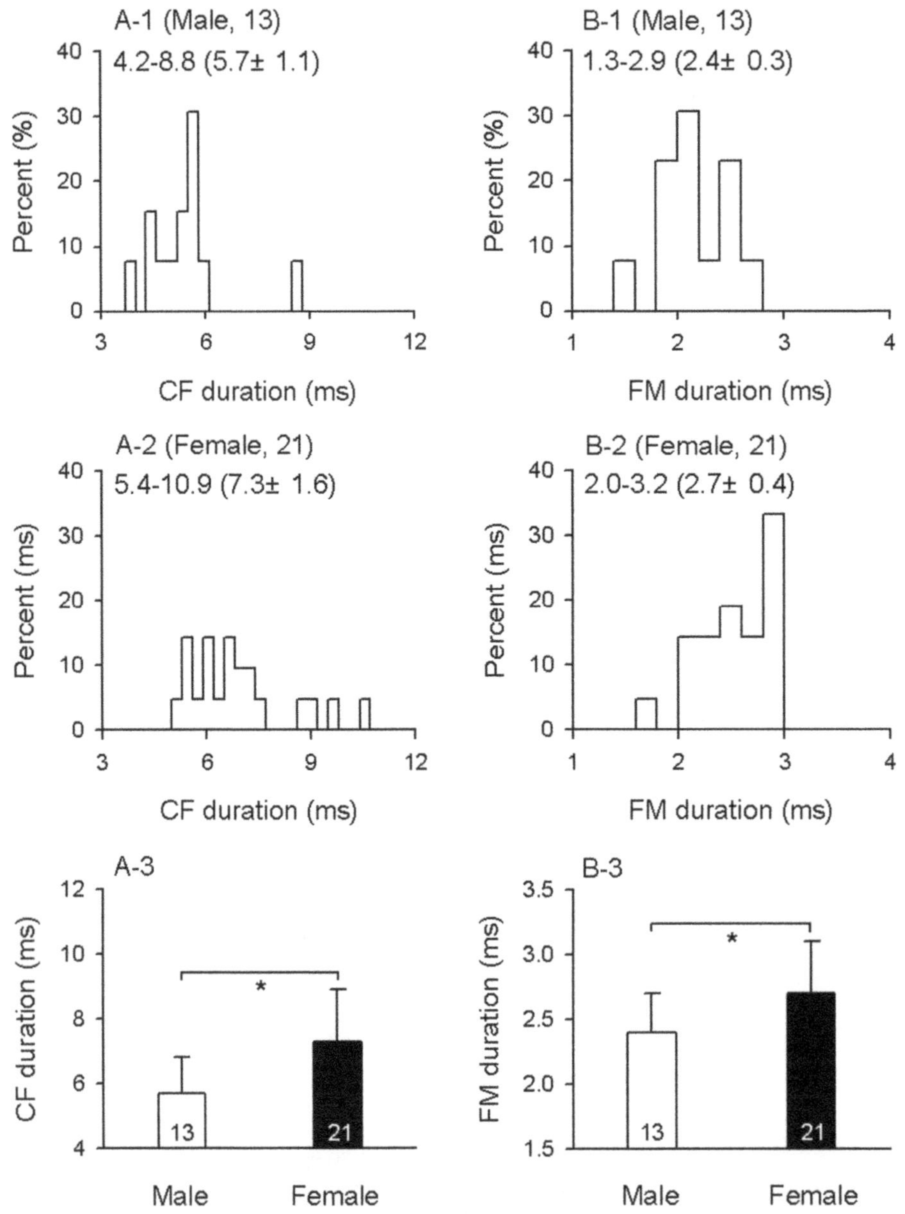

Fig. 3 Sexual dimorphism of the CF and FM duration of echolocation pulses emitted by hanging *Hipposideros pratti*. *A-1, B-1, C-1* vs. *A-2, B-2, C-2*: the distribution of CF and FM duration of echolocation pulses recorded from male and female bats. *A-3, B-3, C-3*: statistical comparison of these two temporal parameters of echolocation pulses recorded from male and female bats (see Fig. 2 for legends)

Possible biological significance of this study

Although there have been studies on sexual dimorphism of bat communication sounds (Kanwal et al. 1994; Liu et al. 2013; Ma et al. 2006), many studies have been mainly focused on the sexual dimorphism of orientation pulses in the CF-FM bat by analyzing the predominant CF (Jones et al. 1992, 1993; Knörnschild et al. 2012; Russo et al. 2001; Feng et al. 2003; Siemers et al. 2005; Suga et al. 1987). In the present study, we have clearly shown that sexual dimorphism of echolocation pulses in the CF-FM bat can also be found in other multiple pulse parameters. For this reason, the present study not only complements but also advances all previous studies with more detailed analysis of sexual dimorphic differences in the echolocation pulses of the CF-FM bat.

While the pulse energy is unevenly distributed on the multiple harmonics of the echolocation pulses of the CF-FM bat, the predominant H_2 has been shown to be essential for successful echolocation in behavioral and neurophysiological studies (Neuweiler et al. 1987; O'Neill and Suga 1982; Schnitzler and Henson 1980; Schnitzler and Denzinger 2011; Suga et al. 1987; Suga 1990). Because

Table 3 Sexual dimorphism in the CF in echolocation pulses of CF-FM bats

Species	Male	Female	Significances
Rhinolophidae			
Rhinolophinae euryale			
South Bulgaria (Siemers et al. 2005)	105.56 ± 0.43	106.63 ± 0.62	s
North Bulgaria (Siemers et al. 2005)	106.21 ± 0.84	105.83 ± 0.41	s
Italy (Russo et al. 2001)	104.32 ± 0.39	104.35 ± 0.33	n. s.
Rhinolophinae cornutus (Feng et al. 2003)	103.13 ± 1.95	106.71 ± 2.30	s
Rhinolophinae mehelyi (Russo et al. 2001)	108.10 ± 0.88	107.48 ± 0.94	n. s.
Hipposiderinae			
Hipposideros rubber (Jones et al. 1993)	134.13 ± 1.84	131.85 ± 0.99	s
Hipposideros caffer (Jones et al. 1993)	138.83 ± 8.25	149.32 ± 4.35	s
Asellia tridens (Jones et al. 1993)	116.37 ± 1.42	118.77 ± 1.59	s
Hipposideros armiger (Feng et al. 2003)	77.52 ± 0.11	76.27 ± 0.78	s
Aselliscus wheeleri (Feng et al. 2003)	125.53 ± 0.22	123.33 ± 1.59	s
Hipposideros pratti (present study)	59.91 ± 0.43	59.28 ± 0.56	s
Mormoopidae			
Pteronotus parnellii (Suga et al. 1987)	61.25 ± 0.534	62.29 ± 0.539	s

Data displayed in the table are expressed as the mean ± standard deviation. Recording condition of the bat was hand-held (Jones et al. 1993; Russo et al. 2001; Siemers et al. 2005), flying (Feng et al. 2003), or sedentary (Suga et al. 1987; present study)
s significant difference, *n. s.* not significant difference

of the time constraint, as the first step, we studied the sexual dimorphism in the multiple pulse parameters of the predominant H_2 of echolocation pulses emitted from sedentary bats mimicking the hanging bats in the roosting cave. An analysis of these multiple parameters of echolocation pulses emitted by flying bat during hunting or foraging should be conducted to further study the sexual dimorphism in echolocation pulse parameters of this bat species. Conceivably, sexual dimorphism in the predominant H_2 of echolocation pulse parameters may also be observed under field study.

During hunting, the bat systematically varies the echolocation pulses in multiple parametric domains to obtain as much information as possible from the returning echoes for successful prey capture and avoidance of obstacles (Griffin 1958; Jen and Kamada 1982; Schnitzler and Henson 1980). It has been shown that the dynamic variation in multiple pulse parameters throughout the prey approach sequence improves the response sensitivity of bat auditory neurons in multiple parametric domains and likely increases the success of hunting (Jen et al. 2001; Wu and Jen 2006, 2008). As mentioned in the "Background" section, many studies have shown that the bats use echolocation pulses for prey capture as well as for social signals for communication, individual recognition, threatening, and guidance (Jameson and Hare 2009; Jones and Ransome 1993; Kazial et al. 2008; Knörnschild et al. 2012; Masters et al. 1995; Möhres 1967; Russo et al. 2007; Schuchmann and Siemers 2010; Voigt-Heucke et al. 2010; Yovel et al. 2009). If our observed sexual dimorphism of echolocation pulses also occurs in nature, then the bats may potentially use this sexual dimorphism cues in pulse parameters for social communication and species and sex identification.

Conclusions

The present studies examined the sexual dimorphism in echolocation pulse parameters of the CF-FM bat, *H. pratti*. The data showed that on average the male bat has higher predominant CF, wider FM bandwidth, and higher pulse repetition rate while the female bat has longer duration of the CF and FM components. These observations suggest that bats may potentially use this sexual dimorphism in echolocation pulse parameters for social communication and species and sex identification.

Abbreviations

CF: constant frequency; FM: frequency modulating; H: harmonic; PRR: pulse repetition rate.

Competing interests

The authors declare that they have no competing interests.

Authors' contributions

Q-CC and Z-YF conceived and designed the experiments. Z-YF, X-YD, NX, QS, G-JL, BL, JuL, JiL, and JT performed the experiments. Z-YF, X-YD, NX, QS, G-JL, BL, JuL, JiL, JT, PH-SJ, and Q-CC analyzed the data. Q-CC contributed the reagents/materials/analysis tools. Z-YF, PH-SJ, and Q-CC wrote the paper. All authors read and approved the final manuscript.

Acknowledgements

We thank the anonymous reviewers for their helpful comments on our earlier manuscript. The work was supported by grants (#31200832, #31070971) from the Natural Science Foundation of China and the Fundamental Research Funds of Central China Normal University (CCNU13A05027).

Author details

[1]School of Life Sciences and Hubei Key Lab of Genetic Regulation and Integrative Biology, Central China Normal University, Wuhan 430079Hubei, China. [2]College of Special Education, Binzhou Medical College, Yantai, Shandong, China. [3]Department of Life Science, National Taiwan Normal University, Taipei, China. [4]Division of Biological Sciences, University of Missouri–Columbia, Columbia, MO 65211, USA.

References

Chen M, Feng J, Li ZX, Zhou J, Zhao HH, Zhang SY, Sheng LX (2002) Echolocation sound waves, morphological features and foraging strategies in *Hipposideros pratti*. Chin J Appl Ecol 13:1629–1632 (in Chinese)

Feng J, Chen M, Liu Y, Li ZX, Zhang XC, Zhou J, Zhang SY (2003) Sexual differences in morphology and echolocation calls in five Chinese bat species. Acta Zool Sin 49:742–747 (in Chinese with English abstract)

Fitzpatrick DC, Suga N, Misawa H (1991) Are the initial frequency-modulated components of the mustached bat's biosonar pulses important for ranging? J Neurophysiol 66:1951–1964

Fu ZY, Tang J, Jen PHS, Chen QC (2010) The auditory response properties of single-on and double-on responders in the inferior colliculus of the leaf-nosed bat, Hipposideros armiger. Brain Res 1306:39–52

Fu ZY, Tang J, Li Y, Zeng H, Chen QC (2011) Frequency-modulation component of the mimic echolocation sound can increase the sensitivity of inferior collicular neurons to sound amplitude in the leaf-nosed bat, *Hipposideros armiger*. Zool Stud 50:537–545

Griffin DR (1958) Listening in the dark. Yale University Press, New Haven

Grilliot ME, Burnett SC, Mendoca MT (2009) Sexual dimorphism in big brown bat (*Eptesicus fuscus*) ultrasonic vocalizations is context dependent. J Mammal 90:203–209

Ho YY, Fang YP, Chou CH, Cheng HC, Chang HW (2013) High duty cycle to low duty cycle: echolocation behaviour of the hipposiderid bat *Coelops frithii*. PLoS One 8, e62938

Jakobsen L, Brinkløv S, Surlykke A (2013) Intensity and directionality of bat echolocation signals. Front Physiol 4:89

Jameson JW, Hare JF (2009) Group-specific signatures in the echolocation calls of female little brown bats (*Myotis lucifugus*) are not an artefact of clutter at the roost entrance. Acta Chiropterol 11:163–172

Jen PHS, Kamada T (1982) Analysis of orientation signals emitted by the CF-FM bat, *Pteronotus p. parnellii* and the FM bat, *Eptesicus fuscus* during avoidance of moving and stationary obstacles. J Comp Physiol A 148:389–398

Jen PHS, Zhou XM, Wu CH (2001) Temporally patterned sound pulse trains affect intensity and frequency sensitivity of inferior collicular neurons of the big brown bat, *Eptesicus fuscus*. J Comp Physiol A 187:605–616

Jiang T, Metzner W, You YY, Liu S, Lu GJ, Li S, Wang L, Feng J (2010) Variation in the resting frequency of *Rhinolophus pusillus* in Mainland China: effect of climate and implications for conservation. J Acoust Soc Am 128:2204–2211

Jones G, Ransome RD (1993) Echolocation calls of bats are influenced by maternal effects and change over a lifetime. Proc Biol Sci 252:125–128

Jones G, Teeling EC (2006) The evolution of echolocation in bats. Trends Ecol Evol 21:149–156

Jones G, Gordon T, Nightingale J (1992) Sex and age differences in the echolocation calls of the horseshoe bat, *Rhinolophus hipposideros*. Mammalia 56:189–193

Jones G, Morton M, Hughes PM, Budden RM (1993) Echolocation flight morphology and foraging strategies of some West African hipposiderid bats. J Zool 230:385–400

Kanwal JS, Matsumura S, Ohlemiller K, Suga N (1994) Analysis of acoustic elements and syntax in communication sounds emitted by mustached bats. J Acoust Soc Am 96:1229–1254

Kazial KA, Masters M (2004) Female big brown bats, *Eptesicus fuscus*, recognize sex from a caller's echolocation signals. Anim Behav 67:855–863

Kazial KA, Kenny TL, Burnett SC (2008) Little brown bats (*Myotis lucifugus*) recognize individual identity of conspecifics using sonar calls. Ethology 114:469–478

Knörnschild M, Jung K, Nagy M, Metz M, Kalko E (2012) Bat echolocation calls facilitate social communication. Proc Biol Sci 279:4827–4835

Kober R, Schnitzler HU (1990) Information in sonar echoes of fluttering insects available for echolocating bats. J Acoust Soc Am 87:882–896

Lazure L, Fenton MB (2011) High duty cycle echolocation and prey detection by bats. J Exp Biol 214(Pt 7):1131–1137

Liu Y, Metzner W, Feng J (2013) Vocalization during copulation behavior in greater horseshoe bats, *Rhinolophus ferrumequinum*. Chin Sci Bul 58:2179–2184

Luo F, Ma J, Li AA, Wu FJ, Chen QC, Zhang SY (2007) Echolocation calls and neurophysiological correlations with auditory response properties in the inferior colliculus of *Pipistrellus abramus* (*Microchiroptera: Vespertilionidae*). Zool Stud 46:622–630

Luo F, Metzner W, Wu FJ, Zhang SY, Chen QC (2008) Duration-sensitive neurons in the inferior colliculus of horseshoe bats: adaptations for using CF-FM echolocation pulses. J Neurophysiol 99:284–296

Ma J, Kobayasi K, Zhang S, Metzner W (2006) Vocal communication in adult greater horseshoe bats, *Rhinolophus ferrumequinum*. J Comp Physiol A 192:535–550

Macías S, Hechavarría JC, Kossl M, Mora EC (2013) Neurons in the inferior colliculus of the mustached bat are tuned both to echo-delay and sound duration. Neuroreport 24:404–409

Masters WM, Raver KAS, Kazial KA (1995) Sonar signals of big brown bats, *Eptesicus fuscus*, contain information about individual identity, age and family affiliation. Anim Behav 50:1243–1260

Möhres FP (1967) Communicative characters of sonar signals in bats. Pp. 939–945. In: Busnel RG (ed) Animal sonar systems: biology and bionics, 2nd edn. Centre National de Recherches Zootechniques de Institut National de la Recherche Agronomique, Jouy-en-Josas, France, pp 719–1233

Neuweiler G (2003) Evolutionary aspects of bat echolocation. J Comp Physiol A 189:245–256

Neuweiler G, Metzner W, Heilmann U, Rübsamen R, Eckrich M, Costa HH (1987) Foraging behaviour and echolocation in the rufous horseshoe bat (*Rhinolophus rouxi*) of Sri Lanka. Behav Ecol Sociobiol 20:53–67

O'Neill WE, Suga N (1982) Encoding of target range and its representation in the auditory cortex of the mustached bat. J Neurosci 2:17–31

Racey PA (1988) Reproductive assessment in bats. In: Kunz TH (ed) Ecological and behavioral methods for the study of bats. Smithsonian Institution Press, Washington D.C., pp 31–45

Russo D, Jones G, Mucedda M (2001) Influence of age, sex and body size on echolocation calls of Mediterranean and Mehely's horseshoe bats, *Rhinolophus euryale* and *R. mehelyi* (Chiroptera: Rhinolophidae). Mammalia 65:429–436

Russo D, Mucedda M, Bello M, Biscardi S, Pidinchedda E, Jones G (2007) Divergent echolocation call frequencies in insular rhinolophids (Chiroptera): a case of character displacement? J Biogeogr 34:2129–2138

Schnitzler HU, Denzinger A (2011) Auditory fovea and Doppler shift compensation: adaptations for flutter detection in echolocating bats using CF-FM signals. J Comp Physiol A 197:541–559

Schnitzler HU, Henson OW Jr (1980) Performance of airborne animal sonar systems: I. Microchiroptera. In: Busnel RG, Fish JF (eds) Animal sonar systems. Plenum Press, New York, pp 109–181

Schuchmann M, Siemers BM (2010) Behavioral evidence for community-wide species discrimination from echolocation calls in bats. Am Nat 176:72–82

Siemers BM, Beedholm K, Dietz C, Dietz I, Ivanova T (2005) Is species identity, sex, age, or individual quality conveyed by echolocation call frequency in European horseshoe bats? Acta Chiropterol 7:259–274

Simmons JA, Vernon JA (1971) Echolocation: discrimination of targets by the bat, *Eptesicus fuscus*. J Exp Zool 176:315–328

Suga N (1990) Biosonar and neural computation in bats. Sci Am 262:60–68

Suga N, Niwa H, Taniguchi I, Margoliash D (1987) The personalized auditory cortex of the mustached bat: adaptation for echolocation. J Neurophysiol 58:643–654

Tang J, Fu ZY, Jen PHS, Chen QC (2011) Recovery cycles of single-on and double-on neurons in the inferior colliculus of the leaf-nosed bat, Hipposideros armiger. Brain Res 1385:114–126

Ulanovsky N, Moss CF (2008) What the bat's voice tells the bat's brain. Proc Natl Acad Sci U S A 105:8491–8498

Vanderelst D, Lee YF, Geipel I, Kalko EKV, Kuo YM, Peremans H (2013) The noseleaf of *Rhinolophus formosae* focuses the Frequency Modulated (FM) component of the calls. Front Physiol 4:191

Voigt-Heucke SL, Taborsky M, Dechmann DNK (2010) A dual function of echolocation: bats use echolocation calls to identify familiar and unfamiliar individuals. Anim Behav 80:59–67

Wu CH, Jen PHS (2006) The role of GABAergic inhibition in shaping duration selectivity of bat inferior collicular neurons determined with temporally patterned sound trains. Hear Res 215:56–66

Wu CH, Jen PSH (2008) Echo frequency selectivity of duration-tuned inferior collicular neurons of the big brown bat, *Eptesicus fuscus*, determined with pulse-echo pairs. Neuroscience 156:1028–1038

Yovel Y, Melcon ML, Franz MO, Denzinger A, Schnitzler HU (2009) The voice of bats: how greater mouse-eared bats recognize individuals based on their echolocation calls. PLoS Comput Biol 5, e1000400

Zhang LB, Jones G, Zhang JS, Zhu GJ, Parsons S, Rossiter SJ, Zhang SY (2009) Recent surveys of bats (Mammalia: Chiroptera) from China. I. Rhinolophidae and Hipposideridae. Acta Chiropterol 11:71–88

7

Differentiation of visual spectra and nuptial colorations of two *Paratanakia himantegus* subspecies (Cyprinoidea: Acheilognathidae) in response to the distinct photic conditions of their habitats

Chia-Hao Chang[1,2], Yi Ta Shao[3,4], Wen-Chung Fu[3], Kazuhiko Anraku[5], Yeong-Shin Lin[1,6] and Hong Young Yan[3,7*]

Abstract

Background: Vision, an important sensory modality of many animals, exhibits plasticity in that it adapts to environmental conditions to maintain its sensory efficiency. Nuptial coloration is used to attract mates and hence should be tightly coupled to vision. In Taiwan, two closely related bitterlings (*Paratanakia himantegus himantegus* and *Paratanakia himantegus chii*) with different male nuptial colorations reside in different habitats. We compared the visual spectral sensitivities of these subspecies with the ambient light spectra of their habitats to determine whether their visual abilities correspond with photic parameters and correlate with nuptial colorations.

Results: The electroretinogram (ERG) results revealed that the relative spectral sensitivity of *P. h. himantegus* was higher at 670 nm, but lower at 370 nm, than the sensitivity of *P. h. chii*. Both bitterlings could perceive and reflect UV light, but the UV reflection patterns differed between genders. Furthermore, the relative irradiance intensity of the light spectra in the habitat of *P. h. himantegus* was higher at long wavelengths (480–700 nm), but lower at short wavelengths (350–450 nm), than the light spectra in the habitats of *P. h. chii*.

Conclusions: Two phylogenetically closely related bitterlings, *P. h. himantegus* and *P. h. chii*, dwell in different waters and exhibit different nuptial colorations and spectral sensitivities, which may be the results of speciation by sensory drive. Sensory ability and signal diversity accommodating photic environment may promote diversity of bitterling fishes. UV light was demonstrated to be a possible component of bitterling visual communication. The UV cue may assist bitterlings in gender identification.

Keywords: Bitterling; Electroretinogram; Sensory drive; UV reflection pattern

Background

Vision is an important sensory modality in fish, with the exception of some fishes, such as cave-dwelling species without eyes. Almost all fishes, from those inhabiting shallow and clear waters to those in the aphotic zone in the deep ocean, can perceive visual signals (Pitcher 1986) in order to communicate, find mates, forage for food, avoid predators, camouflage, or navigate (Hawryshyn 1992; Losey et al. 1999).

Light spectra in aquatic environments are distinct from those on land. Not just water itself, but also suspended materials (e.g., mineral particles, detritus, and planktons), can selectively absorb or scatter different wavelengths of light (Litjens et al. 1999). Since the underwater photic environments may be affected by depth and suspended materials as well as unique features of the habitats, fish under different photic conditions should physiologically adjust its spectral sensitivity in order to enhance detection ability by maximizing the visual contrast (Endler 1992; Lythgoe

* Correspondence: hyyan@gate.sinica.edu.tw
[3]Sensory Physiology Laboratory, Marine Research Station, Academia Sinica, I-Lan, Taiwan
[7]Hanse-Wissenschaftskolleg Institute of Advanced Study, Delmenhorst, Germany
Full list of author information is available at the end of the article

1984). For example, the spectral sensitivity of milkfish (*Chanos chanos*), humpback salmon (*Oncorhynchus gorbuscha*), and yellowfin tuna (*Thunnus albacares*) ontogenetically change to fit the photic environment during each developmental stage (Chang et al. 2009a; Cheng and Flamarique 2004; Loew et al. 2002). Moreover, it has been reported that spectral sensitivity has evolved to adapt to environmental light parameters in cichlids, seabreams, and moray eels (Seehausen et al. 2008; Wang et al. 2011; Wang et al. 2009), especially, the adaptations of RH1 and LWS opsins to light environments which are documented in cichlid fishes (Sugawara et al. 2005; Terai et al. 2006), as well as in different populations of three-spined stickleback (*Gasterosteus aculeatus*) and guppy (*Poecilia reticulata*) (McDonald and Hawryshyn 1995; Tezuka et al. 2014). The retinal photoreceptor cells of fishes utilize visual pigments, which contain a chromophore (11-*cis*-retinal, A1 retinal, or 11-*cis*-3,4-dehydroretinal, A2 retinal) and opsin, to capture photons. Variations in spectral sensitivity mainly result from the alternative usage of different opsin genes (Flamarique 2013; Johnson et al. 2013; Kasagi et al. 2015; Kondrashev et al. 2013; Valen et al. 2013). Furthermore, sexual selection favors individuals that can display visual signals that are readily distinguishable from the background (Hurtado-Gonzales et al. 2014; Miyagi et al. 2012; Morrongiello et al. 2010; Selz et al. 2014). Therefore, in addition to their link with visual properties, photic environments may also influence overall body coloration of fish.

A good example of the interactions among the photic environment, spectral sensitivity, and nuptial coloration is observed in two cichlids inhabiting Lake Victoria, *Pundamilia nyererei* and *Pundamilia pundamilia*. These two cichlids are closely related and *P. nyererei* is sympatric with *P. pundamilia* over its distribution range; moreover, they are morphologically and anatomically similar. Females of these two species are hard to tell from each other, but males of them are obviously different with distinct coloration: male *P. pundamilia* is metallic blue-grey and male *P. nyererei* is bright red dorsally and yellow laterally (Maan et al. 2004). Also, *P. nyererei* lives in deeper water than *P. pundamilia* (Seehausen and Bouton 1997), and so the former is relatively exposed to more red light with a proportionally greater wavelength (Maan et al. 2006). In response to variations in ambient light environments, these two species have evolved distinct spectral sensitivities and nuptial colorations (Maan et al. 2006; Miyagi and Terai 2013).

Bitterling (Acheilognathidae, Cyprinoidea) are cyprinids, distributed over East and Southeast Asia and Europe. They inhabit various lowland freshwater habitats, such as lakes, rivers, and ditches (Kottelat and Freyhof 2007), and have evolved a symbiotic relationship with freshwater mussels (family Unionidae) for reproduction. The reproductive success of male bitterlings (which establish a territory and guard mussels from rivals) is mainly dependent on body size (Casalini et al. 2009; Reichard et al. 2008). However, a recent study demonstrated that nuptial coloration is a critical factor for female mate choice and male–male competition (Smith et al. 2014).

In Taiwan, two native *Paratanakia* bitterlings, *Paratanakia himantegus himantegus* and *Paratanakia himantegus chii* (Fig. 1), are phylogenetically related (Chang et al. 2014; Chang et al. 2009b), and their distributions largely overlap on a geographical scale; however, they are rarely sympatric. Usually, the two bitterlings are found in separate microhabitats: *P. h. chii* mainly dwells in shallow floodplain habitats, such as irrigation channels of rice paddy fields, whereas *P. h. himantegus* is commonly found in lakes and ponds (Chen and Chang 2005; Chen et al. 2012). The most obvious morphological difference between the two bitterlings is the nuptial coloration (mating signals). Breeding males of *P. h. himantegus* possess a red stripe on their dorsal fin, red patches on the operculum and pectoral fin, and red irises, whereas *P. h. chii* breeding males have a yellow-white stripe on the dorsal fin, and their irises are not red (Chen and Chang 2005).

The perception of visual signals is determined by two components: the physical characteristics of a habitat and the sensory properties of a receiver (Endler 1990); therefore, a change in one of these two components tends to drive the development of substitute signals in response to variable habitats (Chunco et al. 2007; Endler 1980), which has been demonstrated by several studies (Dalton et al. 2010; Fuller et al. 2005; Gray et al. 2008; Hoffmann et al. 2007). Based on the observation that the two aforementioned and closely related bitterlings exhibit distinct nuptial colorations, it is hypothesized that the differences in their nuptial colorations may be related to variations in photic properties of its habitats and/or spectral sensitivities of the fish. To test this hypothesis, the spectral sensitivities of these two bitterlings were recorded with electrophysiological method and the light spectra of their habitats were measured and compared.

Methods

Subjects

Fish were caught from Chinlung Lake, Hsichih, New Taipei City (*P. h. himantegus*) and Dashi, Taoyuan County (*P. h. chii*). These two sites belong to the Tamsui River basin. All fish were collected during the spawning season (July 2013). The males exhibited nuptial coloration and the females had long ovipositors. Fish were 25 to 40 mm in standard length. Fish were kept in the laboratory in two 36-L aquaria (separated by taxa) under a natural photoperiod, with water temperature maintained at 25 ~ 28 °C. They were fed ad libitum with artificial fish feed or frozen *Artemia* twice a day. Specimen collection and experiments

Fig. 1 Photos of studied species taken in an indoor aquarium. **a** Male *P. himantegus himantegus*; **b** female *P. h. himantegus*; **c** male *Paratanakia himantegus chii*; **d** female *P. h. chii*. [Photos were taken by Dr. Yun-Ching Chang and Mr. Ta-Ching Chang]

were performed with permission (RFiZOOHY20060701) from the Institutional Animal Care and Use Committee (IACUC) of Academia Sinica.

Electroretinogram recording

Electroretinogram (ERG) recordings were mainly modified from Shao et al. (2014). Fish were anesthetized with 0.025 % buffered MS-222 (ethyl 3-aminobenzoate, methanesulfonic acid salt), and then immobilized through an injection of gallamine triethiodide (Flaxedil; Sigma, St. Louis, MO, USA.), a neuromuscular junction blocking agent, into the dorsal muscle (0.4–0.6 μg per gram of body weight). Each fish was wrapped in wet KimWipes (Kimberley-Clark Taiwan, Taipei) to prevent skin from scraping or drying, and a small tube was inserted into the mouth to deliver well-oxygenated water to the gills. Immobilized fish were kept under dark conditions for at least 1.5 h for dark adaptation before the ERG recording. ERG tests were performed as described previously (Morita et al. 1997; Seeliger et al. 2002; Shao et al. 2014). A 100-W halogen lamp (Osram 64637 100 W/12 V) in a lamp housing device (Nikon C-FI 115, Japan) was used as the light source for tests of visible spectra (400–670 nm). For the shorter wavelengths (≤370 nm), a custom-made UV curing device (Great Lighting Corp. 100 W/365 nm) was used as the light source. A series of neutral density (ND) filters (Andover, NH, USA) were used to reduce the light intensity in 0.3 log unit steps, from −0.3 to −4.8 log units. Peak photon fluxes at each wavelength were determined using a USB 2000 spectrophotometer (Ocean Optics, USA), calibrated by LC-1-cal (Ocean Optics, USA) with OODBase32 software (Ocean Optics, USA), at the fish eye surface. A shutter (Sutter Instrument Lambda SC, USA) was used to provide fixed durations of light stimuli: 10 stimuli, each of 20 ms in duration, with an inter-stimuli interval of 5 s. The ERG tests started from the lowest light intensity, and ND filters were adjusted step-by-step to increase stimuli intensity (16 steps to reach the maximum). ERG signals were recorded using Teflon-coated insulated silver electrodes (63.5 μm in diameter), and the Telfon on the tip end (5 mm) was removed and chlorinated to enhance picking up of evoked potentials. The recording electrode was positioned on the cornea, and the reference electrode was placed on the head. The signals from both electrodes were amplified 20 k-fold and filtered with a high–low band pass filter set between 1 and 3000 Hz (ER-1, Cygnus Technology Inc. USA). The signals were digitized at a 2-kHz sampling rate with PowerLab 4/25 (ADInstruments, Inc. ML845, Australia) and recorded with Scope 3 data acquisition software (ADInstruments, Inc. Australia). The recorded signals were analyzed using custom-made software (programmed by K. Anraku). Signals recorded at each step (100 to 700 ms after stimuli) were compared to the model ERG trace obtained at maximum light stimuli using the correlation method. The replicates of waveforms from each test were compared using Spearman's correlation test. Our pilot study revealed that when a correlation coefficient (r) between two replicates was less than 0.7, the two ERG traces showed little resemblance. Hence, ERG traces were considered valid when r was greater than 0.7 and the *b*-wave latency time delay was less than 100 ms (Seeliger et al. 2002). Deflection of the ERG trace was taken as an indication of the light intensity threshold of the particular wavelength (Morita et al. 1997). Sensitivity was defined as − log10 of the threshold for each tested wavelength. To remove variations between trials, values for each individual were standardized to a 0–100 % scale, where 100 % indicated the maximum sensitivity spectrum (Lisney et al. 2010). After standardization, the sensitivity values of the two bitterlings for each light wavelength stimulus were compared using Tukey's pairwise comparisons (one-way ANOVA). In this study, eight *P. h. himantegus* individuals (four males and four females) and nine *P. h. chii* individuals (six males and three females) were used for recordings.

Body UV reflection pattern

Fish were anesthetized with buffered MS-222 before being photographed. The body UV reflection pattern was recorded using a Nikon D70 camera equipped with a Nikon 105-mm micro lens. The whole process was completed inside a dark room, where fish bodies were illuminated using two UV lamps (UVGL-58 P/N 95-0007-05, long wave mode, 367 nm). A UV passing filter (B + W 403 52 mm) and an IR cut filter (B + W 489 52 mm) were mounted on the lens to block any possible ambient light leakage. The spectral transmission ranges of these two filters are available online (http://diglloyd.com/articles/Filters/spectral-B+W-403.html). The UV reflection patterns of one male and one female of each subspecies were photographed.

Ambient light spectra

The ambient light spectra of four habitats, including the only two known habitats of *P. h. chii* in Taiwan (Dashi Town, Taoyuan County and Longtan Township, Taoyuan County), and two typical habitats of *P. h. himantegus* (Chinlung Lake, Hsichih, New Taipei City and Jiaoxi Township, Yilan County), were measured using a USB-2000 spectrometer with a waterproof fiber optic cable (QP1000-2-UV–VIS) (Ocean Optics, Dunedin, FL, USA), 1000 μm in diameter and 2 m in length; absolute spectral irradiance was measured at three to five sites in each sampling location. Before each measurement, USB-2000 was calibrated with calibration light sources, LS-1-cal (Ocean Optics, Dunedin, FL, USA), with OODBase32 software (Ocean Optics, Dunedin, FL, USA). The light in air was measured upward above the water surface, and the light underwater was measured both upward and sideway at a depth of 25 cm from the water surface at all locations. The underwater sideway light spectrum was the arithmetic mean of two scans in the sun and anti-sun directions (Flamarique et al. 2013). Habitat photic parameters were measured between 10:00 and 14:00 in October, 2014. The 25-cm depth is the averaged depths of the trenches or the near-shore area of the pond, where bitterling breeding pairs and hosting mussels were found. The relative spectral irradiance was smoothed by a factor of 15 using a simple moving average (McLean et al. 2007), and then normalized to a 0–100 % scale, where 100 % indicated the maximum intensity spectrum. Light spectra of the typical habitats of *P. h. chii* and *P. h. himantegus* were the geometric means of the relative spectral irradiance measured in the respective locations. Moreover, the light spectra curves of the typical habitats were compared by Spearman's correlation coefficients with two-tailed tests, which were performed with SPSS v.14 (SPSS Inc., Chicago, IL, USA).

Results

The relative visual spectra of these two *Paratanakia* subspecies measured with ERG recording is shown in Fig. 2, which showed that bitterlings of both subspecies are able to detect light within a range of 370 nm to 670 nm under dark adaptation conditions. The highest sensitivity (lowest threshold) was found at the wavelength of 440 nm for *P. h. himantegus* and 460 nm for *P. h. chii*. Both fishes were less sensitive (higher threshold) to light at a wavelength of 670 nm as compared with other tested wavelengths. No differences in thresholds of spectral sensitivity were observed between gender for either species. To obtain relative spectral sensitivity values, the sensitivity of *P. h. himantegus* was normalized to the stimulation intensity of the threshold at 440 nm (which was set as 100 %), and the sensitivity of *P. h. chii* was normalized to the stimulation intensity of the threshold at 460 nm. The relative visual spectra of the two subspecies are of inverted U shapes, with some variations (Fig. 2). The relative spectral sensitivity curve of *P. h. himantegus* displayed a single peak at 440 nm, but that of *P. h. chii* exhibited two peaks; a maximum sensitivity peak at 460 nm and a secondary peak at 400 nm (99.73 %). Compared to the most sensitive wavelength for the respective fish, the

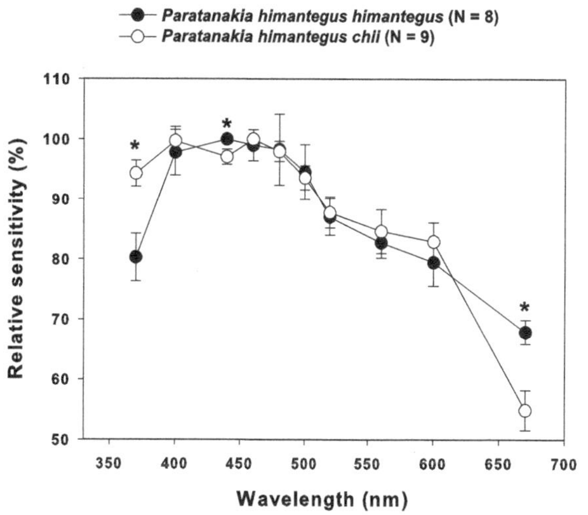

Fig. 2 Relative visual spectra of *Paratanakia himantegus himantegus* and *P. h. chii*. The asterisks indicate wavelengths at which the relative sensitivity of the two species is significantly different ($p < 0.05$). Values are the mean ± S.E.

relative sensitivity of the least sensitive wavelengths (670 nm) was only 70 % in *P. h. himantegus* and 55 % in *P. h. chii*. Examining the two ends of the visual spectra revealed that the relative sensitivity at 670 nm is roughly the same as that at 370 nm for *P. h. himantegus*, but in *P. h. chii*, the relative sensitivity at 670 nm is much lower than that at 370 nm (Fig. 2). Tukey's pairwise (one-way ANOVA) tests were performed to demonstrate that *P. h. himantegus* was more sensitive at 670 and 440 nm but less sensitive at 370 nm ($p < 0.05$) than *P. h. chii* (Fig. 2).

Examination of body UV reflection patterns showed that both species reflect UV light (Fig. 3). The dorsal and anal fins of both *P. h. himantegus* and *P. h. himantegus* males under visible light were more vividly colored (Fig. 3(A-VIS

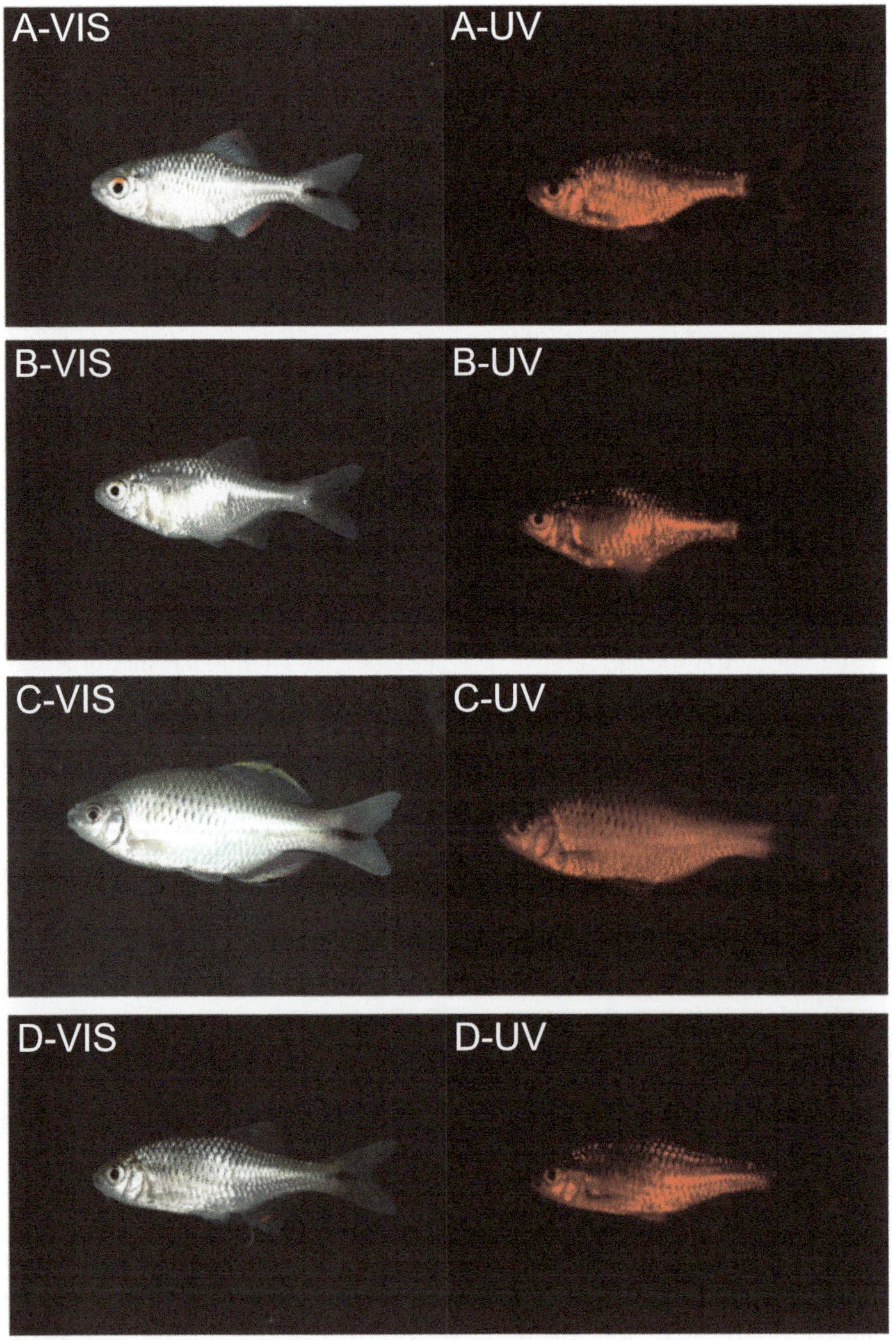

Fig. 3 Photos of body reflection patterns of *Paratanakia himantegus himantegus* and *P. h. chii* under UV light (*UV*) and visual light (*VIS*). (*A-UV*) Male *P. himantegus himantegus*; (*A-VIS*) male *P. h. himantegus*; (*B-UV*) female *P. h. himantegus*; (*B-VIS*) female *P. h. himantegus*; (*C-UV*) male *P. h. chii*; (*C-VIS*) male *P. h. chii*; (*D-UV*) female *P. h. chii*; (*D-VIS*) female *P. h. chii*

and C-VIS)) than those of females (Fig. 3(B-VIS and D-VIS)). Nevertheless, the fins of neither males nor females reflected UV (Fig. 3(A-UV, B-UV, C-UV and D-UV)). There is no clear difference in UV reflection pattern between subspecies. Males of both fishes reflected UV throughout the body (Fig. 3(A-UV, and C-UV)), whereas females had a non-reflecting region above the pelvic fin. UV reflection by the ventral part was stronger than reflection by the dorsal part of the body (Fig. 3(B-UV and D-UV)).

The original data of relative ambient spectra gathered from four habitats were shown in Additional file 1. The light spectra in the air in the sampled habitats were largely consistent (Spearman's rho $r = 0.781$, $p < 0.001$) (Fig. 4a); however, significant differences of light transmission underwater were observed upon comparing spectral irradiance from various habitats. The irradiance of upward light in the habitat of *P. h. himantegus* was higher at relatively long wavelengths (480–700 nm), but lower at shorter wavelengths (350–450 nm), than in the habitats of *P. h. chii* (Fig. 4b). Furthermore, Spearman's correlation coefficients were used to show that the upward underwater spectral curve in the habitats of *P. h. himantegus* and *P. h. chii* were not correlative ($r = 0.161$, $p < 0.001$). At 670 nm, i.e., red light, the relative irradiance in the habitat of *P. h. himantegus* was more than double that in the habitat of *P. h. chii*. However, the relative irradiance in the UV light spectrum, i.e., 370 nm, was far higher in the *P. h. chii* habitat than that in *P. h. himantegus* habitat (Fig. 4b). The sideways light spectra of the *P. h. himantegus* habitat were shifted toward longer wavelengths as compared to the spectra of the habitat of *P. h. chii* (Spearman's rho $r = 0.559$, $p < 0.001$) (Fig. 4c). The highest relative irradiance in the *P. h. himantegus* habitat was observed at 560 nm, whereas the relative irradiance in the *P. h. chii* habitat peaked at a shorter wavelength, 455 nm (Fig. 4c).

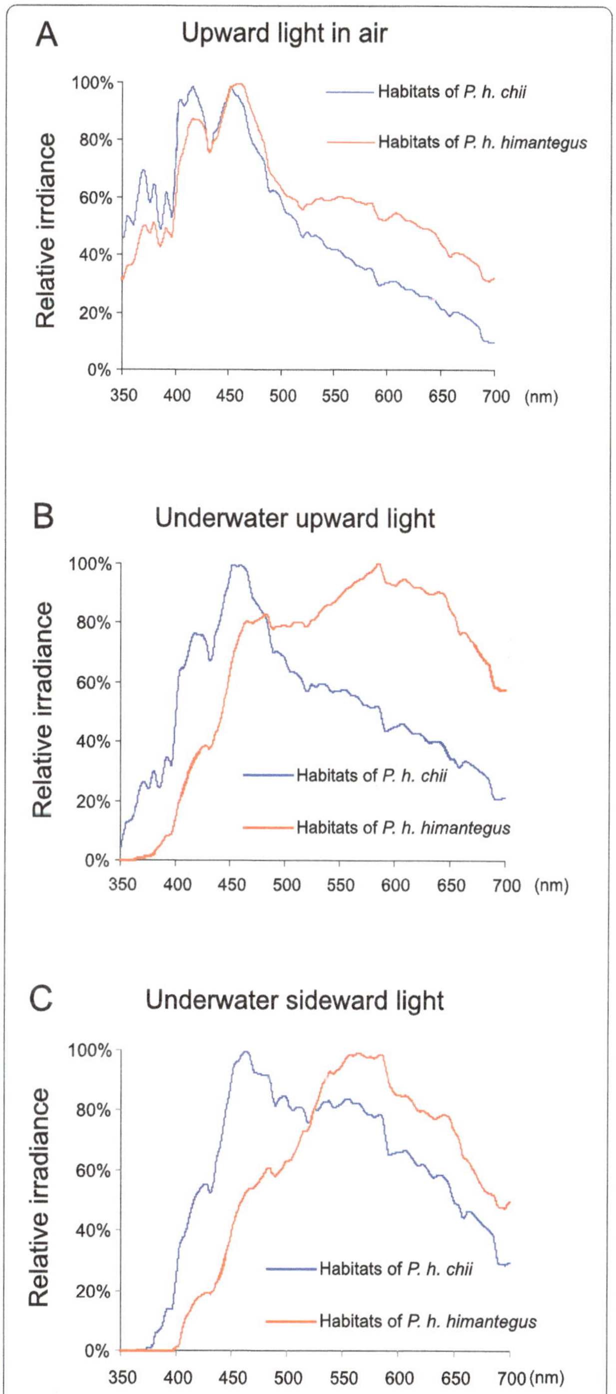

Fig. 4 Light spectra of the sampled habitats of studied species. **a** Upward light in air; **b** underwater upward light; **c** underwater sideward light. Light spectra shown are the geometric means of the measurements at the respective collection sites. Red: habitats of *P. h. himantegus*; Blue: habitats of *P. h. chii*. The original data are provided in Additional file 1

Discussion

According to the surveys carried out by Taiwan Conservation Association of Native Fishes (TCANF), the two bitterlings examined here live in different habitats. *P. h. chii* prefers shallow habitats, such as irrigation channels of rice paddy fields, whereas *P. h. himantegus* prefers deeper water, including lakes and ponds (TCANF unpublished data). Therefore, although these two subspecies can co-exist in one large drainage system, they are seldom sympatric in distribution. These two typical habitats, ditch and pond, not only differ in terms of depth and current but may also differ in their photic environments. Due to phytoplankton, the ambient light spectra of turbid or eutrophic waters may have higher relative irradiance at longer wavelengths (Bowling et al. 1986; Jerlov 1976). As compared with the habitats of *P. h. himantegus*, the habitats of *P. h. chii* usually have a lower density of planktonic algae and clear/running water. Consequently, it has been suggested that *P. h. himantegus* dwells in locations where longer wavelength light predominates to a greater extent than that in the habitats of *P. h. chii* (Fig. 4).

The visual system is responsible not only for recognizing color signals but also for foraging for food or detecting predators. The light environment is a factor known to drive both the reception and production of visual signals of fishes (Hornsby et al. 2013; Hurtado-Gonzales et al. 2014; Shin and Choi 2014; Terai et al. 2006; Tezuka et al. 2014). Indeed, the present study showed that the spectral sensitivities of *P. h. himantegus* and *P. h. chii* are distinct: *P. h. himantegus* is more sensitive than *P. h. chii* to longer wavelength light, while it is less sensitive than *P. h. chii* at shorter wavelengths. As such, the spectral sensitivities of these subspecies are suitable for their respective ambient light conditions. Furthermore, the diverse nuptial colorations of these two bitterling subspecies may also result from the variation in the photic environments and/or the distinction of sensory properties.

Fishes commonly use UV signals in communication (Losey et al. 1999; Witzany 2014). In this study, ERG data demonstrated that both subspecies can perceive UV light, i.e., 370 nm, and *P. h. chii* was more sensitive to UV light than *P. h. himantegus*, which is consistent with the light spectra of their respective habitats. However, since the body UV reflection patterns cannot be quantitatively compared between species, it is not possible to conclude whether the differences in UV sensitivity between the species are related to their body UV reflection patterns. However, the UV reflection patterns of the bitterlings were sexually dimorphic, suggesting that the UV pattern may be a secondary sexual characteristic in these fishes. Furthermore, the nuptial coloration of the males is a crucial signal for the mate choice of females, and the UV cue has been demonstrated to be a factor for nuptial coloration in many fishes (Garcia and Perera 2002; Kodric-Brown and Johnson 2002; Palmer and Hankison 2014; Rick et al. 2006; Smith et al. 2002). Our results imply that the UV reflection pattern of bitterlings may be an important characteristic for consideration in future taxonomic or ecological studies.

The diversification of *P. nyererei* and *P. pundamilia* cichlids is considered to be a good example of speciation caused by sensory drives (Miyagi and Terai 2013). A hypothesis posits that closely related and morphologically similar species should differ in mating signals, which is mainly engendered by three processes: habitat transmission, perceptual tuning, and signal matching. Habitat transmission is important when the habitat physically degrades the intensity of the signal during transmission; therefore, it is advantageous for signals to have a long transmitting distance (Bradury and Vehrencamp 2011; Morrill et al. 2013). Perceptual tuning is the phenomenon whereby some individuals may be inherently more sensitive to certain stimuli, e.g., a specific wavelength of light or frequency of sound, than other organisms, which may be a consequence of adaptation to their habitat (Boughman 2002). Signal matching refers to intimate coupling between the evolution of sexual display in the signal sender and the development of sensory capabilities in the signal receiver (Fisher 1915; Kirkpatrick 1982). Usually, mating signals are compatible with the receiver's perception capabilities, because a more detectable mating signal can increase the reproductive opportunities of the sender and reduce the search cost of the receiver (partner) (Dawkins and Guilford 1996), thereby improving reproductive success. Sensory drive speciation suggests that the gene flow between two sibling species is constrained during contact (Boughman 2002). Habitat transmission, perceptual tuning, and signal matching are all apparent in the two closely related *Paratanakia* subspecies examined in this study. Moreover, earlier genetic studies suggested that there is no gene flow between *P. h. himantegus* and *P. h. chii* (Chang et al. 2014; Chang et al. 2009b), and therefore, it is possible that speciation by sensory drive promotes differentiation of these two bitterlings.

Matsumoto et al. (2014) suggested that speciation by sensory drive most likely occurs when the sensory trait is determined by a small number of loci. Indeed, the visual spectral sensitivity is related to the opsin genes, e.g., long-wavelength-sensitive opsin gene (*lws*) in cichlids, zebrafish (*Danio rerio*), and three-spined stickleback (Seehausen et al. 2008; Shao et al. 2014), or short-wavelength-sensitive opsin gene 2a (*sws2a*) in cichlids (Seehausen et al. 2008). Moreover, body coloration is also likely to be determined by a limited number of genes. For example, the male body coloration of East African cichlid, i.e., *Pseudotropheus saulosi*, may be controlled by only five genes (Gunter et al. 2011). Allopatric speciation (Bohlen et al. 2006) and switches in mussel host preference (Kitamura et al. 2012) have been suggested to be the main factors of bitterling diversification. Future studies of the genes that affect spectral sensitivity and nuptial coloration of these two *Paratanakia* bitterlings are needed to clarify the interactions between natural selection and sexual selection (Maan and Seehausen 2011). Such studies promise to clarify whether speciation by sensory drive promoted the speciation of these two *Paratanakia* bitterling subspecies.

Conclusions

We report that two morphologically similar and phylogenetically close bitterlings, *P. h. himantegus* and *P. h. chii*, have different nuptial colorations and spectral sensitivities and reside in habitats with different environmental photic conditions; these findings support the hypothesis that differences in nuptial coloration may be a consequence of variations in photic properties and spectral sensitivities and imply that speciation could be driven by the sensory system and related environmental conditions (Endler and McLella 1988), thereby promoting bitterling diversity. The spectral sensitivity and body UV reflection pattern data also support the hypothesis that the UV cue is a major component of the visual signals of bitterlings. While the UV cue may not be utilized in species recognition, it may assist bitterlings in gender identification.

Competing interests

The authors declare that they have no competing interests.

Authors' contributions

CHC and YTS carried out the ERG measurements and UV reflection photography experiments, and wrote the first draft of the manuscript. KA designed and wrote the codes for the ERG recording and data analysis systems. WCF assisted in ERG recordings. YSL made comments on the manuscript. HYY provided logistical and financial support during the study, and revised the manuscript. All authors read and approved the final manuscript.

Acknowledgements

This study was supported by intramural research grants from Academia Sinica to HYY. During the course of manuscript preparation, CHC was supported by an overseas study fellowship from the Ministry of Science & Technology, Taiwan (103-2917-I-009-003). YTS was supported by a postdoctoral fellowship from Academia Sinica. Dr. Duncan Wright performed English text editing.

Author details

[1]Department of Biological Science and Technology, National Chiao Tung University, Hsinchu, Taiwan. [2]Department of Biology, St. Louis University, St. Louis, MO, USA. [3]Sensory Physiology Laboratory, Marine Research Station, Academia Sinica, I-Lan, Taiwan. [4]Present Address: Institute of Marine Biology, National Taiwan Ocean University, Keelung, Taiwan. [5]Faculty of Fisheries, Kagoshima University, Kagoshima, Japan. [6]Institute of Bioinformatics and Systems Biology, National Chiao Tung University, Hsinchu, Taiwan. [7]Hanse-Wissenschaftskolleg Institute of Advanced Study, Delmenhorst, Germany.

References

Bohlen J, Šlechtová V, Bogutskaya N, Freyhof J (2006) Across Siberia and over Europe: phylogenetic relationships of the freshwater fish genus *Rhodeus* in Europe and the phylogenetic position of *R. sericeus* from the River Amur. Mol Phylogenet Evol 40:856–865

Boughman JW (2002) How sensory drive can promote speciation. Trends Ecol Evol 17:571–577

Bowling LC, Steane MS, Tyler PA (1986) The spectral distribution and attenuation of underwater irradiance in tasmanian inland waters. Freshw Biol 16:313–335

Bradury JW, Vehrencamp SL (2011) Principles of animal communication. Sinauer Associates, Inc., Sunderland, Massachusetts

Casalini M, Agbali M, Reichard M, Konečná M, Bryjová A, Smith C (2009) Male dominance, female mate choice, and intersexual conflict in the rose bitterling (*Rhodeus ocellatus*). Evolution 63:366–376

Chang C-H, Chiao C-C, Yan HY (2009a) Ontogenetic changes in color vision in the milkfish (*Chanos chanos* forsskål, 1775). Zool Sci 26:349–355

Chang C-H, Lin W-W, Shao Y-T, Arai R, Ishinabe T, Ueda T, Matsuda M, Kubota H, Wang FY, Jang-Liaw NH, Kao HW (2009b) Molecular phylogeny and genetic differentiation of the *Tanakia himantegus* complex (Teleostei: Cyprinidae) in Taiwan and China. Zool Stud 48:839–850

Chang C-H, Li F, Shao K-T, Lin Y-S, Morosawa T, Kim S, Koo H, Kim W, Lee JS, He S, Smith C, Reichard M, Miya M, Sado T, Uehara K, Lavoué S, Chen WJ, Mayden RL (2014) Phylogenetic relationships of Acheilognathidae (Cypriniformes: Cyprinoidea) as revealed from evidence of both nuclear and mitochondrial gene sequence variation: evidence for necessary taxonomic revision in the family and the identification of cryptic species. Mol Phylogenet Evol 81.182–194

Chen I-S, Chang Y-C (2005) A photographic guide to the inland-water fishes of Taiwan. The SueiChan Press, Keelung

Chen I-S, Tzeng C-S, Shao K-T (2012) Red data book of freshwater fishes in Taiwan. Forestry Bureau, Council of Agriculture, Executive Yuan, Taipei, Taiwan

Cheng CL, Flamarique IN (2004) Opsin expression: new mechanism for modulating colour vision. Nature 428:279

Chunco AJ, McKinnon JS, Servedio MR (2007) Microhabitat variation and sexual selection can maintain male colour polymorphisms. Evolution 61:2504–2515

Dalton BE, Cronin TW, Marshall NJ, Carleton KL (2010) The fish eye view: are cichlids conspicuous? J Exp Biol 213:2243–2255

Dawkins MS, Guilford T (1996) Sensory bias and the adaptiveness of female choice. Am Nat 148:937–942

Endler JA (1980) Natural selection on color patterns in *Poecilia reticulata*. Evolution 34:76–91

Endler JA (1990) On the measurement and classification of color in studies of animal color patterns. Biol J Linn Soc 41:315–352

Endler JA (1992) Signals, signal conditions, and the direction of evolution. Am Nat 139: Supplement:125–153

Endler JA, McLella T (1988) The processes of evolution: toward a newer synthesis. Annu Rev Ecol Syst 19:395–421

Fisher RA (1915) The evolution of sexual preference. Eugen Rev 7:184–192

Flamarique IN (2013) Opsin switch reveals function of the ultraviolet cone in fish foraging. Proc R Soc B Biol Sci 280:20122490

Flamarique IN, Cheng CL, Bergstrom C, Reimchen TE (2013) Pronounced heritable variation and limited phenotypic plasticity in visual pigments and opsin expression of threespine stickleback photoreceptors. J Exp Biol 216:656–667

Fuller RC, Carleton KL, Fadool JM, Spady TC, Travis J (2005) Genetic and environmental variation in the visual properties of bluefin killifish, *Lucania goodei*. J Evol Biol 18:516–523

Garcia CM, Perera T (2002) Ultraviolet-based female preferences in a viviparous fish. Behav Ecol Sociobiol 52:1–6

Gray SM, Dill LM, Tantu FY, Loew ER, Herder F, McKinnon JS (2008) Environment-contingent sexual selection in a colour polymorphic fish. Proc R Soc B Biol Sci 275:1785–1791

Gunter HM, Clabaut C, Salzburger W, Meyer A (2011) Identification and characterization of gene expression involved in the coloration of cichlid fish using microarray and QRT-PCR approaches. J Mol Evol 72:127–137

Hawryshyn CW (1992) Polarization vision in fish. Am Sci 80:164–175

Hoffmann M, Tripathi N, Henz SR, Lindholm AK, Weigel D, Breden F, Dreyer C (2007) Opsin gene duplication and diversification in the guppy, a model for sexual selection. Proc R Soc B Biol Sci 274:33–42

Hornsby MAW, Sabbah S, Robertson RM, Hawryshyn CW (2013) Modulation of environmental light alters reception and production of visual signals in Nile tilapia. J Exp Biol 216:3110–3122

Hurtado-Gonzales JL, Loew ER, Uy JAC (2014) Variation in the visual habitat may mediate the maintenance of color polymorphism in a poeciliid fish. PLos One 9, e101497

Jerlov NG (1976) Marine optics. Elsevier Amsterdam, Kingdom of the Netherlands

Johnson AM, Stanis S, Fuller RC (2013) Diurnal lighting patterns and habitat alter opsin expression and colour preferences in a killifish. Proc R Soc B Biol Sci 280:20130796

Kasagi S, Mizusawa K, Murakami N, Andoh T, Furufuji S, Kawamura S, Takahashi A (2015) Molecular and functional characterization of opsins in barfin flounder (*Verasper mosen*). Gene 556:182–191

Kirkpatrick M (1982) Sexual selection and the evolution of female choice. Evolution 36:1–12

Kitamura J, Nagata N, Nakajima J, Sota T (2012) Divergence of ovipositor length and egg shape in a brood parasitic bitterling fish through the use of different mussel hosts. J Evol Biol 25:566–573

Kodric-Brown A, Johnson SC (2002) Ultraviolet reflectance patterns of male guppies enhance their attractiveness to females. Anim Behav 63:391–396

Kondrashev SL, Miyazaki T, Lamash NE, Tsuchiya T (2013) Three cone opsin genes determine the properties of the visual spectra in the Japanese anchovy, *Engraulis japonicus* (Engraulidae, Teleostei). J Exp Biol 216:1041–1052

Kottelat M, Freyhof J (2007) Handbook of European freshwater fishes. Kottelat Freyhof Cornol, Switzerland and Berlin, Germany

Lisney TJ, Studd E, Hawryshyn CW (2010) Electrophysiological assessment of spectral sensitivity in adult nile tilapia *Oreochromis niloticus*: evidence for violet sensitivity. J Exp Biol 213:1453–1463

>Litjens RAJ, Quickenden TI, Freeman CG (1999) Visible and near-ultraviolet absorption spectrum of liquid water. Appl Opt 38:1216–1223

Loew ER, McFarland WN, Margulies D (2002) Developmental changes in the visual pigments of the yellowfin tuna, *Thunnus albacares*. Mar Freshw Behav Physiol 35:235–246

Losey GS, Cronin TW, Goldsmith TH, Hyde D, Marshall NJ, McFarland WN (1999) The UV visual world of fishes: a review. J Fish Biol 54:921–943

Lythgoe JN (1984) Visual pigments and environmental light. Vis Res 24:1539–1550

Maan ME, Seehausen O (2011) Ecology, sexual selection and speciation. Ecol Lett 14:591–602

Maan ME, Seehausen O, Söderberg L, Johnson L, Ripmeester EAP, Mrosso HDJ, Taylor MI, van Dooren TJ, van Alphen JJ (2004) Intraspecific sexual selection on a speciation trait, male coloration, in the Lake Victoria cichlid *Pundamilia nyererei*. Proc R Soc Lond Ser B 271:2445–2452

Maan ME, Hofker KD, van Alphen JJM, Seehausen O (2006) Sensory drive in cichlid speciation. Am Nat 167:947–954

Matsumoto T, Terai Y, Okada N, Tachida H (2014) Sensory drive speciation and patterns of variation at selectively neutral genes. Evol Ecol 28:591–609

McDonald CG, Hawryshyn CW (1995) Intraspecific variation of spectral sensitivity in threespine stickleback (*Gasterosteus aculeatus*) from different photic regimes. J Comp Physiol A 176:255–260

McLean IS, Prato L, McGovern MR, Burgasser AJ, Kirkpatrick JD, Rice EL, Kim SS (2007) The nirspec brown dwarf spectroscopic survey. II. High-resolution *j*-band spectra of m, l, and t dwarfs. Astrophys J 658:1217–1235

Miyagi R, Terai Y (2013) The diversity of male nuptial coloration leads to species diversity in Lake Victoria cichlids. Genes Genet Syst 88:145–153

Miyagi R, Terai Y, Aibara M, Sugawara T, Imai H, Tachida H, Mzighani SI, Okitsu T, Wada A, Okada N (2012) Correlation between nuptial colors and visual sensitivities tuned by opsins leads to species richness in sympatric Lake Victoria cichlid fishes. Mol Biol Evol 29:3281–3296

Morita Y, Meyer-Rochow VB, Uchidaa K (1997) Absolute and spectral sensitivities in dark- and light-adapted pagothenia borchgrevinki, an antarctic nototheniid fish. Physiol Behav 61:159–163

Morrill RJ, Thomas AW, Schiel N, Souto A, Miller CT (2013) The effect of habitat acoustics on common marmoset vocal signal transmission. Am J Primatol 75:904–916

Morrongiello JR, Bond NR, Crook DA, Wong BBM (2010) Nuptial coloration varies with ambient light environment in a freshwater fish. J Evol Biol 23:2718–2725

Palmer MS, SJ Hankison (2014) Use of ultraviolet cues in female mate preference in the sailfin molly, *Poecilia latipinna*. Acta Ethol (early online)

Pitcher TJ (1986) The behaviour of teleost fishes. Springer US, New York, USA

Reichard M, Smith C, Bryja J (2008) Seasonal change in the opportunity for sexual selection. Mol Ecol 17:642–651

Rick IP, Modarressie R, Bakker TCM (2006) UV wavelengths affect female mate choice in three-spined sticklebacks. Anim Behav 71:307–313

Seehausen O, Bouton N (1997) Microdistribution and fluctuations in niche overlap in a rocky shore cichlid community in Lake Victoria. Ecol Freshw Fish 6:161–173

Seehausen O, Terai Y, Magalhaes I, Carleton KL, Mrosso HDJ, Miyagi R, van der Sluijs I, Schneider MV, Maan ME, Tachida H, Imai H, Okada N (2008) Speciation through sensory drive in cichlid fish. Nature 455:620–627

Seeliger MW, Rilk A, Neuhauss SCF (2002) Ganzfeld erg in zebrafish larvae. Doc Ophthalmol 104:57–68

Selz OM, Pierotti MER, Maan ME, Schmid C, Seehausen O (2014) Female preference for male color is necessary and sufficient for assortative mating in 2 cichlid sister species. Behav Ecol 25:612–626

Shao YT, Wang F-Y, Fu W-C, Yan HY, Anraku K, Chen I-S, Borg B (2014) Androgens increase *lws* opsin expression and red sensitivity in male three-spined sticklebacks. PLos One 9, e100330

Shin HS, Choi CY (2014) The stimulatory effect of led light spectra on genes related to photoreceptors and skin pigmentation in goldfish (*Carassius auratus*). Fish Physiol Biochem 40:1229–1238

Smith EJ, Partridge JC, Parsons KN, White EM, Cuthill IC, Bennett ATD, Church SC (2002) Ultraviolet vision and mate choice in the guppy (*Poecilia reticulata*). Behav Ecol 13:11–19

Smith C, Phillips A, Polačik M, Reichard M (2014) Male coloration signals direct benefits in the European bitterling (*Rhodeus amarus*). Environ Biol Fish 97:335–341

Sugawara T, Terai Y, Imai H, Turner GF, Koblmüller S, Sturmbauer C, Shichida Y, Okada N (2005) Parallelism of amino acid changes at the RH1 affecting spectral sensitivity among deep-water cichlids from Lakes Tanganyika and Malawi. Proc Natl Acad Sci U S A 102:5448–5453

Terai Y, Seehausen O, Sasaki T, Takahashi K, Mizoiri S, Sugawara T, Sato T, Watanabe M, Konijnendijk N, Mrosso HDJ, Tachida H, Imai H, Shichida Y, Okada N (2006) Divergent selection on opsins drives incipient speciation in Lake Victoria cichlids. PLos Biol 4, e433

Tezuka A, Kasagi S, Oosterhout CV, McMullan M, Iwasaki WM, Kasai D, Yamamichi M, Innan H, Kawamura S, Kawata M (2014) Divergent selection for opsin gene variation in guppy (*Poecilia reticulata*) populations of Trinidad and Tobago. Heredity 113:381–389

Valen R, Edvardsen RB, Søviknes AM, Drivenes Ø, Helvik JV (2013) Molecular evidence that only two opsin subfamilies, the blue light- (SWS2) and green light-sensitive (RH2), drive color vision in Atlantic cod (*Gadus morhua*). PLos One 9, e115436

Wang FY, Yan HY, Chen JS-C, Wang TY, Wang D (2009) Adaptation of visual spectra and opsin genes in seabreams. Vis Res 49:1860–1868

Wang FY, Tang MY, Yan HY (2011) A comparative study on the visual adaptations of four species of moray eel. Vis Res 51:1099–1108

Witzany G (2014) Biocommunication of animals. Springer, Dordrecht, Netherlands

8

Long-term monitoring of Gastropoda (Mollusca) fauna in planted mangroves in central Vietnam

Sofya Zvonareva[1*], Yuri Kantor[1], Xinzheng Li[2] and Temir Britayev[1]

Abstract

Background: The aim of the research was to study the gastropod assemblage associated with a planted mangrove consisting of *Rhizophora apiculata* in Dam Bay (Tre Island, Nha Trang Bay, central Vietnam). Long-term changes in species composition, density, and biomass of gastropod assemblages in a planted mangrove were compared to those associated with natural mangroves.

Results: The gastropod assemblages in one planted and two natural mangrove associations were studied. Long-term monitoring (2005 to 2013) of assemblages associated with planted mangroves was also carried out. In total, 53 gastropod species in 21 families were recorded. This is comparable or even higher than in other areas of South-East Asia. In planted mangroves, *R. apiculata* trees were planted in mid (in 2004) and lower (2007) intertidal zones. Composition of gastropod fauna during the early period of observations (2005 to 2006) differed significantly in species richness, biomass, and abundance from the later period (2008 to 2013). Initially, the values of these parameters were low in the upper, and mid-intertidal zones and negligible in low intertidal zone. In 2008 to 2013, the values increased in the upper and mid-intertidal zones. *R. apiculata* planting in mid-intertidal zone and changes in its gastropod assemblage strongly affected those of upper intertidal zone as well. Comparison of gastropod fauna in planted mangroves and natural associations revealed significant differences. In planted mangroves, gastropods were mostly represented by eurybiotic species, also found in other not mangrove intertidal habitats, while in natural association many species were predominantly mangrove inhabitants, e.g., Potamididae, Littorinidae, and Ellobiidae. The fact that mangrove plantations in Dam Bay are highly dominated by opportunistic species, together with a paucity of predominantly mangrove-associated gastropods, suggest that the whole ecosystem is not balanced yet.

Conclusions: The gastropod assemblage found in a 9-year old mangrove plantation differed from natural mangrove associations in that the former is dominated by opportunistic eurybiotic gastropod species. This could mean that the ecosystem of planted mangroves is unbalanced and is still in a transitional state.

Keywords: Biodiversity; Biomass; Abundance; Faunal dynamics; Opportunistic species

Background

Mangroves are among the world's most productive ecosystems (Berger et al. 2008). A diverse invertebrate assemblage is associated with the mangal, with molluscs being one of the dominant components in terms of species richness, biomass, and abundance (Cantera et al. 1983, Nagelkerken et al. 2008, Sasekumar 1974). Species composition, diversity, biomass, and abundance of invertebrates change under disturbance, exploitation, or rehabilitation of mangrove association (Skilleter 1996, Skilleter and Warren 2000, Bosire et al. 2004). Therefore, such parameters may reflect the status of mangrove ecosystem and can be used as an indicator of changes in both natural and planted mangroves. Comparison of invertebrate fauna of different types of mangrove stands revealed that mangrove vegetation diversity is positively correlated with associated fauna diversity, and in mature mangrove stands indexes of invertebrate fauna diversity are higher, whereas abundance and biomass are lower compared to newly planted stands (Macintosh et al. 2002).

The mangal is characterized by a high level of heterogeneity. Several studies were dedicated to horizontal distribution of invertebrate fauna in general and of particular

* Correspondence: sonyazv@mail.ru
[1]Laboratory of Ecology and Morphology of Marine Invertebrates, A.N. Severtsov Institute of Ecology and Evolution, 33 Leninskij prosp, Moscow 119071, Russia
Full list of author information is available at the end of the article

groups in mangroves. General faunal zonation in the mangrove ecosystem was provided by Berry (1963), who distinguished a bivalve zone, a *Nerita* zone, an *Uca* zone, a *Littoraria* (as *Littorina*) zone, and burrower zone. Studies of distribution of macro- and mesofauna in tidal flats of northeast Australia revealed that the highest abundance and species diversity was in the mid-intertidal zone and lowest in the lower intertidal zone (Dittmann 2000). On the contrary, the abundance of polychaetes in the mangroves of northern Australia was highest in the lower intertidal zone and lowest in upper intertidal zone (Metcalfe and Glasby 2008).

Species composition also depends on the position of mangroves in relation to the sea shore, e.g., in the landward zone of mangrove vegetation - the most common gastropods belonged to the families Ellobiidae, Assimineidae, and Neritidae, while in the seaward fringe of mangroves - Potamididae (Macnae 1968).

Molluscs and crustaceans play important roles in the mangrove ecosystem, processing mangrove-derived and algal detritus through their feeding and bioturbation activities (Bouillon et al. 2002; Kristensen et al. 2008; Lee 2008).

Gastropod molluscs in mangrove habitats show different levels of fidelity to this ecosystem: some species are predominantly recorded in mangrove habitats (although can be occasionally found in other environments), while others dwell in mangroves, as well as on intertidal and subtidal areas and mudflats (Lozouet and Plaziat 2008; Reid et al. 2008).

Despite the great importance of molluscs in the functioning of mangrove ecosystems, their diversity, biomass, and abundance in mangroves are in need of additional studies (Printrakoon et al. 2008).

Mangroves in Vietnam for centuries played a significant socioeconomic role in the lives of coastal inhabitants. They were employed as source of building material, fuel, food, and other useful products. Herbicide usage during 1962 to 1971 destroyed 104,939 hectares or 36% of the mangrove forests of South Vietnam (Hong and San 1993). Population and economic growth after the war led to even more rapid reduction of mangrove areas. Shrimp and fish ponds, agriculture, charcoal production, and coastal water pollution caused degradation of mangrove associations and their conversion into wastelands (Hong and San 1993). In order to improve this situation, the Vietnamese government created a number of national parks and established protected areas with the aim to conserve and rehabilitate mangrove thickets, including reforestation and afforestation programs (Binh et al. 2005).

There are numerous publications on mangroves of Vietnam mainly related to socioeconomic aspects of the mangal ecosystem (Binh et al. 1997, De Graaf and Xuan 1998, Alongi et al. 2000, Johnston et al. 2000, Thu and Populus 2007, Christensen et al. 2008, Amano et al. 2009). At the same time, the diversity and quantitative characteristics of macrobenthic fauna remain insufficiently studied (Le Vay et al. 2001, Tue et al. 2012). Only two species lists of gastropods in mangroves of Vietnam were found in the published literature: 15 species were reported in Can Gio District (Hong 2004) and 70 species in mangroves along the north coast of Vietnam (Hong and San 1993).

Geographically, the mentioned publications are dedicated to northern and southern Vietnam, while the mangrove associations of central Vietnam remain poorly examined. The peculiarity of mangroves in central Vietnam is that due to the high wave activity that affects the coast, mangroves are located only in well-protected lagoons and estuaries (Hong and San 1993). Often, they do not form a mangrove forest but are represented by a narrow belt of mangrove shrubs along the water's edge or fresh water channels.

In the vicinities of Nha Trang City in the Nha Trang Bay (central Vietnam), the program of mangrove vegetation planting started a decade ago. One of the areas was the Dam Bay on the Tre Island, where planting started in 2004. This provided a unique opportunity to examine the formation and dynamics of benthic community associated with the newly planted mangrove plants, and we launched the program of mangrove macrozoobenthos monitoring of the plantation, started in 2005 and continuing until the present. The aims of this present work were to study the species composition of gastropod assemblages associated with planted mangroves in Dam Bay, describe long-term changes (2005 to 2013) in the species composition, determine the density and biomass of gastropod assemblages in planted mangroves, and compare gastropod assemblages associated with planted mangroves with that of adjacent intertidal areas and natural mangroves.

Methods

Study area

Nha Trang (NT) Bay and Nha Phu (NP) Bay are both located in Khanh Hoa Province in the southern part of central Vietnam (Figure 1a). The areas around the two bays are densely populated, with about 400,000 inhabitants in Nha Trang city and several smaller settlements. Correspondingly, the coastline and aquatic areas of the bays are under strong anthropogenic impact in the form of wastewater discharge, pollutants, fishery, intensive tourist activities, and shipping.

Local climate is influenced by two monsoons determining wet (September to December) and dry seasons (January to August). The tidal regime of NT Bay is irregular diurnal with a spring tide/neap tide cycle of 14 days and a range of 1.5 to 2.0 m. Monthly, there are around 18 to 22 days of diurnal tide and the rest are of mix-semidiurnal. During neap tide days, there is daily additional small high

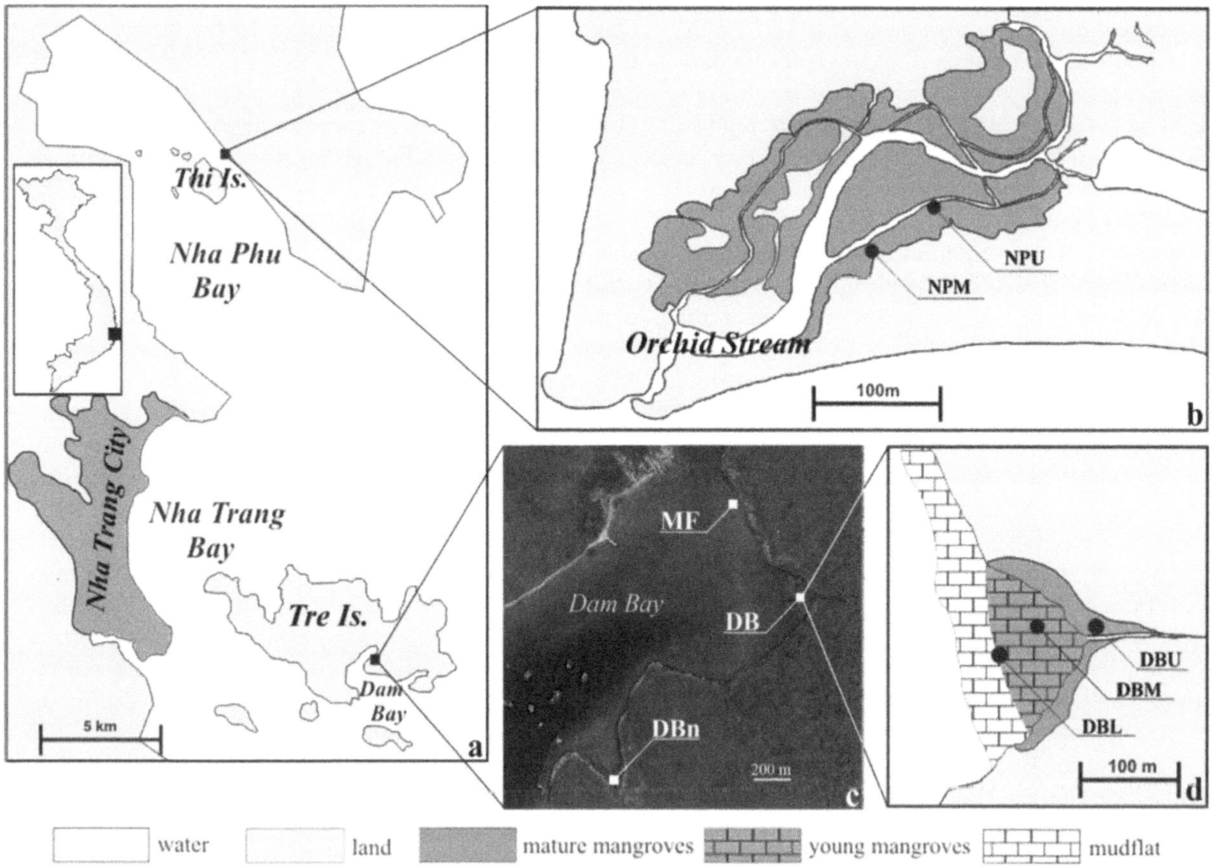

Figure 1 Location of studying sites. **(a)** Map of Nha Trang and Nha Phu Bays; **(b)** location of studied stations in mangrove association in Nha Phu Bay; **(c)** location of studied sites in Dam Bay; **(d)** location of studied stations in mangrove plantation. Abbreviations: NPU, upper station in natural mangrove association in Nha Phu; NPM, lower station in natural mangrove association in Nha Phu; DB, mangrove plantation; DBn, natural mangrove association; MF, unvegetated mudflat; DBU, upper intertidal zone in mangrove plantation in Dam Bay; DBM, mid-intertidal zone in mangrove plantation in Dam Bay; DBL, lower intertidal zone in mangrove plantation in Dam Bay.

tide. During neap tide, the amplitude is about 0.5 m (Lan et al. 2012).

Natural mangrove stands in NT, and NP Bays are scarce and degraded due to anthropogenic impact. They still persist around shrimp and fish ponds in the mouth of the river Be within the Nha Trang city borders and in the inner part of NP Bay. The largest native mangrove association without shrimp and fish ponds is located on the north-eastern coast of NP Bay (Figure 1a).

Natural mangrove vegetation is arranged in narrow belts along the shores of small islets in NP Bay and in the inner part of Dam Bay which are located on the south side of the Tre Island (Figure 1a). These associations experience negligible freshwater intrusions in wet season. Two separated stands of artificial mangrove plantations are present in the inner part of Dam Bay (Figure 1c).

Several small settlements are located on the coast of Dam Bay, as well as numerous fish-farming floats. Despite the fact that Dam Bay area is under program of protection of marine and coastal resources, local people are fishing and collecting crabs and bivalves.

Four studying sites in different types of mangrove vegetation were selected, three in Dam Bay and one in NP Bay. The first, main study site was the mangrove plantation (below referred to as DB site, see Figure 1c,d) of *Rhizophora apiculata* Bl. located in the inner east coast of Dam Bay on silty-sand tidal flat with a width of 200 to 250 m near the belt of mature *Rhizophora* trees mixed with sparse *Avicennia marina* (Forssk.) Vierh. trees. The plantation occupies an area of approximately 1.4 ha. Temporary freshwater streams enter the inland part of plantation for a short time during the wet season, and salinity in the Bay does not deviate significantly from 33‰. The sampling stations were arranged in each intertidal zone in relation to the zones of trees of different age (Figure 1d). At each station, several standard frames (three to eight) were taken (each individual one is referred below as sample). The uppermost station (DBU) was located in the upper intertidal zone occupied by mature trees (central point of this zone is at 12°12′7″N; 109°18′6″E). The second station (DBM) was in the mid-intertidal zone occupied by *R. apiculata* trees planted at a distance of about 1 m in 2004 on sandy-silt mudflat (12°12′5″N; 109°18′12″E). The third station (DBL) was in the lower intertidal zone occupied by *R. apiculata* planted in 2007 (12°12′5″N; 109°18′9″E). Samples were taken in 2009, 2010, and 2011 in the most seaward part of the lower intertidal zone that dries out only during low spring tide where seedlings were planted in 2009. At present, the planted trees are mostly dead in that zone.

The second study site (with a single station DBn, Figure 1c) was located in a small cove with rocky shore and muddy bottom, close to *Rhizophora* plantation. There, mature *R. apiculata* trees form a narrow belt along the water edge. It is wider in the inner part of the basin with finer sediments, where there is a small stream (12°11′50″N; 109°17′57″E). This small natural association corresponds to the mature trees' belt in upper intertidal zone in plantation area. We investigated this site as a control (below referred to as DBn site).

The third study site is located on a non-vegetated mudflat in proximity of mangrove plantation (12°12′14″ N; 109°18′6″E) and was selected as a control (below referred to as MF site, see Figure 1c). We use it as a control because it is similar in terms of sediment and relief as to the area where mangrove seedlings were planted, therefore its fauna should be identical to the fauna of our DBM and DBL stations before mangroves were planted.

The study area in NP Bay (12°24′22″N; 109°14′24″E) was located in the central part of the north-east coast of the Bay opposite to Thi island along Orchid Stream channels. Although this area is subjected to strong anthropogenic impact, it is still the least disturbed mangrove association in the vicinity of Nha Trang City. Recreational zones are located here, and runoff in fresh water channels is produced by eutrophic ponds. This site (below referred to as NP site, see Figure 1b) was examined in order to compare it with the mangrove plantation and natural mangrove association in Dam Bay. This mangrove association is mainly formed by *Rhizophora stylosa* mixed with *Avicennia* sp., *Lumnitzera racemosa* Willd., *Excoecaria agallocha* L., *Aegiceras* sp., and *Bruguiera gymnorhiza* (L.). Samples were taken at two stations, both located in the secondary channel of the stream. The first station was located near the place where this small stream falls into the main stream, and the second station was located little upstream, and they both have full salinity during high tide. Since the structure of this mangrove association is completely different from Dam Bay, it was impossible to associate these two stations with intertidal zones of mangrove sites in Dam Bay, so we analyzed them together, as a single station.

Data collection

Quantitative samples were taken during spring (March to May) and autumn (September to October) once or twice a year, except 2007, when no samples were taken (Table 1). Samples were taken from the ground surface and did not include the molluscs, which live exclusively on tree stems and leaves. This protocol was designed, since otherwise it was impossible to obtain any comparable data.

Sampling frame (in most cases 0.1 m^2) was placed randomly on each studying site during the low tide. Number and size of frames varied from year to year (Table 1). Also, samples were collected non-quantitatively by hand from roots, stems, and branches of mangrove vegetation, sediment surface, rocks, and dead wood when present. Inside the frame, the sediment was withdrawn to the depth of 15 to 20 cm. It was placed in a bucket, filled with water, and stirred up with several water changes. Water was poured through the mesh with openings each diameter 0.5 mm for separating light fraction - detritus, plant remains, and animals, which mostly comprised polychaetes and other worms as well as some crustaceans. The light fraction was collected from the mesh, while the remaining sediment was washed through the sieve with 3-mm mesh. The residues were sorted manually in the laboratory. Living molluscs and other animals were preserved in 70% alcohol. The number of individuals of each species in quantitative samples was counted; excess liquid was removed and wet weight of each species (with the shells in the case of molluscs and brachiopods) was measured.

Table 1 Number and size (m^2) of quantitative and non-quantitative samples by years and site

Site and year of sampling	Samples in U	Samples in M	Samples in L
DB, 2005	3 × 0.1 + n. s.	3 × 0.1 + n. s.	3 × 0.1 + n. s.
DB, 2006	5 × 0.0225	5 × 0.0225	5 × 0.0225
DB, 2008	3 × 0.1 + n. s.	2 × 0.1; 3 × 0.025 + n. s.	3 × 0.025 + n. s.
DB, 2009	7 × 0.1 + n. s.	7 × 0.1 + n. s.	7 × 0.1 + n. s.
DB, 2010	8 × 0.1 + n. s.	8 × 0.1 + n. s.	7 × 0.1 + n. s.
DB, 2011	8 × 0.1 + n. s.	8 × 0.1 + n. s.	8 × 0.1 + n. s.
DB, 2012	7 × 0.1 + n. s.	8 × 0.1 + n. s.	4 × 0.1 + n. s.
DB, 2013	8 × 0.1 + n. s.	7 × 0.1 + n. s.	4 × 0.1 + n. s.
NP, 2012	2 × 0.1 + n. s.	2 × 0.1 + n. s.	-
NP, 2013	n. s.	n. s.	-
MF, 2012	-	2 × 0.1 + n. s.	-
MF, 2013	-	n. s.	-
DBn, 2013	7 × 0.1 + n. s.	-	-

DB, samples in Dam Bay (planted area); MF, mudflat in Dam Bay; NF, samples in Nha Phu; L, lower intertidal zone; M, mid-intertidal zone; U, upper intertidal zone, n. s., non-quantitative samples.

Since the study sites differed significantly in terms of length of observations and collecting efforts, the direct comparison may be misleading. Therefore, we made a comparison of data, obtained within the single year of observations in planted mangrove association (2013) with those from two natural associations in Dam Bay (2013) and NP (2012 to 2013 combined).

Data analysis

The gastropod species were identified to species level in most cases, and information on their ecology and habitats was obtained from Reid (1986), Poppe (2008a,b, 2010), Ng and Sivasothi (1999a,b), Martins (1996), Houbrick (1985, 1992), Hylleberg and Kilburn (2003), Lozouet and Plaziat (2008), and Okutani (2000). For convenience of analysis, gastropods were conditionally divided in two major ecological groups basing on published data (Houbrick 1985; Reid 1986; Houbrick 1992; Ellison et al. 1999; Okutani 2000; Lozouet and Plaziat 2008; Reid et al. 2008): predominantly mangrove-associated and eurybiotic species. Species that are normally found in mangrove ecosystem, although can rarely be recorded in other environments, are attributed to the first group. These are represented in our material by Littorinidae, Potamididae, and Ellobiidae. Second group includes species, found by us in mangrove associations but according to published data are also common in different intertidal and subtidal habitats - mudflats, rocky shores, and dead wood.

Biomass and abundance were calculated for quantitative samples; average biomass and abundance (with standard deviation) was determined for all samples on the sampling station in a year.

Species richness is employed here as the total number of species on a station.

Average species richness per sample - average number of species, collected in each individual sample on a given station, was also calculated.

Results

Mangrove vegetation growth

Seedlings planted in mid-intertidal zone in 2004 by the autumn of 2005 reached about 0.5 m in height and 1.5 m by the autumn of 2008, producing three to five branches and aerial roots. By the autumn of 2013, trees reached height of up to 3 m. In some places, plantings of year 2004 became almost impassible due to aerial roots. Changes in vegetation are shown in Figure 2. Seedlings planted in lower intertidal zone in 2007 have not grown sufficiently during 6 years; in autumn of 2013, they were about 1 m high in the lowest part of lower intertidal zone, but they were mostly dead by 2013, although some of them still have green leaves.

Gastropod assemblage composition in the Nha Trang Bay

Fifty-three gastropod species belonging to 21 families were found in the bay altogether (most common and characteristic species are represented in Figures 3 and 4). Most of them belong to eurybiotic species, while 13 species can be considered as predominantly mangrove associated (Table 2).

In planted mangrove association in Dam Bay, 30 species belonging to 14 families were found during entire period of observations, while 15 species only were found in 2013. The most diverse families were Neritidae (five species), Cerithiidae (four species), and Littorinidae (four species). The eurybiotic species were dominant (see Table 2), while only five species, *Littoraria* spp. were predominantly mangrove inhabitants (Figure 5). The most abundant were *Clithon oualaniensis*, *Cerithideopsilla cingulata*, *Clypeomorus*

Figure 2 Mangrove vegetation growth in mid-intertidal zone of mangrove plantation in Dam Bay, Nha Trang, central Vietnam. In 2005 (left), the height of seedlings was about 0.5 m and in 2013 (right), the height of young trees was about 3 m.

Figure 3 Shells of common eurybiotic (**3** to **7**, **9**) and predominantly mangrove-associated species (**1** to **2**, **8**, **10**). **1** *Littoraria lutea*, **2** *Littoraria intermedia*, **3** *Batillaria zonalis*, **4** *Clypeomorus bifasciata*, **5** *Cerithium coralium*, **6** *Clithon oualaniense*, **7** *Clithon faba*, **8** *Cassidula nucleus*, **9** *Cerithideopsilla cingulata*, **10** *Cerithidea quoyii*.

bifasciata, and *Batillaria zonalis*. Both groups - inhabitants of intertidal soft bottoms (such as *C. oualaniensis*, *B. zonalis*, and *C. cingulata*), and species typical for intertidal rocky bottoms (e.g., *Nerita* spp., *Drupella margariticola*, *Semiricinula turbinoides*) were found.

No obvious differences were observed in species composition or species richness between upper and mid-intertidal zones in DB (21 and 20 species, respectively). The same situation was observed in 2013 (10 and 11 species). The lower intertidal zone differs significantly both in terms of species composition and species richness, where only eight species were found here (two - in 2013), mostly once during entire observation period. Among them, two species were common with upper and mid-intertidal zones, namely *Batillaria zonalis* and *Clypeomorus bifasciata*. At the station nearest to the sea where samples were taken in 2009, 2010, and 2011, no gastropod species were found at all.

In 2013, the difference between upper and mid-intertidal zones was not large in terms of average species richness (six species per sample in both intertidal zones) and abundance (1,429.5 m^{-2} and 1,440.3 m^{-2}, respectively), while biomass was higher in mid-intertidal zone (170 g/m^2 and 210 g/m^2, respectively).

On adjacent unvegetated mudflats, only two species, namely *Milda* sp. and *Monetaria moneta*, were found. The latter species was not found in mangrove associations. In the natural *Rhizophora* association in Dam Bay, 19 species were collected and Cerithiidae (five species) was the most diverse family. As well as in planted mangroves, most of species were eurybiotic ones. The most abundant species were *Batillaria zonalis* and *Clypeomorus bifasciata* which are common on intertidal soft bottoms. *Clypeomorus pellucida*, *Drupella margariticola*, and *Pictocolumbella ocellata* were common on intertidal hard substrata. Only *Littoraria scabra* and *L. intermedia*, both of which are predominantly mangrove-associated species, were found here. The average species richness, abundance, and biomass were substantially lower (1.9 species per sample, 192.2 m^{-2} and 121.6 g/m^2, respectively) than in the plantation.

In NP, 28 species belonging to 11 families were found; the most diverse family was Potamididae (six species). The most abundant species were *Clithon oualaniense* and *Cerithideopsilla cingulata*. Among species found were species typical to intertidal soft bottoms, such as *C. oualaniense*, *Clithon faba*, *Clypeomorus bifasciata*, and *C. cingulata* as well as species typical for intertidal rocky bottom e.g., *Nerita auriculata*, *Nerita cornucopia*, and

Figure 4 Some species of mangal gastropods shown alive. **(a)** *Cassidula nucleus* on *Rhizophora* leave in natural mangrove association in Nha Phu, **(b)** *Terebralia sulcata* on leaf litter near the roots of *Rhizophora* in natural mangrove association in Nha Phu, **(c)** *Littoraria pallescens* on the leaf of *Rhizophora* planted in 2004 in Dam Bay, **(d)** aggregation of *Batillaria zonalis* and *Clithon oualaniensis* on the ground between *Rhizophora* trees planted in 2004 in Dam Bay, **(e)** *Cerithidea quoyii* on the root in natural mangrove association in Nha Phu.

Clypeomorus pellucida. NP mangroves have higher proportion of predominantly mangrove associated species, e.g., *Littoraria* spp., *Cerithideopsilla microptera*, *Cerithidea quoyii*, and *Cassidula nucleus*. Species abundance (1,715.9 m^{-2}) was higher than in both associations in Dam Bay, while the average species richness and biomass were lower than in plantation and higher than in the natural mangrove association in Dam Bay (2.8 species per sample and 148.8 g/m^2).

Long-term changes in gastropod assemblage in Dam Bay mangrove plantation

In the mangrove plantation in Dam Bay, the total number of species varied significantly from year to year due to the occurrence of rare species, which were not recollected every year. The lowest number was in 2006 (three species) and the highest in 2011 (19 species). Predominantly mangrove-associated species first appeared in 2008 and their number grew every year (Figure 5), but even in 2013, species diversity (five species) was much lower than in the natural NP mangrove association (12 species). The number of eurybiotic gastropods changed from year to year, with highest number (15 species) recorded in 2011. This is slightly lower, than in DBn, but almost the same as in NP (Figure 5).

Over the years of monitoring, the steady growth of average species richness per sample was observed (except 2012) with the lowest in 2005 (0.7 species per sample) and the highest in 2013 (6 species per sample). There was little difference between upper and mid-intertidal zones (Figure 6).

In the upper intertidal zone of the mangrove plantation in Dam Bay species richness, biomass, and abundance of gastropods were very low in the first 2 years of monitoring, 2005 and 2006 (two species; 26.3 ± 45.5 g/m^2; 87.9 ± 152.2 m^{-2} in 2005). Then, all parameters increased substantially, the species richness and biomass reached their maxima in 2011 (11 species; 222.4 ± 123.5 g/m^2 respectively) and abundance was highest in 2013 (1429.9 ± 908.7 m^{-2}) (Table 3, Figure 7). In that zone, *Clithon oualaniensis* was dominant in terms of biomass and abundance in 2005 to 2006, starting 2010 the biomass and abundance of *Cerithideopsilla cingulata* was highest, and in latest years (2010 to 2013) *Batillaria zonalis* became the dominant species.

In the mid-intertidal zone of mangrove plantations, species richness, biomass, and abundance of gastropods were very low in 2005 (one species; 3 ± 2.7 g/m^2; 18.2 ± 24.1 m^{-2}) (Table 4). Then, these parameters increased even more dramatically than in the upper intertidal zone and reached the highest values registered in Dam Bay in 2011 (12 species; 315.6 ± 216.6 g/m^2, and 1,523.9 ± 1313.8 m^{-2}, respectively). In 2012, a considerable decrease in abundance and biomass was observed, but in 2013, preceding values were almost reached for abundance but not for biomass (Table 3, Figure 7). High biomass was mainly due to *Batillaria zonalis* and *Cerithideopsilla cingulata*, while density was contributed mostly by *B. zonalis* and by *Clithon oualaniensis*.

In the lower intertidal zone, few molluscs were found both in 2010 (one species; 1.8 ± 4.8 g/m^2; 1.8 ± 6.8 m^{-2}) and 2011 (three species; 0.3 ± 0.5 g/m^2, 4.5 ± 9.1 m^{-2}) in

Table 2 A list of found gastropod molluscs

	DB, 2005	DB, 2006	DB, 2008	DB, 2009	DB, 2010	DB, 2011	DB, 2012	DB, 2013	DBn	NP
FAMILY: Lottiidae										
Lottiidae gen. sp.				+	+	+	+	+		
FAMILY: Trochidae										
Monodonta sp.				+						
FAMILY: Neritidae										
Clithon oualaniense (Lesson, 1831)	+++	+++	++	+++	+++	+++	++	+++		+++
Clithon faba (Sowerby I, 1836)	++				+	+	++	+++		++
Neritina cornucopia (Benson, 1836)										+
Neritina auriculata Lamarck, 1816									+	++
Neritina sp.				+						+
Nerita albicilla Linnaeus, 1758									+	
Nerita undata Linnaeus, 1758						+				
Nerita squamulata Le Guillou, 1841	+			+		+	+		+	
FAMILY: Potamididae										
Cerithideopsilla cingulata (Gmelin, 1791)		+	+++	++	+++	+++	+++	+++		+++
Cerithideopsilla djadjariensis (K. Martin, 1899)[a]										+
Cerithideopsilla microptera (Kiener, 1842)[a]										+
Cerithidea quoyii (Hombron et Jacquinot, 1848)[a]										++
Terebralia sulcata (Born, 1774)[a]										++
Terebralia palustris (Linnaeus, 1767)[a]										++
FAMILY: Cerithiidae										
Cerithium coralium Kiener, 1841				+++	+	+	++	++	+	++
Cerithium traillii (G.B. Sowerby II, 1855)									+	
Clypeomorus bifasciata (G.B. Sowerby II, 1855)			+++	+++	+++	+++	+	++	++	++
Clypeomorus pellucida (Hombron et Jacquinot, 1852)			+	+	++	++	++	++	++	++
Clypeomorus batillariaeformis Habe et Kosuge, 1966									+	
Rhinoclavis vertagus (Linnaeus, 1767)	+					+				
FAMILY: Planaxidae										
Planaxis sulcatus (Born, 1778)									+	
FAMILY: Batillariidae										
Batillaria zonalis (Bruguière, 1792)			+++	+++	+++	+++	+++	+++	+++	+
FAMILY: Thiaridae										
Sermyla riqueti (Grateloup, 1840)										+
FAMILY: Littorinidae										
Littoraria ardouiniana (Heude, 1885)[a]								+		
Littoraria carinifera (Menke, 1830)[a]										+
Littoraria intermedia (Philippi, 1846)[a]									+	+
Littoraria lutea (Philippi, 1847)[a]			++	++		++	+++	+++		++
Littoraria pallescens (Philippi, 1846)[a]						+	+++	++		++
Littoraria scabra (Linnaeus, 1758)[a]					+	+	+	++	+	
Littoraria articulata (Philippi, 1846)[a]										+

Table 2 A list of found gastropod molluscs *(Continued)*

FAMILY: Assimineidae										
Assimineidae gen. sp.			+			+	+			++
Metassiminea philippinica (O. Böettger, 1887)										+
FAMILY: Stenothyridae										
Stenothyra sp.										+
FAMILY: Cypraeidae										
Monetaria annulus (Linnaeus, 1758)	+									
FAMILY: Naticidae										
Notocochlis gualteriana (Recluz, 1844)		+			+	+		+	+	
Naticidae gen. sp.				+	+					+
FAMILY: Ranellidae										
Monoplex pilearis (Linnaeus, 1758)							+			
Gutturnium muricinum (Röding, 1798)								+		
FAMILY: Epitoniidae										
Epitonium sp.			+							
FAMILY: Muricidae										
Drupella margariticola (Broderip, in Broderip et Sowerby, 1833)								+	++	
Tenguella musiva (Kiener, 1835)									+	
Semiricinula turbinoides (Blainville, 1832)				+	+	+			+	
FAMILY: Columbellidae										
Pictocolumbella ocellata (Link, 1807)									++	
FAMILY: Nassariidae										
Nassarius olivaceus (Bruguière, 1789)					+					
FAMILY: Pyramidellidae										
Milda sp.						+			+	
Otopleura auriscati (Holten, 1802)									+	
Pyramidellidae gen. sp.				+		+				
FAMILY: Onchidiidae										
Onchidiidae gen. sp.	+		+							++
FAMILY: Ellobiidae										
Cassidula nucleus (Gmelin, 1791)[a]										++
Laemodonta siamensis (Morelet, 1875)										+
Total:	6	3	9	14	13	19	14	15	19	28

A list of alive gastropod molluscs found in each year in monitored mangrove plantation (DB), natural mangrove association in Dam Bay (DBn) and Nha Phu (NP) with comparative abundance: + singletons and rare species (one to three individuals were found in sampling year); ++ common species (from 3 to 20 individuals were found in sampling year); +++ species with high abundance (more than 20 individuals were found in sampling year).
[a]Predominantly mangrove-associated species.

quantitative samples. No molluscs were found in the most seaward part of the lower intertidal zone.

Discussion

Vegetation growth

According to Hong and San (1993), the growth rate of *Rhizophora apiculata* in Vietnam is highest between 10 and 15 years. Growth performance depends on a number of factors such as soil texture, tide amplitude, salinity and tree density. Growth was higher in areas with salinities 10‰ to 20‰, and the growth rate declines in sites with salinities 25‰ to 35‰. It was reported that in Ca Mau (southern Vietnam), *R. apiculata* younger than 10 years were replanted in the intertidal zones with silt-clay soil; the annual diameter and height increments were 0.74 cm and 0.81 m, respectively. We did not measure the annual growth of planted seedlings but it is clear that the growth rate of *R. apiculata* planted in 2004 in

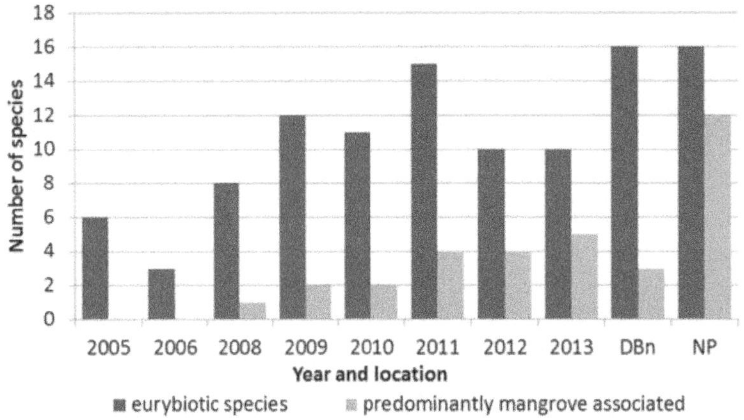

Figure 5 Proportion of predominantly mangrove-associated and eurybiotic gastropod species over the years (2005 to 2013) in mangrove plantations in Dam Bay, Nha Trang, central Vietnam, and in natural mangrove associations in Dam Bay (2013) and Nha Phu (2012 and 2013 combined).

Dam Bay was lower, since they reached only 1.5 m in 4 years and about 3 m in 10 years.

Growth rate of *Rhizophora stylosa* and *Kandelia candel* planted on sandy-clay mudflats in Ha Thin Province (central Vietnam) was low (only 0.40 to 0.87 m in 3-year period) due to the unfavorable soil texture and climatic conditions (Hong and San 1993). So the low growth rate of seedlings planted in 2004 in Dam Bay may be caused by the same reasons as in Ha Thin.

Seedlings planted in 2007 in lower intertidal zone in Dam Bay have not grown sufficiently during 6 years. The reasons of their poor condition may be due to the almost permanent inundation by sea water, so this site is not suitable for planting this mangrove species.

Gastropod species composition

Fifty-two gastropod species were found in three mangrove associations. It is difficult to assess species richness of studied associations due to the lack of information about faunal richness in mangroves of central Vietnam. For other parts of Vietnam, the list is available only for a natural mangrove association in northern Vietnam, where 70 species were recorded (Hong and San 1993). Analysis of this list revealed that several included species that were collected in the lower intertidal zone, e.g., *Conus striatus* and *C. coronatus*, both predatory species in the subtidal zone, Turbinidae, that are typical to coral reefs, one species of Viviparidae, that inhabit only fresh water, and some trochids such as *Angaria delphinus* (=*laciniata*), which normally inhabit subtidal habitats, may not have been collected alive and that these empty shells were brought to the mangroves by hermit crabs or currents. If we exclude from this list the species that are unlikely to belong to the mangrove ecosystem, we have nearly the same species number (53 species) as in our study. Therefore, the observed diversity of molluscs in mangroves of central Vietnam appears similar to that of northern Vietnam.

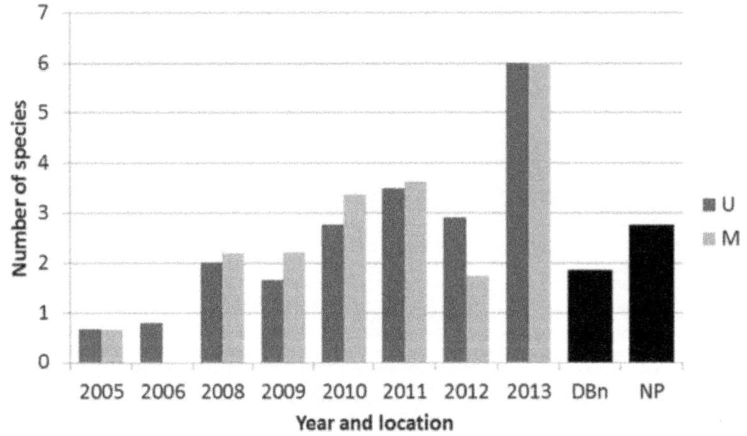

Figure 6 Average species richness per sample in every year in upper (U) and mid-intertidal zones (M) in mangrove plantation in Dam Bay, Nha Trang, central Vietnam, and two natural mangrove associations in Dam Bay (DBn) and Nha Phu (NP) which are in black.

Table 3 Average abundance and biomass with standard deviation of the most common gastropod species in upper intertidal zone

	Number of samples and the year of collection								
	3	5	3	7	8	8	7	8	7
	2005	**2006**	**2008**	**2009**	**2010**	**2011**	**2012**	**2013**	**DBn**
Clithon oualaniense	69.7±120.7 14.3±24.8	346.7±530 30.6±48.8	39.4±53.3 2.5±4.3	74.3±146 4.1±6.92	17.0±32.3 1.8±3.34	46.6±73.4 46.6±73.4	13.6±21.2 0.5±0.8	226.1±242.5 14.5±14.3	
Clithon faba	18.2±31.5 12.0±20.7						1.1±3.2 0.4±1.12	28.4±38.1 2.5±3.5	
Cerithideopsilla cingulata			60.6±105 16.4±28.3	52.9±139.8 4.3±11.3	183±289.1 16.3±13.3	223.9±229.7 62.5±50.9	190.9±150.9 68.5±47.3	206.8±262 47.9±58	
Batillaria zonalis			90.9±142 45.5±66	55.71±147.4 2.7±7.3	239.8±255.3 30.1±21.2	443.2±525.6 109.5±111.1	944.3±650.6 102.1±30.9	136.4±241.3 84.7±184.4	
Cerithium coralium				127.1±183 22.8±39.9	1.1±3.2 0.5±1.5		7.9±19.1 2.6±4.3	5.7±12.8 0.5±1.15	19.5±51.5 13.9±43.5
Clypeomorus bifasciata			3.0±5.2 1.2±2.1	95.71±244.5 21.4±54.2	6.8±9.7 1.0±1.1	23.9±29.9 5.1±7.4	4.5±6.9 1.3±2.3	14.8±16.7 1.6±1.9	28.6±37.3 9.0±16.7
Total for common species	87.9±152.2 26.3±45.5	364.7±530 30.6±48.8	193.9±153.3 65.5±58.8	405.7±579 55.4±94.8	447.7±524.8 49.7±35	950±411.5 218.9±115.7	660.2±595.3 169.3±136.2	1426.1±908.9 169.2±61.5	184.4±224.2 120.7±154.5
Total for all species	87.9±152.2 26.3±45.5	364.7±527.2 34±47	193.9±153.3 65.5±58.8	408.6±583.7 55.4±94.8	452.3±495.7 52.9±39.1	955.7±421.5 222.4±123.5	660.2±595.3 169.3±136.2	1429.5±908.7 170±57.5	192.2±224.1 121.6±154.5

Average abundance (m^{-2}, above line) and biomass (g/m^2, below line) with standard deviation of the most common gastropod species (found in half or more sampling seasons) in quantitative samples and total abundance and biomass of gastropods in upper intertidal zone in each year of monitoring in mangrove plantations in Dam Bay and natural mangrove association in Dam Bay (DBn).

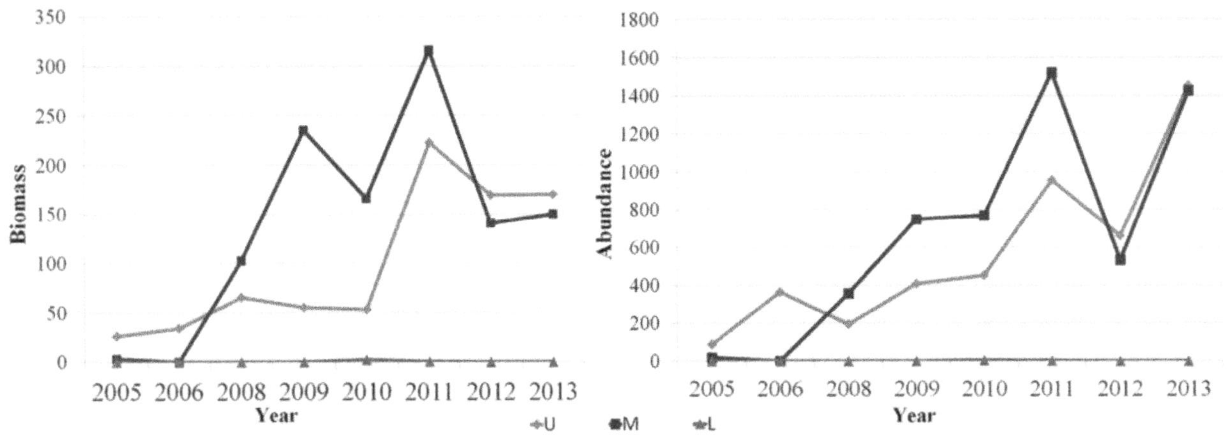

Figure 7 Biomass and abundance dynamics in upper (U), mid (M) and low (L) intertidal zones in mangrove plantation in Dam Bay, Nha Trang, central Vietnam.

The number of species recorded in this study is similar or even higher than in other areas of south-east Asia, for example, approximately 50 species were found in Malay peninsula (Ashton et al. 2003), 56 species were found in the Indian Sundarbans including areas adjacent to mangrove trees biotopes (Dey 2006), 30 gastropod species were found in natural and replanted mangroves on the west coast of Thailand (Macintosh et al. 2002), and 32 species were collected on the coast of south Thailand (Sri-aroon et al. 2005).

Probably the highest number, some 151 gastropod species in total, were found in mangroves of Abatan river on Bohol Island, Philippines (Lozouet and Plaziat 2008). The mentioned locality is characterized by a wide variety of habitats - from nearly freshwater, with *Nypa* palms, to typically full salinity marine areas. The authors restricted the number of species, which have been collected exclusively within mangroves to 65, although due to a large number of singletons, this number is definitely an underestimation.

Species richness in planted mangroves in Dam Bay alone (30 species) increased significantly during monitoring years, with appearance in 2008 and slow growth of number of predominantly mangrove-associated species (*Littoraria*). Nevertheless, most of predominantly mangrove-associated species are still absent in Dam Bay plantation (e.g., Potamididae - *Cerithideopsilla djadjariensis, C. microptera, Cerithidea quadrata, Terebralia* spp., some others) despite their presence in natural association in Nha Phu and pelagic development (Houbrick 1991) that should facilitate the species dispersal. We are unable to account for the absence of these species in Dam Bay at the moment. It should be mentioned that the relatively high number of recorded species is due to the continuous observations, and in any given year, the number of collected species was much lower and never exceeded 19 (in 2011), thus making it the least rich in gastropods among all studied sites in the Nha Trang area.

In the natural mangrove association in Dam Bay, we found 19 gastropod species. This was 11 species less than the total number observed in the mangrove plantation, but the sampling efforts in the latter were considerably higher. The main difference in species composition between mangrove plantations and the natural mangrove association in Dam Bay was that species associated with intertidal rocks were absent in mangrove plantations, such as *Pictocolumbella ocellata, Tenguella musiva, Drupella margariticola, Planaxis sulcatus,* and *Nerita undata.* On the contrary, some species that were common in the plantations were absent in the natural association, such as *Clithon oualaniensis* and *Cerithideopsilla cingulata.* This could not be explained only by differences in abiotic factors between the sites, and this difference in species composition may be possibly related to the ecology of these species. Amongst the predominantly mangrove-associated gastropods, only two *Littoraria* species were found here.

Species diversity of Nha Phu (28 species in single collecting year) is much higher than in other sites and is comparable with the total number of species in mangrove plantation (collected in 2005 to 2013), though species composition is different here. The most diverse gastropod family in Nha Phu was the Potamididae, while in the mangrove plantation, only one species in this family was found. Most of the species of this family except *Cerithideopsilla cingulata* are considered as predominantly mangrove associated (Reid et al. 2008). Some species of Ellobiidae are considered as one of the main components of mangrove-associated gastropods (Reid et al. 2008, Reid et al. 2013), and they were present only in Nha Phu. These predominantly mangrove-associated species are mostly arboreal, and this can explain the fact that NP mangroves have lower average species richness, as shown by quantitative

Table 4 Average abundance and biomass with standard deviation of the most common gastropod species in mid intertidal zone

	Number of samples and the year of collecting								
	3	5	5	7	8	8	8	7	2
	2005	**2006**	**2008**	**2009**	**2010**	**2011**	**2012**	**2013**	**NP**
Clithon oualaniense			35.6±37.1 2.7±3.3	165.7±267.4 11.1±17.2	117.0±130.9 8.7±9.8	404.5±625.6 21.5±29.1	13.6±35 1.0±1.8	374±186.5 22.7±11.4	309.1±500.2 23.0
Clithon faba					1.1±3.2 0.2±0.7	5.7±10.8 1.0±2.4		98.7±117.7 5.5±8.3	
Cerithideopsilla cingulata					6.8±12.6 1.2±2.2	50.0±98.6 19.3±39.8	18.2±41.5 6.2±15.8	10.4±12.2 4.1±4.2	1379.55±2070.6 124.4
Batillaria zonalis			247.8±378.2 85.8±123.7	478.6±449.3 192.3±189.4	576.1±421.5 137.6±127.1	946.6±724.7 256.3±197.3	484.1±634.2 125.1±150.8	868.8±470.6 142.9±75.4	2.27±4.5 0.5
Cerithium coralium					10.2±28.9 10.2±27.1		10.2±25.5 7.5±19.1		
Clypeomorus bifasciata			70.8±80.4 13.5±12.6	114.3±198.6 25.2±50	54.5±73.4 7.1±7.3	112.5±123.7 17.4±26.6			
Total for common species	0	0	354.2±485.5 102±136.4	758.6±509.4 223±203.5	765.9±493.1 165.1±130.2	1519.3±1311.5 315.5±216.6	528.4±688.6 140±163.3	1416.9±614.1 186.1±76	1690.9±2000.2 147.9±157.3
Total for all species	18.2±24.1 3±2.7	0	357.9±482.4 102.4±136	762.9±507.3 230.1±199	768.2±494 165.8±130.2	1523.9±1313.8 315.6±216.6	530.7±688.7 140.4±163.6	1440.3±606 210±80	1715.9±1993.7 148.8±157.3

Average abundance (m^{-2}, above line) and biomass (g/m^2, below line) with standard deviation of the most common gastropod species (found in half or more sampling seasons) in quantitative samples and total abundance and biomass of gastropods in mid intertidal zone in each year of monitoring in mangrove plantations in Dam Bay and natural mangrove association in Nha Phu (NP).

samples taken on the surface of sediments (only 2.8 species per sample) comparing with DB (six species per sample in year 2013).

A possible explanation of the absence of key groups of predominantly mangrove-associated gastropod species in both study areas in Dam Bay may be due to differences in the types of vegetative associations in Dam Bay and Nha Phu. In Dam Bay, the mangrove trees comprise small monospecific associations, while at Nha Phu the mangrove association has more complex vegetation structure, which is also much larger than both associations in Dam Bay and experiences a wider range of hydrological conditions. This confirms the statement of Macintosh et al. (2002) about the correlation between the diversity of mangrove vegetation and mangal-associated fauna.

Biomass and abundance of Gastropoda

Only gastropods that inhabit the surface of the substratum were used for analysis of biomass and abundance and therefore the following discussion has some limitations, since most of predominantly mangrove-associated species (mostly living on the roots and branches of the mangrove trees) were not quantitatively sampled.

The mangrove plantation in Dam Bay is characterized by high level of micro heterogeneity. This is confirmed by very high values of standard deviation in biomass and abundance indexes. The abundance of gastropods strongly depends on proximity of the frame to the nearest tree. In general, the samples, collected in close proximity, were poorer than samples taken at some distance from the trees (pers. obs.). This heterogeneity obscures the general tendencies. In the discussion below, we used the average values of biomass and abundance.

In the upper intertidal zone, biomass and abundance varied from year to year and both parameters significantly increased during monitoring period (Figure 7). In later years, biomass and abundance of *Batillaria zonalis* and *Cerithideopsilla cingulata* were relatively high, while in the early years (2005 to 2006), the two species were absent. One might expect that the parameters would remain stable, since no changes in vegetation happened in the upper intertidal zone over the period of observation. Thus, the changes in gastropod population parameters in the upper intertidal zone may have been caused by changes in the mid-intertidal zone after planting the mangrove seedlings, and where biomass and abundance of gastropods, especially of *B. zonalis* and *C. cingulata*, increased even more dramatically in the later period of monitoring. Low biomass and abundance or even absence of gastropod molluscs in mid-intertidal zone in 2005 to 2006 can be explained by the fact that molluscs just started to populate the young plantation.

Both in upper and mid-intertidal zones in late period (2008 to 2013), biomass and abundance of gastropods increased due to *Batillaria zonalis* and *Clithon oualaniensis. Batillaria zonalis*, *Cerithideopsilla cingulata*, and *C. oualaniensis* occur in different habitats in addition to the mangal - in fish ponds, estuarine systems, intertidal muddy, and sand banks sometimes in high densities (Vohra 1970; Grüneberg and Nugaliyadde 1976; Kamimura and Tsuchiya 2004; Fujioka et al. 2007; Tan and Clements 2008). In the mangrove plantation, these species can be considered as opportunistic. They penetrated the mid-intertidal zone in Dam Bay with the appearance of mangrove vegetation providing enrichment of detritus and conditions suitable for the growth of micro-algae on which these species feed (Kamimura and Tsuchiya 2004; Vohra 1970).

The facts that the mangrove plantation in Dam Bay is highly dominated by opportunistic species together with paucity of predominantly mangrove-associated gastropods suggest that the whole ecosystem is not balanced yet. Macintosh et al. (2002) obtained a similar result - the species diversity indexes were lower, and abundance was higher in planted mangroves than in mature mangroves. Therefore, we can expect that biomass and abundance of these species eventually will reach a plateau and may even decrease as the whole system reaches an equilibrium.

In the natural mangrove association in Dam Bay, the dominant species was the same as in the plantation - *Batillaria zonalis.* Although this species occurred in comparatively lower density in the natural mangrove, its biomass was comparable with that of the mangrove plantation, due to larger sizes of individuals in DBn. Two other most abundant species in the plantation, namely *Cerithideopsilla cingulata* and *Clithon oualaniensis* were absent in the natural mangrove. This difference can probably be explained by unequal amount of organic matter in the substratum of these two associations. Since the natural association in Dam Bay is represented only by a narrow belt of trees, large amounts of organic matter are washed away by tides, while in thicker mangrove plantations, it is reposed, thus allowing opportunistic species to attain high densities. Since this natural association corresponds to the upper intertidal zone of the mangrove plantation area with mature *Rhizophora* trees, the low abundance of opportunistic species confirms the fact that mid-intertidal zone is strongly affecting the fauna of upper intertidal zone. *Cerithideopsilla cingulata* and *Clithon oualaniensis* also dominated quantitative samples collected in Nha Phu. This suggests that the natural mangrove association despite its richness in gastropods is unbalanced too. One of the reasons may be the inputs from eutrophicated ponds through the channels in the natural mangrove vegetation.

Conclusions

Nine-year-old mangrove plantation differed from natural mangrove association in biomass and abundance. Species diversity in plantations increased through the monitoring period but it is dominated by opportunistic eurybiotic gastropod species. This suggests that the ecosystem is still in a transitional state.

This investigation is the first study of gastropod molluscs in mangroves of central Vietnam, and it is one of the first long-term monitoring studies in the newly planted mangroves. The findings of this study potentially can make a contribution to understanding of establishing of mangrove ecosystem and highlighted the parameters in evaluating the state of the mangrove ecosystem.

Competing interests
The authors declare that they have no competing interests.

Authors' contributions
ZSS, KYI, BTA, and XL collected and processed the samples, analyzed the results, and wrote the manuscript. All authors read and approved the final manuscript.

Authors' information
ZSS is a PhD student in the Laboratory of Ecology and Morphology of Marine Invertebrates; this study is a part of PhD thesis.
KYI is a PhD, DSc, leading researcher, and scientific advisor of ZSS.
BTA is a PhD, DSc, Prof., and director of the Laboratory of Ecology and Morphology of Marine Invertebrates.
XL, with a degree in PhD, is a research professor.

Acknowledgements
We would like to thank our colleagues from A.N. Severtsov Institute of Ecology and Evolution Yu.V. Deart, Yu. Dgebuadze, A.E. Fedosov, I.N. Marin, P, E.S. Mekhova, and A.V. Zykova; colleagues from Moscow State University E.V. Vortsepneva and A.E. Zhadan; B.I. Sirenko from Zoological Institute; and A.A. Udalov from P.P. Shirshov Institute of Oceanology for their assistance in field work and data analysis. The fieldwork in Vietnam was supported by the Coastal Branch of the Russian Vietnamese Tropical Center (Nha Trang, Vietnam), and we also thank colleagues from Tropical Centre Vo Thi Ha, Nguyen Thi Hai Thanh, Tran Quoc Hoan, and Nguyen Van Quang for their assistance in organizing and conducting material collecting. The work was partially supported by the grants of Russian Foundation for Basic research 13-04-91175-GFEN_a 'Mangrove ecosystems of South China Sea: degradation and rehabilitation' and 14-04-00481-a.

Author details
[1]Laboratory of Ecology and Morphology of Marine Invertebrates, A.N. Severtsov Institute of Ecology and Evolution, 33 Leninskij prosp, Moscow 119071, Russia. [2]Institute of Oceanology, Chinese Academy of Sciences, 7 Nanhai Road, Qingdao, 266071, China.

References
Alongi D, Johnston D, Xuan T (2000) Carbon and nitrogen budgets in shrimp ponds of extensive mixed shrimp-mangrove forestry farms in the Mekong Delta, Vietnam. Aquac Res 31:387–399

Amano T, Yoshinaga I, Yamagishi T (2009) Contribution of anammox bacteria to benthic nitrogen cycling in a mangrove forest and shrimp ponds, Haiphong, Vietnam. Microbes Environ 26:1-6, doi:10.1264/jsme2.ME10150

Ashton EC, Macintosh DJ, Hogarth P (2003) A baseline study of the diversity and community ecology of crab and molluscan macrofauna in the Sematan mangrove forest, Sarawak, Malaysia. J Trop Ecol 19:127–142

Berger U, Rivera-Monroy VH, Doyle TW, Dahdouh-Guebas F, Duke NC, Fontalvo-Herazo ML, Hildenbrandt H, Koedam N, Mehlig U, Piou C, Twilley RR (2008) Advances and limitations of individual-based models to analyze and predict dynamics of mangrove forests: a review. Aquat Bot 89:260–274, doi:10.1016/j.aquabot.2007.12.015

Berry AJ (1963) Faunal zonation in mangrove swamps. Bull Natl Museum State Singapore 32:90–98

Binh CT, Phillips MJ, Demaine H (1997) Integrated shrimp-mangrove farming systems in the Mekong delta of Vietnam. Aquac Res 28:599–610

Binh TNKD, Vromant N, Hung NT, Hens L, Boon EK (2005) Land cover changes between 1968 and 2003 in Cai Nuoc, Ca Mau Peninsula, Vietnam. Environ Dev Sustain 7:519–536, doi:10.1007/s10668-004-6001-z

Bosire JO, Dahdouh-Guebas F, Kairo JG, Cannicci S, Koedam N (2004) Spatial variations in macrobenthic fauna recolonisation in a tropical mangrove bay. Biodivers Conserv 13:1059–1074, doi:10.1023/B:BIOC.0000018149.88212.2d

Bouillon S, Koedam N, Raman A, Dehairs F (2002) Primary producers sustaining macro-invertebrate communities in intertidal mangrove forests. Oecologia 130:441–448, doi:10.1007/s004420100814

Cantera J, Arnaud PM, Thomassin BA (1983) Biogeographic and ecological remarks on molluscan distribution in mangrove biotopes. 1. Gastropods. J Molluscan Stud Suppl 12A:10–26

Christensen S, Tarp P, Hjortsø C (2008) Mangrove forest management planning in coastal buffer and conservation zones, Vietnam: a multimethodological approach incorporating multiple stakeholders. Ocean Coast Manag 51:712–726, doi:10.1016/j.ocecoaman.2008.06.014

De Graaf G, Xuan T (1998) Extensive shrimp farming, mangrove clearance and marine fisheries in the southern provinces of Vietnam. Mangroves Salt Marshes 2:159–166

Dey A (2006) Handbook on mangrove associate molluscs of Sundarbans. Zoological survey of India, Kolkata

Dittmann S (2000) Zonation of benthic communities in a tropical tidal flat of north-east Australia. J Sea Res 43:33–51, doi:10.1016/S1385-1101(00)00004-6

Ellison AM, Farnsworth EJ, Merkt RE (1999) Origins of mangrove ecosystems and the mangrove biodiversity anomaly. Global Ecol Biogeogr 8:95–115, doi:10.1046/j.1466-822X.1999.00126.x

Fujioka Y, Shimoda T, Srithong C (2007) Diversity and community structure of macrobenthic fauna in shrimp aquaculture ponds of the Gulf of Thailand. Japan Agric Res 41:163–172

Grüneberg H, Nugaliyadde L (1976) Population studies on a polymorphic prosobranch snail (*Clithon* (*Pictoneritina*) *oualaniensis* Lesson). Philos Trans R Soc Lond B Biol Sci 275:385–437

Hong PN (2004) Effects of mangrove restoration and conservation on the biodiversity and environment in Can Gio District. In: Vanucci M (ed) Mangrove Management and Conservation: Present and Future. NY United Nations University Press, New York, pp 111–137

Hong PN, San HT (1993) Mangroves of Vietnam. IUCN, Bangkok

Houbrick RS (1985) Genus *Clypeomorus* Jousseaume (Cerithiidae - Prosobranchia). Smithson Contrib Zool 403:1–131

Houbrick RS (1991) Systematic review and functional morphology of the mangrove snails *Terebralia* and *Telescopium* (Potamididae; Prosobranchia). Malacologia 33:289–338

Houbrick RS (1992) Monograph of the Genus *Cerithium* Bruguière in the Indo-Pacific (Cerithiidae - Prosobranchia). Smithson Contrib Zool 510:1–211

Hylleberg J, Kilburn RN (2003) Marine molluscs of Vietnam (Tropical marine mollusc program). Phuket Mar Biol Cent Spe Pub 28:1–300

Johnston D, Trong NV, Tien DV, Xuan TT (2000) Shrimp yields and harvest characteristics of mixed shrimp-mangrove forestry farms in southern Vietnam: factors affecting production. Aquaculture 188:263–284

Kamimura S, Tsuchiya M (2004) The effect of feeding behavior of the gastropods *Batillaria zonalis* and *Cerithideopsilla cingulata* on their ambient environment. Mar Biol 144:705–712, doi:10.1007/s00227-003-1238-x

Kristensen E, Bouillon S, Dittmar T, Marchand C (2008) Organic carbon dynamics in mangrove ecosystems: a review. Aquat Bot 89:201–219, doi:10.1016/j.aquabot.2007.12.005

Lan T, Trang C, Huong D, Cuong C, Vinh V (2012) Sustainability in coastal urban environment: thematic profiles of resources and their users. Case studies of Hai Phong and Nha Trang City-Vietnam. SECOA FP7 Res. Proj. Vol. 3. Sapienza Universita Editrice, Rome, pp 301–343

Le Vay L, Ut V, Jones D (2001) Seasonal abundance and recruitment in an estuarine population of mud crabs, *Scylla paramamosain*, in the Mekong Delta, Vietnam. Hydrobiologia 449:231–239

Lee SY (2008) Mangrove macrobenthos: assemblages, services, and linkages. J Sea Res 59:16–29, doi:10.1016/j.seares.2007.05.002

Lozouet P, Plaziat J-C (2008) Mangrove environments and mollusks Abatan River, Bohol and Panglao Islands. ConchBooks, Hackenheim, Central Philippines

Macintosh DJ, Ashton EC, Havanon S (2002) Mangrove rehabilitation and intertidal biodiversity: a study in the Ranong mangrove ecosystem, Thailand. Estuar Coast Shelf Sci 55:331–345, doi:10.1006/ecss.2001.0896

Macnae W (1968) A general account of the fauna and flora of mangrove swamps and forests in the Indo-West-Pacific region. Adv Mar Biol 6:73–270
Martins AM d F (1996) Relationships within the Ellobiidae. In: Taylor J (ed) Orig. Evol. Radiat. Mollusca. Oxford University Press, Oxford, pp 285–294
Metcalfe KN, Glasby CJ (2008) Diversity of Polychaeta (Annelida) and other worm taxa in mangrove habitats of Darwin Harbour, northern Australia. J Sea Res 59:70–82, doi:10.1016/j.seares.2007.06.002
Nagelkerken I, Blaber SJM, Bouillon S, Green P, Haywood M, Kirton LG, Meynecke JO, Pawlik J, Penrose HM, Sasekumar A, Somerfield PJ (2008) The habitat function of mangroves for terrestrial and marine fauna: a review. Aquat Bot 89:155–185, doi:10.1016/j.aquabot.2007.12.007
Ng PKL, Sivasothi N (1999a) A guide to the mangroves of Singapore 1. Singapore Science Centre, Singapore
Ng PKL, Sivasothi N (1999b) A guide to the mangroves of Singapore 2. Singapore Science Centre, Singapore
Okutani T (2000) Marine mollusks in Japan. Tokai University Press, Tokyo
Poppe GT (2008a) Volume I: Philippine marine mollusks. ConchBooks, Hackenheim
Poppe GT (2008b) Volume II: Philippine marine mollusks. ConchBooks, Hackenheim
Poppe GT (2010) Volume III: Philippine marine mollusks. ConchBooks, Hackenheim
Printrakoon C, Wells F, Chitramvong Y (2008) Distribution of molluscs in mangroves at six sites in the upper Gulf of Thailand. Raffles Bull Zool 18:247–257
Reid DG (1986) The littorinid molluscs of mangrove forests in the Indo-Pacific region:the genus *Littoraria*. British Museum (Natural History), London
Reid DG, Dyal P, Lozouet P, Glaubrecht M, Williams ST (2008) Mudwhelks and mangroves: the evolutionary history of an ecological association (Gastropoda: Potamididae). Mol Phylogenet Evol 47:680–699, doi:10.1016/j.ympev.2008.01.003
Reid DG, Claremont M, Smith L, Shamoto M, Glaubrecht M, Ozawa T (2013) Mosaics in the mangroves: allopatric diversification of tree-climbing mudwhelks (Gastropoda: Potamididae: Cerithidea) in the Indo-West Pacific. Biol J Linn Soc 110:564–580, doi:10.1111/bij.12151
Sasekumar A (1974) Distribution of macrofauna on a Malayan Mangrove Shore. J Anim Ecol 43:51–69
Skilleter GA (1996) Validation of rapid assessment of damage in urban mangrove forests and relationships with molluscan assemblages. J Mar Biol Assoc UK 76:701–716
Skilleter GA, Warren S (2000) Effects of habitat modification in mangroves on the structure of mollusc and crab assemblages. J Exp Mar Bio Ecol 244:107–129, doi:10.1016/S0022-0981(99)00133-1
Sri-aroon P, Lohachit C, Harada M (2005) Brackish-water mollusks of Surat Thani Province, southern Thailand. Southeast Asian J Trop Med Public Health 36:180–188
Tan SK, Clements R (2008) Taxonomy and distribution of the Neritidae (Mollusca: Gastropoda) in Singapore. Zool Stud 47:481–494
Thu PM, Populus J (2007) Status and changes of mangrove forest in Mekong Delta: case study in Tra Vinh, Vietnam. Estuar Coast Shelf Sci 71:98–109, doi:10.1016/j.ecss.2006.08.007
Tue NT, Hamaoka H, Sogabe A, Quy TD, Nhuan MT, Omori K (2012) Food sources of macro-invertebrates in an important mangrove ecosystem of Vietnam determined by dual stable isotope signatures. J Sea Res 72:14–21, doi:10.1016/j.seares.2012.05.006
Vohra FC (1970) Some studies on *Cerithidea cingulata* (Gmelin 1790) on a Singapore sandy shore. J Molluscan Stud 39:187–202

9

A new species of fish-eating rat, genus *Neusticomys* (Sigmodontinae), from Ecuador

J Delton Hanson[1*], Guillermo D'Elía[2], Sheri B Ayers[3], Stephen B Cox[1], Santiago F Burneo[4] and Thomas E Lee Jr[5]

Abstract

Background: In this study, the genetic substructure and morphology of the species *Neusticomys monticolus* was evaluated. A nuclear marker and mitochondrial maker were used to examine phylogeographic structure and to estimate genetic distances. Two statistical measurement analyses were applied to morphological data.

Results: These data recovered two morphologically distinct phylogeographic groups corresponding to populations on the eastern and western slopes of the Andes. Further, these eastern and western Andean slope populations of *N. monticolus* are 8.5 % divergent using sequence data from cytochrome-*b* (0.8 % divergent in the interphotoreceptor retinoid-binding protein gene).

Conclusions: Populations currently assigned to *N. monticolus* constitute a species complex. The name *N. monticolus* is here restricted to western Andean slope populations. Populations on the eastern slope of the Andes are assigned to a new species, to which the authors assign the name *Neusticomys vossi* sp. nov.

Keywords: Andes, Cricetidae; Ecuador; Ichthyomyini; Muroidea; *Neusticomys monticolus*; *Neusticomys vossi* sp. nov

Background

Sigmodontine rodents constitute a diverse group of New World rodents; extant diversity encompasses ca. 400 species arranged in 86 genera (D'Elía and Pardiñas 2015). Traditionally, sigmodontine taxa have been arranged into tribes (Reig 1980; Musser and Carleton 2005; D'Elía et al. 2007), one of which is the tribe Ichthyomyini comprising five genera and 17 species ranging from Bolivia, central Brazil, and the Guianas to southern Mexico (Jenkins and Barnett 1997; Musser and Carleton 2005). Ichthyomyines are small to medium-sized semiaquatic animalivorous rodents (Voss 1988), constituting a remarkably distinct group within the sigmodontine radiation. Several morphological features differentiate them from other sigmodontines and as such have been interpreted as supporting the monophyly of the tribe (Voss 1988). Recent phylogenetic analyses, based on nuclear DNA sequences which include representatives of only two ichthyomyine genera (*Rheomys* and *Neusticomys*), have questioned the monophyly of the tribe (Martínez et al. 2012; Parada et al. 2013; Salazar-Bravo et al. 2013). However, a recent inspection of the sequence of *Neusticomys* (EU649036) indicates that it may be a chimeric sequence that includes a fragment retrieved from an oryzomyine. This would therefore weaken the argument against a monophyletic Ichthyomyini.

The ichthyomyine genus that deviates the least from the general sigmodontine body plan is *Neusticomys* (Voss 1988); for example, on its hind feet, it lacks the stiff hairs (presumably an adaptation for swimming) found in other members of the tribe. This genus was nominated by Anthony (1921) to contain the Andean species *Neusticomys monticolus*. Almost 7 decades later, Voss (1988) subsumed the predominantly lowland *Daptomys* Anthony (1929) under *Neusticomys*. Recently, Percequillo et al. (2005) described another lowland species of *Neusticomys* from central Brazil increasing the known diversity of the genus to six species. Of these, *N. monticolus* is the most widely distributed species, being endemic to the Andes Mountains of Colombia and Ecuador, at elevations between 1800 and 3750 m (Lee et al. 2006a, 2006b; Musser and Carleton 2005). The lifestyle of the species (living near fast-moving streams) has made it difficult to collect. Individuals either are caught in pitfall traps set next to streams, or Sherman traps set in the

* Correspondence: j.delton.hanson@gmail.com
[1]RTLGenomics, 4321 Marsha Sharp Fwy Door 2, Lubbock, TX 79407, USA
Full list of author information is available at the end of the article

stream, and they do not seem to be attracted to typical baits, rather those that are caught seem to have been attracted due to curiosity or by insects gathered around typical rodent bait (Personal observation of TEL). These characteristics result in few museum specimens of *N. monticolus* available for study, with still fewer available tissue samples. In 1988, there were ~50 known specimens of *N. monticolus* (Voss 1988), none of which had associated frozen tissue that we could identify. An additional six specimens were collected during field work conducted in Ecuador during 2003, 2005, 2007, and 2008 (Lee et al. 2008; M. Pinto personal communication).

The only taxonomic assessment of *N. monticolus* to date is that presented by Voss (1988), and no published taxonomic oriented study based on genetic data has focused on this species or any other *Neusticomys*. Here we attempt to characterize the genetic variation of *N. monticolus* by examining the nucleotide sequence of the mitochondrial cytochrome-*b* gene (*Cytb*) and nuclear IRBP gene (*Rbp3*) of individuals from populations on both sides of the Andes. Given the already noticed morphological differences among the phylogeographic units here uncovered, we describe a new species of *Neusticomys*.

Methods

Sampling

Four field trips were conducted to survey the mammalian fauna of Ecuador. Animals were collected following methods approved by the American Society of Mammalogists Animal Care and Use Committee (Sikes 2011). Two specimens of *Neusticomys* were collected on the western side of the Andes near the type locality of *N. monticolus* in 2003 as well as an additional two from a different locality in 2008. One specimen each was collected in 2005 and in 2007 at two different localities in the eastern Andes. These specimens are housed at various natural history collections (Appendix). Further, the tissue was obtained for another member of Ichthyomyini (*Rheomys raptor*), and sequences were obtained from GenBank for eight additional members of the subfamily Sigmodontinae (one representative of seven tribes within Oryzomalia, one member of Sigmodontini) and four non-sigmodontine members of the family Cricetidae (Appendix).

Morphological analysis

The following measurements (Table 1, defined in Voss 1988) were examined for 41 specimens of *Neusticomys*

Table 1 Mean measurements and ranges (in mm) for distinct sets of specimens of *Neusticomys*; specimen QCAZ 7830 is the holotype of *N. vossi* sp. nov

	N. monticolus ♀	*N. monticolus* ♂	*N. monticolus* Antiquia	*N. vossi* sp. nov. ♀	*N. vossi* sp. nov. ♂	QCAZ 7830
TL	216.56 (196–287)	226.44 (187–313)	206.8 (191–217)	199.5 (188–211)	205.57 (193–221)	211
HBL	115.67 (95–195)	122.88 (96–211)	106.8 (100–112)	102.82 (97–110.99)	105.17 (101–108)	110.99
LT	100.89 (90–109)	103.56 (87–114)	100 (91–105)	101.25 (96–108)	99 (92–108)	108
HFC	25.56 (25–26)	26.25 (24–36)	24.8 (24–26)	26 (25–27)	25.43 (24–27)	27
EC	10 (6–12)	10.25 (8–14)	11 (11–11)	10.25 (8–13)	9.86 (9–11)	13
CIL	24.36 (23.1–25)	24.96 (24.2–25.8)	23.75 (22.7–24.9)	23.7 (23.2–25.01)	24.13 (23.4–24.9)	25.01
LD	6.01 (5.5–6.2)	6.09 (5.4–6.6)	5.76 (5.5–6.1)	5.66 (5.34–6.1)	6.04 (5.7–6.4)	6.1
LM	4.16 (4–4.91)	4.07 (3.9–4.4)	4.06 (3.9–4.1)	4.12 (4–4.2)	4.09 (4–4.3)	4.17
LIF	4.74 (4.3–5.9)	4.69 (4.4–5.06)	4.46 (4.2–4.7)	4.39 (4.11–4.7)	4.86 (4.7–5.2)	4.45
BIT	1.44 (1.3–1.75)	1.44 (1.3–1.6)	1.42 (1.4–1.5)	1.28 (1.1–1.47)	1.27 (1.2–1.4)	1.47
BIF	2.11 (1.9–2.4)	2.09 (1.9–2.3)	2.16 (2.1–2.2)	1.99 (1.97–2)	2.15 (2–2.3)	1.99
BPB	2.93 (2.7–3.1)	2.96 (2.7–3.4)	2.88 (2.5–3.3)	2.87 (2.7–3.14)	2.85 (2.6–3.1)	3.14
LN	9.65 (8.6–11.24)	9.48 (8.5–10.2)	9.74 (9.2–10.7)	9.1 (8.81–9.5)	9.46 (8.9–10.1)	8.89
BN	2.89 (2.6–3.2)	2.99 (2.7–3.2)	2.94 (2.7–3.1)	2.37 (2.27–2.52)	2.7 (2.5–3)	2.52
LIB	4.81 (4.6–5)	4.85 (4.6–5.2)	4.84 (4.6–5.1)	4.59 (4.5–4.67)	4.61 (4.4–4.9)	4.57
ZB	12.83 (12.2–13.3)	13.16 (12.3–13.4)	12.1 (12.1–12.1)	12.03 (11.73–12.49)	12.54 (11.9–13)	12.49
BB	12.13 (11.8–12.3)	12.16 (11.7–12.5)	11.38 (11.2–11.5)	11.73 (11.52–11.91)	12.13 (11.9–12.5)	11.91
BZP	1.31 (1–2.06)	1.14 (1–1.3)	1.02 (0.9–1.1)	1.17 (1.1–1.25)	1.13 (1.1–1.3)	1.25
BM1	1.45 (1.4–1.57)	1.44 (1.3–1.5)	1.44 (1.3–1.6)	1.22 (1.01–1.4)	1.38 (1.3–1.5)	1.16
HI	4.3 (2.36–4.9)	4.68 (4.1–5.2)	4.36 (4–4.7)	3.77 (3.03–4.3)	4.49 (4.2–5)	3.03
DI	1.33 (1–1.5)	1.42 (1.3–1.5)	1.42 (1.4–1.5)	1.21 (1.2–1.23)	1.3 (1.2–1.4)	1.23
BOC	7.09 (6.7–7.39)	7.04 (6.6–7.5)	6.75 (6.6–7)	6.65 (6.47–6.8)	6.85 (6.8–7)	6.63

Measurements acronyms and definitions are found in Voss (1988)

(including four specimens not presented in Voss (1988); Appendix): TL—total length, HBL—head body length, LT—length of tail, HF.C—length of hindfoot collected, E.C—length of ear collected, CIL—condylo-incisive length, LD—length of diastema, LM—length of maxillary molars, LIF—length of the incisive foramina, BIT—-breadth of the incisor tips, BIF—breadth of the incisive foramina, BPB—breadth of the palatal bridge, LN—length of nasals, BN—breadth of nasals, LIB—least interorbital breadth, ZB—zygomatic breadth, BB—breadth of braincase, BZP—breadth of the zygomatic plate, BM1—breadth of M1, HI—height of incisor, DI—depth of incisor, and BOC—breadth of occipital condyle. The combined measurement data were first examined with principle components analysis (PCA) in the R programming package (R Development Core Team 2009) to determine if distinct groups were identified. Defined groups were then analyzed with linear discriminant analysis (LDA) using the R programming package (R Development Core Team 2009). LDA was used to graphically explain differences between phylogeographic groups using multiple variables.

Molecular methods

Genomic DNA was isolated from approximately 0.1 g of either the liver or muscle, using a phenol extraction method (Longmire et al. 1997). The polymerase chain reaction (PCR) was used to amplify the complete 1143 bp of *Cyt*b for the seven specimens and up to 1266 bp of *Rbp*3. Reaction concentrations (25 μl volume) included ≤300 ng genomic DNA, 0.07 mM dNTPs, 2.86 mM MgCl, 5 μl 10× buffer, 1.25 U Taq (Go Taq, Promega, Madison, Wisconsin), and 0.286 μM of primers L14115 and H15288 (*Cyt*b; Martin et al. 2000) or A1 and B2 then A1 F and E2 B2 (*Rbp*3; Stanhope et al. 1992; Weksler 2003). Thermal profiles for PCR included an initial denaturation step at 95 °C (2 min), 30 to 40 cycles with denaturation at 95 °C (45 s), annealing at 48 °C-*Cyt*b or 54 °C-*Rbp*3 (1 min), extension at 72 °C (1 min 30 s), and a final extension cycle of 72 °C (8 min). Amplicons were purified using the QIAquick PCR purification kit (Qiagen, Inc., Valencia, California) and then sequenced using ABI Prism Big Dye Terminator v3.1 ready reaction mix (Applied Biosystems, Foster City, California) and a 3100-*Avant* automated sequencer (Applied Biosystems, Foster City, California). *Cyt*b primers L14115 and H1528 and the internal primers O400R (Hanson and Bradley 2008), F1 (Whiting et al. 2003), O700H (Hanson and Bradley 2008), and 700 L (Peppers and Bradley 2000) and *Rbp*3 primers A1 and B2 and the internal primers F and E2, 395R (Hanson et al. 2010), and C and D2 (Jansa and Voss 2000; Weksler 2003) were used for cycle sequencing at 95 °C (30 s) denaturing, 50 °C (20 s) annealing, and 60 °C (4 min) extension. Following 30 to 40 cycles, reactions were purified and precipitated in isopropanol. Sequencher 4.1.4 (Gene Codes, Ann Arbor, Michigan) was used to proof sequences. DNA sequences were deposited in GenBank, and accession numbers are listed in Appendix.

Genetic analyses

ClustalW, MUSCLE, and manual approaches were used in MEGA4 (Tamura et al. 2007) to align nucleotide sequences and gave identical alignments. As no additional ichthyomyine *Cyt*b or *Rbp*3 sequence were available in GenBank (the one sequence of *Rbp*3 available for *Neusticomys* has been identified as chimeric and was regenerated in this paper), newly obtained sequences of *Neusticomys* and *Rheomys* were integrated into a matrix with sequences gathered from eight Sigmodontinae tribes (one representative of seven tribes within Oryzomalia, one member of Sigmodontini) and four non-Sigmodontinae tribes which were used to form the out-group. When available, we included full-length sequences gathered from a specimen of the type species of the type genus of each tribe. The matrices were analyzed using a Bayesian approach (Rannala and Yang 1996) in MrBayes version 3.1.2 (Ronquist and Huelsenbeck 2003). MrModeltest (Nylander 2004) was used to estimate the most appropriate model of sequence evolution (GTR + I + G). Bayesian analysis was performed with sequences partitioned by codon using site-specific gamma distribution allowing for a proportion of invariable sites. Runs, consisting of four Markov chains, were allowed to proceed for ten million generations and were sampled every 1000 generations. The first 1000 trees were discarded as "burnin" based on stabilization of likelihood scores; the remaining trees were used to compute a 50 % majority rule consensus tree and obtain posterior probability (PP) estimates for each clade. Additionally, *Cyt*b pairwise genetic distances were calculated between the recovered phylogroups using MEGA4 (Tamura et al. 2007) and the Kimura two-parameter model (Kimura 1980).

Results

Morphological data

Results of the PCA of morphometric data (Fig. 1a, b) are congruent with the genetic data in identifying two putatively allopatric groups in Ecuador, one (V) formed by specimens collected on the eastern slopes of the Andes and the other (M) formed by specimens collected on the western slopes of the Andes. In addition, a second western group, distributed north near Antioquia, Colombia, was also recovered (there is currently no genetic data available for this group). The eastern and western groups marginally overlap. This overlap decreased when age groups and sexes were examined separately (Fig. 1a, b, respectively). The two western subgroups overlap more than the eastern and western groups but still are well separated. When the three groups were examined using

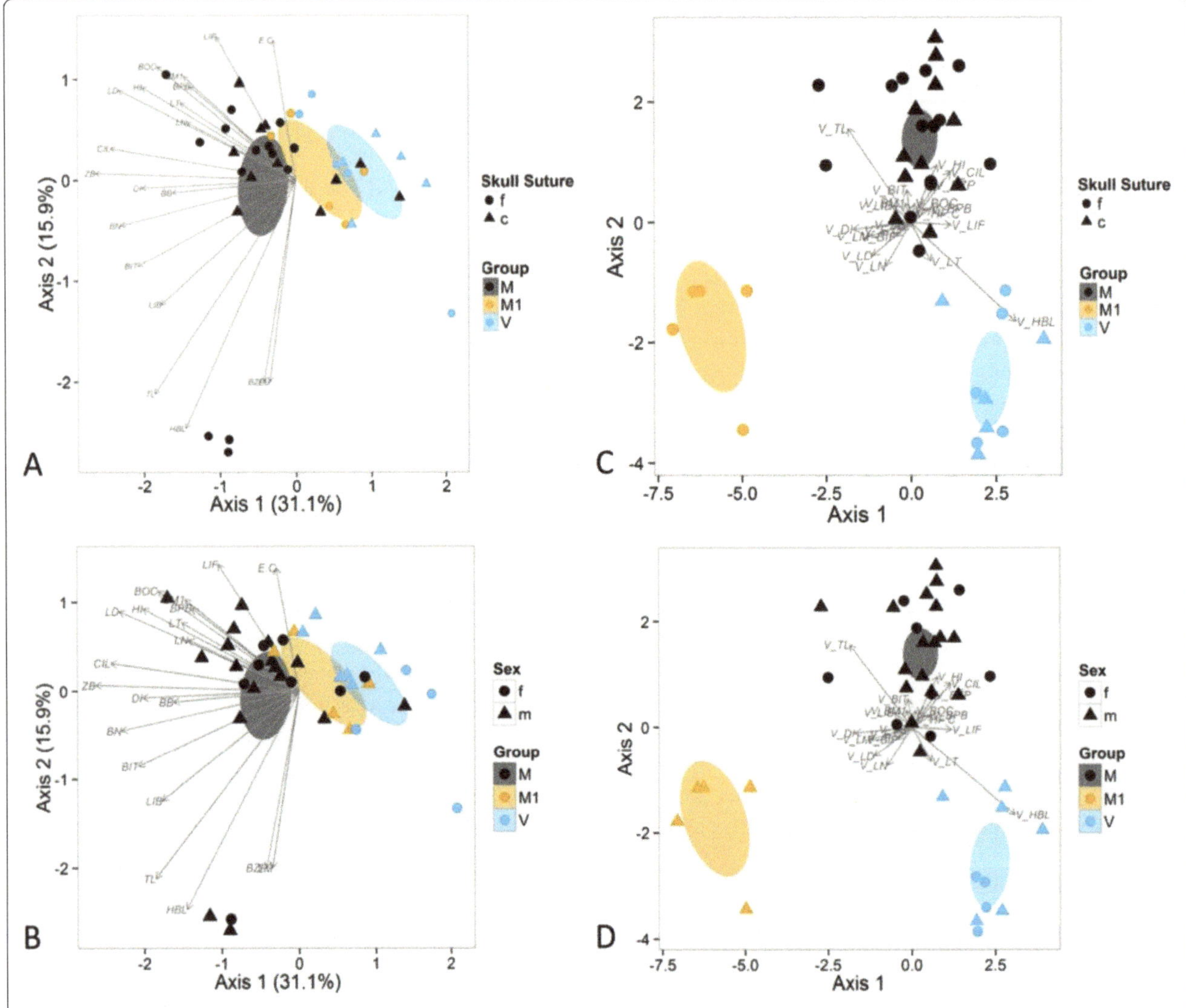

Fig. 1 Graphical representation of the PCA analyses (**a**, **b**) and LDA analysis (**c**, **d**). Vectors labeled as in Table 1. **a**, **c** Samples identified by locality and age. Eastern samples (*N. vossi* sp. nov.: *V*) are *light gray*, western samples (*N. monticolus*: *M*) are *dark gray*, specimens from Antioquia (*N. monticolus*: *M1*) are *mid-gray*. Adults (fused craniosutures) are *circles*, and subadults (closed craniosutures) are *triangles*. **b**, **d** Samples identified by locality and sex. Location symbols same as above, females—*circles*, males—*triangles*

LDA (Fig. 1c, d), the separation was complete in both age and sex analyses (Fig. 1c, d).

Genetic data

Both phylogenetic analyses of mitochondrial and nuclear DNA sequences (Fig. 2) depict a strongly monophyletic Sigmodontinae (clade I; PP = 1.00). Importantly, the tribe Ichthyomyini also is recovered as monophyletic (clade II; PP = 1.00); *Rheomys* appears sister to *Neusticomys*. Within *Neusticomys*, two distinct groups are indicated, each highly supported (PP = 1.00). The groups appear to be allopatric; clade V is distributed east of the Andes and clade M west of the Andes (Fig. 3). Average genetic distances among the eastern and western groups were 8.5 % (*Cyt*b) and 0.8 % (*Rbp*3); a much higher value than that observed within each group (eastern: 0.4 and 0.0 %; western: 0.1 and 0.1 %, respectively).

Taxonomic implications

Despite the small sample size examined, we consider that when taken together, available data provide sufficient evidence to justify recognition of an additional species of *Neusticomys*. As no name is available, we name and describe it below.

Neusticomys vossi sp. nov.

Voss' fish-eating rat urn:lsid:zoobank.org:act:FB716D66-3D57-40A0-9E48-0E8A5A9B402E

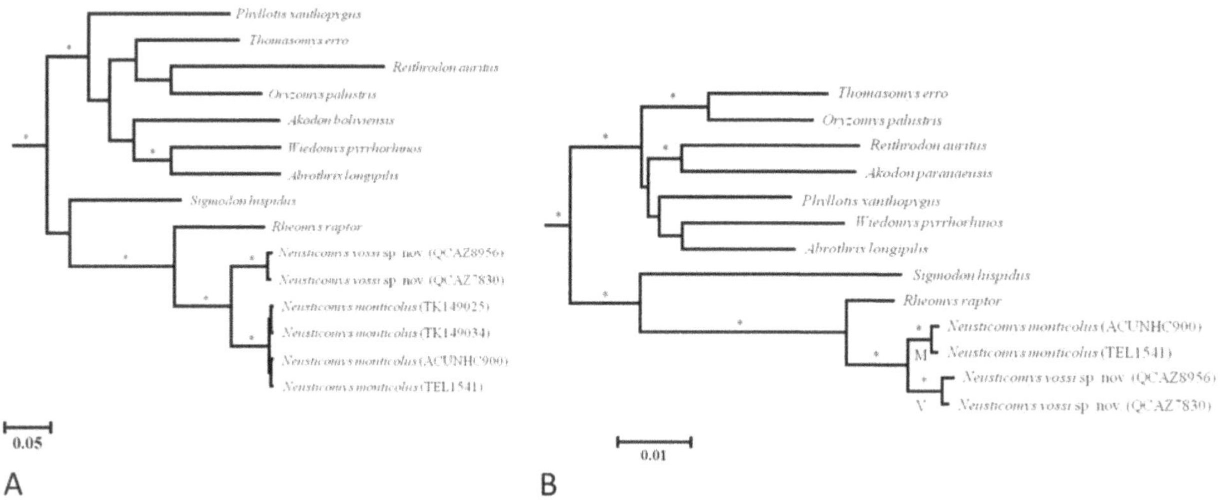

Fig. 2 Phylogenetic tree of sigmodontine rodents based on *Cytb* (**a**) and *Rbp3* (**b**) DNA sequences. Bayesian posterior probability values greater than 0.95 are represented by asterisks placed above the branch

Holotype—QCAZ 7830, an adult lactating female collected by T. E. Lee (personal catalog number TEL1846) in August 2005 with a Sherman trap placed in a forested mountain stream near a small (1 m tall) waterfall (Fig. 4). The specimen is preserved as the skin, skull and skeleton, and frozen tissue and deposited at El Museo de Zoología, Pontificia Universidad Católica del Ecuador, Quito, Ecuador.

Type locality—12 km by road northwest of Cosanga (0° 31′ 70″ S, 77° 52′ 99″ W), Napo Province, Ecuador (1900 m).

Diagnosis—As stated by Percequillo et al. (2005), species of *Neusticomys* are difficult to diagnose using presumptive autapomorphies; rather, unique combinations of character states are operationally useful for species recognition. *Neusticomys vossi* sp. nov. is a species of Ichthyomini, that can be distinguished from other ichthyomyine species by its smaller size, reduced hallux, single cusped last molar, large interparietal bone, and an apparent less advanced aquatic specialization, and can be distinguished from other congeners (except *N. monticolus*) by having dull grayish rather than brownish pelage and by occipital condyles not projecting posteriorly beyond rest of occiput. *N. vossi* sp. nov. can be distinguished from *N. monticolus* by its smaller overall size and by its narrower incisors, braincase, occipital condyles, and rostrum as well as shorter condylo-incisive length, length of diastema, length of maxillary molars, zygomatic breadth, breadth of braincase, and breadth of occipital condyles.

Holotype measurements—Head body length (103 mm), tail length (108 mm), hindfoot (27 mm), ear (13 mm), breadth of nasals (2.52 mm), length of nasals (8.89 mm), depth of incisor (1.23 mm), length of incisors (3.03 mm), breadth of M1 (1.16 mm), and the length of the incisive foramina (4.45 mm).

Description—Dull grayish black pelage (Fig. 5). Ventral pelage color often slightly lighter than dorsal pelage. Mystacial vibrissae and oral margins usually silver. Tail uniformly colored with whitish hairs along ventral surface and occasionally a white tip. Some individuals may have white midpectoral blazes and irregular whitish dorsal spotting. Manus and pes covered with whitish silver hairs turning to dark at the wrist, ankles, and metapodials.

Rostrum slender with threadlike zygomatic arch (Fig. 6). Infraorbital foramina very large. Palatal foramina very large and medially broad, reaching the anterior lobe of large palate, well surpassing the posterior end of the toothrow. Square mesopterygoid fossa lacking a median palatine process. Braincase broad, flat, and smooth. Coronoid process long and narrow. No ridges on frontals or parietals. Small depression at base of nasals, plane of nasals continuous with frontals. Large interparietal bone. Occipital condyles not projecting posteriorly beyond rest of occiput, and not exposed dorsally. Well-developed gnathic process. Peglike process on maxillary roots of zygomata external to molars. Upper incisors ungrooved and orthodont. Parallel upper molar rows. Molar cusps nearly opposite. High crowned molars with deep flexes almost touching in the middle line of molars. M1 much larger than M2 and M3. M1 with three similar-sized lobes M2 with paracone-protocone pair much wider than the metacone-hypocone pair, M3 reduced.

Comparisons—Voss (1988) noted that specimens from Papallacta in the eastern Andes, referred here to *N. vossi* sp. nov., average slightly smaller than *N. monticolus* from Guarumal in the western Andes; the same is true for the

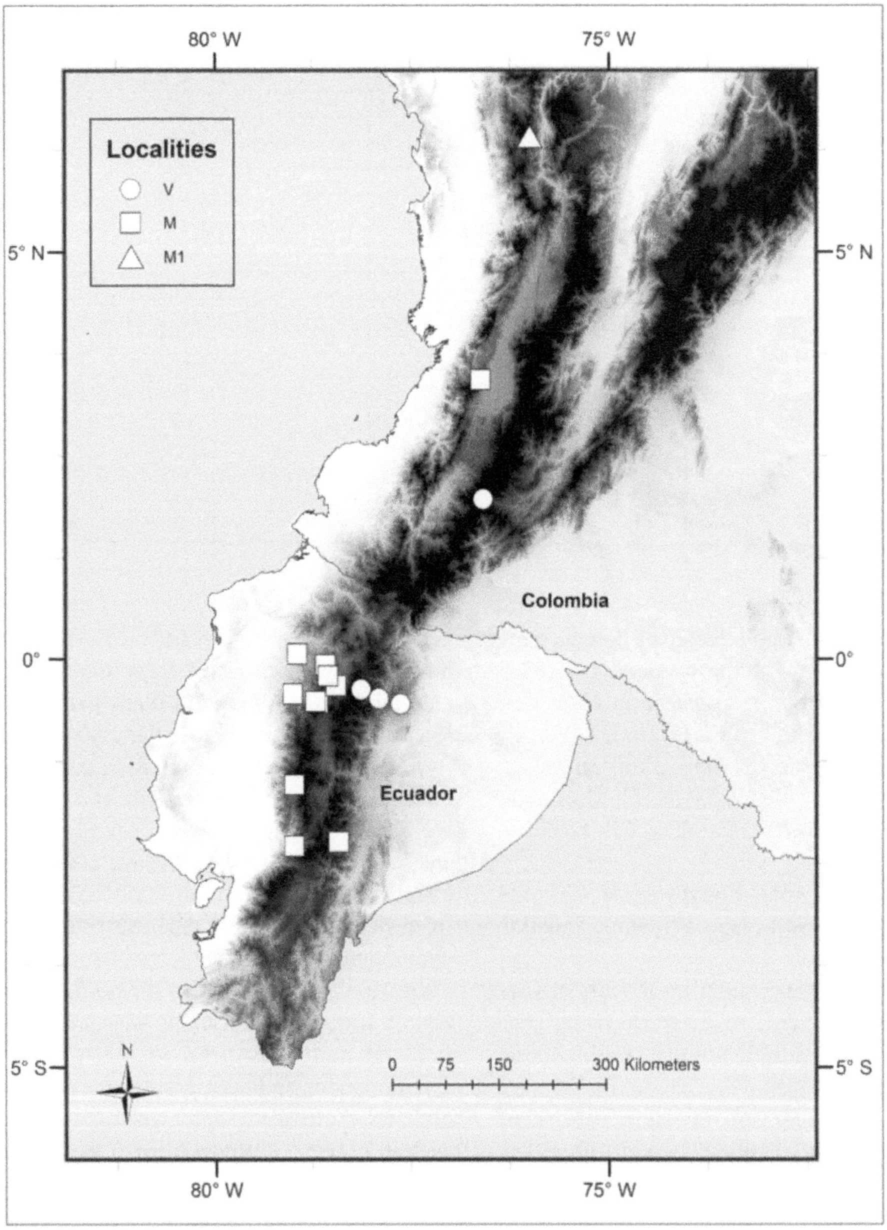

Fig. 3 Map of recording localities of specimens of *Neusticomys* analyzed in the present study. Eastern samples (*N. vossi* sp. nov.: *V*) are *circles*, western samples (*N. monticolus: M*) are *squares*, and specimens from Antioquia (*N. monticolus: M1*) are *triangles*

specimens, including the holotype of *N. vossi* sp. nov., collected by us. The breadth of the occipital condyles appears nearly diagnostic between the two groups in Voss (1988) study. BOC in *N. vossi* sp. nov. samples is ≤6.9 mm with females being smaller than males. For the most part, *N. monticolus* shows a BOC of ≥6.9 mm with females being smaller than males.

Similar to *Neusticomys venezuelae* and *Neusticomys mussoi*, Voss' fish-eating rat is differentiated from *Neusticomys ferreirai*, *Neusticomys oyapocki*, and *Neusticomys peruviensis* in that the posterior edge of the inferior zygomatic root lies above the anterecone of M1. *N. vossi* sp. nov. differs from *N. oyapocki* and *N. ferreirai* in having three upper and lower molars.

Distribution—Known from four sites on the eastern slopes of the Andes in northern Ecuador and southern Colombia ranging from 1° 58′ N, 76° 35′ W in the north to 0° 33′ N, 77° 36′ W and from 1900 to 3750 m.

Etymology—*Neusticomys vossi* sp. nov. is named to honor Dr. Robert S. Voss of the American Museum of Natural History. Rob Voss is the author of a large series of key contributions towards the understanding of South

Fig. 4 View of the type locality of *Neusticomys vossi* sp. nov

American mammals; in particular, he authored a now classical monograph on ichthyomyine rodents. As such, he was the first to recognize the morphological differences between the new species described here and *N. monticolus*. As part of an ichthyomyine review, Voss (1988) wrote the following as a comment on *N. monticolus*, "The series from Guarumal together with those from nearby Las Machinas and from the Rio Pita closely resemble the Volcan Pichincha specimens, but the Papallacta series exhibits some metric differences."

Natural history—One *N. vossi* sp. nov. was recorded at Papallacta with two large embryos on the 15th of May (Voss 1988). The type collected in August is a lactating female. The type was taken by a small waterfall about 1 m tall, with traps exposed to the spray of water from the fall, which is congruent with the description by Tate (1931) of *Neusticomys* habitat. The stream was rocky and about 1 m across with fast rapids (Fig. 4). This specimen represents a low elevation record in Ecuador at 1900 m for the eastern Andes (Lee et al. 2006a). *Oreoryzomys balneator* and *Thomasomys erro* were collected in the same trap line or in nearby forests, as the type specimen (Lee et al. 2006a). Vegetation along the stream consisted of plants with large waxy leaves. Plants of the families Araceae, Arecaceae, Cecropiaceae, Chloranthaceae, Cyatheaceae, Cyclanthaceae, Flacourtiaceae, Lauraceae, Lobeliaceae, Melastomataceae, Meliaceae, Moraceae, Piperaceae, and Poaceae were found along the stream bank (Lee et al. 2006a).

Discussion

Samples from eastern and western localities were sequenced to examine phylogeographic structure of *N. monticolus*. Two highly differentiated and potentially allopatric groups were recovered; these groups are congruent with the morphological forms previously identified by Voss (1988). Externally, the two groups do not appear to present differences besides pelage color in a

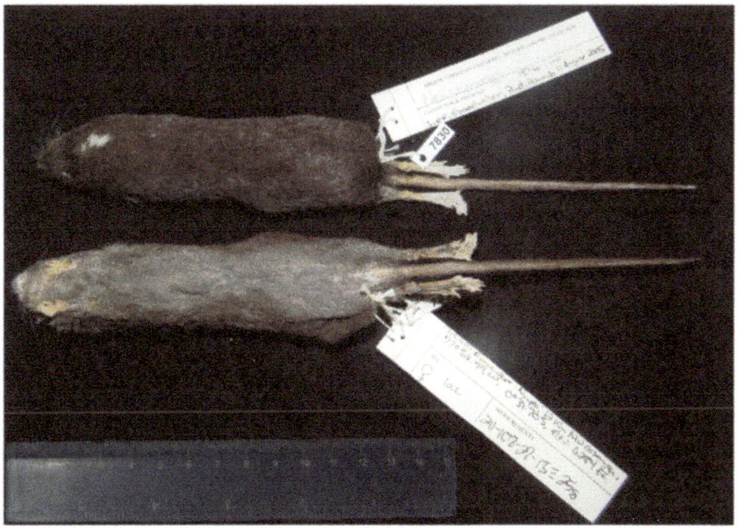

Fig. 5 Skin of the holotype of *Neusticomys vossi* sp. nov. (QCAZ 7830)

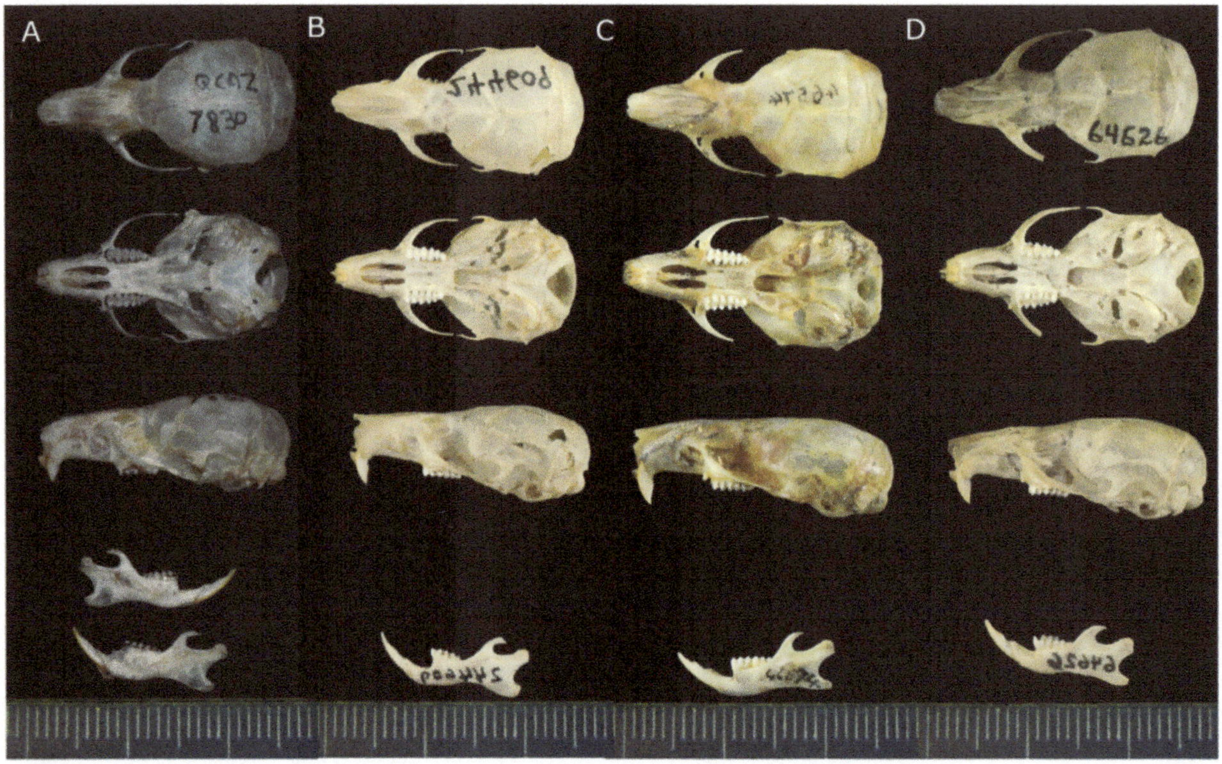

Fig. 6 Skulls and jaw for *Neusticomys vossi* sp. nov. (**a** QCAZ 7830 and **b** AMNH 244609) and *N. monticolus* (**c** AMNH 46574 and **d** AMNH 64626). **a**, **c** are the respective type specimens. All skulls are from adult females with closed cranial sutures except for **c** which is a juvenile male with open sutures

few individuals but show differences in certain cranial characteristics as well as an overall smaller size for the eastern animals (Voss 1988). However, Voss (1988) considered these differences insufficient to consider both forms as distinct species.

Genetic and genealogical results combined with morphological differences presented herein suggest that populations currently allocated to *N. monticolus* east and west of the Andes represent two distinct biological entities. Mean *Cyt*b genetic distance between the two *Neusticomys* haplogroups (8.5 %) is in the order of mean genetic distances shown between species in other groups of sigmodontine rodents (e.g., *Abrothrix* ca. 5 to 10 % [Feijoo et al. 2010; D'Elía et al. in press]; species of the *Akodon boliviensis* species group: 2.8–7.7 % [Jayat et al. 2010]; *Eligmodontia*: 4.6–11.4 % [Mares et al. 2008]; *Juliomys*: ca. 12 % [Pardiñas et al. 2008]; *Melanomys*: 4.5–7.6 % [Hanson and Bradley 2008]; *Nectomys*: 7.36 % [Hanson and Bradley 2008]; *Oligoryzomys*: 4.45–15 % [Hanson et al. 2011; Palma et al. 2010; Richter et al. 2010; Rogers et al. 2009]; *Oryzomys*: 4.5–12.1 % [Hanson et al. 2010]; *Oxymycterus*: 2.5–9.6 % [Jayat et al. 2008]); *Rhipidomys*: 4.1–12.4 % [Costa et al. 2011]; *Scapteromys*: ca. 4.5 % [D'Elía and Pardiñas 2004]; *Sigmodon*: 8.5–20.8 % [Henson and Bradley 2009]). Although the species used for comparison are not the closest possible relatives to *Neusticomys*, all belong to other sigmodontine tribes and provide a diverse array of reference points. Further, the *Rbp*3 genetic distance between the two *Neusticomys* groups (0.8 %) is significant when compared to the 1.5 % difference between them and *Rheomys* (the only other Ichthyomini examined genetically). For a frame of reference, the *Rbp*3 genetic distance between *Neusticomys* and other members of the Sigmodontinae is only 6–8 % (compared to 15–20 % in *Cyt*b). Eastern and western Andean clades of *Neusticomys* previously assigned to *N. monticolus* showed enough genetic differentiation—despite the low sample size—to warrant further examination using other source of evidence (e.g., karyological, morphological, ecological studies; see Baker and Bradley 2006; Bradley and Baker 2001). In this particular situation, the additional investigation was mostly conducted some 21 years (Voss 1988) prior to the discovery of the highly differentiated haplogroups, as well as complemented here.

In addition to the new species described above, the morphometric data further support a unique group near Antiquia, Colombia, which Voss (2015) identified in his account of *N. monticolus* saying, "The few available specimens from the western Andes (Cordillera Occidental) of

Colombia exhibit morphological differences from Ecuadorean material and may represent a distinct species." We identify this subgroup as one requiring further molecular examination.

Conclusions

Ichthyomyini is the least studied tribe of sigmodontine rodents; studies on these mice are rare in the literature, limiting in the current century to the reporting of new geographic records (e.g., Leite et al. 2007; Miranda et al. 2012; Nunes 2002; Pacheco and Ugarte-Núñez 2011), the presentation of new data in reports of mammalian surveys (e.g., Lee et al. 2006a, 2006b, 2008; Voss et al. 2001), synthesis of available knowledge for one species (Packer and Lee 2007), and the description of a new species (e.g., Percequillo et al. 2005). As such, aside from Voss (1988) study, no comprehensive phylogenetic hypothesis is available for the tribe. Therefore, currently, it is not feasible to advance either a robust biogeographic hypothesis accounting for ichthyomyine diversification or one centered on *Neusticomys*. Similarly, given the scarcity of ichthyomyines in natural history collections and the difficulty of collecting them in the field (but see Pacheco and Ugarte-Núñez 2011), the task of performing an exhaustive phylogeographic study of any ichthyomyine species may be difficult to accomplish.

The present study is the first study of ichthyomyine alpha taxonomy since Percequillo et al. (2005) described *N. ferrerai* and the first taxonomic study focused on members of the tribe Ichthyomyini using sequence data. Given the results of our morphological assessment (and that of Voss 1988), it would be informative to perform an analysis of DNA sequences from specimens collected near Antioquia, Colombia.

Resumen

En este estudio se evaluó la estructura genética y morfológica de la variación de la especie *Neusticomys monticolus*. Distancias genéticas y estructura filogeográfica fueron estimadas en base a secuencias de un gen nuclear y otro mitocondrial. Los datos morfológicos fueron analizados estadísticamente. Los datos indican la existencia de dos grupos filogeográficos que difieren morfológicamente y corresponden a poblaciones de las vertientes este y oeste de los Andes. Estos grupos difieren en promedio en 8.5 % en el gen del citocromo b (0.8 % en el gen de la proteína de unión interfotorreceptor retinoide). Las poblaciones actualmente asignadas a *N. monticolus* constituyen un complejo de especies. El nombre *N. monticolus* es aquí restringido a las poblaciones de la vertiente occidental de los Andes. Las poblaciones de las estribaciones orientales de los Andes son asignadas a una nueva especies, a la cual es dado el nombre *Neusticomys vossi* sp. nov.

Appendix

Specimens examined—Specimens are arranged by species and locality. Specific identification numbers (museum or collector numbers) and GenBank accession numbers (*Cyt*b/*Rbp*3; for specimens included in molecular analysis) are listed in parentheses, respectively. Museum acronyms are as follows: Abilene Christian University Natural History Collections (ACUNHC); El Museo de Zoología, Pontificia Universidad Católica del Ecuador (QCAZ); Departamento de Microbiología, Universidad del Valle (HTC); Field Museum of Natural History (FMNH); Museum of Southwestern Biology (MSB); Museum of Vertebrate Zoology (MVZ); The Museum of Texas Tech University (TTU, TK—tissue collection); Thomas E. Lee (TEL; vouchers available at ACUNHC or QCAZ); University of Kansas Museum of Natural History (KU); and University of Michigan Museum of Zoology (UMMZ)

Akodon boliviensis—BOLIVIA: Tarija; 4.5 km E of Iscayachi (MSB68571, *Cyt*b [GenBank:KC841367]).

Akodon paranaensis—PARAGUAY (TTU108159, *Rbp*3 [EU649035]).

Abrothrix longipilis—ARGENTINA: Río Negro, La Veranda (MVZ1554494, *Rbp*3 [AY163557]). CHILE: Araucania; Fundo Hermanos Garcia (MSB205660, *Cyt*b [GenBank:GU564083]).

Arvicola terrestri—SWITZERLAND (MVZ155884, [GenBank:AY275106/AY277407]).

Cricetus cricetus—AUSTRIA: Niederosterreich; 1 km NE Gutramsdorf (MVZ155880, [GenBank:AY275109/AY277410]).

Neusticomys monticolus—COLOMBIA: Antioquia, Santa Bárbara (FMNH71218; FMNH71220; FMNH71221; FMNH71222; FMNH71223). COLOMBIA: Valle, Pichindé (HTC1372; HTC2180; HTC2237; HTC3365). ECUADOR: Chimborazo, Pauchi (AMNH66848). ECUADOR: Cotopaxi, Reserva Otonga (TK149025, [GenBank:KF359517]; TK149034, [GenBank:KF359518]). ECUADOR: Pichincha, Guarumal (UMMZ126298; UMMZ126299; UMMZ126298; UMMZ126299; UMMZ155789; UMMZ155790; UMMZ 155791; UMMZ155793; UMMZ155794). ECUADOR: Pichincha, Las Machinas (AMNH64634; AMNH64636; AMNH64637; AMNH64638; AMNH64639). ECUADOR: Pinchincha, Rio Pita (AMNH64627; AMNH64628; AMNH64629; AMNH64631; AMNH64632). ECUADOR: Pinchincha, San Ignacio (AMNH64625; AMNH64626). ECUADOR: Pichincha, Tandayapa Valley (QCAZ6446, [GenBank:KF359515/KR105606]; ACUNHC900/QCAZ 6531, [GenBank:KF359516/KR105605]).

Neusticomys vossi sp. nov.—COLOMBIA: Huila, San Antonio (FMNH71224; FMNH71225). ECUADOR: Bolivar, Sinche (AMNH62920). ECUADOR: Napo; 11 km SE of Baeza (QCAZ7830, [GenBank:KF359513/KR105608]). ECUADOR:Napo, Papallacta (AMNH244608;

AMNH244609; UMMZ126297; UMMZ155604; UMMZ 155605; UMMZ155606). ECUADOR: Napo; Volcan Sumaco (QCAZ8956, [GenBank:KF359514/KR105607]).

Oryzomys palustris—USA: Texas; Galveston, Virginia Point (TTU82920, [GenBank:DQ185382/EU273431]).

Phyllotis xanthopygus—ARGENTINA: Río Negro, Pilcanieyeu (MVZ182703, *Rbp*3 [GenBank:AY163632]). CHILE: Tarapaca; Parinacota; Arica *Cyt*b (FMNH133830, [GenBank:U86831]).

Reithrodon auritus—ARGENTINA: Rio Negro, Las Victorias (MVZ182704, [GenBank:EU579474/AY163634]).

Rheomys raptor—COSTA RICA: Puntarenas; Monteverde Cloud Forest Reserve, Quebrada Cuecha (KU159017, [GenBank:KF359512/AY163635]).

Sigmodon hispidus—USA: Texas; Cameron, Brownsville (TK32481, *Cyt*b [GenBank:AF425199]). USA: Kansas; Ellis, Hays (OK5840, *Rbp*3 [EU635707]).

Thomasomys erro—ECUADOR: Napo; 12 km N Cosanga (ACUNHC1137/QCAZ7649, [GenBank:EU579476/EU649057]).

Wiedomys pyrrhorhinos—BRAZIL: Minas Gerais; Ponte do Colatino (MVZ197566, *Cyt*b [GenBank:EU579477). BRAZIL: Bahia, Jaborandi (CRB1839, *Rbp*3 [AY163644]).

Competing interests

The authors declare that they have no competing interests.

Authors' contributions

JDH supervised the molecular work, drafted the manuscript, and processed the measurements. GD helped draft the manuscript, and proofed and modified the species description. SBA performed the molecular work and helped drafting the manuscript. SBC performed the statistical analyses. TEL and SB collected the specimens and measurements and provided guidance on the natural history of the genus. All authors read and approved the final version of the manuscript.

Acknowledgements

An Abilene Christian University Math/Science Research Council grant funded the trip to Cosanga Ecuador. M. Pinto, D. Wilson, and K. Helgen graciously loaned tissue from their field work. M. Pinto and A. Camacho provided pictures of the type and reference individuals. M. Mauldin assisted with the map edits. R. Voss allowed access to the original data sheets for his 1988 manuscript. Abilene Christian Natural History Collections, El Museo de Zoología, Pontificia Universidad Católica del Ecuador, The Museum of Texas Tech Natural Science Research Laboratory, and the University of Kansas Museum of Natural History for facilitating the tissue loans. Ecuadorian specimens were collected under the following research permits by the Ecuadorian Ministry of Environment: 012-IC-FAU-DNBAPVS/N (July 15th, 2005); 020-IC-FAU-DNBAPVS/MA (July 20th, 2006); 002-IC-FAU/FLO-DRFN-P/MA (July 30th, 2007); and 02-12 IC-FAU-FLO-DPAC/MA (May 18th, 2012).

Author details

[1]RTLGenomics, 4321 Marsha Sharp Fwy Door 2, Lubbock, TX 79407, USA. [2]Instituto de Ciencias Ambientales y Evolutivas, Universidad Austral de Chile, Valdivia, Chile. [3]True Health Diagnostics, Frisco, TX, USA. [4]Museo de Zoología Pontificia Universidad Católica del Ecuador, Quito, Ecuador. [5]Abilene Christian University, Abilene, TX, USA.

References

Anthony HE (1921) Preliminary report on Ecuadorean mammals. No 1. Am Mus Novit 20:1–6

Anthony HE (1929) Two genera of rodent from South America. Am Mus Novit 383:1–6

Baker RJ, Bradley RD (2006) Speciation in mammals and the genetic species concept. J Mammal 87:643–662

Bradley RD, Baker RJ (2001) A test of the genetic species concept: cytochrome-*b* sequences and mammals. J Mammal 82:960–973

Costa BMA, Geise L, Pereira LG, Costa LP (2011) Phylogeography of *Rhipidomys* (Rodentia: Cricetidae: Sigmodontinae) and description of two new species from southeastern Brazil. J Mamm 249:945–962

D'Elía G, Pardiñas UFJ (2015) Subfamily Sigmodontinae Wagner, 1843. In: Patton JL, Pardiñas UFJ, D'Elía G (eds) Mammals of South America, vol 2, Rodents. The University of Chicago Press, Chicago, Illinois, pp 63–70

D'Elía G, Pardiñas UFJ, Teta P, Patton JL (2007) Definition and diagnosis of a new tribe of sigmodontine rodents (Cricetidae: Sigmodontinae), and a revised classification of the subfamily. Gayana 71:187–194

D'Elía G, Teta P, Upham NS, Pardiñas UFJ, Patterson BD (in press) Description of a new soft-haired mouse, genus *Abrothrix* (Sigmodontinae), from the temperate Valdivian rainforest. J Mamm.

D'Elía G, Pardiñas UF (2004) Systematics of argentinean, Paraguayan, and uruguayan swamp rats of the genus *Scapteromys* (Rodentia, Cricetidae, Sigmodontinae). J Mammal 85:897–910

R Development Core Team (2009) R: a language and environment for statistical computing. R Foundation for Statistical Computing, Vienna, Austria, http://www.R-project.org. ISBN 3-900051-07-0

Feijoo M, D'Elía G, Pardiñas UFJ, Lessa EP (2010) Systematics of the southern Patgonian-Gueguian endemic *Abrothrix lanosus* (Rodentia: Soigmodontinae): phylogenetic position, karyotypic and morphological data. Mamm Biol 75:122–137

Hanson JD, Bradley RD (2008) Molecular diversity within *Melanomys caliginosus* (Rodentia: Oryzomyini): evidence for multiple species. Occas Papers Mus Texas Tech Univ 275:1–11

Hanson JD, Indorf JL, Swier VJ, Bradley RD (2010) Molecular diversity within the *Oryzomys palustris* complex: evidence for multiple species. J Mammal 91:336–347

Hanson JD, Utrera A, Fulhorst CF (2011) The delicate pygmy rice rat (*Oligoryzomys delicatus*) is the principal host of Maporal virus (family Bunyaviridae, genus *Hantavirus*). Vector Borne Zoonotic Dis 11:691–696

Henson DD, Bradley RD (2009) Molecular systematics of the genus *Sigmodon*: results from mitochondrial and nuclear gene sequences. Can J Zool 87:211–220

Jansa SA, Voss RS (2000) Phylogenetic studies on didelphid marsupials I. Introduction and preliminary results from nuclear IRBP gene sequences. J Mamm Evol 7:43–77

Jayat JP, D'Elía G, Pardiñas UFJ, Miotti MD, Ortiz PE (2008) A new species of the genus *Oxymycterus* (Mammalia: Rodentia: Cricetidae) from the vanishing Yungas of Argentina. Zootaxa 1911:31–51

Jayat JP, Ortiz PE, Pardiñas UFJ, Salazar-Bravo J, D'Elía G (2010) The *Akodon boliviensis* species group (Rodentia: Cricetidate: Sigmodontinae) in Argentina: species limits and distribution, with the description of a new entity. Zootaxa 2409:1–61

Jenkins PD, Barnett AA (1997) A new species of water mouse, of the genus *Chibchanomys* (Rodentia, Muridae, Sigmodontinae) from Ecuador. Bull Nat His Mus Lond (Zool) 63:123–128

Kimura M (1980) A simple method for estimating evolutionary rate of base substitutions through comparative studies of nucleotide sequences. J Mol Evol 16:11–120

Lee TE, Alvarado-Serrano D, Platt RN, Goodwiler GG (2006a) Report on a mammal survey of the Cosanga River drainage, Ecuador. Occas Papers Mus Texas Tech Univ 260:1–10

Lee TE, Packer JB, Alvarado-Serrano D (2006b) Results of a mammal survey of the Tandayapa Valley, Ecuador. Occas Papers Mus Texas Tech Univ 290:1–14

Lee TE, Burneo SF, Marchán MR, Roussos SA, Sebastian Vizcarra-Vàscomez R (2008) The mammals of the temperate forests of Volcán Sumaco, Ecuador. Occas Papers Mus Texas Tech Univ 276:1–10

Leite RN, Da Silva MNF, Gardner TA (2007) New records of *Neusticomys oyapocki* (Rodentia, Sigmodontinae) from a human-dominated forest landscape in northeastern Brazilian Amazonia. Mastozool Neotrop 14:257–261

Longmire JL, Maltbie M, Baker RJ (1997) Use of lysis buffer in DNA isolation and its implications for museum collections. Occas Papers Mus Texas Tech Univ 163:1–4

Mares MA, Braun JK, Coyner BS, Van den Bussche RA (2008) Phylogenetic and biogeographic relationships of gerbil mice *Eligmodontia* (Rodentia, Cricetidae) in South America, with a description of a new species. Zootaxa 1753:1–33

Martin Y, Gerlach G, Schlotter C, Meyer A (2000) Molecular phylogeny of European muroid rodents based on complete cytochrome-*b* sequences. Mol Phylogenet Evol 16:37–47

Martínez JJ, Ferro LI, Mollerach MI, Barquez RM (2012) The phylogenetic relationships of the Andean swamp rat genus *Neotomys* (Rodentia, Cricetidae, Sigmodontinae) based on mitochondrial and nuclear markers. Acta Theriol 57:277–287

Miranda CL, Rogério RV, Semedo TBF, Flores TA (2012) New records and geographic distribution extension of *Neusticomys ferreirai* and *N. oyapocki* (Rodentia, Sigmodontinae). Mammalia 76:335–339

Musser GG, Carleton MD (2005) Family Cricetidae. In: Wilson DE, Reeder DM (eds) Mammal species of the world: a taxonomic and geographic reference, vol 2, 3rd edn. Johns Hopkins University Press, Baltimore Maryland, pp 894–1522

Nunes A (2002) First record of *Neusticomys oyapocki* (Muridae, Sigmodontinae) from the Brazilian Amazon. Mammalia 66:445–447

Nylander JAA (2004) Mr.Modeltestv2. Program distributed by the author. Evolutionary Biology Centre, Uppsala University, Sweden

Pacheco V, Ugarte-Núñez J (2011) New records of Stolzmann's fish-eating rat *Ichthyomys stolzmanni* (Cricetidae, Sigmodontinae) in Peru: a rare species becoming a nuisance. Mamm Biol 76:657–661

Packer JB, Lee TE (2007) *Neusticomys monticola*. Mammal Species 805:1–3

Palma RE, Rodríguez-Serrano E, Rivera-Milla E, Hernandez CE, Salazar-Bravo J, Carma MI, Belmar-Lucero S, Gutierrez-Tapia P, Zeballos H, Yates TL (2010) Phylogenetic relationships of the pygmy rice rats of the genus *Oligoryzomys* Bangs, 1900 (Rodentia: Sigmodontinae). Zool J Linn Soc 160:551–566

Parada A, Pardiñas UF, Salazar-Bravo J, D'Elía G, Palma RE (2013) Dating an impressive Neotropical radiation: molecular time estimates for the Sigmodontinae (Rodentia) provide insights into its historical biogeography. Mol Phyl Evol 66:960–968

Pardiñas UF, Teta P, D'Elia G, Galliari C (2008) Rediscovery of Juliomys pictipes (Rodentia: Cricetidae) in Argentina: emended diagnosis, geographic distribution, and insights on genetic structure. *Zootaxa* 1758:29–44.

Peppers LL, Bradley RD (2000) Molecular systematics of the genus *Sigmodon*. J Mammal 81:332–343

Percequillo AR, Carmignotto AP, de Silva MJ (2005) A new species of *Neusticomys* (Ichthyomyini, Sigmodontinae) from central Brazilian Amazonia. J Mammal 86:873–880

Rannala B, Yang Z (1996) Probability distribution of molecular evolutionary trees: a new method of phylogenetic inference. J Mol Evol 43:304–311

Reig OA (1980) A new fossil genus of South American cricetid rodents allied to *Wiedomys*, with an assessment of the Sigmodontinae. J Zool 192:257–281

Richter MH, Hanson JD, Cajimat MN, Milazzo ML, Fulhorst CF (2010) Geographical range of Rio Mamore virus (family Bunyaviridae, genus *Hantavirus*) in association with the small-eared pygmy rice rat (*Oligoryzomys microtis*). Vector Borne Zoonotic Dis 10:613–620

Rogers DS, Arenas EA, González-Cózatl FX, Hardy DK, Hanson, JD, Lewis-Rogers N (2009) Molecular phylogenetics of *Oligoryzomys fulvescens* based on cytochrome *b* gene sequences, with comments on the evolution of the genus *Oligoryzomys*. Cervantes, F, ed, 60 Años de la Colección Nacional de Mamíferos del Instituto de Biología, Unam. Aportaciones al conocimiento y conservacíon de los mamíferos Mexicanos.

Ronquist F, Huelsenbeck JP (2003) MRBAYES 3: Bayesian phylogenetic inference under mixed models. Bioinformatics 19:1572–1574

Salazar-Bravo J, Pardiñas UF, D'Elía G (2013) A phylogenetic appraisal of Sigmodontinae (Rodentia, Cricetidae) with emphasis on phyllotine genera: systematics and biogeography. Zool Scripta 42:250–261

Sikes RS, Gannon WL, the Animal Care and Use Committee of the American Society of Mammalogists (2011) Guidelines of the American Society of Mammalogists for the use of wild mammals in research. J Mammal 92:235–253

Stanhope MJ, Czelusniak J, Si JS, Nickerson J, Goodman M (1992) A molecular perspective on Mammalian evolution from the gene encoding interphotoreceptor retinoid binding protein with convincing evidence for bat monophyly. Mol Phylogenet Evol 1:148–160

Tamura K, Dudley J, Nei M, Kumar S (2007) MEGA4: Molecular Evolutionary Genetics Analysis (MEGA) software version 4.0. Mol Biol Evol 24:1596–1599

Tate GHH (1931) Random observations on habits of South American mammals. J Mammal 12:248–256

Voss RS (1988) Systematics and ecology of ichthyomyine rodents (Muroidea): patterns of morphological evolution in a small adaptive radiation. Bull Am Mus Nat His 188:259–493

Voss RS (2015) Tribe Ichthyomyini Vorontsov, 1959. In: Patton JL, Pardiñas UFJ, D'Elía G (eds) Mammals of South America, vol 2, Rodents. The University of Chicago Press, Chicago, Illinois, pp 279–291

Voss RS, Lunde DP, Simmons NB (2001).The mammals of Paracou, French Guiana: a neotropical lowland rainforest fauna part 2. Nonvolant species. Bull. Am. Mus. Nat. His 3-236.

Weksler M (2003) Phylogeny of Neotropical Oryzomyine rodents (Muridae: Sigmodontinae) based on the nuclear IRBP exon. Mol Phylogenet Evol 29:331–349

Whiting AS, Bauer AM, Sites JW (2003) Phylogenetic relationships and limb loss in sub-Saharan African scincines lizards (Squamata: Scincidae). Mol Phylogenet Evol 29:582–598

10

The research on the formation mechanism of extraordinary oxidative capacity of skeletal muscle in hibernating ground squirrels (*Spermophilus dauricus*)

Shanfeng Jiang[1], Yunfang Gao[1*], Yangmei Zhang[1], Kun Liu[1], Huiping Wang[1] and Nandu Goswami[2]

Abstract

Background: Previous studies indicate that hibernating animals, under conditions of torpor for long periods, show an increased oxidative muscle fibers (type I) ratio and a decreased glycolytic muscle fibers (type II) ratio in skeletal muscle and accompanied by extraordinary oxidative ability. This observation is completely contrasted with non-hibernators, which show a shift of oxidative muscle fibers (type I) to glycolytic muscle fibers (type II). Presently, the mechanisms by which these changes occur remain unclear. To investigate the mechanism of high oxidative capacity of the skeletal muscles in hibernating ground squirrels, capillary density (CD), and capillary/fiber (C/F) were measured by immunohistochemistry. mRNA expression levels of hypoxia-inducible factor-1α (HIF-1α) and vascular endothelial growth factor (VEGF) were determined using real-time quantitative PCR assay. Spectrophotometry was applied to determine the activities of hexokinase (PK), pyruvate kinase (HK), and cytochrome c oxidase (CcO).

Results: In the soleus muscle (SOL), mRNA expression levels of HIF-1α and VEGF in torpor became slightly lower but were not statistically significant; they were, however, significantly higher in the arousal group. In hibernating animals, no significant change occurred in CD but C/F increased by 15 %. CcO showed the highest activity in torpor. There were no significant differences in the activities of HK and PK between the torpid animals and summer active animals in SOL. However, PK activity increased by 34 % after hibernation.

Conclusions: Oxidative capacity may be ensured by an increase of capillary supply of skeletal muscle in hibernating animals.

Keywords: Muscle disuse atrophy; Hibernation; C/F; *HIF-1α*; VEGF

Background

Muscle atrophy induced by disuse is an important clinical problem. Countermeasures that can prevent muscle atrophy as well as restore the exercise capacity (Boonyarom and Inui 2006; Williams et al. 2009) are regularly sought after. Despite long periods of hibernation, associated with muscle disuse, arousal of hibernators is associated with normal motion. During hibernation, hibernators are in a torpid state except for interbout-arousals. So skeletal muscle unloading causes its disuse. This model is applied to study treatment for muscle atrophy due to disuse (Ivakine and Cohn 2014; Gao et al. 2012). Their skeletal muscle mass may decrease slightly during hibernation, but limited or no changes in contractile performance have been demonstrated (Bodine 2013; James et al. 2013). This awakening with normal muscle function is important for obtaining food or escaping from predators. The mechanism by which these animals are able to preserve muscle function during hibernation is not clear.

* Correspondence: gaoyunf@nwu.edu.cn
[1]Key Laboratory of Resource Biology and Biotechnology in Western China (Northwest University), Ministry of Education, Xi'an 71069, China
Full list of author information is available at the end of the article

Mammalian skeletal muscle fibers can be classified into two kinds: type I and type II. Fibers of type I mainly rely on aerobic metabolism while fibers of type II mainly rely on anaerobic metabolism. Slow-twitch muscles (e.g., SOL) are mainly composed of type I muscle fibers. The transformation of muscle fiber from type I to type II, corresponding with a reduced oxidative capacity, is one of the mechanisms that have been identified to contribute to muscle atrophy in simulated microgravity (using hindlimb-unloading in animals or in bed rested subjects) as well as in spaceflight (Bricout et al. 1999). There is a decrease in the ability of SOL to oxidize long chain fatty acids during space flight and a decrease in respiratory capacity of whole SOL during muscle atrophy (Fell et al. 1985; Grichko et al. 2000). Although oxidative capacity reduced for whole SOL, oxidative enzyme activity in individual fibers seemed unchanged (Manchester et al. 1990) and glycolytic enzyme activity increased (Stein et al. 2002). Reduced oxidative capacity for whole SOL in hind limb suspended animals or simulated spaceflight may be attributed to the transformation of muscle fiber types. Inadequate oxygen supply, which occurred in skeletal muscle induced by the cephalad blood redistribution in these models (McDonald et al. 1992; Yamasaki et al. 2004) may play an important role in this transformation.

In contrast, although oxygen consumption of torpid animals is only 5–10 % of active animals (Muleme et al. 2006), it is reported that many hibernating species in steady-state torpor are not hypoxic. During torpor, Arctic ground squirrels (*Spermophilus parryii*), as well as other species of ground squirrels, are well oxygenated with normal to above normal arterial oxygen pressures (Drew et al. 2004; Frerichs et al. 1994; Ma et al. 2005). And previous studies indicated that in torpid animals, despite a long period of muscle disuse, skeletal muscle fibers showed a shift of type II to type I as well as increased oxidative muscle fibers ratio accompanied by decreased glycolytic muscle fibers ratio (Gao et al. 2012).

It appears that torpid animals have remarkable aerobic oxidation ability and why this occurs was explored in this paper. In skeletal muscles, capillary density (CD) and capillary/fiber (C/F) are both commonly used indicators for the evaluation of skeletal muscle oxygen supply (Duscha et al. 1999; Egginton et al. 2001). CD is the ratio of capillaries number to fibers area; C/F is the ratio between the number of capillaries and the number of muscle fibers. In non-hibernators, atrophy of skeletal muscle due to disuse leads to a relative increased CD because of the decreased cross section area. Though the CD increased, the significantly decreased ratio of C/F indicates a reduced oxygen supply instead of a real capillary proliferation in skeletal muscle (Desplanches et al. 1991; Kvist et al. 1995; Qin et al. 1997). Hypoxia-inducible factor-1 (HIF-1) is the nuclear factor of induction and regulation to hypoxia (Majmundar et al. 2010). Under hypoxic conditions, it can induce transcription of vascular endothelial growth factor (VEGF) (Wang et al. 2007). And VEGF is the most important angiogenic stimulators which can stimulate angiogenesis to increase local oxygen supply (Dai and Rabie 2007). Thus, changes in HIF-1α and VEGF are important indicators to determine the extent of hypoxia, and they could also be used as sensitive indicators to judge whether the organ is hypoxia.

Aerobic oxidation of skeletal muscle mainly depends on the mitochondria activity. Cytochrome c oxidase (CcO), as a mitochondrial marker enzyme, will change its activity with mitochondrial oxidative capacity. In the glycolysis process, the main functional enzymes are hexokinase (HK) and pyruvate kinase (PK). Their performance reflects the anaerobic metabolism ability of skeletal muscle.

Therefore, to study the possible mechanisms that could contribute to the switch in metabolism of skeletal muscle fiber types in torpid animals, the indictors for oxygen supply, such as CD and C/F, HIF-1α, and VEGF mRNA, the indictors for metabolic capability such as CcO, PK, and HK were detected in different periods of hibernation. To our knowledge, this is the first study that examines systematically these switch mechanisms from the perspective of oxygen supply and metabolism. We hypothesized that the adequate oxygen supply is a pre-requisite to assure a switch in skeletal muscle fiber type and that this would be accompanied by a corresponding changes in metabolic enzymes.

Methods

Animals

Daurian ground squirrels (*S. dauricus*) from the Weinan region, Shaanxi province of China, were matched for body mass and assigned to four groups of eight animals each: autumn group/pre-hibernation group (Pre-H), torpor group, interbout-arousal group, aroused from hibernation for more than 3 months group (Post-H).

The animals were raised individually in conventional plastic cages and provided with standard laboratory rat

Table 1 CD and C/F in soleus muscle (mean ± SD)

	Pre-hibernation	Torpor	Interbout-arousal	Posthibernation
CD (/mm^2)	14.16 ± 1.36	17.556 ± 2.60	13.96 ± 0.446	14.24 ± 1.17
C/F	2.89 ± 0.23	3.31 ± 0.40*	2.89 ± 0.16	2.92 ± 0.50

*$P < 0.05$, compared with pre-hibernation

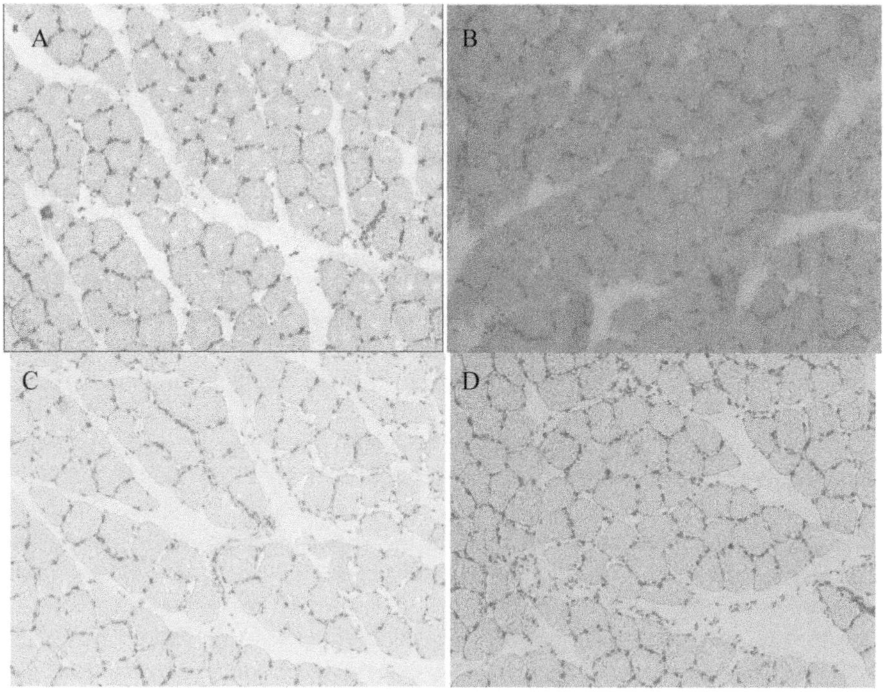

Fig. 1 Capillaries and fibers visualized by alkaline phosphatase staining. Numbers of capillaries (*black dots*) and fibers (*irregular circle*) were calculated. Fibers area was obtained by summing each fiber area. **a** pre-hibernation; **b** torpor; **c** interbout-arousal; **d** posthibernation

chow and water ad libitum and were housed in an environment with natural light, following international guidelines for the care and use of laboratory animals. The Laboratory Animal Care Committee of the People's Republic of China Ministry of Health approved all procedures. The ground squirrels were brought to a dark, hibernation room (4–6 °C) at a time in year when animals usually start to hibernate.

Sample preparation

Animals were anesthetized with sodium pentobarbital (45 mg/kg body weight i.p.) and SOL was removed and weighed to 0.1 mg accuracy (Sartorius, BS210S, Germany). Muscle samples were stored at −80 °C before being used for enzymatic analyses.

CD and C/F of the SOL

Muscle samples of 5 mm were cut at the SOL and then dipped in paraformaldehyde solution of 4 % for 12 h. Samples were then transferred into a sucrose solution of 30 %, embedded vertically with an Optimal Cutting Temperature gum (Tissue-Tek®, CA 90501, Sakura Finetek USA, Inc., Torrance, USA), and serial slices of 10 μm were cut in a cryostat (Leica, CM1850, Wetzlar, Germany) at −25 °C. Slices were placed on microscope slides which were pretreated with Poly-L-Lysine (10 g/l). The frozen sections were fixed for 3 min and then incubated darkly for 15 min at 37 °C. Sections were pictured with a 20× magnification microscope. CD and C/F were calculated by dividing the number of capillaries in a given area by the number or area of fibers, respectively (Egginton et al. 2001).

mRNA expression levels of HIF-1α and VEGF

Total RNA was isolated using RNAiso Plus (TaKaRa bio, Dalian, China) according to the manufacturer's protocol. All materials used for RNA preparation were treated with 0.1 % (v/v) diethylpyrocarbonate (DEPC) and autoclaved. The OD values of 260 and 280 nm were measured to determine the RNA content. Reverse transcription was performed according to manufacturer's protocol of TAKARA reagent (TaKaRa bio, Dalian,

Table 2 Relative expression amount of HIF-1α mRNA and VEGF mRNA in soleus muscle (mean ± SD)

	Pre-hibernation	Torpor	Interbout-arousal	Posthibernation
HIF-1α mRNA	0.0110600 ± 0.00418182	0.0066100 ± 0.00130571	0.0057983 ± 0.00130259	0.0196200 ± 0.00710753#
VEGF mRNA	0.0198600 ± 0.00462688	0.0134717 ± 0.00294716	0.0115067 ± 0.00259340*	0.0345333 ± 0.01308872#

*$P < 0.05$, compared with Interbout-arousal; #$P < 0.05$, compared with torpor

China). Primers for real-time PCR were designed as follows:

HIF-1α forward: 5′-TTTTAATACCCTCTGATTTAGCA TGT-3′
HIF-1α reverse: 5′-AGCTCTGAGTAATTCTTCACCC-3′
VEGF mRNA forward: 5′-CTGTACCTCCACCATGCC AAGT-3′
VEGF mRNA reverse: 5′-ACGCACTCCAGGGCTTCA T-3′
GAPDH forward: 5′-GACAACTTTGGCATCGTGGA-3′
GAPDH reverse: 5′-ATGCAGGGATGATGTTCTGG-3′

Two microliter of 100 ng DNA template was added into fluorescent quantitation PCR reaction system (0.8 μL primers each, 10 μL SYBR Green, and 6.4 μL sterile purified water). Final concentration of primers is 100 ng and 0.4 μmol/L. PCR reaction parameters were set according to the instructions. To analyze the relative concentration of mRNA, $2^{-\Delta\Delta ct}$ method was used (Yu et al. 2012).

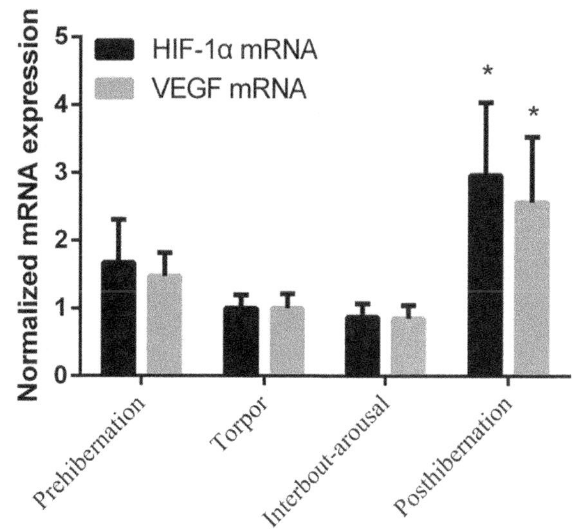

Fig. 2 Normalized mRNA expression of HIF-1α and VEGF. *$P < 0.05$, compared with torpor

Enzyme activities of HK, PK, and CcO

Muscle samples were homogenized and then centrifuged at 2500 rpm for 10 min before supernatant collection. Supernate was separately processed to detect HK and PK. Enzyme activities of HK, PK, and CcO were obtained using Hexokinase Assay Kit, Pyruvate Kinase Assay Kit, and Cytochrome c Oxidase Assay Kit, respectively (NJJCBIO; Nanjing, China).

Statistical analyses

SPSS 16.0 was used for all statistical analyses. Data are indicated as means ± SD. A one-way ANOVA was used to determine overall differences; Fisher's LSD post hoc test was used to determine group differences. ANOVA-Dunnett's T3 method was used when no homogeneity was detected. Significance was assumed with $P < 0.05$.

Results

Compared with the Pre-H group, CD of SOL increased slightly in the hibernating group but not statistically significant ($P > 0.05$). C/F significantly increased by 15 % ($P < 0.05$). In the interbout-arousal group, the CD and the ratio of C/F did not significantly changed ($P > 0.05$) (Table 1, Fig. 1).

Compared with the Pre-H group, expression levels of HIF-α1 and VEGF mRNA decreased in the hibernating group, albeit not significantly ($P > 0.05$). The Post-H group which had the highest expression levels of two genes, increased by about 190 and 150 % in HIF-1α and VEGF mRNA, respectively, compared with the hibernating group ($P < 0.05$) (Table 2, Fig. 2).

CcO activity of hibernating group achieved the highest level (Table 3). It increased by 72.03 % when compared with that of Pre-H group. Following hibernation it decreased by 61.69 %.

In SOL, there was no significance of difference of the activity of HK among the groups (Table 4).

Activity of the PK appeared to decrease in torpor when compared with the pre-hibernation group but not significantly. Compared with the hibernating group, PK activity increased by 34 % in the post-H group.

Discussion

The most important finding of our study is that capillary supply and oxidative enzyme activity increases, and glycolytic enzyme seems unchanged in the slow-twitch soleus muscle of the hibernating animals.

From the perspective of capillary supply, CD showed no significant changes in the hibernating group, but the C/F increased 15 %. This suggests that vascular proliferation occurred in SOL, which, in turn, could have enhanced the blood oxygen supply thus ensuring that there was enough oxygen for consumption to meet up the demands for increased aerobic oxidation. A limit of this measurement is unable to ascertain whether any change in capillary supply was predominantly associated

Table 3 Changes of CcO activity in different periods (mean ± SD)

	Summer active	Pre-hibernation	Torpor	Posthibernation
CcO activity in SOL (U/ml)	0.9721 ± 0.1949##	0.4382 ± 0.0734	0.7538 ± 0.2404##	0.2888 ± 0.0426**

##$P < 0.01$, ##compared with pre-hibernation; **$P < 0.01$, **compared with torpor

Table 4 The changes of glycolytic enzyme activity in soleus of ground squirrels (mean ± SD)

	Pre-hibernation	Torpor	Interbout-arousal	Posthibernation
HK activity (U/gprot)	31.08752 ± 8.56072	32.88304 ± 16.82466	38.85082 ± 7.63251	36.8637 ± 13.96182
PK activity (U/gprot)	28.15717 ± 4.40807	22.22285 ± 3.29632	29.12844 ± 8.54316	33.76566 ± 6.7182**

**$P < 0.01$, compared with pre-hibernation

with a particular muscle fiber. Considering spatial heterogeneity (Egginton et al. 2001), it is possible that vascular proliferation occurred nearby slow-twitch muscle fiber. But global oxygen supply to SOL should be increased since hibernating animals are well oxygenated (Drew et al. 2004; Frerichs et al. 1994; Ma et al. 2005; Maginniss and Milsom 1994).

The expression of HIF-1α mRNA decreased by 50 % during the entire torpor period in medial gastrocnemius and SOL (slow-twitch) in ground squirrel compared with that in September. For lateral gastrocnemius (fast twitch), mRNA expression levels of HIF-1α of torpor period increased by 400 % compared with before hibernation, indicating that lateral gastrocnemius muscle is in a state of hypoxia (Nowell et al. 2011). The results of our study show a similar decreasing trend for HIF-1α mRNA in SOL. According to previous studies, HIF-1α protein levels increased significantly in skeletal muscle (Maginniss and Milsom 1994; Morin and Storey 2005). But no significant changes were seen in HIF-1α mRNA expression levels in hibernating thirteen-lined ground squirrels. Thirteen-lined ground squirrels may be similar to golden-mantled ground squirrels (*S. lateralis*) which become hypoxic during torpor owing to long periods of apnea (Drew et al. 2004).

Previous studies have shown a reduction in the oxidative capacity of skeletal muscle fibers and decreases in CcO activity, in 14-day tail-suspended animals (a model for disuse atrophy). Additionally, it was associated with transformation from type I muscle fibers to type II muscle fibers (Roy et al. 1996). But in our study, we observed an elevation in the CcO levels during hibernation disuse, which is consistent with a switch from fibers type II fibers to type I muscle fibers.

Hexokinase is a rate-limiting enzyme in the glycolytic pathway. Its activity has been shown to be significantly reduced during torpor in ground squirrel (Abnous and Storey 2008). Our results show that compared with the Pre-H group, HK and PK activities of the hibernating groups did not change significantly. But PK activity during Post-H increased significantly ($P < 0.01$). According to the work of Dr. Matt Andrew's lab, brown adipose tissue may play critical roles in high efficiency fat catabolism, non-shivering thermogenesis, and transitions into and out of the torpid state (Hampton et al. 2013). Thus, skeletal muscle may experience a shift in substrate utilization away from glucose towards fat. That may result in a normal HK and PK level.

Activity changes of oxidative enzyme and glycolytic enzyme are in consistent with the transformation of muscle fiber from type II to type I. But it is completely contrasted to that of disuse muscle atrophy, whose glycolytic enzyme activity increased and oxidative enzyme decreased or unchanged. This suggested that metabolism enzyme may be important indictors or inducers for different muscle types' shift.

Taken together, all above findings imply that although the hibernating ground squirrels are in a low oxygen environment, they have sufficient capillary supply to maintain daily oxidative metabolism and even to provide some additional oxygen for arousal. It appears that the skeletal muscles of torpid animals possess adequate blood oxygen supply.

Conclusions

Our results show that capillary supply of skeletal muscle increases in torpid animals. This may ensure large increases in oxidative capacity during hibernation which may play important roles in allowing the skeletal muscle to achieve the transformation of type I muscle to type II muscle.

Abbreviations

C/F: capillary/fiber; CcO: cytochrome c oxidase; CD: capillary density; HIF-1: hypoxia-inducible factor-1; HK: hexokinase; PK: pyruvate kinase; SOL: soleus muscle; VEGF: vascular endothelial growth factor.

Competing interests

The authors declare that they have no competing interests.

Authors' contributions

YFG and HPW conceived and designed the experiments. SFJ, YMZ, and KL performed the experiments and analyzed the data. SFJ, YFG, HPW, and NG wrote the manuscript. All authors read and approved the final manuscript.

Acknowledgements

This work was supported by funds from the National Natural Science Foundation of China (Grant No. 31270455), the Specialized Research Fund for the Doctoral Program of Higher Education of China (Grant No.20116101110013), and the International Scientific and Technological Cooperation Projects in Shaanxi Province of China (Grant No. 2013KW26-01).

Author details

[1]Key Laboratory of Resource Biology and Biotechnology in Western China (Northwest University), Ministry of Education, Xi'an 71069, China. [2]Institute of Physiology, Center of Physiological Medicine, Medical University Graz, Graz, Austria.

References

Abnous K, Storey KB (2008) Skeletal muscle hexokinase: regulation in mammalian hibernation. Mol Cell Biochem 319(1–2):41–50. doi:10.1007/s11010-008-9875-5

Bodine SC (2013) Hibernation: the search for treatments to prevent disuse-induced skeletal muscle atrophy. Exp Neurol 248:129–135. doi:10.1016/j.expneurol.2013.06.003

Boonyarom O, Inui K (2006) Atrophy and hypertrophy of skeletal muscles: structural and functional aspects. Acta Physiol (Oxford, England) 188(2):77–89. doi:10.1111/j.1748-1716.2006.01613.x

Bricout VA, Serrurier BD, Bigard AX, Guezennec CY (1999) Effects of hindlimb suspension and androgen treatment on testosterone receptors in rat skeletal muscles. Eur J Appl Physiol Occup Physiol 79(5):443–448. doi:10.1007/s004210050535

Dai J, Rabie AB (2007) VEGF: an essential mediator of both angiogenesis and endochondral ossification. J Dent Res 86(10):937–950

Desplanches D, Mayet MH, Ilyina-Kakueva EI, Frutoso J, Flandrois R (1991) Structural and metabolic properties of rat muscle exposed to weightlessness aboard Cosmos 1887. Eur J Appl Physiol Occup Physiol 63(3–4):288–292

Drew KL, Harris MB, LaManna JC, Smith MA, Zhu XW, Ma YL (2004) Hypoxia tolerance in mammalian heterotherms. J Exp Biol 207(Pt 18):3155–3162. doi:10.1242/jeb.01114

Duscha BD, Kraus WE, Keteyian SJ, Sullivan MJ, Green HJ, Schachat FH, Pippen AM, Brawner CA, Blank JM, Annex BH (1999) Capillary density of skeletal muscle: a contributing mechanism for exercise intolerance in class II-III chronic heart failure independent of other peripheral alterations. J Am Coll Cardiol 33(7):1956–1963

Egginton S, Fairney J, Bratcher J (2001) Differential effects of cold exposure on muscle fibre composition and capillary supply in hibernator and non-hibernator rodents. Exp Physiol 86(5):629–639

Fell RD, Steffen JM, Musacchia XJ (1985) Effect of hypokinesia-hypodynamia on rat muscle oxidative capacity and glucose uptake. Am J Physiol 249(3 Pt 2):R308–R312

Frerichs KU, Kennedy C, Sokoloff L, Hallenbeck JM (1994) Local cerebral blood flow during hibernation, a model of natural tolerance to "cerebral ischemia". J Cereb Blood Flow Metab 14(2):193–205. doi:10.1038/jcbfm.1994.26

Gao YF, Wang J, Wang HP, Feng B, Dang K, Wang Q, Hinghofer-Szalkay HG (2012) Skeletal muscle is protected from disuse in hibernating dauria ground squirrels. Comp Biochem Physiol A Mol Integr Physiol 161(3):296–300. doi:10.1016/j.cbpa.2011.11.009

Grichko VP, Heywood-Cooksey A, Kidd KR, Fitts RH (2000) Substrate profile in rat soleus muscle fibers after hindlimb unloading and fatigue. J Appl Physiol 88(2):473–478

Hampton M, Melvin RG, Andrews MT (2013) Transcriptomic analysis of brown adipose tissue across the physiological extremes of natural hibernation. PLoS One 8(12), e85157. doi:10.1371/journal.pone.0085157

Ivakine EA, Cohn RD (2014) Maintaining skeletal muscle mass: lessons learned from hibernation. Exp Physiol 99(4):632–637. doi:10.1113/expphysiol.2013.074344

James RS, Staples JF, Brown JC, Tessier SN, Storey KB (2013) The effects of hibernation on the contractile and biochemical properties of skeletal muscles in the thirteen-lined ground squirrel, Ictidomys tridecemlineatus. J Exp Biol 216(Pt 14):2587–2594. doi:10.1242/jeb.080663

Kvist M, Hurme T, Kannus P, Jarvinen T, Maunu VM, Jozsa L, Jarvinen M (1995) Vascular density at the myotendinous junction of the rat gastrocnemius muscle after immobilization and remobilization. Am J Sports Med 23(3):359–364

Ma YL, Zhu X, Rivera PM, Toien O, Barnes BM, LaManna JC, Smith MA, Drew KL (2005) Absence of cellular stress in brain after hypoxia induced by arousal from hibernation in Arctic ground squirrels. Am J Physiol Regul Integr Comp Physiol 289(5):R1297–R1306. doi:10.1152/ajpregu.00260.2005

Maginniss LA, Milsom WK (1994) Effects of hibernation on blood oxygen transport in the golden-mantled ground squirrel. Respir Physiol 95(2):195–208

Majmundar AJ, Wong WJ, Simon MC (2010) Hypoxia-inducible factors and the response to hypoxic stress. Mol Cell 40(2):294–309. doi:10.1016/j.molcel.2010.09.022

Manchester JK, Chi MM, Norris B, Ferrier B, Krasnov I, Nemeth PM, McDougal DB Jr, Lowry OH (1990) Effect of microgravity on metabolic enzymes of individual muscle fibers. FASEB J 4(1):55–63

McDonald KS, Delp MD, Fitts RH (1992) Effect of hindlimb unweighting on tissue blood flow in the rat. J Appl Physiol (1985) 72(6):2210–2218

Morin P Jr, Storey KB (2005) Cloning and expression of hypoxia-inducible factor 1α from the hibernating ground squirrel, Spermophilus tridecemlineatus. Biochim Biophys Acta—Gene Struct Expr 1729(1):32–40, http://dx.doi.org/10.1016/j.bbaexp.2005.02.009

Muleme HM, Walpole AC, Staples JF (2006) Mitochondrial metabolism in hibernation: metabolic suppression, temperature effects, and substrate preferences. Physiol Biochem Zool 79(3):474–483. doi:10.1086/501053

Nowell MM, Choi H, Rourke BC (2011) Muscle plasticity in hibernating ground squirrels (Spermophilus lateralis) is induced by seasonal, but not low-temperature, mechanisms. J Comp Physiol B 181(1):147–164. doi:10.1007/s00360-010-0505-7

Qin L, Appell HJ, Chan KM, Maffulli N (1997) Electrical stimulation prevents immobilization atrophy in skeletal muscle of rabbits. Arch Phys Med Rehabil 78(5):512–517

Roy RR, Baldwin KM, Edgerton VR (1996) Response of the neuromuscular unit to spaceflight: what has been learned from the rat model. Exerc Sport Sci Rev 24:399–425

Stein TP, Schluter MD, Galante AT, Soteropoulos P, Tolias PP, Grindeland RE, Moran MM, Wang TJ, Polansky M, Wade CE (2002) Energy metabolism pathways in rat muscle under conditions of simulated microgravity. J Nutr Biochem 13(8):471–478, http://dx.doi.org/10.1016/S0955-2863(02)00195-X

Wang Y, Wan C, Deng L, Liu X, Cao X, Gilbert SR, Bouxsein ML, Faugere MC, Guldberg RE, Gerstenfeld LC, Haase VH, Johnson RS, Schipani E, Clemens TL (2007) The hypoxia-inducible factor alpha pathway couples angiogenesis to osteogenesis during skeletal development. J Clin Invest 117(6):1616–1626. doi:10.1172/jci31581

Williams D, Kuipers A, Mukai C, Thirsk R (2009) Acclimation during space flight: effects on human physiology. Can Med Assoc J 180(13):1317–1323. doi:10.1503/cmaj.090628

Yamasaki M, Shimizu T, Katahira K, Waki H, Nagayama T, O-ishi H, Katsuda S, Miyake M, Miyamoto Y, Wago H, Okouchi T, Matsumoto S (2004) Spaceflight alters the fiber composition of the aortic nerve in the developing rat. Neuroscience 128(4):819–829, http://dx.doi.org/10.1016/j.neuroscience.2004.07.022

Yu H, Zhu M, Qin Y, Zhong Y, Yan H, Wang Q, Bian H, Li Z (2012) Analysis of glycan-related genes expression and glycan profiles in mice with liver fibrosis. J Proteome Res 11(11):5277–5285. doi:10.1021/pr300484j

11

The poorly known genus *Ventrifurca* Roewer, 1913 revisited (Opiliones: Cranaidae)

Osvaldo Villarreal M[1*], Adriano B Kury[1] and Ricardo Pinto-da-Rocha[2]

Abstract

Background: The distinction of genera in Cranaidae (as in many other Gonyleptoidea) is traditionally made on the basis of a combination of armature of scutal areas, tarsomere counts, and some secondary sexual structures (which may vary among males). This historically led to the creation of non-natural groups of species as meaningless genera. Here some nominal genera of Cranaidae are addressed, and this paucity of diagnostic characters is alleviated with more detailed diagnoses and descriptions.

Results: The poorly known genus *Ventrifurca* Roewer, 1913, hitherto monotypic, is revisited. The genera *Microcranaus* Roewer, 1913 (with three species), *Cayabeus* Roewer, 1932 (monotypic), and *Angistrius* (monotypic) are here proposed as junior subjective synonyms of *Ventrifurca*. Both type species of *Microcranaus* and *Cayabeus* are considered junior subjective synonyms of *Ventrifurca albipustulata*; thus, *Cayabeus perlatus* Roewer, 1932 and *Microcranaus pustulatus* Roewer, 1913 = *Ventrifurca albipustulata* Roewer, 1913. *Ventrifurca abnormis* comb. nov. is proposed for *Angistrius abnormis* Roewer, 1932. *Rhopalocranaus dybasi* Goodnight & Goodnight, 1947, currently *Neocranaus dybasi*, is transferred to *Ventrifurca*, and the new combination *Ventrifurca dybasi* comb. nov. is herein proposed for it. *Ventrifurca albipustulata* is redescribed, and for the first time, the male genitalia is described and illustrated, and SEM pictures are offered. The new species *Ventrifurca caffeinica* sp. nov. is described from Quindío department, Colombia. A new diagnosis to *Ventrifurca* is proposed. The relationships of *Ventrifurca* are discussed. The remaining two species of *Microcranaus* are transferred to the family Manaosbiidae: (1) *Microcranaus columbianus* newly combined as *Rhopalocranaus columbianus* (Roewer, 1963) and (2) *Microcranaus gracilis* newly combined as *Camelianus gracilis* (Roewer, 1913). New diagnoses are given for both species.

Conclusions: A non-monotypic genus of Andean Cranaidae is defined, associated with a well-defined geographic region, and unloaded of extraneous species which belong to another family. Recognition of variation in heteromorphic males reveals past errors of assignment when only limited samples were available for study.

Background

The family Cranaidae is a group of mainly Andean-Amazonian harvestmen, only recognized in the end of the twentieth century (Kury 1994) and whose diversity is underestimated. Its taxonomy is unsatisfactory and plagued by many monotypic and poorly known genera. The monophyly of this taxon has been recovered in some phylogenetic studies (Orrico and Kury 2009; Sharma and Giribet 2011; Pinto-da-Rocha et al. 2014). Nevertheless, a molecular study by Pinto-da-Rocha et al. (2014) considered it as a member of the Gonyleptidae. However, because Orrico and Kury (2009) found that of the four traditional subfamilies of cranaids, two are microdiverse and the other two are paraphyletic, they proposed to abandon the use of subfamilies at least as currently defined. On the other hand, Kury (2012b) proposed the inclusion of Zamorinae as a fifth subfamily wich, posteriorly was transferred to Nomoclastidae by Kury & Villarreal (2015). In view of this uncertainty in recent literature, the use of subfamilies in Cranaidae is meant here only for convenience of reference of traditional assemblages of species. A phylogenetic analysis project to understand the internal and external relationships of Cranaidae/Cranainae is ongoing by the first author. Cranaidae (as a family) is used here as a neutral term.

Recently, some works have been published addressing varied taxonomy and systematic issues in the family,

* Correspondence: osvaldovillarreal@gmail.com
[1]Departamento de Invertebrados, Museu Nacional/UFRJ, Quinta da Boa Vista, São Cristóvão, 20.940-040 Rio de Janeiro, RJ, Brazil
Full list of author information is available at the end of the article

primarily three systematic works, two papers focusing on the relationship of the genus *Phareicranaus* Roewer, 1913 (Pinto-da-Rocha and Kury 2003; Pinto-da-Rocha and Bonaldo 2011) and another focused on testing the monophyly of Stygnicranainae (Orrico and Kury 2009). Subsequently, some contributions on the alpha taxonomy of Cranaidae were published, either as (1) transfer of taxa from other families to Cranaidae (Pinto-da-Rocha and Hara 2009; Pinto-da-Rocha et al. 2012), (2) description of new taxa (Kury 2012a, b), (3) detection of 'false cranaids' (Villarreal and Kury 2012), or (4) synonymy or revalidation and redescription of some genera (Hara et al. 2014). Those contributions tried to clarify the definition of the genera composing this family, but the taxonomy of cranaids is still confused and most genera are poorly known.

In this paper, the genus *Ventrifurca* Roewer, 1913 is revisited, its type species is redescribed, a new species is described from Quindío department, Colombia, and a new generic diagnosis is offered. *Cayabeus* Roewer, 1932 and *Microcranaus* Roewer, 1913 are considered junior subjective synonyms of *Ventrifurca*. Two spurious species of *Microcranaus* are transferred to Manaosbiidae. Comments about the relationships of the genus *Ventrifurca* within the family are presented, and a diagnostic character in the genital macrosetae is detected for a group of cranaid genera.

Historical background

Roewer (1913b: 349) described the new subfamily Cranainae in Gonyleptidae with some old and many new genera and species. Of relevance here, he described the following as new: (1) the genus *Microcranaus*, based on two new species - *Microcranaus pustulatus* from 'Ecuador: Cachab' and *Microcranaus gracilis* from 'Surinam: Saramacca'. He designated *Microcranaus pustulatus* Roewer, 1913 as type species of the genus; (2) the monotypic genus *Ventrifurca* based on the new species *Ventrifurca albipustulata* from 'Columbien: Camelia'; and (3) the genus *Rhopalocranaus* with four new South American species, designating as type species *Rhopalocranaus marginatus* Roewer, 1913.

Posteriorly, Roewer (1932: 337 and 346) described in Gonyleptidae Cranainae two new monotypic genera - *Cayabeus*, based on the new species *Cayabeus perlatus* from 'Ecuador (Cayabé)', which is a misspelling of the town Cayambe (Pichincha), and *Angistrius*, based on the new species *Angistrius abnormis* from Alausi.

Goodnight and Goodnight (1947: 40) described in the wastebasket genus *Rhopalocranaus* (then in Gonyleptidae Cranainae) the new species *Rhopalocranaus dybasi* from 'mountains west of Cali, Department of Cauca, Colombia'. Soares and Soares (1948: 593, 609) perpetrated an unlucky synonymy, seemingly based on nothing: (1) *Rhopalocranaus* Roewer, 1913 = *Cranaus* Simon, 1879.

Kury (1997: 4) transferred *Rhopalocranaus* to Manaosbiidae. Kury (2003) in his catalog of New World Laniatores proposed many changes and updates, among which are the following: (1) updated and completed details of locality records of old literature and (2) transferred *Rhopalocranaus dybasi* to *Neocranaus* Roewer, 1913, creating the new combination *Neocranaus dybasi* (Goodnight and Goodnight, 1947).

Methods

The standard of description and morphological nomenclature follows Villarreal and Kury (2012) and Kury (2012a, b) and outlines of the dorsal scutum as described in Kury et al. (2007). MS = macrosetae of penis ventral plate. Other abbreviations used: CL = carapace length, CW = carapace width, AL = abdominal scutum length, AW = abdominal scutum width, DSL = dorsal scutum length, Tr = trochanter, Fe = femur, Pa = patella, Ti = tibia, Mt = metatarsus, Ta = tarsus. All measurements are in mm unless otherwise noted. Abbreviations of the repositories cited are FMNH (Field Museum of Natural History, Chicago), IAvH (Instituto Alexander von Humboldt, Colombia), MNHN (Muséum National d'Histoire Naturelle, Paris), MNRJ (Museu Nacional, Rio de Janeiro), QCAZ (Museo de Zoologia, Pontificia Universidad Católica de Quito), and SMF (Senckenberg Museum, Frankfurt). The A-E system of naming the macrosetae of ventral plate follows Kury and Villarreal (2015).

Results

In the following list, literature entries are not exhaustive (these can be found in Kury 2003), but rather restricted to the critical contributions.

Cranaidae Roewer, 1913

Ventrifurca Roewer, 1913

Ventrifurca Roewer, 1913b: 382; Kury, 2003: 100 (type species *Ventrifurca albipustulata* Roewer, 1913, by monotypy).

Microcranaus Roewer, 1913b: 352; Soares and Soares, 1948: 608; Kury, 2003: 95 (type *Microcranaus pustulatus* Roewer, 1913, by original designation). Syn. nov.

Cayabeus Roewer, 1932: 337; Kury, 2003: 91 (type species *Cayabeus perlatus* Roewer, 1932, by monotypy). Syn. nov.

Angistrius Roewer 1932: 346; Kury 2003: 90 [type species: *Angistrius abnormis* Roewer, 1932, by monotypy]. Syn. nov.

Diagnosis. Dorsal scutum type beta, elongate pyriform in dorsal view (Figures 1A, 2A, and 3A). Dorsal scutum sexually dimorphic in lateral view, in males frontal hump higher, reaching eye height and abdominal scutum flat, resulting in spine of area III erect (in females frontal

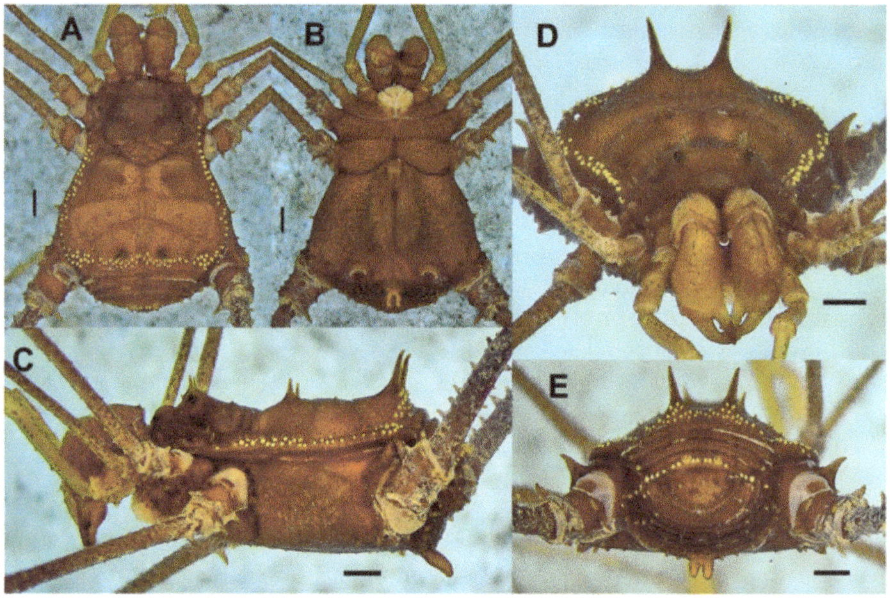

Figure 1 *Ventrifurca albipustulata* (MNRJ 19197) male. **(A)** Habitus, dorsal view. **(B)** Ventral view. **(C)** Lateral view. **(D)** Dorsofrontal view. **(E)** Posterior view. Scale bars = 1 mm.

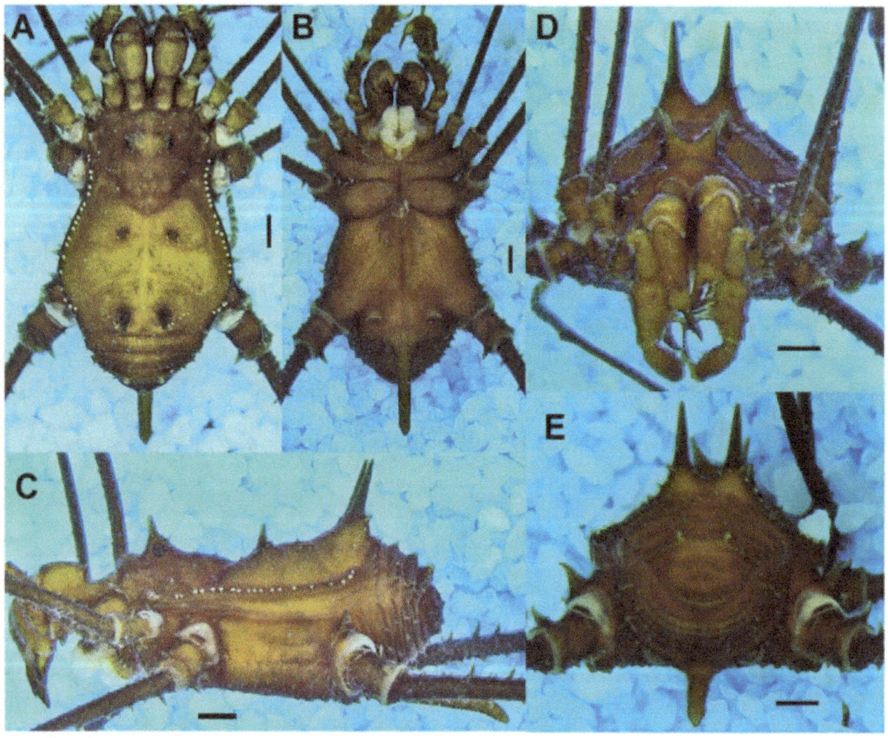

Figure 2 *Ventrifurca caffeinica* sp. nov. (MNRJ 8351) male holotype. **(A)**. Habitus, dorsal view. **(B)** Ventral view. **(C)** Lateral view. **(D)** Frontal view. **(E)** Posterior view. Scale bars = 1 mm.

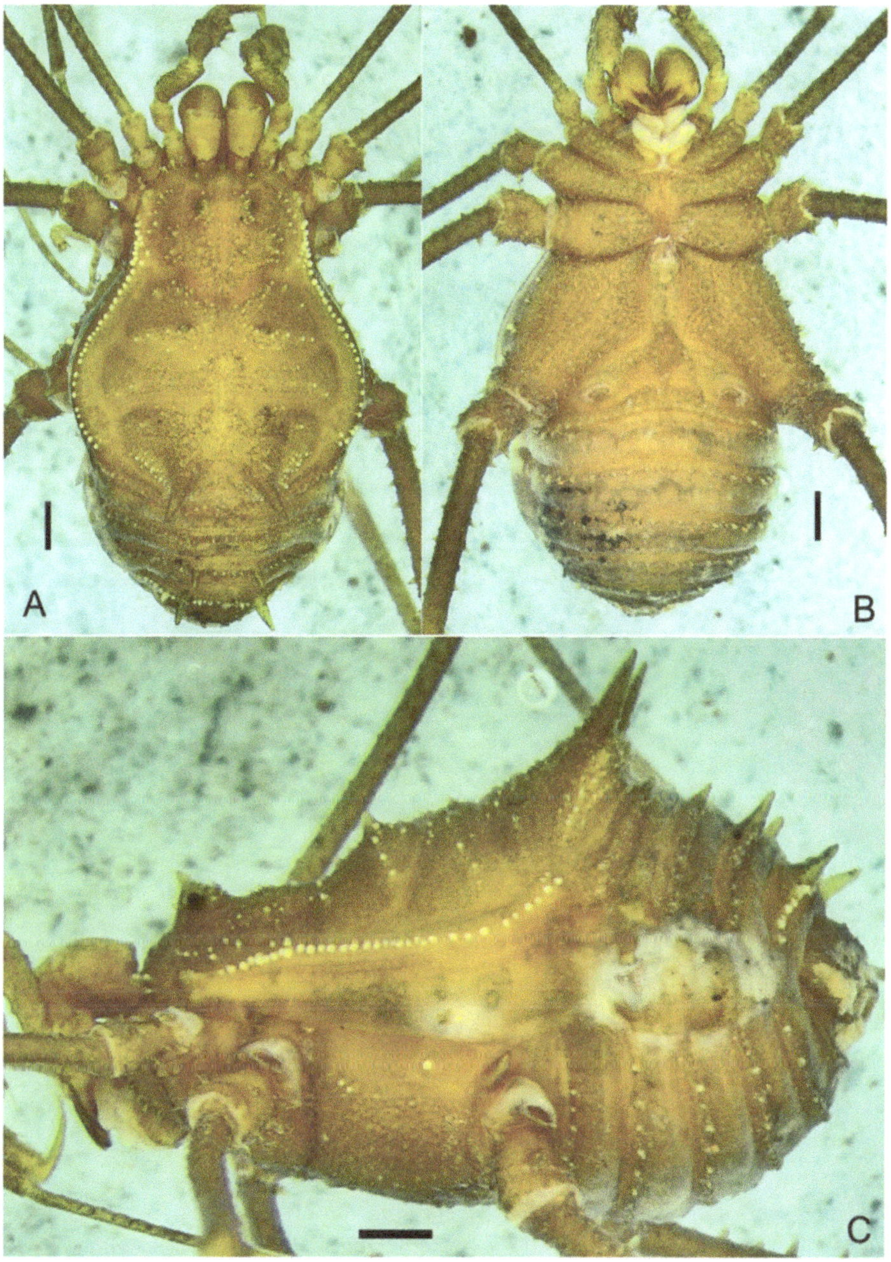

Figure 3 *Ventrifurca dybasi* (FMNH-INS 000 011 250) female holotype. **(A)** Habitus dorsal view. **(B)** Ventral view. **(C)** Lateral view. Scale bars = 1 mm.

hump is very low and abdominal scutum is bulged, resulting in spines of area III inclined backwards) (e.g., Figure 4E,F). Mesotergum divided into 3 well-marked areas, area I divided into left and right halves by longitudinal straight groove (Figures 1A, 2A, 5A and 6A). Mesotergum smooth with a pair of high spines on area I (higher in males), a row of granules on II and a pair of very high spines on area III (Figures 1A, 2A,C,D, 5A and 6A). Lateral margin of scutum with irregular rows of yellow tubercles, extending (completely or not) onto posterior border of area III (Figures 1A,C,D,E, 2A,C,D, 3A,C and 7A,C,E). Free tergite III with a pair of lighter spines, larger in females (Figures 5A,B and 6A,E). Stigmatic area of males with variably developed process on the medial zone of the posterior border, from 2 small buttons to 2 conspicuous parallel processes fused at base or isolated (Figures 1B,E and 5C,E) or a large cylindrical process (Figures 2B,C and 6A,D,F). Stigmatic area sexually dimorphic, shorter in females, elongate in males. Pedipalpi slightly elongated with femur and patella cylindrical and unarmed, tibia enlarged with few ventrolateral spiniform tubercles (Figure 8D,E) or with little tubercles

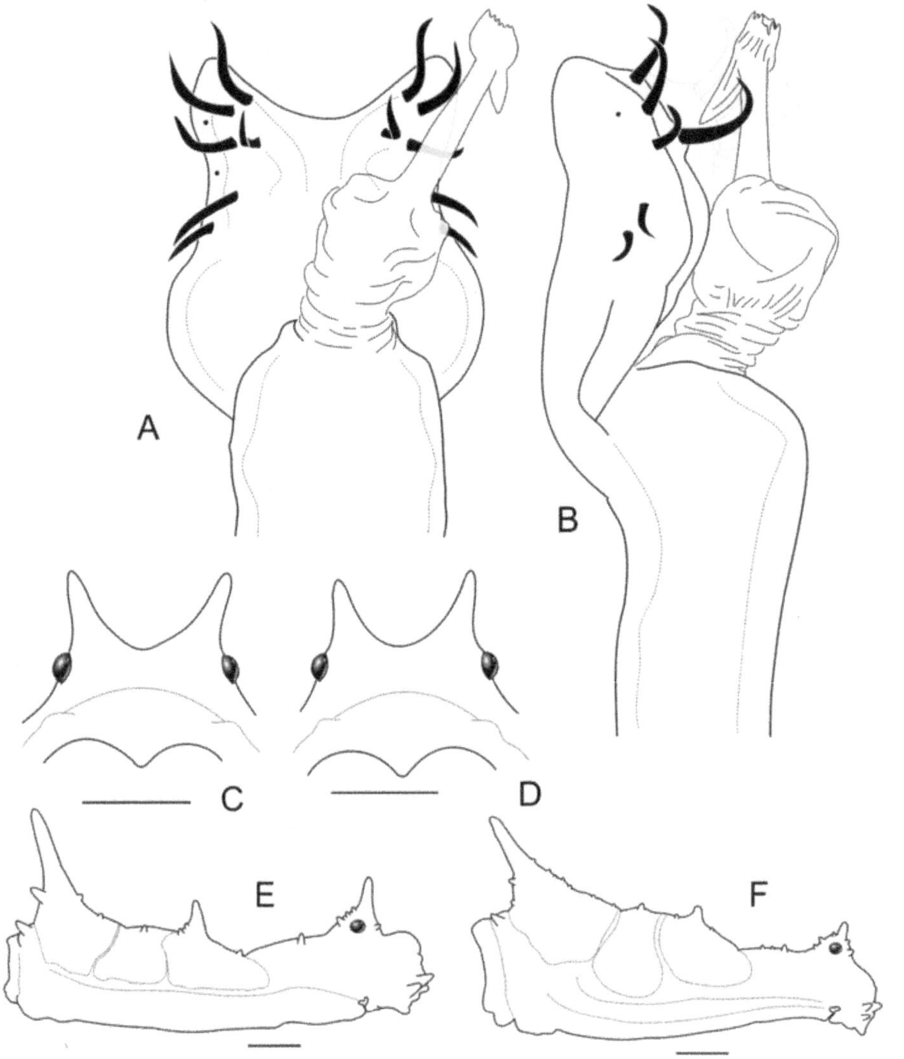

Figure 4 *Ventrifurca caffeinica* sp. nov. (MNRJ 8351) male holotype. Apical portion of the penis: **(A)** Dorsal view. **(B)** Lateral view. **(C-D)** Intraspecific variation on the ocularium of the males (MNRJ 8351 and MNRJ 7038) in frontal view. Sexual dimorphism in dorsal scutum. **(E)** Male. **(F)** Female. Scale bars = 1 mm.

but with strong sexual dimorphism (Figure 9A,B,C,I,J). Leg IV of males armed with many spiniform tubercles. Femur proximally armed with four longitudinal rows of spiniform tubercles, distally with 1 retrolateral ventrally curved tubercle, 1 retrodorsal distal and 1 conspicuous retrolateral subdistal tubercle (Figure 8F,G). Penis. Ventral plate of penis in the same direction of the longitudinal axis of the penis, trapezoid and guitar shaped, with distal border concave and laterodistal corner rounded (Figures 4A,B, 10A,B,C, 11A,C and 12A,B,C). Stylus surpassing ventral plate, without dorsal process, with little stylar caps (Figures 11C,D,E and 12C,D,E). Ventral plate chaetotaxy: laterodistal group (MS C group), with 3 curved setae grouped together on the distal corner of the plate (Figures 10A, 11A, and 12A,B,C); dorsomedial setae (MS D) (largest among those of ventral plate), represented only by 1 pair of large and curved setae pointing distally, located near of the medial line of the plate, close to distal group (Figures 4B, 10A,B, 11A,B, and 12A); lateroproximal group (MS A) placed medially conformed by a pair separated from the rest by a gap (MS B absent) (Figures 4A,B, 10A,B, 11A,B,C, and 12B). Lateroventral group (MS E) conformed by two pairs of short and rounded setae located on the laterodistal keel of ventral plate (VP; Figures 4A,B, 10A,B, and 12B).

Included species (Figure 13). (1) *Ventrifurca abnormis* (Goodnight and Goodnight, 1947) comb. nov. - Ecuador, Chimborazo (Northwestern Andean montane forests NT0145); (2) *Ventrifurca albipustulata* Roewer, 1913 - Ecuador, Esmeraldas; Pichincha (Northwestern Andean montane forests NT0145 and Western Ecuador moist forests NT0178); (3) *Ventrifurca dybasi* (Goodnight and

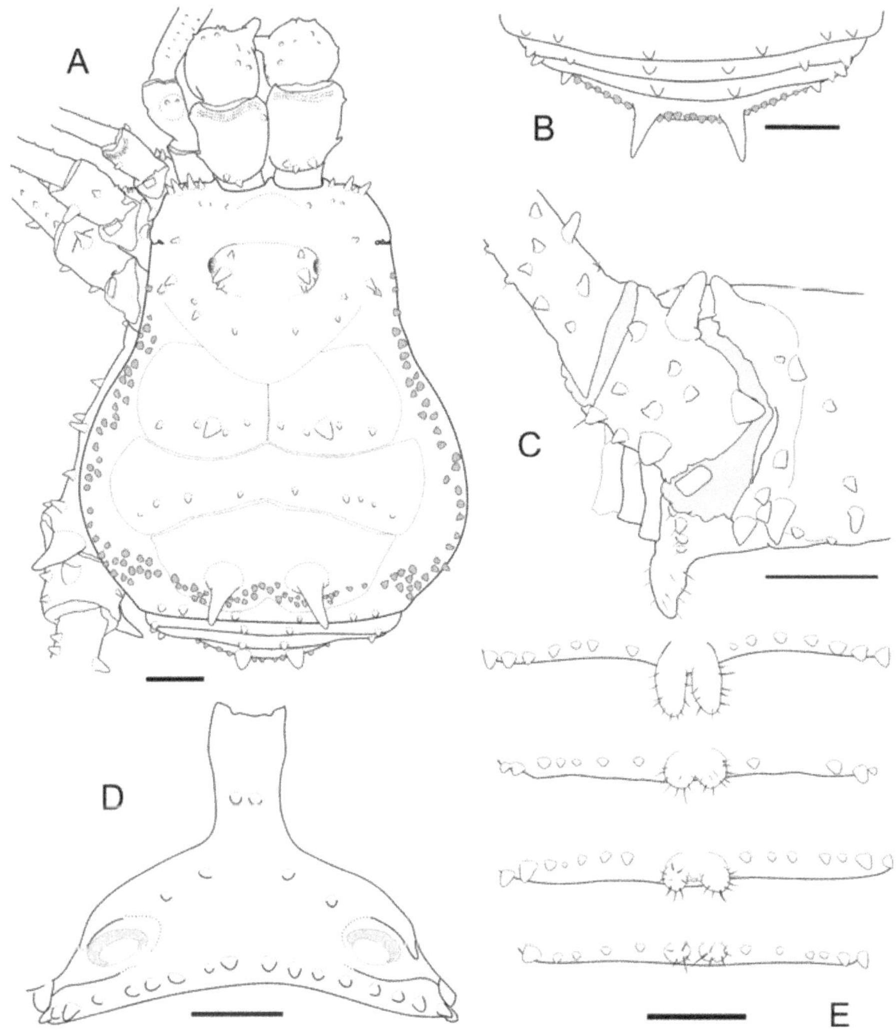

Figure 5 *Ventrifurca albipustulata* (MNRJ 19197). **(A)**. Habitus, dorsal view (male). **(B)** Free tergites, dorsal view (female). **(C)** Coxa and trochanter IV in lateral view (male). **(D)** Stigmatic area, ventral view (female). **(E)** Variation of posterior border of the stigmatic area in ventral view (males). Scale bars = 1 mm.

Goodnight, 1947) comb. nov. - Colombia, Valle del Cauca (Cauca Valley montane forests NT0109), and (4) *Ventrifurca caffeinica* sp. nov. - Quindío (Cauca Valley montane forests NT0109).

Ventrifurca abnormis (Goodnight and Goodnight, 1947) comb. nov.

Figures 13 and 14

Angistrius abnormis Roewer 1932: 346, figs. 63-64; Kury 2003: 90.

Type data

♂ holotype (SMF RII 261/5), from Ecuador, Chimborazo, Alausi, WWF ecoregion NT0145 (Northwestern Andean montane forests).

Diagnosis

Lateral and posterior rows of yellow granules of scutum continuous all along, but with 2 clusters at scutal groove and on the lateral margins of area III. Free tergite III with scattered yellow granules. Scutal area I with contrasting yellow tubercles in the middle and posteriorly. Ocularium, carapace, scutal area II with yellow tubercles (Figure 14A). Basichelicerite short. Pedipalpal femur ectally unarmed (Figure 14B). Stigmatic area on males with robust uniramous median apophysis (Figure 14B,C). Ratio Fe IV of ♂/scutum = 2.3.

Ventrifurca albipustulata Roewer, 1913

Figures 1, 5, 7, 8, 10, 11, and 13; Tables 1 and 2

Ventrifurca albipustulata Roewer, 1913b: 383, figs. 149-150; Kury, 2003: 100.

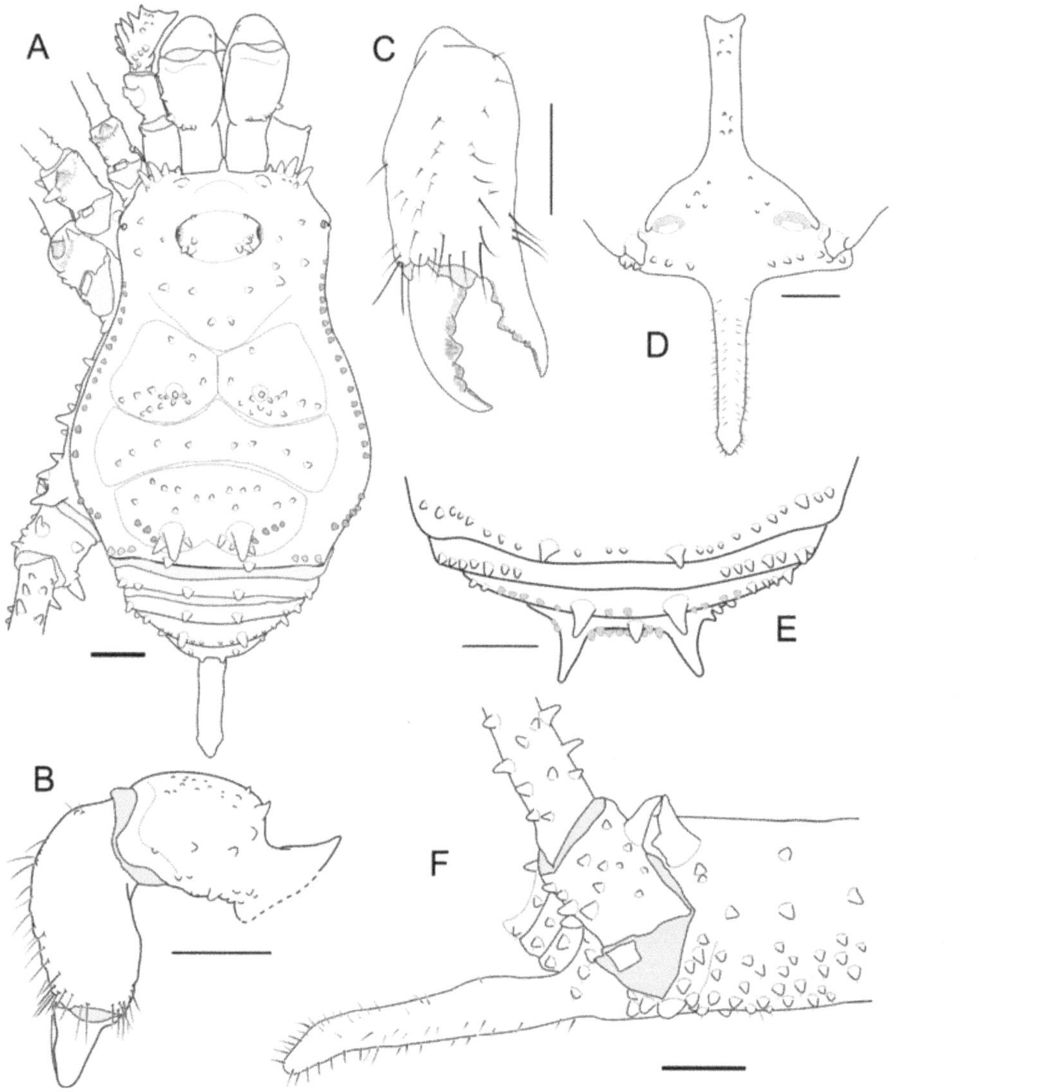

Figure 6 *Ventrifurca caffeinica* sp. nov. **(A)** Habitus, dorsal view (male holotype). **(B)** Left Ch, ectal view. **(C)** Left Ch, frontal view. **(D)** Stigmatic area, ventral view (male holotype). **(E)** Free tergites in dorsal view (female MNRJ 19387). **(F)** Cx-Tr IV and apophysis of stigmatic area, lateral view (male holotype). Scale bars = 1 mm.

Microcranaus pustulatus Roewer, 1913b: 353, fig. 137; Kury, 2003: 93. Syn. nov.

Cayabeus perlatus Roewer, 1932: 337, fig. 53; Kury, 2003: 91. Syn. nov.

Type data

(1) 1 ♂ 1 ♀ syntypes of *Ventrifurca albipustulata* (SMF RI 846, photographed, Roewer cited only 1 male 'type', but the type series consists of 1 male and 1 female) from 'Columbien' without further locality data, while the publication has 'Columbien Camelia' and Kury (2003) extrapolated 'Cafetal Camelia, Antioquia' based on the locality of Swiss Führmann and Mayor 1910 Expedition. Roewer studied the material from this expedition in other contributions, but there is no evidence that this specific material is prevenient from that. There are many places called 'La Camelia' in Colombia, and we do not have evidence to pinpoint one as the type locality, especially because the actual label bears nothing further than 'Colombia'. (2) ♂ holotype of *Microcranaus pustulatus* (SMF RI 851, photographed), from Ecuador, 'Cachab'. There is a Cachabé railway station of collecting by Rosenberg of BMNH, in the Esmeraldas Province at 500 ft. This must correspond to Esmeraldas, San Javier de Cachaví, ca. 138 m, 1.06611/−78.77722, WWF ecoregion NT0178 (Western Ecuador moist forests). There is a railway station on the Cachabí River at 150 m, Esmeraldas, currently known as Urbina, very close to San Javier. Kury (2003) interpolated 'Carchí Province' Cachabí River, but these localities are more to the west, already in the

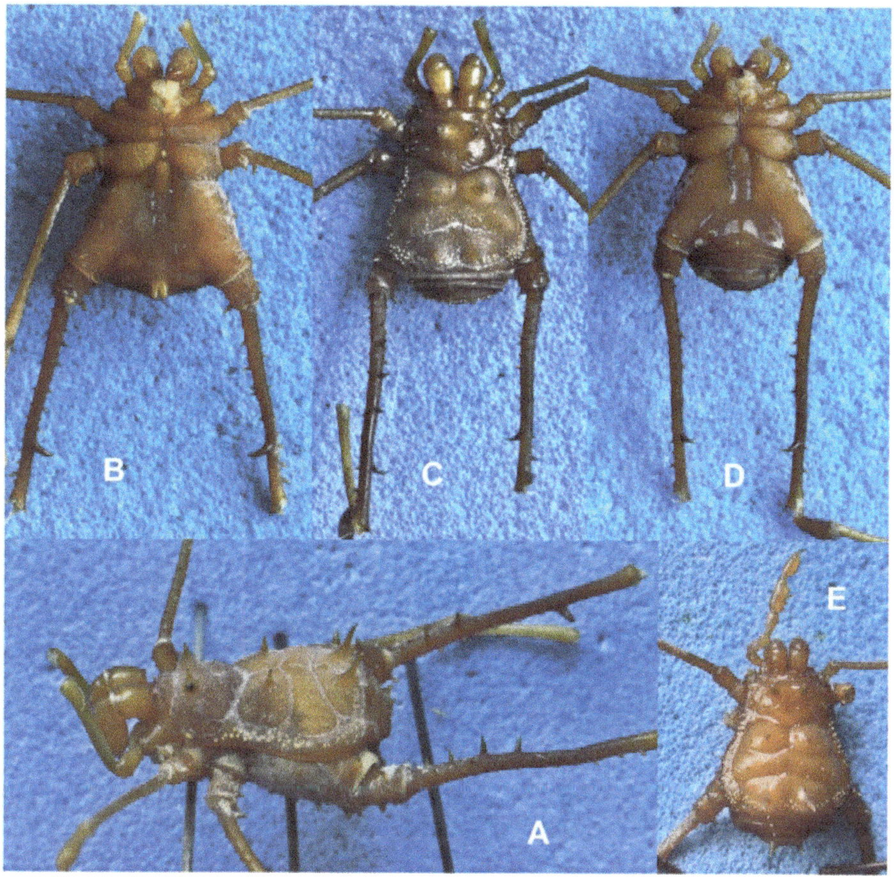

Figure 7 Type material of nominal species here synonymized under *Ventrifurca albipustulata*. *Ventrifurca albipustulata* (SMF RI 846) male syntype: **(A)** Habitus, dorsolateral view. **(B)** Same, ventral view. *Microcranaus pustulatus* (SMF RI 851) male holotype: **(C)** Habitus, dorsal view. **D.** Habitus, ventral view. *Cayabeus perlatus* (MNHN Coll. Simon 20319) female syntype: **(E)** Habitus, dorsal view.

Esmeraldas province. (3) 2 ♀ [wrongly reported as ♂] syntypes of *Cayabeus perlatus* (SMF RI 1423 and MNHN Coll. Simon 20319, Roewer number 5377, both photographed), Ecuador, Pichincha, Cayambe [misspelled as 'Cayabé'], WWF ecoregion NT0145 (Northwestern Andean montane forests).

Other material examined

4 ♂ 1 ♀ (MNRJ 19197) Ecuador, Esmeraldas, Chuchuvi, Cumbres Andinas, N 00.88404°, W 078.51437°, alt: 742 m, 16.iii.2011, leg. A Chagas, A Giupponi, A Kury and M Vega, WWF ecoregion NT0145 (Northwestern Andean montane forests). 1 ♂ (QCAZ AK 0262), same data. 1 ♂ 1 ♀ (MNRJ 19386) Ecuador, Esmeraldas, Tundaloma Lodge, near Calderón. N 01.18277°; W 078.75259°. Alt. 55 m. 8-9.ii.2014, leg. A Kury and A Giupponi, WWF ecoregion NT0178 (Western Ecuador moist forests).

Diagnosis

Lateral and posterior rows of yellow granules of scutum irregularly doubled or tripled and continuous all along. Free tergite III with row of yellow granules along its extension. Scutal area I without contrasting yellow tubercles. Ocularium, carapace, scutal area II without yellow tubercles (Figures 1A,C and 5A). Basichelicerite short. Pedipalpal femur ectally unarmed (Figure 8D, J). Stigmatic area on males with short bifurcated median apophysis, which may be extremely reduced (Figures 1B,C,E, 5C,E, and 7B,D). Ratio Fe IV of ♂/scutum = 1.4. Stylar cap short (Figure 11E).

Redescription (♂ MNRJ 19197)

Body and appendage measurements are in Table 2.

Dorsum. Dorsal scutum type beta without conspicuous coda (convergent-sided terminal area) (Figures 1A and 5A). Scutum rather flattened in lateral view (Figure 1C,D,E). Abdominal scutum widest at groove III level; lateral margin of dorsal scutum with yellow rounded tubercles from carapace to area III, following outline of scutum (Figure 5A). Ocularium without median depression, armed with a paramedian pair of acuminate tubercles set widely apart (Figure 1D) and some small tubercles. Mesotergum well delimited, divided into 3 well-marked areas: area I divided into left and right halves by longitudinal straight groove,

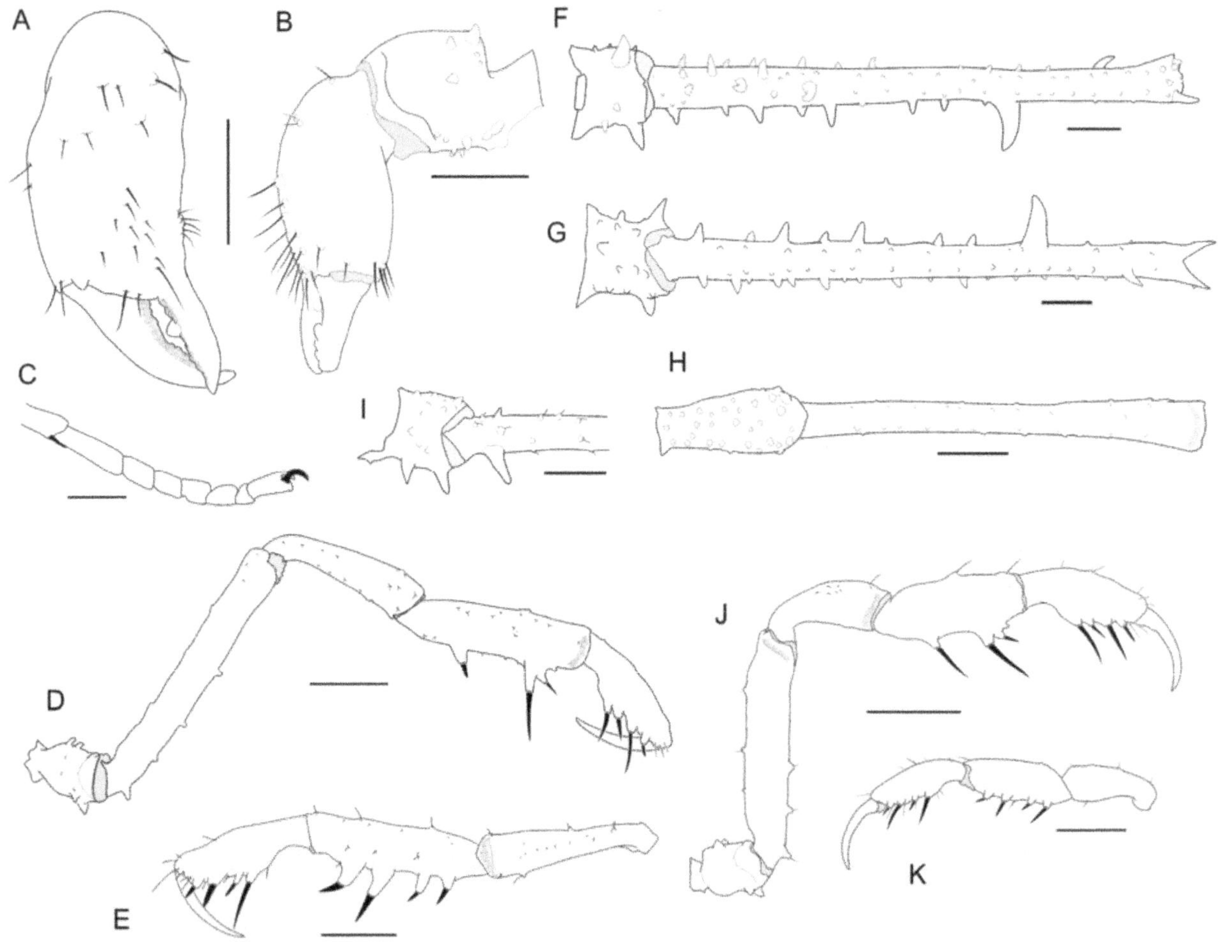

Figure 8 *Ventrifurca albipustulata* (MNRJ 19197) male. **(A)** Right chelicera, frontal view. **(B)** Left chelicera, ectal view. **(C)** Right Ta IV, retrolateral view. **(D)** Right Pp, ectal view. **(E)** Right Pp, dorsomesal view. **(F)** Right Tr and Fe IV, dorsal view. **(G)** Same, ventral view. **(H)** Right Pa and Ti IV, dorsal view. **(I)** Left Tr and base of Fe III, ventral view. Female: **(J)** Right Pp, ectal view. **(K)** Right Pp, dorsomesal view. Scale bars = 1 mm.

and with 1 conspicuous spine on each side and four little granules (Figure 5A); area II penetrating subtly into area I, with a median transverse row of small tubercles; area III with a pair of paramedian acuminate erect high subparallel spines, located near posterior margin, posterior half of this area densely covered with yellow and rounded tubercles (Figures 1A and 5A). Posterior border of scutum straight, armed with a row of granules. Free tergites I-II with a pair of paramedian granules and 1 slightly larger laterodistal on each side, contrasting yellow; free tergite III armed with a posterior row of 16 yellow conspicuous tubercles, the paramedian pair much stouter and concolorous with background (Figure 1A,E).

Venter. Stigmatic area with 3 pairs of tiny granules aligned in two longitudinal and divergent rows and 2 enlarged parallel apophyses (formed by multiple tubercles fused) with common base and rounded apex on the medial area of the posterior border (Figures 1B and 5D,E).

Chelicera (Figure 8A,B). Chelicera slightly swollen, bulla with a group of 2 to 3 proximal and 1 ectoproximal tubercles; hand enlarged, with some frontal tubercles and a patch of about 5 to 6 mesal setiferous tubercles. Fixed finger with a dentate lamella and 1 conspicuous tooth on the proximal half.

Pedipalpi (Figure 8D,E). Femur, patella and tibia elongate; trochanter with 2 dorsal tubercle, 2 minute dorsodistal granules; ventrally with 2 tubercles. Femur cylindrical with a discrete row of ventral spaced granules, about 40% DSL. Patella unarmed, slightly swollen distally, about 1/4 of DSL. Tibia cylindrical, about 1/4 of DSL, ectal with 3 spines (iIi) and mesal with four (iiII). Tarsus without coarse dorsal granulation, ectal with 4 spines (IiIi) and mesal with 4 spines (IiIi).

Legs (Figure 8 C,F,G,H,I). Trochanter II with conspicuous dorsal tubercle, 2 retroventral and another retrodorsal. Trochanter III with 3 prolateral tubercles, 1 prodorsal and 1 retrodorsal and 2 large retrolateral and 2 proventral and retroventral. Trochanter IV with 1 large prodorsal tubercle, 2 retrolateral tuber-cles, distal 1 much larger, 1 ventral blunt tubercle. Femora I-III straight, with numerous rows of granules. Femur III

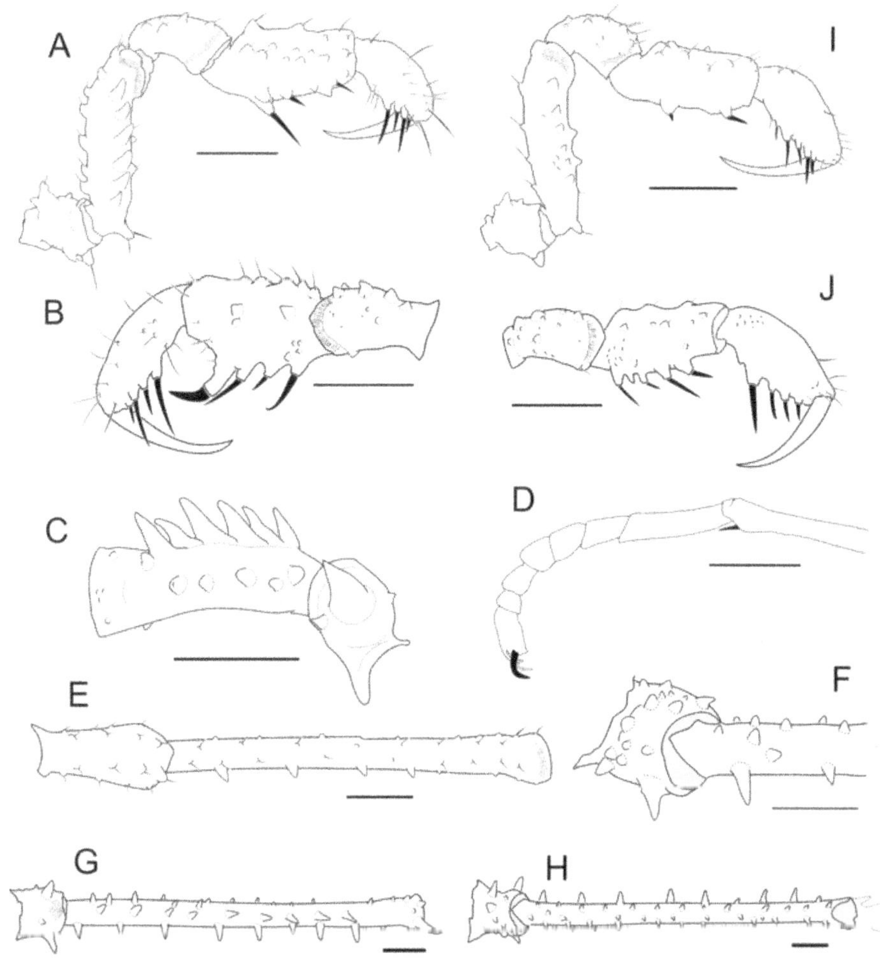

Figure 9 *Ventrifurca caffeinica* sp. nov. male and female. Male holotype (MNRJ 8351): **(A)** Right Pp, ectal view. **(B)** Right Pp, mesodorsal view. **(C)** Right Tr and Fe of Pp, dorsal view. **(D)** Right Ta IV, prolateral view. **(E)** Pa-Ti IV, dorsal view. **(F)** Tr- basal Fe III, ventral view. **(G)** Tr-Fe IV, dorsal view. **(H)** Same, ventral view. Female paratype (MNRJ 19387): **(I)** Right Pp, ectal view. **(J)** Left Pp, mesodorsal view. Scale bars = 1 mm.

furthermore with 1 retrodorsal distal spine and 1 proximal retroventral tubercle. Femur IV slightly sinuous with 6 longitudinal rows of granules; retrodorsal row with 3 larger proximal and 1 distal; those of both ventral rows very small; retroventral row with 1 curved subdistal spines and proventral row with a large curved perpendicular spine. Ratio Fe IV/scutum = 1.4. Tarsal counts in Table 1.

Genitalia (Figures 10 and 11). Ventral plate (VP) subrectangular with mid-constriction and apical parabolic concavity. Ventral surface of VP with 2 lung-shaped fields of needle microsetae each with a distal patch of asparagus-like microsetae. All macrosetae (MS) displaced to distal. Set MS A-B composed of 2 pairs or short lateral setae. 3 to 4 pairs of short curved MS C with helycoidal apex. MS D: 1 pair, strongly displaced to dorsal and curved apically. MS E: 2 pairs, very small, located on laterodistal flange, between groups A and C. Glans sac columnar elongate. Stylus substraight, without processes. Capellum formed by fused dorsal and ventral stylar caps with a pair of acuminate lateral prongs and short acuminate mid prong.

Variation (Figure 5E). The apophyses of stigmatic area may be either isolated or fused at base, as well as buttons or longer finger-like projections, with combinations between these.

Female. Similar to the males. Differential characters are as follows: without frontal hump and abdominal areas high giving the body a non-flattened appearance; spines of the abdominal area I and ocularium short. Paramedian tubercles of free tergite III larger than the males (Figure 5B). Stigmatic area short (non-elongated) without posterior process (Figure 5D). Chelicera not swollen. Pedipalpi sexually dimorphic, slightly elongated, with spines on the tibia and tarsus smaller than the male (Figure 8J,K). Legs with the longitudinal row of tubercles less conspicuous, without none dorsal or lateral tubercle conspicuously greater.

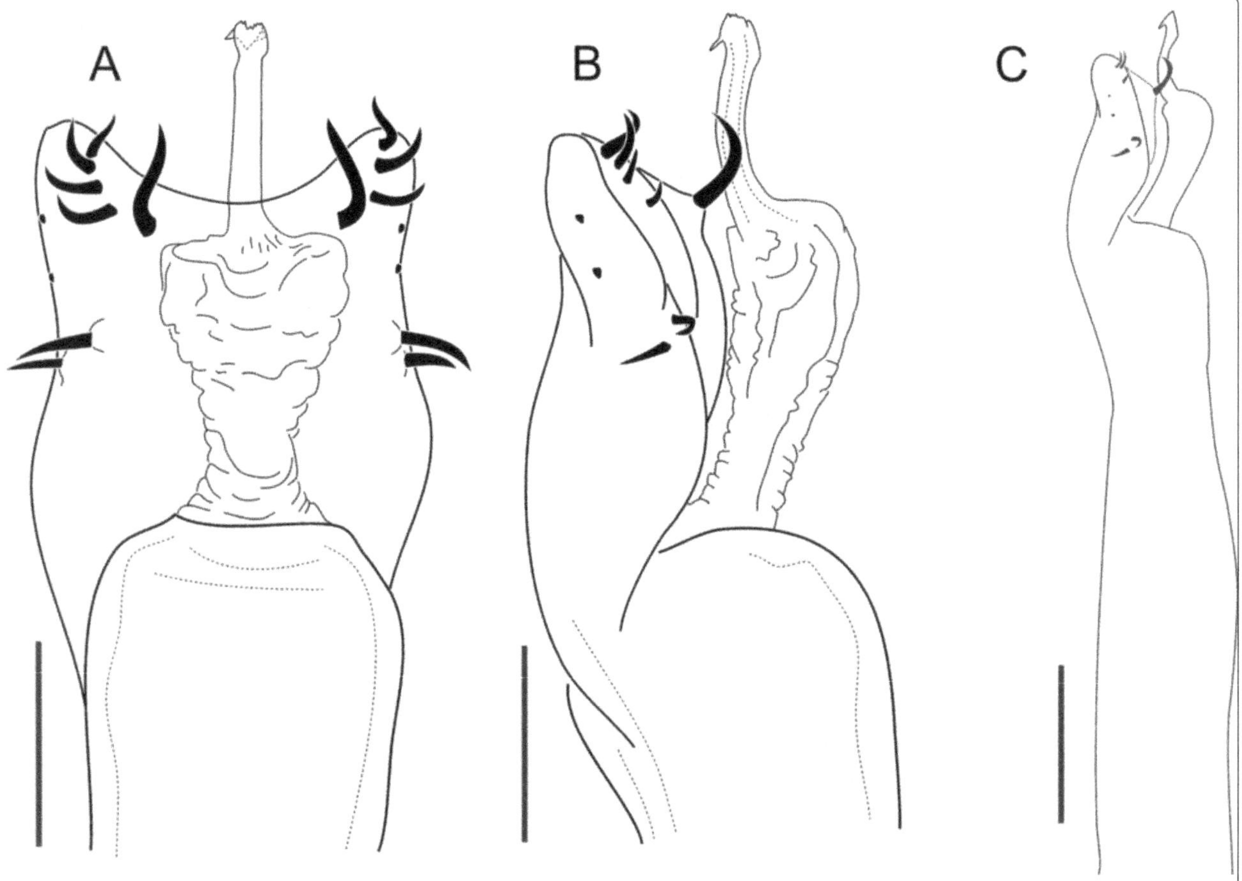

Figure 10 *Ventrifurca albipustulata* (MNRJ 19197). Distal portion of the penis: **(A)** Dorsal view. **(B)** Lateral view. **(C)** Apical half of the penis, lateral view. Scale bars (A, B) = 0.01 and (C) = 0.1 mm.

Ventrifurca caffeinica **sp. nov.**
Figures 2, 4, 6, 12, and 13; Tables 2 and 3

Etymology

Species name is a neo-Latin adjective from the substance caffeine, referring to the international fame of Quindío department as a coffee producer.

Type data

Holotype: ♂ (MNRJ 8351) Colombia, Quindío, Filandia, b. Palacio, v. Biscocho 1,700 m 30.xi.2002 leg. L Mendoza. Paratypes: 1 ♂ (MNRJ 7038) Colombia, Quindío, Bosque de Bremen, Filandia x.1996 leg. C Rodríguez; 1 ♂ (IAvH 300053) Colombia, Quindío, Municipio. Filandia, Vda. Cruces, Finca Bosquinza, 1,900 m 4°42′15″ N 75°37′57″ W 04.vi.2002 Plantación de pino 6 leg. G Jiménez & MF Reina; 1 ♀ (MNRJ 19387) Colombia, Quindío, b. Silencio ix.1998 leg. G Ramírez.

Diagnosis

Lateral and posterior rows of yellow granules of scutum as a single line and interrupted posteriorly with only a few granules behind spines of area III. Free tergite III without row of yellow granules. Scutal area I with only a few contrasting yellow tubercles around paramedian spines. Ocularium, carapace, scutal area II without yellow tubercles (Figures 2A,C and 6A). Basichelicerite robust and long (Figures 2A and 6A). Pedipalpal femur with ectal row of stout spines (Figure 9A,I). Stigmatic area on males with very long cylindrical uniramous median acuminate apophysis (Figures 2B,E and 6A,D,F). Ratio Fe IV of ♂/scutum = 1.4. Stylar cap much elongate (Figure 12C,D,E).

Body and appendage measurements are in Table 2.

Dorsum

Dorsal scutum type beta elongate without conspicuous coda (convergent-sided terminal area) (Figures 2A and 6A). Scutum rather flattened in lateral view, excepted for base of spines of area III (Figure 2C). Abdominal scutum widest at groove III level; lateral margin of dorsal scutum with yellow rounded tubercles from carapace to area III, following outline of scutum as a discontinuous row (Figures 2A and 6A). Ocularium without median

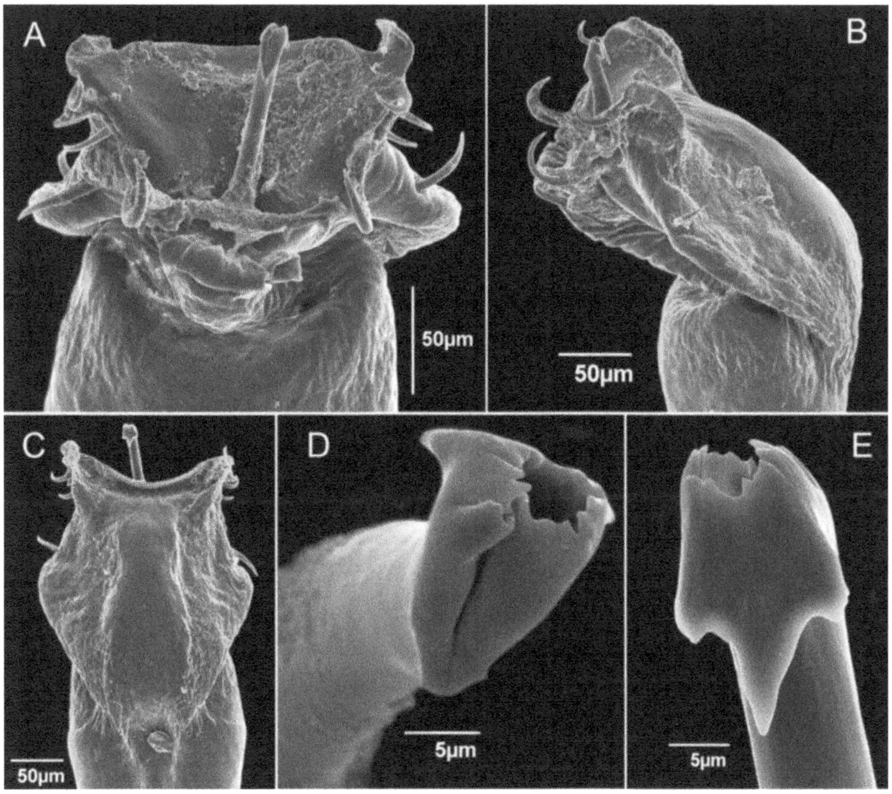

Figure 11 *Ventrifurca albipustulata* (MNRJ 19197). Penis, distal part. **(A)** Dorsoapical view. **(B)** Lateral view. **(C)** Ventral view. **(D)** Stylar caps, dorsoapical view. **(E)** Stylar caps, ventral view.

depression, armed with a paramedian pair of acuminate tubercles set widely apart (Figures 2D and 4C, D) and some small tubercles. Mesotergum well delimited, divided into 3 well-marked areas: area I divided into left and right halves by longitudinal straight groove, and with 1 conspicuous spine on each side and 14 little granules of which a few are contrasting yellow (Figure 6A); area II penetrating subtly into area I, with a median transverse row of small tubercles; area III with a pair of paramedian acuminate erect high subparallel spines, located near posterior margin, each with 2 to 3 accessory spines; posterior half of this area with a row of yellow tubercles (Figures 2A and 6A). Posterior border of scutum straight, armed with 3 yellow granules each side plus a pair of paramedian larger ones. Free tergites I-II with a pair of paramedian granules and 1 to 2 smaller laterodistal on each side; free tergite III armed with a posterior row of 13 tubercles, the paramedian pair much stouter (Figure 6A).

Venter

Stigmatic area with 2 groups of tiny anterior granules; 2 clusters of granules near of stigmata and 5 to 6 large granules each side of the posterior border. Posteroventral process cylindrical and large, located on the medial area of the posterior border (Figures 2B,C and 6D,F).

Chelicera (Figure 6A,B,C)

Chelicera slightly swollen, bulla with 2 proximal and 1 to 2 ectoproximal tubercles; hand enlarged, with some frontal tubercles arranged in 2 rows and a patch of about 3 mesal setiferous tubercles. Fixed finger with a dentate lamella and 1 conspicuous tooth on the proximal half.

Pedipalpus (Figure 9A,B,C)

Trochanter with 2 dorsal tubercles, 2 minute dorsodistal granules; ventrally with 2 tubercles. Femur cylindrical, slightly elongated, armed with longitudinal rows of tubercles, the dorsal and mesal rows are larger, and with a ventroproximal large tubercle; patella and tibia normal (not unusually elongated as in other members of the genus); patella curved, with some dorsal granules; tibia dorsally granulated, ectal with 3 spines (Iii) and mesal with four (IiII), the most distal curved and strong; tarsus without coarse granulation, ectal with 4 spines (IiIi) and mesal with 4 spines (IiIi).

Legs (Figure 9D,E,F,G,H)

Trochanter II with 2 + 1 + 2 ventral (3 rows, middle larger), 1 prodorsal distal, 2 retrodorsal (distal larger), 2 retrolateral, 1 retroventral tubercles. Trochanter III with 2 + 1 + 2 ventral (3 rows, middle larger), 2 proventral, 2 retroventral,

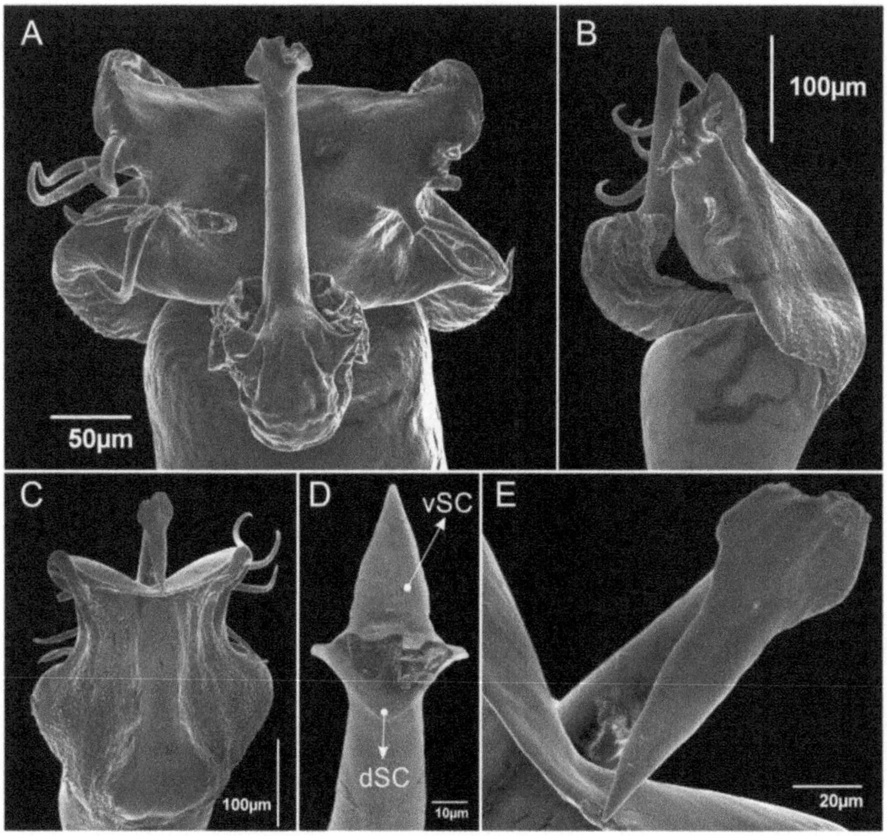

Figure 12 *Ventrifurca caffeinica* sp. nov. (MNRJ 7038). Penis, distal part. **(A)** Dorsoapical view. **(B)** Lateral view. **(C)** Ventral view. **(D)** Stylar caps, dorsoapical view. **(E)** Stylar caps, ventral view. vSC: ventral stylar cap; dSC: dorsal stylar cap.

2 (1 much larger) retrolateral, 2 prodorsal, 2 (distal larger) retrodorsal and 2 small prolateral tubercles. Trochanter IV with 2 + 1 + 2 ventral (3 rows, middle larger), 1 large prodorsal, 3 retrodorsal clustered, 2 small prolateral, 2 retrolateral distal 1 much larger, tubercles. Femora I-III straight, with numerous rows of granules. Femur III furthermore with 1 retrolateral proximal and 1 retrodorsal distal tubercles. Femur IV straight, with 6 longitudinal rows of granules; ones in retrolateral row larger, 1 retrodorsal distal tubercle. Tibia I-IV straight, with numerous rows of granules. Tibia IV with a retrolateral row of tubercules. Ratio Fe IV/scutum = 1.4. Tarsal counts in Table 3.

Genitalia (Figures 4A, B and 12)

Ventral plate (VP) subrectangular with deep mid-constriction and apical parabolic concavity. Ventral surface of VP with 2 lung-shaped fields of needle microsetae each with a distal patch of asparagus-like microsetae. All macrosetae (MS) displaced to distal. Set MS A-B composed of 2 pairs or short lateral setae. 3 pairs of short curved MS C with helycoidal apex. MS D: 1 pair, strongly displaced to dorsal and curved apically. MS E: 2 pairs, very small, located on laterodistal flange, ventrally at group C. Glans sac columnar elongate. Stylus straight, without processes. Capellum formed by marginally fused dorsal and ventral stylar caps with pair of acuminate lateral prongs and very stout acuminate mid prong.

Variation

There are only two beta males in the lots examined; one of them has apophyses of stigmatic area reduced. Spines of the ocularium can be subparallel or slightly divergent in frontal view.

Female

Similar to the males. Differential characters are as follows: without frontal hump and abdominal areas high, giving the body a non-flattened appearance; spines of the abdominal area I and ocularium short (Figure 4E,F). Paramedian tubercles of free tergite III larger than in the males (Figures 6E). Stigmatic area short (non-elongated) without posterior process. Chelicera not swollen. Pedipalpi sexually dimorphic, with tubercles smaller than in the male (Figures 9I, J), and more distal ectal spine of the tibia straight and normal. Legs with the longitudinal row of tubercles less conspicuous.

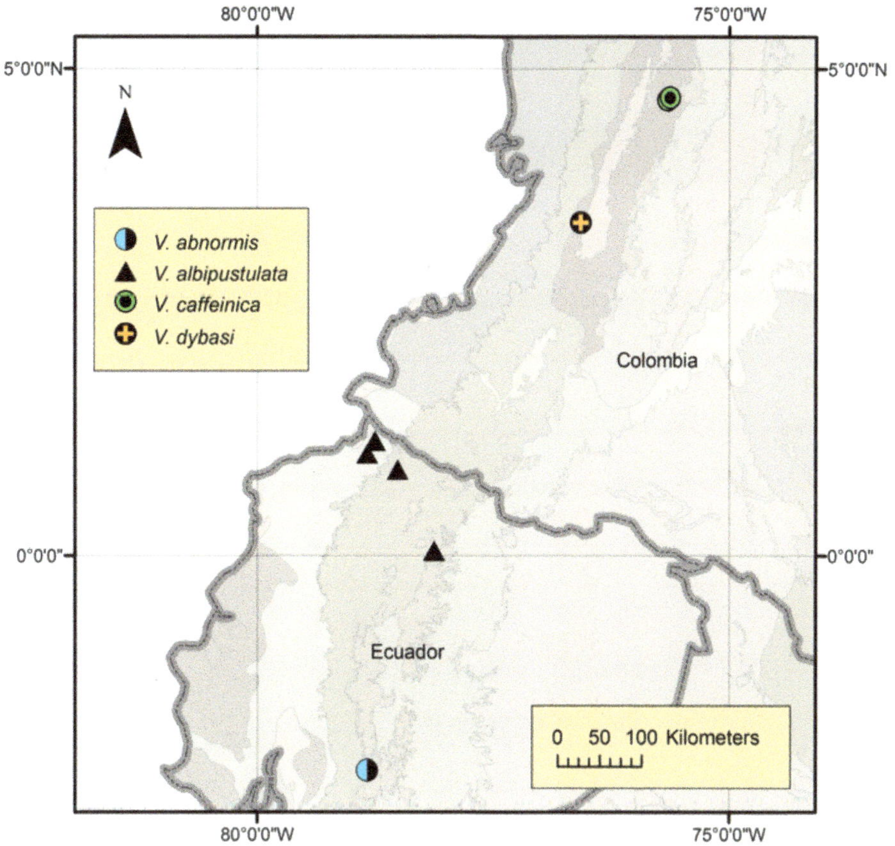

Figure 13 Northwestern South America, showing the distribution of the species of *Ventrifurca*. Shaded areas in the background are the WWF ecoregions.

Ventrifurca dybasi (Goodnight and Goodnight, 1947) comb. nov.

Figure 3A,B,C

Rhopalocranaus dybasi Goodnight and Goodnight, 1947: 40, fig. 20.

Neocranaus dybasi: Kury 2003: 95.

Type data

♀ holotype (FMNH), from Colombia, Valle del Cauca, Mountains W Cali, WWF ecoregion NT0109 (Cauca Valley montane forests). Examined.

Diagnosis (Figure 3)

Lateral and posterior rows of yellow granules of scutum as a single line (excepted at carapace where they are clustered) and interrupted posteriorly with numerous granules postero-lateral on area III. Free tergite III with row of yellow granules along its extension. Scutal area I with an arch of yellow tubercles interrupted in laterals. Ocularium, carapace, scutal area II with scattered yellow tubercles. Pedipalpal femur with ectal row of stout spines.

Manaosbiidae Roewer, 1943.

Rhopalocranaus columbianus (Roewer, 1963) comb. nov., new familial assignment

Figure 15A,C,D

Microcranaus columbianus Roewer, 1963: 64, figs 34-35; Kury 2003: 95.

Type data

♂ holotype 1 ♂ 2 ♀ paratypes (SMF 12707), from Colombia, Cauca, Monterredondo (03°13′N - 76°12′W), 1,400 m.

Diagnosis

Chelicerae, pedipalpi and legs yellowish brown, dorsal scutum dark brown. Scutal areas higher than ocularium in lateral view; ocularium with 2 high and acuminate tubercles; area I, III with a pair of high and acute tubercles. Free tergites I-III each with a pair of paramedian tubercles (higher on III). Penis: ventral plate in the same direction of the penis main axis, subrectangular in dorsal view, with lateral margins slightly divergent and the distal border slightly concave. Chaetotaxy: with only 2 pairs of macrosetae (MS) A, straight or slightly curved, cylindrical and acuminated, with a gap between MS A1-A2. One pair distally positioned, near to laterodistal corner

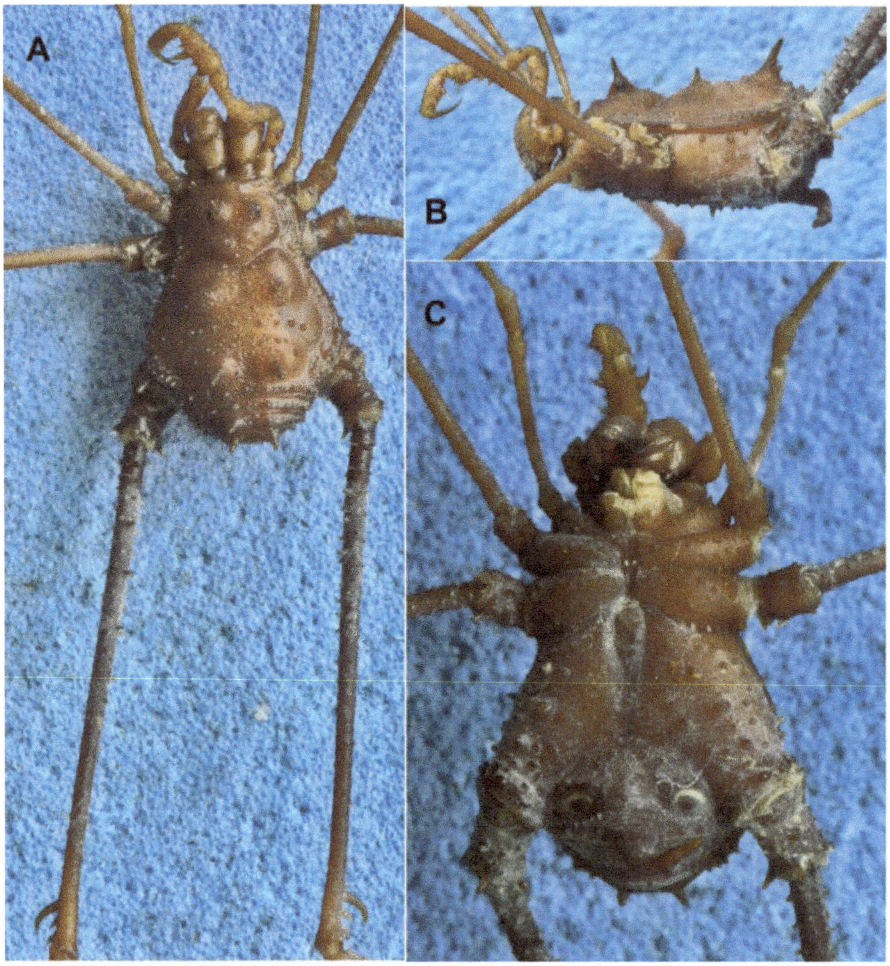

Figure 14 *Ventrifurca abnormis* male holotype (SMF RII 261/5). **(A)** Habitus, dorsal view. **(B)** Habitus, lateral view. **(C)** Habitus, ventral view.

of VP, the other 1 pair, basally positioned, together but dorsally to MS B. Macrosetae C with 3 pairs of curved and long setae distally located (the basal most slightly separate from the distal pair). MS D with only 1 pair of short setae, located dorsally to MS C. MS group by MS E group composed by 2 pairs reduced and rounded setae. Stylus with globose glans, apex straight and cylindrical, without stylar caps or other structures. Male basitarsus I moderately spindled and undivided.

Table 1 Tarsal formula of *Ventrifurca albipustulata* (numerals in parentheses represent number of distitarsomeres)

		I	II	III	IV
MNRJ 19197	♂	?/7(3)	14(3)	7	7
MNRJ 19197	♂	7(3)	14-16(3)	7	7
MNRJ 19197	♂	7(3)	16(3)	7	7
MNRJ 19197	♂	7-8(3)	15-16(3)	7	7
MNRJ 19197	♀	7(3)	14(3)	7	7
QCAZ AK 0262	♂	7(3)	14(3)	7	7

***Camelianus gracilis* (Roewer, 1913) comb. nov., new familial assignment**

Figure 15B.

Microcranaus gracilis Roewer, 1913b: 355, fig. 138.
'Microcranaus' gracilis: Kury, 2003: 206.

Type data

♀ holotype [wrongly reported as male] (SMF RI-849), from Surinam, Saramacca.

Diagnosis (based on female holotype)

Body russet, scutum, tergites and sternites darker, mottled in black. Appendages pale yellow with loose black reticle (according to Roewer 1913b). Dorsum. Anterior margin of carapace with three tubercles each side. Ocularium with 2 conical dorsal tubercles (about size of eye), 2 anterior tubercles, 2 posterior tubercles. Carapace with 3 to 4 tubercles behind each side of ocularium. Lateral margin with four tubercles near coxa III. Area I with a large apophysis with rounded apex, surrounded by 6 tubercles, 3 tubercles each side; II with 2 to 3

Table 2 *Ventrifurca* spp. body and appendage measurements, expressed in millimeters

	V. albipustulata		*V. caffeinica*	
	♂ (MNRJ 19197)	♀ (MNRJ 19197)	♂ (MNRJ 8351)	♀ (MNRJ 19387)
CL	3.4	2.6	3.5	3.0
CW	3.8	3.5	3.8	3.9
AL	4.2	3.7	4.2	4.1
AW	6.5	3.7	6.4	6.5
IOD	1.7	1.3	1.7	1.2
BaCh	1.6	1.2	1.6	1.4
FePp	2.5	1.8	3.2	2.0
PaPp	1.7	1.2	1.8	1.1
TiPp	1.9	1.3	1.9	1.2
Fe I	5.0	4.2	5.3	3.5
Fe II	9.9	9.9	11.7	-
Fe III	7.1	6.5	8.2	5.9
Fe IV	9.1	8.8	10.2	7.4
Ti I	2.8	2.6	3.5	2.6
Ti II	6.8	6.7	8.1	-
Ti III	3.8	3.3	3.9	3.2
Ti IV	5.8	4.8	5.8	4.6

tubercles each side; III with 2 divergent pointed apophyses, directed backwards, with 2 tubercles each side. Posterior margin with 8 tubercles. Free tergites I with 14 tubercles, 1 pair higher (about eye high); II with 4 to 5 tubercles each side, with a pair of wide and pointed tubercles (larger than tergite width), 2 tubercles between largest ones; III with 10 tubercles, 3 of them larger. Anal operculum with 10 tubercles. Cheliceral segment I with 7 tubercles on lateral margin of bulla. Tarsal formula: 6(3); 11(3); 6; 6. Male unknown.

Discussion and conclusions

Rationale of the synonymies

Roewer (1913a, b, 1932), as typical of his system, used supposed differences (some of them known to be irrelevant, see for example de Piza 1940: 313 ff.; de Piza 1942: 403 ff.; Ringuelet 1959: 188) in the armature of the scutal areas and of the free tergites to separate these genera, recognizing three categories ('spine', 'small spine', and 'tubercle'). Furthermore, *Ventrifurca* was isolated from the other two genera by having a short bifurcate projection on stigmatic area, and *Microcranaus* isolated by having tarsus I, III, and IV hexamerous instead of heptamerous. In our sample of *V. albipustulata* (six specimens), tarsomere range is 7-8/14-16/7/7; curiously none has 6 tarsomeres on leg I. The tarsomere count, said by Roewer (inspired in Sørensen) and followers to be fairly constant (Sørensen 1902: 3; Roewer 1913a: 5; Goodnight and Goodnight 1953: 36), has later proven to be variable, and 5 versus 6 or 6 versus 7 tarsomeres were already found in many species (e.g., Soares 1943: 223; de Mello-Leitão 1949: 3; Muñoz-Cuevas 1973: 230). The paired apophyses of the stigmatic area would be the most significant feature to diagnose *Ventrifurca*. However, this feature is herein demonstrated to be variable intraspecifically, which led different male specimens to be described as different genera (see 'variation' above and Figure 5E).

Table 3 Tarsal formula of *Ventrifurca caffeinica* sp. nov.

		I	II	III	IV
MNRJ 8351	♂	7-6(3)	11-12(3)	7	7
MNRJ 7038	♂	?-6(3)	?	6-7	7
IAvH 300053	♂	6(3)	?	7	?-7
MNRJ 19387	♀	6(3)	12(3)	7	7

Interchange of species to and from *Ventrifurca*/*Rhopalocranaus*

It is not unusual, in works done under the Roewerian paradigm, to join completely unrelated species which happen to coincide in tarsal counts and area armature. *M. columbianus* has been originally included in *Microcranaus* under such constraints. However, examination of the type material by one of us, including male genitalia, led to its placement in Manaosbiidae, where it fits in *Rhopalocranaus* by the shape of the ventral plate of the penis, pattern of their macrosetae, and ornamentation of the dorsal scutum and ocularium (a review of Manaosbiidae by RPR is in course). On the other hand, following the same reasoning of Roewer, Goodnight and Goodnight (1947) discovered a species here included in *Ventrifurca* after revision of the holotype, but described it under the then amorphous assemblage *Rhopalocranaus*, namely *R. dybasi*. Only much later, Kury (1997) transferred *Rhopalocranaus* from Cranainae to Manaosbiidae. However, as demonstrated above, *R. dybasi*, even known only by the female, is a Cranaidae and matches the present diagnosis of *Ventrifurca*.

Distribution of *Ventrifurca*

The distribution of cranaid species is primarily associated with the Andean and Amazonian forest. It is currently difficult to find any pattern of distribution of the genera in this family due to (1) the unsatisfactory taxonomy with many monotypic/meaningless genera, (2) species known from only one locality, and (3) poorly studied collections in the Andean region. A few comments can be made on two monophyletic groups: (1) the 'longipalp cranaids' (formerly Stygnicranainae) which are distributed in the south of Colombia and north of Ecuador associated to Northwestern Andean montane

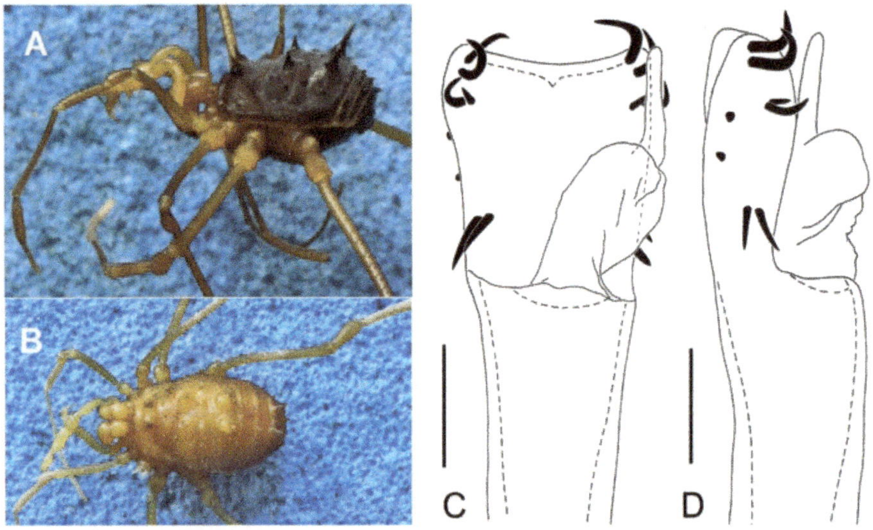

Figure 15 Species of Manaosbiidae transferred here from *Microcranaus*. *Rhopalocranaus columbianus* (SMF 12707) male holotype: **(A)** Habitus, lateral view. *Camelianus gracilis* (SMF RI-849) female holotype: **(B)** dorsal view. *Rhopalocranaus columbianus*. Distal portion of the penis. **(C)** Dorsal view. **(D)** Lateral view. Scale bars (C-D) = 0.1 mm.

forests (Orrico and Kury 2009) and (2) the genus *Phareicranaus* which has some species groups (Villarreal, unpublished data), one of them associated to north of south America (Venezuela, Panama and Trinidad) and another well distributed in the Amazonian Basin (Brazil, Ecuador, Peru).

Although only three species of *Ventrifurca* are known, this could be a more diverse genus and seems to be restricted to the western range of the Colombian Andes and Ecuador, from north of Ecuador about 700 km north into northwestern Colombia. Reliable records of *Ventrifurca* species are only from WWF ecoregions NT0109 and NT0145, both belonging to the larger region Northern Andean montane forests. This corresponds to Morrone's (2014) Cauca Province. On the other hand, the type locality of *Microcranaus pustulatus*, the 'Cachab' - which is here interpreted as Río Cachabí, at much lower altitude, already in the Choco-Darien moist forests region (matching Morrone's Western Ecuador Province) - might be inaccurate, so *Ventrifurca* can be said to be safely recorded only from Northern Andean montane forest. Another interpretation is that the original locality report is accurate, and *Ventrifurca* is much wider spread than we think, descending into the Pacific lowland forest. Only further collecting expeditions will allow putting those hypotheses to test.

Relationships of *Ventrifurca*

A great number of Cranaidae (e.g., *Phalangodus*, *Phareicranaus* in part, *Carsevennia* Roewer, 1913, *Heterocranaus* Roewer, 1913, *Yania* Roewer, 1919) has short, compact pedipalpi, with femur strongly convex dorsally and spined mainly on the males. But a more slender, sexually dimorphic pedipalpus, slightly elongate in males of *V. albipustulata*, hints to some relationship with the three genera of 'longipalp cranaids' which are thought to constitute a clade (see illustrations of all in Orrico and Kury 2009). In *Tryferos*, the pedipalpus of the male is immensely elongate (femur to tibia) and femur unarmed while the female has regular podomeres. *Agathocranaus* is known only from a male, but podomeres are only moderately long. Of special interest in *Agathocranaus* is the circular/elliptical transverse section of tarsus (instead of hemispherical). This feature is consistently associated with elongate pedipalpi and is known in Biantidae, Epedanidae, Stygnidae (Kury and Pérez-G 2007; Pinto-da-Rocha 1997; Kury 2008, 2009), and some Gonyleptidae (Caelopyginae, Sodreaninae, and Progonyleptoidellinae) (Pinto-da-Rocha 2002; Pinto-da-Rocha and Bragagnolo 2011). *Stygnicranaus* appears to represent a step further in the evolution of 'longipalpness,' because it has the same tarsal conformation as *Agathocranaus*, but with greater elongation of Fe-Ti both in males and in females. Even if *Ventrifurca caffeinica* sp. nov. is an exception, frequently species with ornamentation on the posterior edge of the stigmatic area have at least males if not both sexes with elongation of leastwise one of the segments of the pedipalpus, e.g., *Angistrisoma atrolutea* Roewer, 1932, *A. fuscum* Roewer, 1932, *Angistrius abnormis* Roewer, 1932, *Ventripila marginata* Roewer, 1917, and *Ventrisudis mira* Roewer, 1963.

The guitar-shaped penis ventral plate, the distal concentration of macrosetae, and the presence of a stylar cap (thus called by Kury 2012a, also *capellum* in Orrico and Kury 2009), all present in *Ventrifurca*, are quite widespread in some Cranaidae genera (e.g., *Heterocranaus*, *Chiriboga*, and *Metacranaus* in Pinto-da-Rocha and Bonaldo 2011;

Zannicranaus in Kury 2012a). More strikingly similar is the greatly medially shifted pair of dorsal macrosetae present in *Stygnicranaus* and *Metacranaus*. As male genital morphology is largely unknown in Cranaidae, the evolution of their features is hitherto only poorly understood, a scenario that is gradually changing with focused work on this family.

Competing interests

The authors declare that they have no competing interests.

Authors' contributions

OVM led the execution of the project, prepared most photographs and drawings, and wrote the bulk of the text, including diagnoses and descriptions of the species and some of the discussion. ABK conceived the initial, much narrower, idea of the paper, discovered synonymies, traveled to Ecuador where he collected *Ventrifurca* specimens, arranged loan of Chicago type specimens, edited images, created the map, checked descriptions and diagnoses, and redacted the historical background and most of the discussion. RPR traveled to France and Germany in order to study type material of Roewer's species and produced pictures of them, drafted the genitalia of Manaosbiidae, and placed the non-Cranaidae in genera. All authors intensively discussed all parts of the text, genera and species delimitation, and image organization, and all read and approved the final manuscript.

Acknowledgements

This study is co-funded by grants CNPq (477502/2012-1 and 504327/2012-7) and FAPERJ (E-26/111.431/2012 and E-26/111.705/2012) to ABK, CAPES (PEC-PG 5900115) to OVM, and FAPESP (2009/17206-5; BIOTA, 2013/50297-0), NSF (DOB 1343578), and NASA to RPR. We would like to thank Peter Jäger (SMF) for the hospitality in Frankfurt to RPR, Mark Judson and Elise-Anne Leguin (MNHN) idem in Paris, Claudia Medina (IAvH) for hospitality to ABK in Villa de Leyva, Petra Sierwald (FMNH) for sending type material, and Álvaro Barragán and Mauricio Vega (QCAZ) for helping out in ABK's expedition to Ecuador. Maria Cleide de Mendonça (MNRJ) provided optical facilities.

Author details

[1]Departamento de Invertebrados, Museu Nacional/UFRJ, Quinta da Boa Vista, São Cristóvão, 20.940-040 Rio de Janeiro, RJ, Brazil. [2]Departamento de Zoologia, Instituto de Biociências, Universidade de São Paulo, Caixa Postal 11461, 05422-970 São Paulo, SP, Brazil.

References

Goodnight CJ, Goodnight ML (1947) Phalangida from tropical America. Fieldiana Zool 32(1):1–58

Goodnight CJ, Goodnight ML (1953) The opilionid fauna of Chiapas, Mexico, and adjacent areas (Arachnoidea, Opiliones). Am Mus Novit 1610:1–81

Hara MR, Pinto-da-Rocha R, Villarreal MO (2014) Revision of the cranaid genera *Phalangodus*, *Iquitosa* and *Aguaytiella* (Opiliones: Laniatores: Gonyleptoidea). Zootaxa 3814(4):567–580

Kury AB (1994) The genus *Yania* and other presumed Tricommatidae from South American highlands (Opiliones, Cranaidae, Prostygninae). Revue Arachnologique 10(7):137–145

Kury AB (1997) The genera *Saramacia* Roewer and *Syncranaus* Roewer, with notes on the status of the Manaosbiidae (Opiliones, Laniatores). Bol Mus Nac Rio de J (Nov Ser) (Zool) 374:1–22

Kury AB (2003) Annotated catalogue of the Laniatores of the New World (Arachnida, Opiliones). Rev Iber Aracnol Vol Especial Monográfico 1:1–337

Kury AB (2008) On the systematic position of *Dino* Loman and *Toccolus* Roewer (Opiliones, Laniatores, Epedanidae), with the description of a new species from western Java, Indonesia. Zootaxa 1932:61–68

Kury AB (2009) A new genus of Stygninae from a relictual rainforest in Ceará, northeastern Brazil (Opiliones, Laniatores, Stygnidae). Zootaxa 2057:63–68

Kury AB (2012a) A new genus of Cranaidae from Ecuador (Opiliones: Laniatores). Zootaxa 3314:31–44

Kury AB (2012b) First report of the male of *Zamora granulata* Roewer 1928, with implications on the higher taxonomy of the Zamorinae (Opiliones, Laniatores, Cranaidae). Zootaxa 3546:29–42

Kury AB, Pérez-G, A (2007). Biantidae Thorell, 1889. Pp 176-179. In: Pinto-da-Rocha, R., G. Machado & G. Giribet (eds.). Harvestmen: the biology of the Opiliones. Harvard University Press, Cambridge and London. x + 597 pages

Kury AB, Villarreal-M O, Costa CS (2007) Redescription of the type species of *Cynorta* Koch, 1839 (Arachnida, Opiliones, Cosmetidae). J Arachnol 35(2):325–333

Kury AB, Villarreal MO (2015) The prickly blade mapped: establishing homologies and a chaetotaxy for macrosetae of penis ventral plate in Gonyleptoidea (Arachnida, Opiliones, Laniatores). Zool J Linnean Soc 174:1–46.

de Mello-Leitão CF (1949) Famílias, subfamília, espécies generos novos de opiliões e notas de sinonimia. Bol Mus Nac Rio de J (Nov Ser) (Zool) 94:1–33

Morrone JJ (2014) Biogeographical regionalisation of the Neotropical region. Zootaxa 3782(1):001–110, http://dx.doi.org/10.11646/zootaxa.3782.1.1

Muñoz-Cuevas A (1973) Sur les caractères génériques de la familie des Gonyleptidae (Arachnida, Opilions, Laniatores). Bull Mus Nat Hist Nat (3) 87(113):225–234

Orrico VGD, Kury AB (2009) A cladistic analysis of the Stygnicranainae Roewer, 1913 (Arachnida, Opiliones, Cranaidae) – where do longipalp cranaids belong? Zool J Linnean Soc 157:470–494

Pinto-da-Rocha R (1997) Systematic review of the Neotropical family Stygnidae (Opiliones, Laniatores, Gonyleptoidea). Arq Zool 33(4):163–342

Pinto-da-Rocha R (2002) Systematic review and cladistic analysis of the Caelopyginae (Opiliones, Gonyleptidae). Arq Zool 36(4):357–464

Pinto-da-Rocha R, Bonaldo AB (2011) Species relationships in the Neotropical genus *Phareicranaus* Roewer 1913 (Opiliones: Cranaidae): two new species and new data from penial morphology. Zootaxa 3135:1–34

Pinto-da-Rocha R, Bragagnolo C (2011) Systematic revision and cladistic analysis of the Brazilian subfamily Sodreaninae (Opiliones: Gonyleptidae). Invertebr Syst 24 ["2010"]:509–538

Pinto-da-Rocha R, Hara MR (2009) New familial assignments for three species of Neotropical harvestmen based on cladistic analysis (Arachnida: Opiliones: Laniatores). Zootaxa 2241:33–46

Pinto-da-Rocha R, Kury AB (2003) Phylogenetic analysis of *Santinezia* with description of five new species (Opiliones, Laniatores, Cranaidae). J Arachnol 31(2):173–208

Pinto-da-Rocha R, Benedetti AR, Vasconcelos EG, Hara MR (2012) New systematics assignments in Gonyleptoidea (Arachnida, Opiliones, Laniatores). Zookeys 198:25–68

Pinto-da-Rocha R, Bragagnolo C, Marques FPLM, Antunes M Jr (2014) Phylogeny of harvestmen family Gonyleptidae inferred from a multilocus approach (Arachnida: Opiliones). Cladistics 30:519–539. doi:10.1111/cla.12065

de Piza ST Jr (1940) Breves considerações em torno de alguns nóvos Gonyléptidas do Brasil. Rev Agric 15(7–8):312–324

de Piza ST Jr (1942) A respeito da sistemática de alguns opiliões. Rev Bras Biol 2(4):403–416

Ringuelet RA (1959) Los arácnidos Argentinos del orden Opiliones. Rev Mus Argent Cienc Nat Bernardino Rivadavia 5(2):127–439, figs 1–62, plates 1–20

Roewer CF (1913a) Die Familie der Gonyleptiden der Opiliones-Laniatores. Arch Naturgesch (Abt A Original-Arbeiten) 79(4): 1–256, pl. 1a.

Roewer CF (1913b) Die Familie der Gonyleptiden de Opiliones - Laniatores. [part 2]. Arch Naturgesch (Abt A Original-Arbeiten) 79(5): 257–472, pl. 2

Roewer CF (1932) Weitere Weberknechte VII (7. Ergänzung der: "Weberknechte der Erde", 1923) (Cranainae). Arch Naturgesch (NF) 1(2):275–350

Roewer CF (1963) Opiliones aus Peru und Colombien. [Arachnida Arthrogastra aus Peru V]. Senckenb Biol 44(1):5–72

Sharma PP, Giribet G (2011) The evolutionary and biogeographic history of the armoured harvestmen – Laniatores phylogeny based on ten molecular markers, with the description of two new families of Opiliones (Arachnida). Invertebr Syst 25:106–145

Simon E (1879) Essai d'une classification des Opiliones Mecostethi. Remarques synonymiques et descriptions d'espèces nouvelles. Première partie. Annales de la Société Entomologique de Belgique, 22: 183–241.

Soares BAM (1943) Notas sobre opiliões II. Pap Avulsos Dep Zool (Sao Paulo) 3(15):221–227

Soares BAM, Soares HEM (1948) Monografia dos gêneros de opiliões neotrópicos I. Pap Avulsos Dep Zool (Sao Paulo) 5(9):553–636

Sørensen WE (1902) Gonyleptiden (Opiliones, Laniatores). Ergebnisse der Hamburger Magalhaensischen Sammelreise 6(5):1–36

Villarreal MO, Kury AB (2012) *Licornus* Roewer, 1932: newly transferred to Ampycinae and first record of the family Gonyleptidae (Opiliones: Laniatores) from Venezuela. Zootaxa 3544:71–78

12

Biological mosquito control is affected by alternative prey

Ram Kumar[1,2,3], Priyanesh Muhid[4], Hans-Uwe Dahms[5], Jaigopal Sharma[6] and Jiang-Shiou Hwang[2*]

Abstract

Background: Mosquitofish were introduced to several countries of the tropics and subtropics as biological agents for the control of mosquito larvae. Meanwhile, they became a threat to native communities and fish worldwide, similar to other invasive species through resource competition, overexploitation, or habitat alteration. We investigated prey selectivity patterns of *Gambusia affinis* (mosquitofish) preying on larvae of the two Indian major carps (*Catla catla* and *Labeo rohita*) in the presence of varied proportions of alternative prey (rotifers, cladocerans, chironomid and mosquito larvae) under laboratory conditions.

Results: The patterns of prey selectivity in mosquitofish were influenced by the presence of alternative prey and their relative abundance in the environment. Carp larvae, when present in equal proportions, were randomly selected by mosquitofish, however, positively selected when present in higher proportions. In the presence of *Hexarthra mira, Daphnia similoides* or the mosquito larval instar-IV as an alternative prey, the mosquitofish preferred fish larvae regardless of prey proportions. In the medium where either mosquito larval instar-I or chironomid larvae were given as alternative prey, the mosquitofish either rejected or randomly selected the carp larvae. Given a multispecies prey combination, mosquitofish primarily selected the larvae of *L. rohita* and mosquito larval instar-I. We also found a prey switching ability of mosquitofish in relation to varying abundances of prey species in the environment.

Conclusions: The present results suggest that mosquito immatures are not the preferred food of mosquitofish when fish larvae are present in their natural habitats. Since mosquitofish and carp larvae have overlapping natural habitats and prey preferences are the invasive mosquitofish may have a substantial impact on native communities of invertebrates and fish. This way, they are equally important for extensive fisheries and conservation management.

Keywords: Invasive fish; Conservation management; Optimal foraging; Prey selection; Prey switching; Aquaculture

Background

Two species of mosquitofish, *Gambusia affinis* (Baird and Girard 1853) and *Gambusia holbrooki* (Girard 1859), have become examples of invasive species globally with negative impacts on native communities. They have invaded low-order stream sections, most of the wetlands, shallow lakes and almost all ornamental ponds in 67 tropical and subtropical countries (Duryea et al. 1996, Lever 1996, Swanson et al. 1996, Helfman 2007). In addition to competing for resources, interfering with reproduction and introducing parasites (Arthington 1991, Crowl et al. 1992, Kennard et al. 2005, Benejam et al. 2009, Hazelton and Grossman 2009), mosquitofish are quite aggressive and often attack fish more than twice their own body size (Wurtsbaugh et al. 1980, Howe et al. 1997). Because of their higher dispersal tendency, invasiveness (Rehage and Sih 2004) and adverse effects on native aquatic communities (Barrier and Hicks 1994, Webb and Joss 1997, Arthington and Marshall 1999, Smith et al. 2011), the database of the IUCN Invasive Species Specialist Group (www.issg.org/database) classifies mosquitofish, along with eight other species, as among the world's 100 worst invasive species ISSG, 2000; Lowe et al. 2000). Invasive exotic species may displace native species through resource competition or by overexploiting native species to the point of extinction (Dahms 2004, Clavero and García-Berthou 2005) or they may alter the habitat to such an extent that native species are no longer able to exist (Dahms 2005, Gooden et al. 2009).

* Correspondence: Jshwang@mail.ntou.edu.tw
[2]Institute of Marine Biology, National Taiwan Ocean University, 2 Pei Ning Road, Keelung 202, Taiwan
Full list of author information is available at the end of the article

In several studies, mosquitofish have been used as a model organism to investigate pelagic trophic interactions (Hurlbert et al. 1972, Lancaster and Drenner 1990, Peck and Walton 2008), to test optimal foraging theory (Bence and Murdoch 1986, Bence 1988) or to assess genetic diversity (Vidal et al. 2010). Some dietary studies have focussed on the role that mosquitofish play in controlling insect populations (Linden and Cech 1990, Gophen et al. 1998, Hamer et al. 2002). Mosquitofish are perennially abundant in degraded hypertrophic lakes (e.g. Naini Lake, Rajghat Lake and Old Fort Lake in Delhi, India (Kak 1999), that support persistent cyanobacterial blooms (Kak 1999), an environment otherwise uninhabitable for native species. There is indication on mosquitofish-driven displacement of native species (Nagdali and Gupta 2002, Helfman et al. 2009). In India, there is a strong negative association between mosquitofish density and the Indian major carp (Nagdali and Gupta 2002, Jhingran 1991) in the natural settings. A mass-mortality event of *G. affinis* increased the density of larger-sized zooplankton and fish larvae in Lake Nainital, India (Nagdali and Gupta 2002), confirming its negative effect on native aquaculturally important fishes (Arthington 1991, Arthington and Marshall 1999, Marchetti and Moyle 2001).

Several studies have demonstrated the competitive advantage of mosquitofish over small indigenous fishes (Courtenay and Maffe 1989, Howe et al. 1997, Ivantsoff 1999) and amphibians (Vargas and De Sostoa 1996, Goodsell and Kats 1999, Komak and Crossland 2000). Other studies have described the effects of mosquitofish predation on invertebrates such as zooplankton (Margaritora 1990), odonate larvae (Englund 1999) and fairy shrimp (Leyse et al. 2004, Sornsupharp et al. 2013a, b). Despite being generalistic feeders (Rehage et al. 2005), mosquitofish exhibit strong prey selection in an environment with multiple prey types (Hurlbert and Mulla 1981, García-Berthou 1999, Kumar et al. 2008). They are able to attack a variety of prey, and their preferences are determined by several factors including prey size, morphology, biomass (Wurtsbaugh et al. 1980), transparency, relative abundance of prey in the environment (Rao and Kumar 2002, Kumar et al. 2012) and the swimming behaviour of prey (Linden and Cech 1990, Swanson et al. 1996, Alcaraz et al. 2008). This selective predation has a significant effect on the structure and dynamics of freshwater ecosystems (Kennard et al. 2005, Rao and Kumar 2002, Kumar et al. 2008), as the less preferred prey populations obtain an environment free of competition (Peck and Walton 2008, Lee et al. 2010).

Although mosquitofish have often been released in highly disturbed or artificial habitats as biocontrol agents of mosquito immatures, they have the potential to spread into pristine water bodies (Courtenay and Maffe 1989), where they can severely impact native fishes (Howe et al. 1997), amphibians (Webb and Joss 1997, Goodsell and Kats 1999) and invertebrates (Leyse et al. 2004). In a comparative study of *G. holbrooki* and a common native species in eastern Australia, the Pacific Blue-eye, *Pseudomugil signifier* Kner, it was suggested that the native fish *P. signifier* is a potential effective predator of mosquito larvae. In contrast, mosquitofish becomes a threat to native fishes that share similar habitats. This is particularly due to their eco-physiological advantages related to fast growth, early maturity, viviparity and their voracious appetite (Vargas and de Sostoa 1996).

Therefore, it is relevant to elucidate the preference of mosquitofish for the native fish communities from a wide spectrum of natural prey species (Olden et al. 2010). Relatively few studies have analysed the dietary patterns of mosquitofish (Crivelli and Boy 1987, Mansfield and McArdle 1998, García-Berthou 1999) using co-occurring fish larvae and an array of different co-occurring prey types under controlled conditions.

On the Indian subcontinent, the two commercially and aquaculturally important carp species *Catla catla* (Hamilton 1822) and *Labeo rohita* (Hamilton 1822) often co-occur with introduced mosquitofish. These two major carps are native to Indian freshwaters and are extensively used in fish-culture in both lentic and lotic waters (Jhingran 1991, Rao 2003). Due to the invasive nature of mosquitofish and the threat that mosquitofish provide for the indigenous fauna, it has become necessary to evaluate their predation potential on carp larvae. The optimal foraging model (Werner and Hall 1974, Krebs and Kacelnik 1991) predicts that a predator positively selects more profitable prey based on the energy spent in searching, encountering, attacking, capturing and ingesting, compared to the energy gained from digesting other prey (Kumar and Rao 2003, Kumar et al. 2008). Therefore, the availability of less-profitable prey should not influence the predators' decision to select more profitable prey. Hence, with an increasing abundance of profitable prey, the predator's preference should switch from random selection to positive selection (Oaten and Murdoch 1975, Akre and Johnson 1979, Kiørboe et al. 1996). The ability of a predator to switch from one prey species to another depending on their relative abundance saves energy spent in searching for less abundant prey. This is because searching is an adaptive behaviour that not only allows the predator to maximise its net energy intake but may also have a stabilising effect on the prey population (Oaten and Murdoch 1975, Kiørboe et al. 1996). Carp larvae (body length: 3.5–5 mm) have a higher biomass than zooplankton and an inefficient escape response, which makes them more vulnerable to predation. Thus, this study compared the foraging preferences of mosquitofish between fish larvae and a series of invertebrate prey and examined the effects of variations in the relative abundance of alternative prey on fish larval preferences.

Here we pursue our hypotheses that (I) mosquitofish positively selects carp larvae over alternative invertebrate prey, and (II) mosquitofish preference for carp larvae depends on the relative abundance of alternative prey in the environment. To test these hypotheses, we evaluated quantitatively the consumption of fish larvae by the mosquitofish *G. affinis* (Baird and Girard, 1854), in the presence of varied proportions of alternative prey such as rotifers, cladocerans and dipteran (culicid and chironomid) larvae.

Methods

Collection, culture and preparation of experimental animals

All experimental animals, except chironomid larvae, used in this study were collected in the field prior to the experiment and cultured in the laboratory. Table 1 lists the experimental animals and their characteristics. Individuals of each prey species were used only once in an experiment. The predators used in the experiments were exposed to all prey species before testing, because they were cultured using combinations of the same prey species (fish larvae were used only occasionally) as food.

Predator

Similarly sized female mosquitofish (*G. affinis*) were collected with a fish net from an experimental pond in the botanical garden of Sri Guru Tegh, Bahadur Khalsa College, University of Delhi, India. All fish were immediately brought to the laboratory, where they were acclimated before the experiments. Until the experiment began, the fish were maintained in 5 stock aquaria of 15 L each, and provided with natural macrophytes collected from the same pond. All aquaria were cleaned, and the water was changed once a week. Each aquarium contained about 35–40 fish of similar size to be acclimated. A 14-h L:10-h D photoperiod was maintained at ambient room temperature. Mosquitofish were fed with a mixture of zooplankton, mosquito larvae and chironomid larvae. Mosquitofish individuals were used only once in feeding trials. Throughout the experimental period of 12 days, in total 130 individual fish were used for the experiments.

Prey

The experiments in this study involved representative species that were consumed by the predator (García-Berthou 1999, Kumar et al. 2008) and commonly co-occur with mosquitofish in temporary waters, ornamental ponds, pools, lakes, etc. in tropical environments. Before mosquitofish were captured, zooplankton samples were collected by filtering 50–60 L pond water through a 53-μm mesh. The rotifers *Hexarthra mira* (Hudson 1871) and *Asplanchna intermedia* (Hudson 1886) and the cladoceran *Ceriodaphnia cornuta* (Sars 1885) were isolated from zooplankton samples that were collected from the same pond as the mosquitofish (see also Chullasorn et al. 2009). However, the larger cladoceran *Daphnia similoides*

Table 1 Body sizes (mean ± SE, $n = 20$) and dry mass of prey species used in the experiment

Group	Species	Length (μm)	Dry mass (μg)	Micro-distribution and seasonality
Rotifera	*Hexarthra mira* (Hudson, 1871)	186 ± 14.60	0.08	Planktonic; throughout the water column; throughout the year with highest abundance in summer
	Asplanchna intermedia (Hudson, 1886)	986 ± 68.90	1.86	Planktonic; throughout the water column; abundant in lake plankton during summer
Cladocera	*Ceriodaphnia cornuta* (Sars, 1885)	526 ± 16.87	2.6	Planktonic; throughout the water column; perennially abundant
	Daphnia similoides (Hudec, 1991)	1961 ± 34.50	27.2	Planktonic; throughout the water column; occurrence from fall to spring with a peak during winter
Insecta	*Anopheles stephensi* (Liston, 1901) larvae instar-I-II	736 ± 14.50	4.48	Water surface; most from summer to fall
	A. stephensi larvae instar-IV	1896 ± 82.00	185	Water surface; most abundant from summer to fall
	Chironomus larvae	3280 ± 185.54	240	Littoral and benthic
Pisces: Cyprinidae (prey)	*Catla catla* (Hamilton, 1822)	4624 ± 219.00	385	Benthopelagic in fresh to brackish water; spring and fall
	Labeo rohita (Hamilton, 1822)	3410 ± 246.04	310	Throughout the water column
Pisces: Poeciliidae (predator)	*Gambusia affinis* (Baird and Girard, 1854)	4723 ± 458.60	–	Throughout the water column; littoral, higher density in surface waters; perennial, more abundant from April to September

(Hudec 1991) was collected from an ornamental pond with a lower density of mosquitofish. The field-collected animals were slowly acclimated to autoclaved tap water by progressively increasing the proportion of autoclaved tap water to the filtered pond water. The cultures of the rotifer *H. mira* and both cladoceran species were expanded and maintained in autoclaved tap water, using the green alga *Chlorella vulgaris* Beijerinck 1890 at a density of 2–3.5 × 10^6 ml as food. The rotifer *Brachionus calyciflorus* Pallas 1766 was used as food for the culture of the predatory rotifer *A. intermedia*. Polymorphism was observed in the *A. intermedia* culture. Although saccate, cruciform and campanulate morphs appeared in the laboratory cultures, only the campanulate morphs were used for the experiments (see Dahms et al. 2011a, Hwang et al. 2013).

The larvae of the mosquito *Anopheles stephensi* Liston 1901 were obtained from a laboratory culture. Larval cultures were maintained in a separate room at optimal, constant conditions (temperature 25 ± 2 °C, relative humidity 70 ± 5 %, photoperiod 14 h L:10 h D). Mosquito eggs collected from the cage were placed in enamel trays containing dechlorinated water for hatching. A standard diet consisting of yeast extract and finely ground dog biscuits (2:3) was added to this tray to feed the mosquito larvae. *Chironomus* larvae were collected from outdoor concrete tubs as required. The larvae of *C. catla* and *L. rohita* were obtained from the aquaculture laboratory of the Department of Zoology, University of Delhi.

For all cultures and experiments, room temperature was maintained at 25 ± 2 °C. Table 1 lists the experimental animals and their characteristics.

Experimental treatments

All experiments, including five replicates for each treatment, were conducted in 30 × 20 cm glass troughs containing 2 L autoclaved tap water. A series of short-term experiments was conducted to investigate the feeding rates (number of larvae ingested h^{-1}) of the mosquitofish. The initial density (Table 2) for each prey type and test duration was chosen based on preliminary feeding experiments, such that at the end of each experiment a substantial number of prey remained unconsumed. The mosquitofish used in the experiments had sufficient exposure to all prey species prior to testing, since they were cultured using food in different combinations of the same prey species except fish larvae. One individual mosquitofish was transferred from the stock aquarium to the experimental trough 6 h prior to the experiment and deprived of food. After the fish was starved for 6 h, prey was individually counted and placed gently in each experimental trough.

Table 2 Experimental protocols used for tests of *L. rohita* larval consumption and prey selection by mosquitofish

Experiment	Test no.	Prey species	Combined prey density
1. Larval consumption rate			
	1	*C. catla*	60
		L. rohita	60
2. Prey selectivity: selectivity within a taxonomic group			
Fish larvae—fish larvae choice	2	*C. catla—L. rohita*	60
L. rohita larvae—rotifers choice	3	*H. mira—L. rohita*	60
		A. intermedia—L. rohita	60
L. rohita larvae—cladocerans choice	4	*C. cornuta—L. rohita*	60
		D. similoides—L. rohita	60
L. rohita larvae—insect larvae	5	*A. stephensi* larvae instar-I—*L. rohita*	60
		A. stephensi larvae instar-IV—*L. rohita*	60
		Chironomus larvae—*L. rohita*	60
3. Multispecies choice	6		600

Experiment 1: consumption of fish larvae

The ingestion rates of female mosquitofish preying on *C. catla* and *L. rohita* larvae were estimated. Sixty larvae of either *L. rohita* or *C. catla* were offered to previously acclimatised and starved mosquitofish in each experimental trough. The experimental design included two treatments (*L. rohita* + *G. affinis*; *C. catla* + *G. affinis*) × 2 controls (*L. rohita* alone; *C. catla* alone) × 5 replicates, for 20 glass troughs. Mosquitofish were removed from the experimental dishes after 3 h of incubation, and the number of remaining carp larvae was counted.

Experiment 2: prey selection experiment

The following prey combinations were used for the selectivity experiments: (i) *C. catla—L. rohita* larvae, (ii) rotifers (either *H. mira* or *A. intermedia*)—*L. rohita* larvae (iii) cladocerans (either *D. similoides* or *C. cornuta*)—*L. rohita* larvae (iv) mosquito larvae (either larval instars-I-II or instar-IV)—*L. rohita* larvae and (v) chironomid larvae—carp larvae (vi) multispecies prey combination (all abovementioned prey types in the medium). To test the null hypothesis that the preference of the mosquitofish does not change with varying proportions of alternative prey types, two prey-type choice experiments offered prey types in three different proportions (0.33, 0.5 and 0.67), providing a total of 60 prey individuals. Table 2 shows the experimental protocols for the different selectivity tests, and Table 3 shows the prey species

Table 3 Prey species and densities used in multispecies prey selectivity tests

No.	Species	No. 2000 ml^{-1}
1	*L. rohita*	8
2	*C. catla*	8
3	*H. mira*	200
4	*A. intermedia*	60
5	*C. cornuta*	50
6	*D. similoides*	24
7	*A. stephensi* larvae instar-I	150
8	*A. stephensi* larvae instar-IV	50
9	*Chironomus* larvae	50

used in a multispecies choice test. The mosquitofish were allowed to feed for 1 h. The difference between the initial and final numbers of prey in the beaker was taken as the number of prey consumed by the mosquitofish during the test period. Random sampling of some beakers did not reveal any loss of prey due to death by other causes than predation.

Statistical analyses

One-way ANOVA was used to test the differences in the consumption of *C. catla* and *L. rohita,* in terms of the number of individuals and dry mass ingested. For two prey-type choice experiments, the expected consumption (number and dry mass) of one prey species in relation to the other *versus* the actual prey consumption was calculated for each condition. The differences between the actual and expected prey consumption for three proportions were analysed with two-way ANOVA for both of the prey types in two prey choice tests. Nonparametric Mann–Whitney rank tests (Mann and Whitney 1947) were used to determine the differences between expected and observed values of number and dry mass ingested at particular proportions of particular prey types, depending on the two-way ANOVA results (see also Tseng et al. 2008, Kim et al. 2013a, b). The difference between the expected and actual prey consumption for each prey type in a multiple prey choice test was analysed with repeated one-way ANOVA. Prey selectivity was calculated using Manly's selectivity index (α_i) (Manly 1974), modified for a situation in which the predator consumes a substantial portion of the available prey, and hence, prey numbers in the medium decline with time, similar to our experiments. The modified formula (Chesson 1983) is:

$$\alpha_i = \frac{\ln(n_{i0}-r_i)/n_{i0}}{\sum_{j=1}^{m} \ln((n_{i0}-r_i)/n_{i0})}, i = 1, 2, 3 \ldots\ldots\ldots m$$

where α_i is Manly's α (preference index) for prey type I; n_{i0} is the number of items of type i present at the beginning of a foraging bout; r_i is the number of items of food type i in the consumer's diet; and m is the number of prey types.

Manly's α_i values range from 0 to 1. The α_i value for non-selective feeding is 0.5 with two prey-type choice tests. The α_i value for non-selective feeding is 0.11 in a multispecies choice test with nine prey types. Index values above 0.5 (0.11 for multispecies prey choice tests) indicate positive selection (preference), and values below 0.5 indicate negative selection (avoidance). Deviations in selectivity index values from the value for non-selective feeding were tested for statistical significance using Hotelling's T^2 test. Differences in the number and dry mass of the two prey types consumed and the arcsine-transformed selectivity index values (α_i) among three proportions were analysed with MANOVA because the response variables were not independent (Gill 2001). MANOVA results reveal the independent variables (e.g. predators, presence of alternative prey and relative abundance of prey in the present study) that best differentiate mosquitofish preferences (Gill 2001). The actual consumption data were also subjected to a general linear model factorial univariate ANOVA, to assess the interaction of the variables (prey type × proportions) and the effects of the individual factors on fish larval consumption rates. To identify homogeneous subsets of means, the ANOVA was followed by pairwise multiple comparisons using Tukey's test. Using a general linear model (Gill 2001) and a factorial multivariate ANOVA (MANOVA) test (Gotelli and Ellison 2004), this study tested the null hypotheses that the presence of alternative prey and different relative prey proportions do not affect mosquitofish preferences for fish larvae. As Pillai's trace, Wilk's lambda and Hotelling–Lawly's trace converge to a similar P value despite having differences in their degrees of freedom (Scheiner 2001), Pillai's trace was used for further interpretation because it provides the most robust test (Scheiner 2001).

Prior to use of parametric statistics, the w/s normality test was used for all data. w/s normality test requires only the sample standard deviation and the data range (Kanji, 2011). The test is based on the q statistics, which is the 'studentized' range, or the range expressed in standard deviation units. The standardised range $q = \frac{w}{s}$ was estimated, where the sample standard deviation (s) and the range (w). The null hypothesis H_0: prey consumption is not significantly different than normal prey consumption rate in nature. The calculated q values were within the range of critical values, therefore, H_0 is accepted (the data are normally distributed).

The data obtained in proportions (index values, proportion eaten etc.) were arcsine transformed prior to applying parametric statistics. The distribution of proportions is binomial, and arcsine transformation of data makes the distribution normal (Sokal and Rohlf 2012). The observed proportion p is changed to a new form $\theta = \sin^{-1}(\sqrt{p})$.

To test prey-switching mechanisms at all prey combinations, the observed and expected consumption was subjected to regression analyses. The null hypothesis was that the ratio of two alternative prey consumed (P_1:P_2) in a given time period is constant and independent of the relative availabilities of the two prey. The formula for prey switching (Landry 1981, Kiørboe et al. 1996) calculates the proportionality constant (C) formula as follows:

$$\frac{P_1}{P_2} = C(N_1/N_2)$$

where N_1 and N_2 represent the densities of two prey species and P_1 and P_2 are the numbers of the two prey species consumed in a given time period. Increasing the values of C (proportionality constant) with increasing values of N_1/N_2 implies prey switching.

Results

Consumption of fish larvae

Mosquitofish ingested the larvae of both fish species and consumed *L. rohita* larvae at a significantly higher rate than *C. catla* larvae (one-way ANOVA; $P < 0.05$, Table 4, Fig. 1). However, the differences were not statistically significant in terms of larval dry mass of the two fish species ingested (one-way ANOVA, $P = 0.718$, Table 4, Fig. 1).

Prey selection experiment

Gambusia affinis ingested the larvae of native cyprinid fish in all prey combinations and proportions. However, the mosquitofish preference for the fish larvae was a function of the type and relative proportion of alternative prey. The effect of alternative prey type and the interaction of prey type and prey proportion on the consumption and preference for fish larvae were highly significant (MANOVA, $P < 0.01$, Table 4). The effect of prey proportion was dependent on the presence of the alternative prey type. In all test combinations, the prey preference (Manly's α_i) was significantly correlated ($R^2 = 0.53$ to 0.7, $P < 0.05$) with prey ingestion rates (Fig. 2).

Table 4 One-way ANOVA results for differences in consumption of *C. catla* and *L. rohita* by mosquitofish

	Sum of squares	df	Mean square	F value	P value
Dry mass of fish larvae consumed					
Between groups	36,901.389	1	36,901.389	0.135	0.718
Within groups	4,359,138.889	16	272,446.181		
Total	4,396,040.278	17			
Number of fish larvae consumed					
Between groups	14.222	1	14.222	7.062	0.017
Within groups	32.222	16	2.014		
Total	46.444	17			

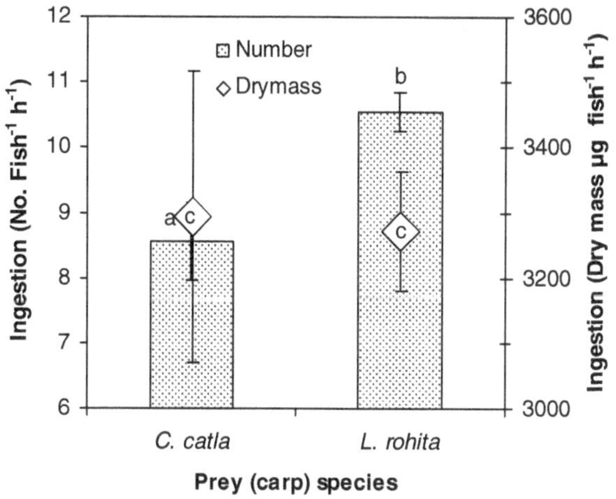

Fig. 1 Per capita and per hour ingestion (mean ± SE) of fish larvae (*C. catla* and *L. rohita*) by mosquitofish female

C. catla—*L. rohita* choice test

When offered a combination of *C. catla* and *L. rohita* larvae, mosquitofish preferred the more abundant prey. Neither the prey type nor their interaction was significant, but the prey proportion was highly significant (MANOVA, $P < 0.001$, Fig. 3, Table 5) for both, prey ingestion rate and the selectivity index (α_i) values. In either case, the number and dry mass of fish larval prey ingested by mosquitofish were significantly less (one-way ANOVA, $P < 0.01$, Fig. 3) than the expected value at the lowest prey proportion (0.33). The number of ingested *L. rohita* was significantly higher ($P < 0.05$) at higher prey proportion, but the ingested dry mass of *L. rohita* did not differ significantly from the expected value. The number and dry mass of either ingested fish larval prey did not differ significantly from the expected value (Mann–Whitney U test; Fig. 3) at equal prey proportions.

Mosquitofish showed differential selection behaviour for the two tested rotifer species. They showed a strong preference for fish larvae, regardless of their proportion against the evasive rotifer *H. mira* (Hotelling's T^2 test, $P < 0.05$, Fig. 4a, Manly's $\alpha = 0.78$–0.96). The selectivity index value was significant (MANOVA, $P < 0.001$, Fig. 4a) for prey type, but neither for proportion nor for interaction ($P = 0.16$). The ingestion of *L. rohita* larvae was significantly higher than the expected value (two-way ANOVA, $P < 0.001$, Fig. 4a). However, the dry mass of ingested fish larvae did not differ significantly from the expected value, regardless of its relative proportion in the medium ($P > 0.9$). In terms of dry mass, the

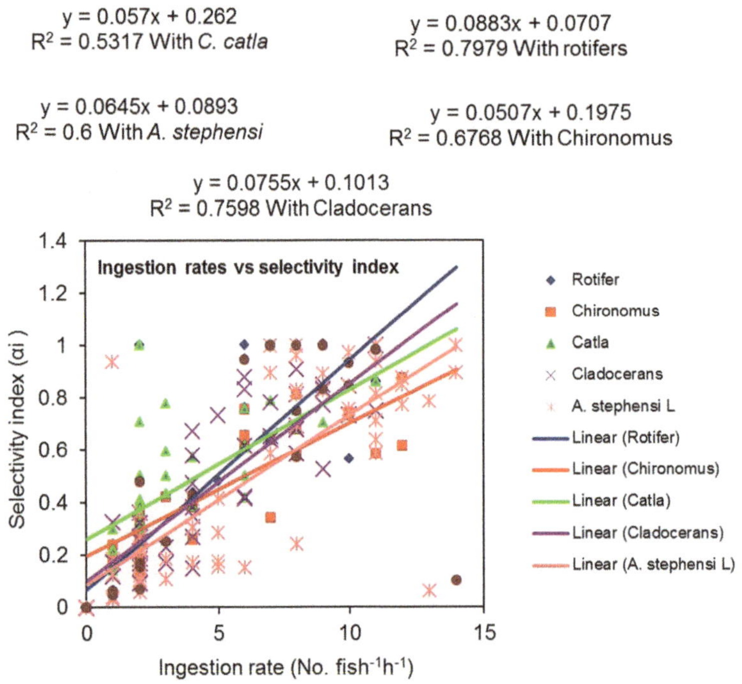

Fig. 2 Correlation between Manly's selectivity index (α_i) and ingestion rates in mosquitofish

mosquitofish positively selected fish larvae against the larger rotifer *A. intermedia* in terms of dry mass regardless of prey proportions (Hotelling's T^2 test, $P < 0.05$; Manly's $\alpha_i = 0.76–0.92$, Fig. 4b). However, in terms of the number of ingested larvae, the fish larvae were selected positively (Hotelling's T^2 test, $P < 0.05$; Manly's $\alpha_i = 0.72–0.9$; Fig. 4b) at higher or equal proportions but were randomly selected at lower proportion. In contrast, *A. intermedia* was selected positively at higher proportion (0.67) (Fig. 4b) and negatively selected at equal or lower proportion when the index value was calculated with the number of ingested prey. From a combination of *L. rohita* larvae and *C. cornuta*, the mosquito fish prey preference depended on the proportion of prey in the environment (MANOVA, $P < 0.001$, Fig. 5a). *C. cornuta* was positively selected when offered at higher proportion whereas fish larvae were positively selected when offered at higher or equal proportion (Fig. 5a).

Regardless of the prey proportion, fish larvae were positively selected (MANOVA, $P < 0.001$; Hotelling's T^2 test, $P < 0.001$; Manly's $\alpha = 0.76$ to 0.84) when the larger cladoceran *D. similoides* was present as an alternative prey; whereas *D. similoides* was not selected against the fish larvae (Fig. 5b). Prey type-related differences in number and dry mass ingestion were highly significant (MANOVA, $P < 0.001$, Table 5), whereas proportion-related differences were neither significant for the dry mass (MANOVA, $P = 0.188$) nor for the selectivity index values calculated with the ingested dry mass (MANOVA, $P = 0.483$). The number and dry mass of ingested *D. similoides* was significantly less (two-way ANOVA, $P <$ 0.01, Fig. 5b) than the respective expected values, regardless of the relative prey proportion. In contrast, the numbers of ingested fish larvae were significantly higher (two-way ANOVA, $P < 0.01$) than the expected value.

The mosquitofish exhibited a strong preference for instar-I stage of *A. stephensi* and negatively selected the fish larvae (Hotelling's T^2 test, $P < 0.05$; Manly's $\alpha = 0.14$ to 0.36, Fig. 6a) regardless of prey proportion in the medium. Dry mass and number of *A. stephensi* instar-I actually ingested by the mosquitofish were significantly higher (two-way ANOVA, $P < 0.001$, Fig. 6a) than the expected values at all proportions (two-way ANOVA, $P = 0.052$). The interaction between prey type and proportion was not significant (two-way ANOVA, $P = 0.557$). The actual dry mass of ingested *L. rohita* did not differ significantly (two-way ANOVA, $P = 0.858$, Fig. 6a) from the expected value but the number of ingested *L. rohita* was significantly lower (two-way ANOVA, $P < 0.001$, Fig. 6a) than the respective expected values. In contrast, mosquitofish strongly avoided (Manly's $\alpha = 0.06–0.18$) *A. stephensi* instar-IV and positively selected fish larvae (Manly's $\alpha = 0.82–0.94$, Fig 6b) regardless of the proportion of prey in the medium. The number and dry mass of actually ingested mosquito larvae instar-IV were significantly lower (two-way ANOVA, $P < 0.001$, Fig. 6b) than the expected values regardless of prey proportions in the environment.

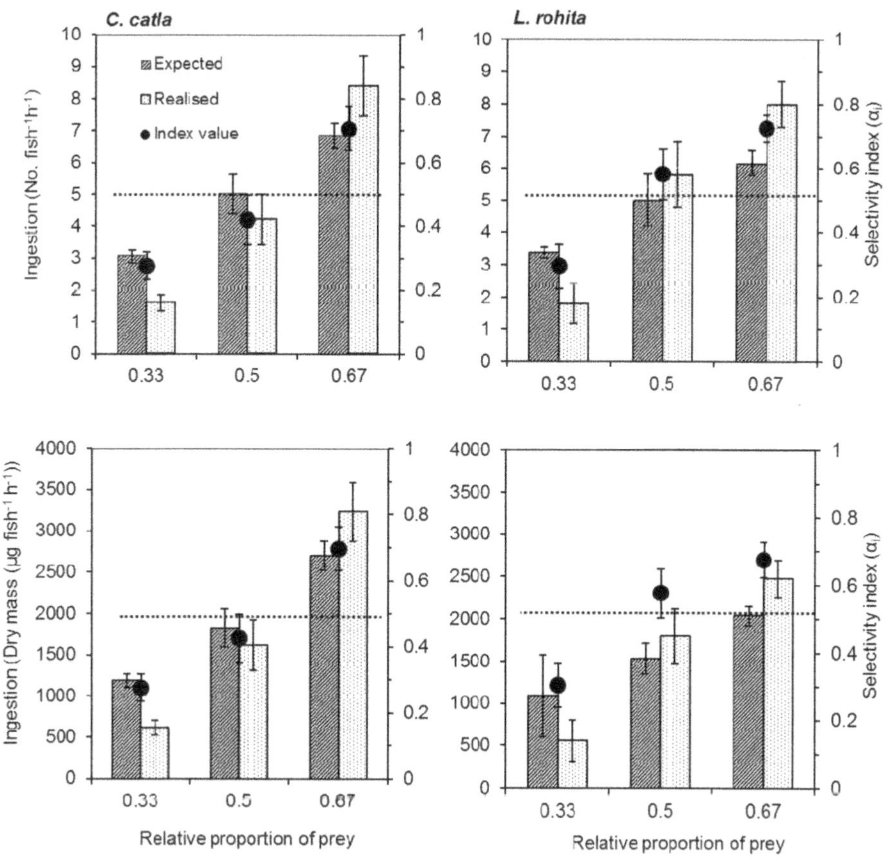

Fig. 3 Expected and observed mean ingestion rates and Manly's selectivity index (α_i) in mosquitofish given a mixture of *C. catla* and *L. rohita* larvae. The *dashed line* in each indicates the α_i value (secondary Y-axis) for non-selective feeding

The actual dry mass of ingested fish larvae did not differ significantly (two-way ANOVA, $P < 0.858$, Fig. 6b) from the expected value but the number of ingested fish larvae was significantly higher (two-way ANOVA, $P < 0.001$, Fig. 6b) than the expected value at a higher proportion. The prey proportion ($P < 0.001$), but not the prey type × proportion interactions were significant ($P > 0.337$).

In the fish larvae-chironomid larvae choice test, prey type-related differences in prey ingestion rates were significant for the dry mass of ingested prey ($P < 0.001$; MANOVA, Table 5) and for the prey selectivity index values calculated with the number of prey (MANOVA, $P < 0.001$, Table 5, Fig. 6c) consumed. Neither the number of prey ($P = 0.24$; MANOVA) nor the selectivity index value calculated with dry mass ($P < 0.067$) was significant. The proportion-related effects on differences in prey ingestion and selectivity index values for number and dry mass were highly significant (MANOVA, $P < 0.001$,

Table 5 Results of a multivariate ANOVA (Pillai's Trace and Roy's Largest Root test) for number and dry mass of prey consumed and for selectivity index value calculated for number and dry mass of prey ingested

Effect		Value	*F* value	Numerator (df)	Denominator (df)	*P* value
Intercept	Pillai's trace	1.000	199,732.51	24	1	0.002
	Roy's Largest Root	4,793,580.3	199,732.51	24	1	0.002
Type	Pillai's trace	1.000	84,979.222	24	1	0.003
	Roy's Largest Root	2,039,501.3	84,979.222	24	1	0.003
Proportion	Pillai's trace	1.920	1.997	48	4	0.265
	Roy's Largest Root	409,866.4	34,155.5	24	2	0
Type × proportion	Pillai's trace	1.997	56.091	48	4	0.001
	Roy's Largest Root	100,356.61	8363.051	24	2	0

Table 6 Between-subjects effects (MANOVA) for multispecies prey choice test

Dependent variable	Type III sum of squares	df	Mean square	*F*	*P* value
Number ingested	122,520.8	8	15,315.1	26.769	0.000
Dry mass ingested	2,764,194.5	8	345,524.312	56.952	0.000
Selectivity index value for number	0.242	8	0.03	41.721	0.000
Selectivity index value for dry mass	0.226	8	0.028	31.878	0.000

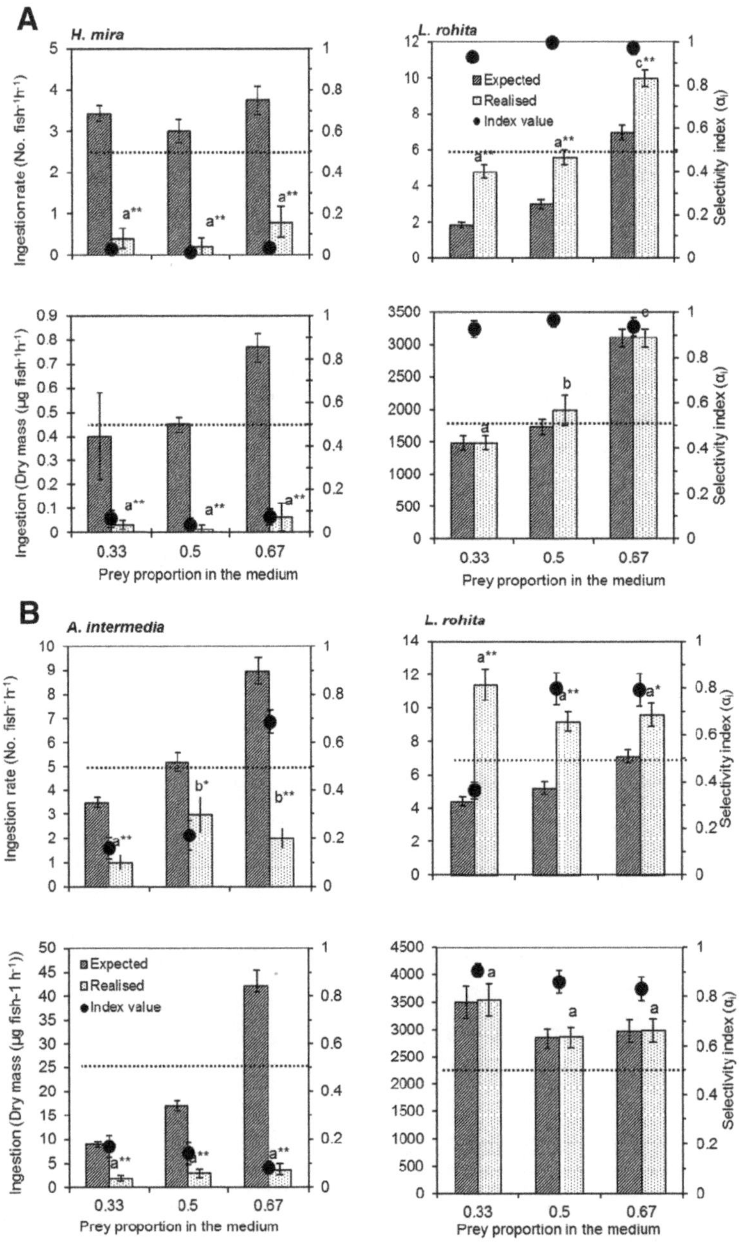

Fig. 4 Expected and observed mean ingestion rates and Manly's selectivity index (α_i) for mosquitofish in Rotifera—*L. rohita* larvae choice test: **a** *H. mira*—*L. rohita* test combinations; **b** *A. intermedia*—*L. rohita* test combinations. The *dashed line* in each indicates the α_i value (secondary Y-axis) for non-selective feeding

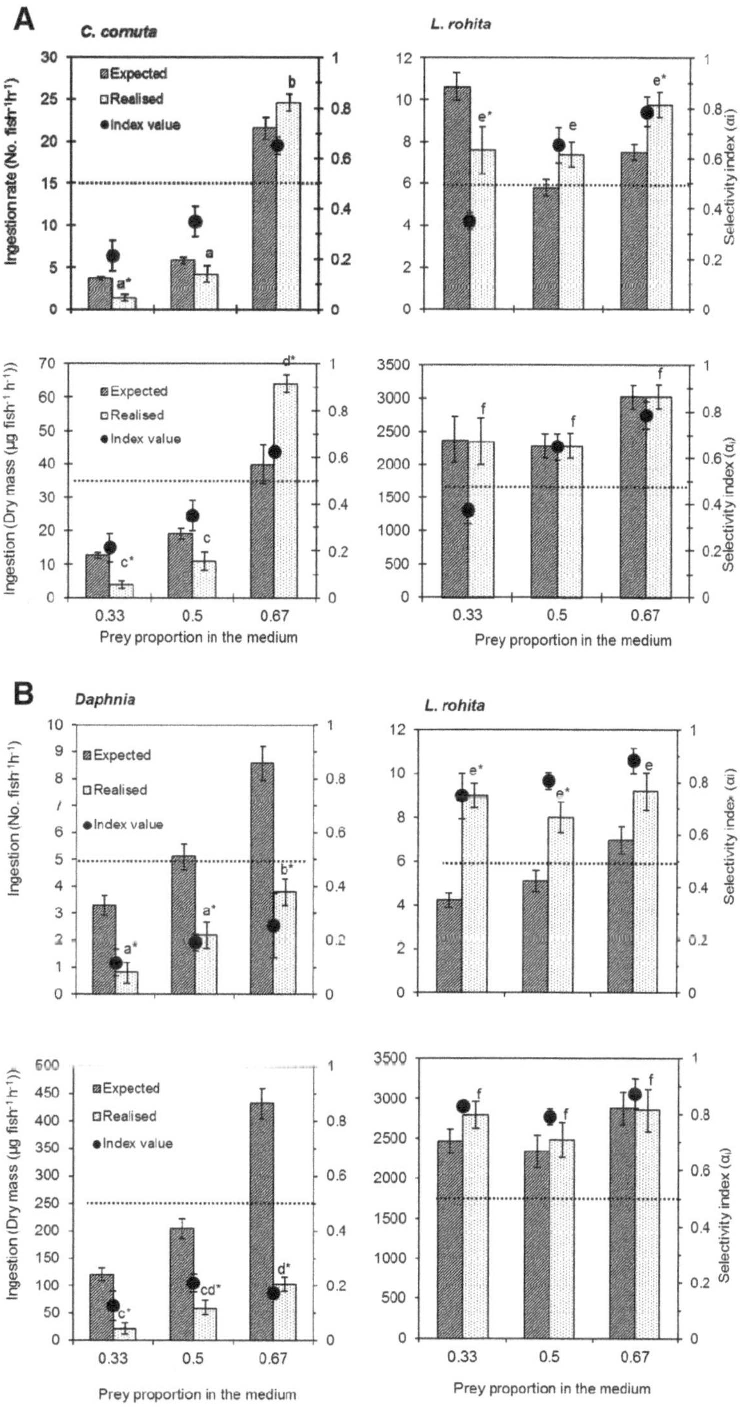

Fig. 5 Expected and observed mean ingestion rates and Manly's selectivity index (a_i) for mosquitofish in Cladocera—*L. rohita* larvae choice test: **a** *C. cornuta*—*L. rohita* test combinations; **b** *D. similoides*—*L. rohita* test combinations. The *dashed line* in each indicates the a_i value (secondary Y-axis) for non-selective feeding

Table 5, Fig. 6c). The interaction effect between prey type and proportion was significant only for the selectivity index value calculated for the number of prey ingested. Number and dry mass of either prey type ingested were significantly lower (Mann–Whitney *U* test, $P < 0.01$) at lower proportion, did not differ significantly at equal proportion (except the number of chironomid larvae ingested) and were significantly higher than the respective expected values at the higher prey proportion.

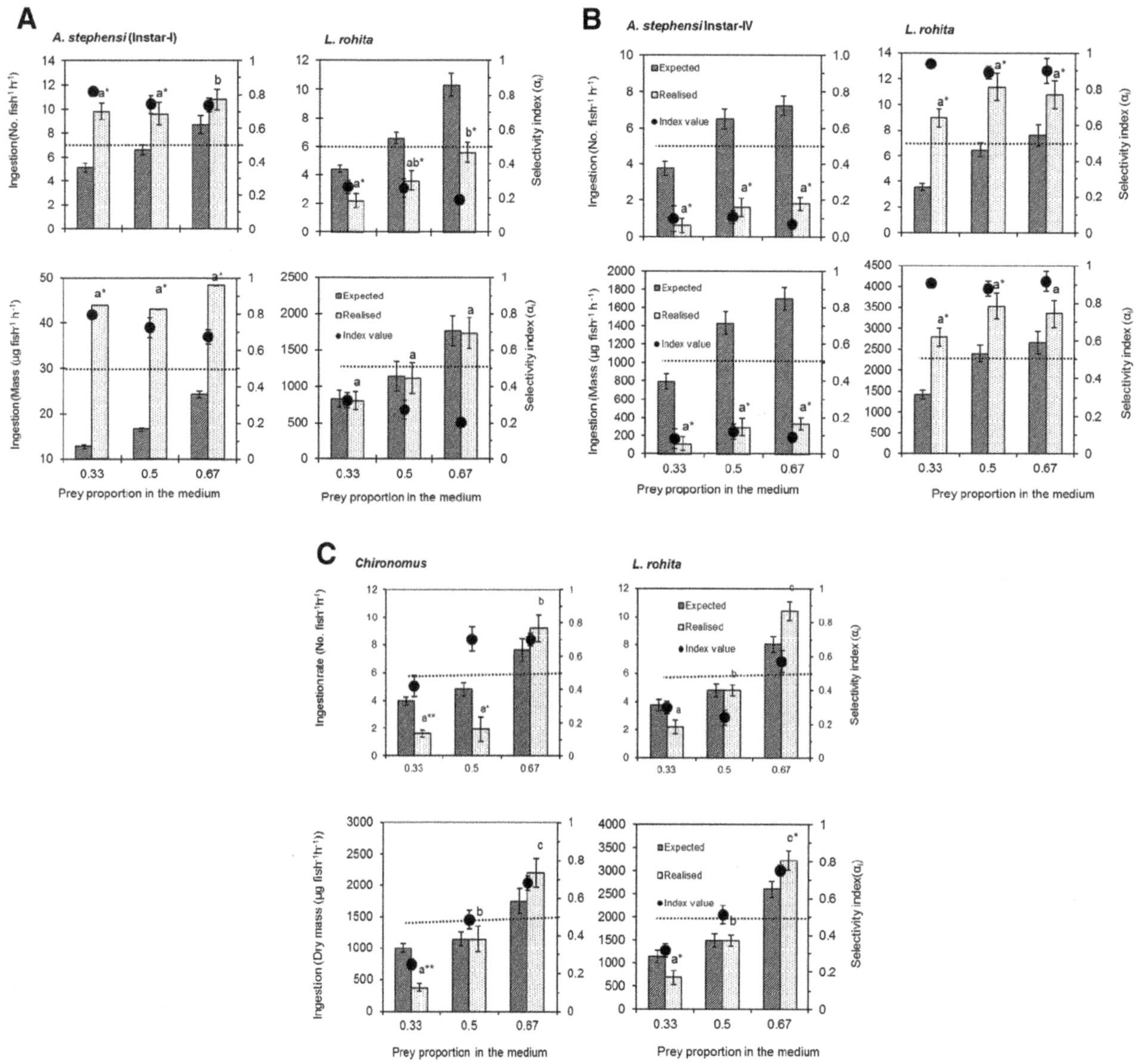

Fig. 6 Expected and observed mean ingestion rates and Manly's selectivity index (α_i) for mosquitofish in Insect larvae—*L. rohita* combination choice test: **a** *A. stephensi* instar-I—*L. rohita* test combinations; **b** *A. stephensi* instar-IV—*L. rohita* test combinations and **c** *Chironomus* larvae—*L. rohita* test combinations. The *dashed line* in each indicates the a value (secondary Y-axis) for non-selective feeding

Given a choice of all nine prey types (Table 3), the mosquitofish exhibited a significant preference for the larvae of *L. rohita* (Hotelling's T^2 test, $P < 0.01$; Manly's $\alpha = 2.46$, Fig. 7) and the mosquito larvae—instar-I (Hotelling's T^2 test, $P < 0.01$; Manly's $\alpha = 2.36$, Fig. 7), strong avoidance of *D. similoides* and mosquito larvae instar-IV and random selection for the remaining prey. With respect to dry mass consumed, both the fish larvae were positively selected; *C. cornuta, D. similoides, A. stephensi* instar–I and the remaining prey types were avoided (Hotelling's T^2 test $\alpha = 2.46$, Fig. 7). In terms of dry mass, the actual ingestion was significantly higher ($P < 0.01$; Mann–Whitney *U* test; MANOVA, Table 6) for larvae of both fish species, *C. cornuta* and *A. stephensi* instar-I, and was significantly lower (Mann–Whitney *U* test, $P < 0.01$, Fig. 8) for *H. mira* and *A. stephensi* instar-IV than the respective expected values.

The prey combination proportion and their interaction effects were highly significant (two-way ANOVA, $P < 0.01$) for the total dry biomass ingested by mosquitofish. Significantly less dry mass was ingested from the mosquito larvae instar-I—*L. rohita* larvae combination at all three proportions (Fig. 8), and from the equal or higher proportions of the *H. mira*—fish larvae combination.

Finally, an analysis of our data from all eight tests with two prey types, using Landry's (1981) formula (see

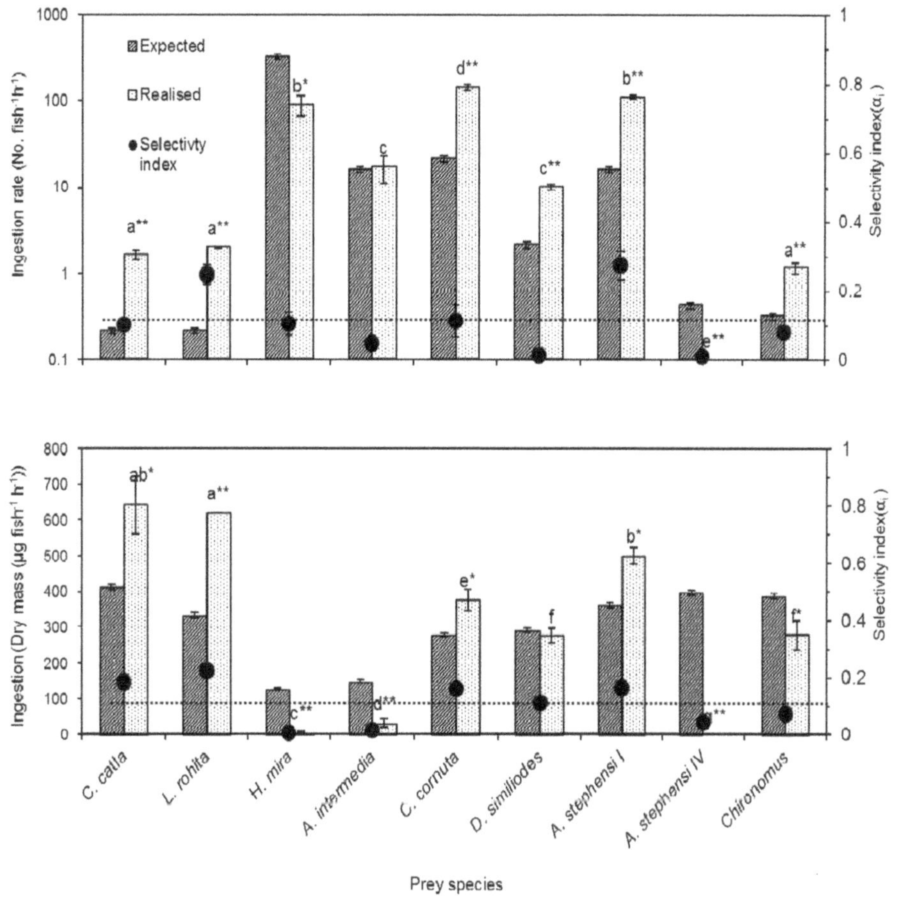

Fig. 7 Expected and observed mean ingestion rates of number (fish^{-1} h^{-1}) and dry (µg fish^{-1} h^{-1}) biomass and Manly's selectivity index (α_i) for mosquitofish in multispecies prey choice test. The *dashed line* indicates the a value (secondary Y-axis) for non-selective feeding

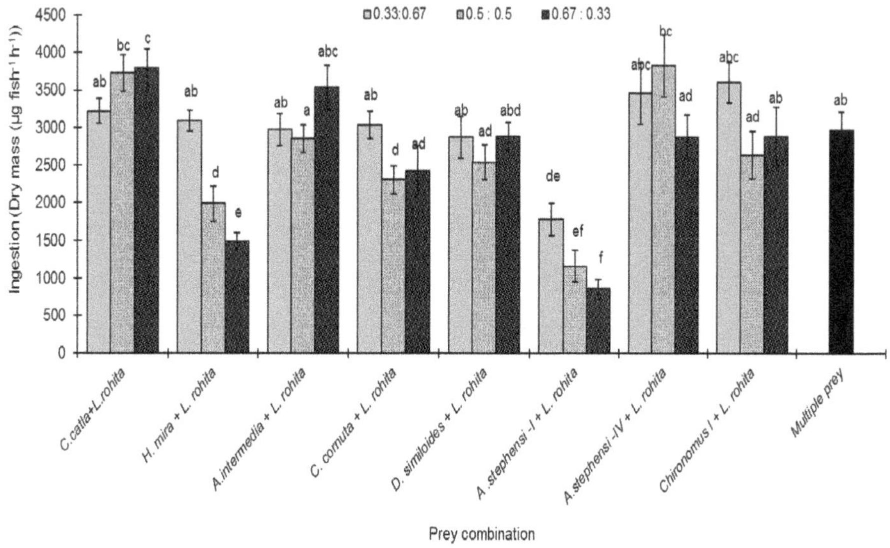

Fig. 8 Total dry mass ingestion (µg Ind^{-1} h^{-1}) in the mosquitofish at different prey combinations and proportions

'Methods'), is as follows: the proportionality constant increased with increasing value of N_1/N_2 in *C. catla—L. rohita* ($R^2 = 0.54$; $P < 0.01$; Fig. 9), *C. cornuta—L. rohita* ($R^2 = 0.54$; $P < 0.01$) and *Chironomus* larvae—*L. rohita* combinations ($R^2 = 0.73$; $P < 0.001$), which indicates switching in the prey preference of mosquitofish in relation to the relative abundance of prey in the environment (Fig. 9).

Discussion

This study demonstrated that the mosquitofish *G. affinis* prefers fish larvae and consumes a substantial number of fish larvae even in the presence of an alternative, naturally occurring prey. Mosquitofish consistently preferred fish larvae in all three test combinations (involving *H. mira*, *D. similoides* and *A. stephensi* instar-IV) across the three relative abundances tested. Mosquitofish preferred fish larvae in three test combinations when present in equal proportions and preferred fish larvae in higher proportions at all prey combinations tested, except with *A. stephensi* instar-I larvae. These results show that the vulnerability of fish larvae to mosquitofish predation depends upon the identity and the relative abundance of alternative prey. Fish larvae were consistently preferred only if an alternative prey was small rotifers, *D. similoides*, or larger instars of mosquito larvae. However, as recorded in Experiment I, the fish larvae were consumed readily in the absence of an alternative prey. Prey profitability varies as a function of their relative abundance and the distinctiveness of alternative prey (Rao and Kumar 2002). The size, morphology and behaviour of an alternative prey, therefore, indirectly influence the predatory removal of carp larvae from a system. Since quantitative data on the abundance of mosquitofish in natural waters are not available, we cannot extrapolate the effect of its predation on natural populations of fish larvae. Since their worldwide introduction, several studies on mosquitofish have confirmed negative effects of their predation on indigenous fish species (Courtenay and Maffe 1989, Howe et al. 1997, Ivantsoff 1999). The role of an invasive mosquitofish in shaping wetland fish assemblage structures has been well established in Australian wetlands (Macdonald et al. 2012). However, no study has been conducted on Indian wetlands. The multispecies prey choice test in this study shows the selective elimination of carp larvae by poeciliids in a natural setting. Note that the feeding data in this study do not include partially eaten or killed but only unconsumed larvae. Therefore, this study provides conservative estimates of total mortality rates. However, it is likely that *Gambusia*-imposed elimination of carp larvae in this study is higher than that in nature, because the prey density used here was higher than that in the field.

The prey selectivity of mosquitofish is influenced by hunger level (which was not investigated here), but more importantly by the size, shape, taste acceptability and behaviour of the prey. For instance, the laboratory experiment in this study showed that larval carp were more vulnerable to mosquitofish predation when either the evasive rotifer, *H. mira*, or the larger cladoceran, *D. similoides*, were also present.

The larvae of *C. catla* and *L. rohita* were similar in many respects, but differed in size. Evidently, either of these larvae was preyed upon randomly by mosquitofish if present in equal proportions. This is because preference is a function of the relative proportion of larvae in the medium. Because of their small size, rotifers are an ideal first food for the larvae of carp species (Lubzens

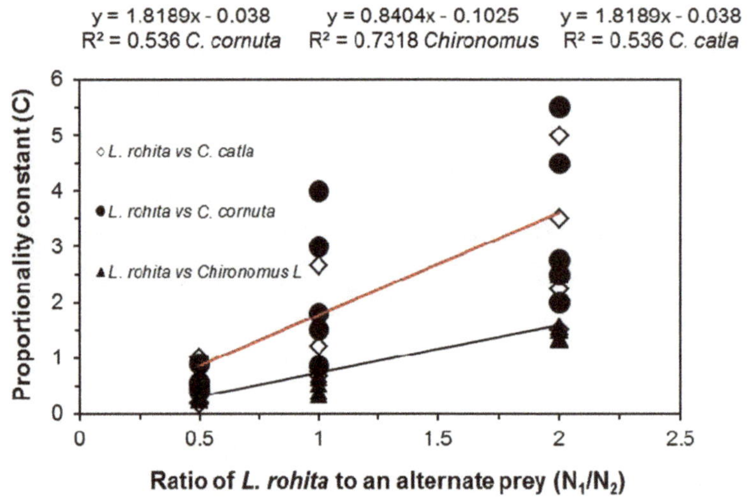

Fig. 9 Prey switching in mosquitofish, i.e. value of *C* (proportionality constant: ratio of the proportion of prey in the diet and the initial prey proportion in the medium) in relation to the initial ratio (N_1/N_2) of the two prey abundances in the medium

et al. 1989, Kumar et al. 2012). However, if mosquitofish are present in the system, they will actively eliminate carp larvae and avoid the rotifer *H. mira*. *H. mira* performs evasive darting movements that reduce the risk of predation, as reported for the predatory rotifer *Asplanchna* (Gilbert and Williamson 1978, Iyer and Rao 1996) and the predatory copepod *Mesocyclops* (Williamson 1983, Kumar 2003, Rao and Kumar 2002). *H. mira* was the smallest alternative prey used in this study. Hence, plausible explanations for its low elimination rates would be the evasive behaviour and smaller size of this rotifer. A similar explanation was provided in an earlier study investigating the interaction between the predatory rotifer *A. intermedia* and *H. mira* (Iyer and Rao 1996). The preference of freshwater poeciliids for prey below the threshold size decreases abruptly (e.g. Werner et al. 1983, Bence 1988, Bence and Murdoch 1986). The larger rotifer, *A. intermedia*, are transparent and swim slowly and smoothly; both traits may reduce their conspicuousness (Gilbert and Williamson 1978, Dahms et al. 2011b). These characteristics may be responsible for the lower vulnerability of *A. intermedia*. Due to their wriggling behaviour and eye spots, carp larvae are more visually apparent to feeding mosquitofish. *A. intermedia* was neither preferred nor avoided when present in equal or higher proportions. The mosquitofish preference pattern was related to the relative abundance of the prey. Due to morphological, physiological and ecological constraints, fish species may not always behave according to the predictions of optimal foraging theory in selecting a particular prey type or size (Pyke 1984, Cabral and Marques 1999, Mookerji and Rao 1993, 1994).

The smaller cladoceran, *C. cornuta*, might be more conspicuous for mosquitofish because of its darker colour and thicker cuticle. The relatively thicker cuticle of *C. cornuta* protects it against predation by *Mesocyclops* (Matsumura-Tundisi et al. 1990, Rao and Kumar 2002, Kumar 2003). For *C. cornuta*, the risk of being conspicuous apparently outweighs any benefit accruing from its thicker cuticle, which may offer some protection against predation by *Mesocyclops* (Matsumura-Tundisi et al. 1990, Rao and Kumar 2002) but not against mosquitofish. Prey visibility and consequently the reactive distance increase with prey contrast (Rao 2003). Since nearly all zooplankton species are transparent, anything that increases their visual contrast (such as the thicker and darker cuticle of *Ceriodaphnia*) will ultimately make them more vulnerable to a visual predator. The negative selection of *D. similoides* against carp larvae must be interpreted with caution. Bence and Murdoch (1986) found that prey size *per se* is not a reliable determinant of mosquitofish preference when its diet choice includes prey species that differ in other attributes. Repeated behavioural observations of *D. similoides* under a stereomicroscope in this study showed avoidance behaviour against mosquitofish. Previous studies indicate that larger-sized *D. similoides* were not a preferred food for mosquitofish (Hurlbert and Mulla 1981, Bence and Murdoch 1986). The body size and escape behaviour of the prey are also determinant factors for the prey preference of mosquitofish: early instars of mosquito larvae were strongly selected over fish larvae. Avoidance of instar-IV mosquito larvae by the mosquitofish in the carp larvae–mosquito larvae selection experiment may not be due only to a difference in size, as both the larvae (mosquito larval instar-IV and carp larvae) were comparable in size. The observed avoidance of larger instars of mosquito larvae could be attributed to their relatively thick exoskeleton and dense setation. As observed through a stereomicroscope, mosquito larvae instar-IV did not elicit an active attack response by the mosquitofish, even after an encounter. The active avoidance of carp larvae in the presence of chironomid larvae as an alternative prey further suggests that the body size of the mosquito larval instar-IV was not a causative factor in its avoidance.

Because they are larger in size and heavier in dry biomass, chironomid larvae were preferred above fish larvae when present in equal or higher proportions. However, mosquitofish preferences switched to carp larvae when larval carp were present in higher proportions. For prey that differ in frequency of encounter and ease of capture but are similar in nutritional quality, optimal foraging theory predicts switching behaviour depending on the relative abundance of prey (Bence and Murdoch 1986, Schoener 1987, Krebs and Kacelnik 1991). Many fish species are able to switch from one feeding mode to another, depending on the size and concentration of the prey (Mookerji and Rao 1993). *Chironomus* larvae that were preferred in our two prey choice tests were not a preferred food types in a multispecies choice test. Note that previous fish experience with prey items may also have affected the preferences of prey. Mosquitofish were thoroughly exposed to the *Chironomus* larvae but not to the fish larvae, as fish larvae were added in the food of mosquitofish only occasionally. In nature, *Chironomus* larvae are commonly located at the bottom of a waterbody during the day but they migrate upwards at night and are not readily encountered by visual predators such as mosquitofish.

The laboratory experiments in this study clearly indicated a strong influence of mosquitofish on commercially important carp species. The results of a multispecies prey choice test further supported this observation. Even in the presence of eight alternative prey types, the mosquitofish showed a strong preference for carp larvae. In the yolk-sac stage, carp larvae are slightly buoyant with less ability to avoid predation, which makes them more vulnerable to mosquitofish predation. Mosquitofish, being voracious feeders, may also have adversely affected carp larvae by eliminating their food from the environment (Arthington 1991, Bøhn and Amundsen 2001). Since vulnerability to predation in many species is size-selective (Paradis et al. 1996), larvae that are larger at hatching or reach a large size more quickly face a lower predation risk. However, continued food deprivation, leading to irreversible starvation effects in carp larvae could act indirectly by increasing predation risk (Rice et al. 1987). This is because weaker fish larvae take longer to grow out of the 'size bottleneck' of high predation susceptibility (Werner et al. 1983, Bøhn and Amundsen 2001). Mosquitofish, and especially *G. affinis*, were introduced worldwide to control mosquito populations. However, mosquitofish have a 'voracious' appetite and feed at rates much higher than most other similar-sized fish (Meffe and Snelson 1989).

Mosquitofish are not very efficient in preying on mosquito larvae in winter (<20 °C). Sexual dimorphism and size differences play important roles in mosquitofish feeding behaviour, with multiple effects on the food web (Cabral et al. 1998, Specziár 2004). Previous studies have also reported gender-specific differences in the extent of carnivory and efficiency of utilising carp larval prey (Cabral and Marques 1999, Specziár 2004).

The single species test with predator (control agent) does not accrue same results in nature because of the presence of vast arrays of prey species. Therefore removals of target populations' depend on dietary niche breadth and preferences of the biocontrol agent. In the present study, environmentally relevant organisms were chosen and realistic concentrations were tested in our experimental set-up. The prey combinations used in the experiment are frequently occurring in natural mosquito breeding habitats and *Gambusia* are introduced for mosquito control. So, the laboratory results of our study can immediately be translated to the field. The mosquitofish form schools that may contain hundreds of fish. The schooling behaviour of mosquitofish leaves predation-free patches. Hence, predation risks may vary locally, depending on overall predator densities and the location of schools (Werner et al. 1983). Female carp could avoid mosquitofish at several spatial scales. Adult carp could use the strength of fish chemicals and exudates to identify areas in ponds where dense mosquitofish schools are present (Griswold and Lounibos 2005) and avoid these areas in selecting their breeding sites. In addition to seasonal and between-site variations, mosquitofish show an ontogenetic diet shift from diatoms, crustacean nauplii and microcrustaceans (particularly cladocerans) to larger prey, such as adult nematocerans (Mansfield and McArdle 1998, García-Berthou 1999). These attributes should be taken into account when extrapolating these laboratory results to natural conditions. In nature, mosquitofish decimate the larvae of their native counterparts not only directly, by predation, but also indirectly, through competition for food (Bence 1988, Arthington 1991, Bøhn and Amundsen 2001, Blaustein and Chase 2007, Leprieur et al. 2009).

The results of this study confirm the previous observations of opportunistic feeding in mosquitofish. The positive selection of fish larvae (when present in higher proportion) against all eight invertebrate prey types suggests heavy predation pressure on carp larvae in a natural setting. This particularly holds for the breeding season, when the relative abundance of fish larvae increases sharply.

Conclusions

Due to their higher dispersal tendency, invasiveness and adverse effects on native aquatic communities, the database of the IUCN Invasive Species Specialist Group classifies mosquitofish, along with eight other species, as among the world's 100 worst invasive species. Several studies have demonstrated the competitive advantage of mosquitofish over small indigenous fishes and amphibians. Their selective predation might have a significant effect on the structure and dynamics of freshwater ecosystems, as the less-preferred prey populations get an environment free of competition.

In order to understand the mechanisms of mosquitofish impact on native fish communities, it will be important to determine the preference of mosquitofish for the larvae of native fish species in relation to other naturally occurring prey. We observed a switch in preferences in relation to the relative abundance of a particular prey type suggesting that mosquitofish use an adaptive feeding strategy to maximise their fitness in an environment where prey are scarce. The selectivity experiments in this study indicate that mosquitofish positively selected fish larvae in all prey combinations tested. These results suggest that mosquito larvae are not the preferred food of mosquitofish when fish larvae are present in the natural habitat. Since mosquitofish and carp larvae have overlapping natural habitats, mosquitofish predation and prey preferences should have a consequence on regulatory management practices in conservation management and aquaculture.

Our results particularly suggest caution in the introduction of mosquitofish as biocontrol agents of mosquitoes in natural waters where carp are important in aquaculture, especially in South Asia. Although it is a threat to indigenous biodiversity and fails to control mosquitoes effectively, stocking and distribution of mosquitofish are still being practiced and encouraged by governmental and non-governmental disease-vector control agencies in several countries.

Competing interests
The authors declare that they have no competing interests.

Authors' contributions
RK made the conception and design of the experiment and analysis of data. PM was acquiring the data and provided some interpretation of the data. JS was involved in experiment with fish larvae. H-UD was revising the manuscript critically for important intellectual content and added substantially to the 'Introduction' and 'Discussion'. J-SH contributed to the 'Discussion' and gave final approval of the version to be published. All authors read and approved the final manuscript.

Acknowledgements
The Indian Council of Medical Research (ICMR, India) is acknowledged for sanctioning a research project (Evaluation of the potential of copepods to control disease vectors; Project No. ICMR-5/8-7 (179)/2004-ECD-II). and Acharya Narendra Dev College, University of Delhi, India, is acknowledged for providing facilities and support to RK. The Ministry of Science and Technology of Taiwan is thanked for a post-doctoral fellowship (NSC 98-2811-B-019-002) to RK and research funding to JSH (NSC 102-2923-B-019-001-MY3, MOST 104-2611-M-019-004 and Most 104-2621-M-019-002). and research funding to T H SHIH (MOST 104-2621-M-019-002 and MOST 104-2621-M-037-001). This work was supported by a grant of KMU (KMU-Q104018/104.01.01-104.12.31). We thank Prof. T. R. Rao for stimulating discussions and suggestions. We also appreciate the assistance of Dr. Poonam for support at various stages of data analyses and manuscript preparation.

Author details
[1]Ecosystem Research Laboratory, Department of Zoology, Acharya Narendra Dev College (University of Delhi), Govindpuri, Kalkaji, New Delhi 110 019, India. [2]Institute of Marine Biology, National Taiwan Ocean University, 2 Pei Ning Road, Keelung 202, Taiwan. [3]Centre for Environmental Sciences, School of Earth Biological and Environmental Sciences, Central University of Bihar, Patna, India. [4]Australian Rivers Institute, Griffith University, Nathan, QLD 4111, Australia. [5]Department of Biomedical Science and Environmental Biology, KMU - Kaohsiung Medical University, No. 100, Shin-Chuan 1st Road, Kaohsiung 80708, Taiwan. [6]Department of Biotechnology, Delhi Technological University, Delhi 110042, India.

References

Akre BG, Johnson DM (1979) Switching and sigmoid functional response curves by damsel fly naiads with alternative prey available. J Anim Ecol 48:703–720

Alcaraz C, Bisazza A, García-Berthou E (2008) Salinity mediates the competitive interactions between invasive mosquitofish and an endangered fish. Oecologia 155:205–213

Arthington AH (1991) Ecological and genetic impacts of introduced and translocated freshwater fishes in Australia. Can J Fish Aquat Sci 48(Suppl 1):33–43

Arthington AH, Marshall CJ (1999) Diet of the exotic mosquitofish, *Gambusia holbrooki*, in an Australian lake and potential for competition with indigenous fish species. Asian Fisheries Society 12:1–16

Barrier RFG, Hicks BJ (1994) Behavioral interactions between black mudfish (*Neochanna diversus*) and mosquitofish (*Gambusia affinis* Baird and Girard 1854). Ecol Freshw Fish 3:93–99

Bence JR (1988) Indirect effects and biological control of mosquitoes by mosquitofish. J Appl Ecol 25:505–521

Bence JR, Murdoch WW (1986) Prey size selection by the mosquitofish: relation to optimal diet theory. Ecology 67:324–336

Benejam L, Alcaraz C, Sasal P, Levert GS, García-Berthou E (2009) Life history and parasites of the invasive mosquitofish (*Gambusia holbrooki*) along a latitudinal gradient. Biol Invasions 11:2265–2277

Blaustein L, Chase JM (2007) Interactions between mosquito larvae and species that share the same trophic level. Annu Rev Entomol 52:489–507

Bøhn T, Amundsen PA (2001) The competitive edge of an invading species. Ecology 82:2150–2163

Cabral JA, Marques JC (1999) Life history, population dynamics and production of eastern mosquitofish, *Gambusia holbrooki* (Pisces, Poeciliidae), in rice fields of the lower Mondego River Valley, western Portugal. Acta Oecologia 20:607–620

Cabral JA, Mieiro CL, Marques JC (1998) Environmental and biological factors influence the relationship between a predator fish, *Gambusia holbrooki*, and its main prey in rice fields of the Lower Mondego River Valley (Portugal). Hydrobiologia 382:41–51

Chesson J (1983) The estimation and analysis of preference and its relationship to foraging models. Ecology 64:1297–1304

Chullasorn S, Yang W-X, Dahms H-U, Kangtia P, Holinska M, Anansatitporn W, Sanoamuang LO, Hwang J-S (2009) Naupliar development of *Eucyclops* cf. *serrulatus tropicalis*, *Euc.* cf. *spatulatus*, and *Ectocyclops medius* Kiefer, 1930 (Copepoda: Cyclopidae). Zool Stud 48(1):12–32

Clavero M, García-Berthou E (2005) Invasive species are a leading cause of animal extinctions. Trends Ecol Evol 20:110–110

Courtenay WR Jr, Maffe GK (1989) Small fishes in strange places: a review of introduced poecilliids. In: Meffe GK, Snelson FF Jr (eds) Ecology and evolution of livebearing fishes (Poeciliidae). Prentice Hall, Englewood Cliffs, New Jersey, pp 319–331

Crivelli AJ, Boy V (1987) The diet of mosquito fish, *Gambusia affinis* (Baird and Girard) (Poeciliidae) in Mediterranean France. Revue d'Ecologie 42:421–435

Crowl TA, Townsend CR, McIntosh AR (1992) The impact of introduced brown and rainbow trout on native fish: the case of Australasia. Rev Fish Biol Fish 2:217–241

Dahms H-U (2004) Copepoda; Cyclopoida – Genera *Mesocyclops* and *Thermocyclops* by Ueda & J. Reid (eds.) Book Review. Hydrobiologia 515:248–249

Dahms H-U (2005) Aquatic ecology of rice fields. C.H. Fernando, F. Goeltenboth and J. Margraf, editors (Volumes Publishing, Kitchener, Ontario, Canada, N2B2B9, A third millenium book 205, first edition, Limnology 6:137–138, book review

Dahms H-U, Dobretsov S, Lee J-S (2011a) Effects of UV radiation on marine ectotherms in polar regions. Comp Biochem Physiol C 153:363–371

Dahms H-U, Hagiwara A, Lee J-S (2011b) Ecotoxicology, ecophysiology, and mechanistic studies with rotifers. Aquat Toxicol 101:1–12

Duryea R, Donnelly J, Guthrie D, O'Malley C, Romanowski M, Schmidt R (1996) *Gambusia affinis* effectiveness in New Jersey mosquito control. In: Proceedings of the Eighty-Third Annual Meeting of the New Jersey Mosquito Control Association, Inc., pp 95–102

Englund RE (1999) The impacts of introduced Poeciliid fish and Odonata on the endemic *Megalagrion* (Odonata) damselflies of Oahu Island, Hawaii. J Insect Conserv 3:225–243

García-Berthou E (1999) Food of introduced mosquito fish: ontogenetic diet shift and prey selection. J Fish Biol 55:135–147

Gilbert JJ, Williamson CE (1978) Predator–prey behaviour and its effect on rotifer survival in associations of *Mesocyclops edax*, *Asplanchna girodi*, *Polyarthra vulgaris* and *Keratella cochlearis*. Oecologia 37:13–22

Gill J (2001) Generalized linear models: a unified approach. Sage University Paper Series on Quantitative Applications in the Social Sciences No. 134. Sage, Thousand Oaks, CA, USA

Gooden B, French K, Turner PJ, Downey PO (2009) Impact threshold for an alien plant invader, *Lantana camara* L., on native plant communities. Biol Conserv 142:2631–2641

Goodsell JA, Kats LB (1999) Effect of introduced mosquitofish on Pacific tree frogs and the role of alternative prey. Conserv Biol 13:921–924

Gophen M, Yehuda Y, Malinkov A, Degani G (1998) Food composition of the fish community in Lake Agmon. Hydrobiologia 380:49–57

Gotelli NJ, Ellison AM (2004) A primer of ecological statistics. Sinauer Associates, Sunderland

Griswold MW, Lounibos LP (2005) Does differential predation permit invasive and native mosquito larvae to coexist in Florida? Ecol Entomol 30:122–127

Hamer AJ, Lane SJ, Mahony MJ (2002) The role of introduced mosquitofish (*Gambusia affinis*) in excluding the native green and golden bell frog (*Litoria aurea*) from original habitats in south-eastern Australia. Oecologia 132:445–452

Hazelton PD, Grossman GD (2009) The effects of turbidity and an invasive species on foraging success of rosyside dace (*Clinostomus funduloides*). Freshw Biol 54:1977–1989

Helfman GS (2007) Fish Conservation: a guide to understanding and restoring global aquatic biodiversity and fishery resources. Publishers Island Press, 1718, Washington, DC

Helfman GS, Collette BB, Facey DE, Bowen BW (2009) The diversity of fishes: biology, evolution, and ecology. Second edition. Wiley-Blackwell

Howe E, Howe C, Lim R, Burchett M (1997) Impact of the introduced poeciliid *Gambusia holbrooki* (Girard, 1859) on the growth and reproduction of *Pseudomugil signifer* (Kner, 1865) in Australia. Mar Freshw Res 48:425–434

Hurlbert SH, Mulla MS (1981) Impacts of mosquitofish (*Gambusia affinis*) predation on plankton communities. Hydrobiologia 83:125–151

Hurlbert SH, Zedler J, Fairbanks D (1972) Ecosystem alteration by mosquitofish (*Gambusia affinis*) predation. Science 175:639–641

Hwang D-S, Dahms H-U, Park HG, Lee J-S (2013) A new intertidal *Brachionus* and intrageneric phylogenetic relationships among *Brachionus* as revealed by allometry and CO1-ITS1 gene analysis. Zool Stud 52(13):1–10

ISSG (2000) 100 of the world's worst invasive species database. Invasive Species Specialist Group. University of Auckland, Auckland

Ivantsoff WA (1999) Detection of predation on Australian native fishes by *Gambusia holbrooki*. Mar Freshw Res 50:467–468

Iyer N, Rao TR (1996) Responses of the predatory *Asplanchna intermedia* to prey species differing in vulnerability: laboratory and field studies. Freshw Biol 36:521–533

Jhingran VG (1991) Fish and fisheries of India. 3rd edition. Hindustan Publishing Corporation, Delhi, India, XIV +645pp

Kak A (1999) Field and laboratory studies on interaction between rotifers and cladocerans. Ph.D Thesis, University of Delhi

Kanji GK (2011) 100 Statistical tests, 3rd edn. Sage publications, New Dehi, 242

Kennard MJ, Arthington AH, Pusey BJ, Harch BD (2005) Are alien fish a reliable indicator of river health? Freshw Biol 50:174–193. doi:10.1111/J.1365[--]2427.2004.01293.X

Kim J-H, Jeon HJ, Baek JM, Han KN, Dahms H-U (2013a) EDCs-induced glucocorticoid receptor related genes expression of the river pufferfish, *Takifugu obscurus*. Aquac Res 44:985–994

Kim J-H, Jung SJ, Lee W-O, Dahms H-U, Han K-N (2013b) Osmoregulation related gene expression of the anadromous river pufferfish *Takifugu obscurus* under sudden salinity changes. J Fish Physiol Biochem. [doi:10.1007/s10695-013-9837-z] ONLINE 02August2013

Kiørboe T, Saiz E, Viitasalo M (1996) Prey switching behaviour in the planktonic copepod *Acartia tonsa*. Mar Ecol Prog Ser 143:65–75

Komak S, Crossland MR (2000) An assessment of the introduced mosquitofish (*Gambusia affinis holbrooki*) as a predator of eggs, hatchlings and tadpoles of native and non-native anurans. Wildl Res 27:185–189

Krebs JR, Kacelnik A (1991) Decision making. In: Krebs JR, Davis NB (eds) Behavioral ecology, an evolutionary approach. Blackwell Scientific Publications, London, pp 105–136

Kumar R (2003) Effects of *Mesocyclops thermocyclopoides* (Copepoda: Cyclopoida) predation on the population growth patterns of different prey species. J Freshw Ecol 18:384–393

Kumar R, Rao TR (2003) Predation on mosquito (*Anopheles stephensi* and *Culex quinquefasciatus*) larvae by *Mesocyclops thermocyclopoides* (Copepoda; Cyclopoida) in the presence of alternate prey. Int Rev Hydrobiol 88:570–581

Kumar R, Dahms H-U, Muhid P, Tseng L-C, Hwang J-S (2008) Potential of three aquatic predators to control mosquitoes in the presence of alternative prey: a comparative experimental assessment. Mar Freshw Res 59:817–835

Kumar R, Souissi S, Hwang J-S (2012) Vulnerability of carp larvae to copepod predation as a function of larval age and body length. Aquaculture 338–341:274–283

Lancaster HF, Drenner RW (1990) Experimental mesocosm study of the separate and interactive effects of phosphorous and mosquitofish (*Gambusia affinis*) on plankton community structure. Can J Fish Aquat Sci 47:471–479

Landry MR (1981) Switching between herbivory and carnivory by the planktonic marine copepod *Calanus pacificus*. Mar Biol 65:77–82

Lee CH, Dahms H-U, Cheng SH, Souissi S, Schmitt FG, Kumar R, Hwang J-S (2010) Predation of *Pseudodiaptomus annandalei* (Copepoda: Calanoida) by the grouper fish fry *Epinephelus coioides* under different hydrodynamic conditions. J Exp Mar Biol Ecol 393:17–22

Leprieur F, Brosse S, García-Berthou E, Oberdorff T, Olden JD, Townsend CR (2009) Scientific uncertainty and the assessment of risks posed by non-native freshwater fishes. Fish Fish 10:88–97

Lever C (1996) Naturalized fishes of the world. Academic Press, London

Leyse KE, Lawler SP, Strange T (2004) Effects of an alien fish, *Gambusia affinis*, on an endemic California fairy shrimp, *Linderiella occidentalis*: implications for conservation of diversity in fishless waters. Biol Conserv 118:57–65

Linden AL, Cech JJJ (1990) Prey selection by mosquitofish (*Gambusia affinis*) in California rice fields: effect of vegetation and prey species. J Am Mosq Control Assoc 6:115–120

Lowe S, Browne N, Boudjelas S (2000) 100 of the world's worst invasive alien species-a selection from the Global Invasive Species Database. Invasive Species Specialist Group, World Conservation Union

Lubzens E, Tandler A, Minkoff G (1989) Rotifers as food in aquaculture. Hydrobiologia 186(187):387–400

Macdonald JI, Tonkin ZD, Ramsey DS, Kaus AK, King AK, Crook DA (2012) Do invasive eastern gambusia (*Gambusia holbrooki*) shape wetland fish assemblage structure in South-eastern Australia? Mar Freshw Res 63:659–671

Manly BFJ (1974) A model for certain types of selection experiments. Biometrics 30:281–294

Mann HB, Whitney DR (1947) On a test of whether one or two random variables is stochastically larger than the other. Ann Math Stat 18:50–60

Mansfield S, McArdle BH (1998) Dietary composition of *Gambusia affinis* (Family: Poeciliidae) populations in the northern Waikato region of New Zealand. N Z J Mar Freshw Res 32:375–383

Marchetti MP, Moyle PB (2001) Keeping alien fishes at bay: effects of flow regime and habitat structure on fish assemblages in a regulated California stream. Ecol Appl 11:75–87

Margaritora FG (1990) Influence of *Gambusia affinis* on the features and dynamics of the zooplankton community in the pools of Castel Porziano (Latium). Riv Idrobiol 29:747–762

Matsumura-Tundisi T, Rietzler AC, Espindola ELG, Tundisi JG, Rocha O (1990) Predation on *Ceriodaphnia cornuta* and *Brachionus calyciflorus* by two *Mesocyclops* species coexisting in Barra Bonita reservoir (SP, Brazil). Hydrobiologia 198:141–151

Meffe GK, Snelson FF (1989) An ecological overview of poeciliid fishes. In: Meffe GK, Snelson FF (eds) Ecology and evolution of livebearing fishes (Poeciliidae). Prentice Hall, Upper Saddle River, New Jersey, pp 13–31

Mookerji N, Rao TR (1993) Patterns of prey selection in rohu (*Labeo rohita*) and singhi (*Heteropneustes fossilis*) larvae under light and dark conditions. Aquaculture 118:85–104

Mookerji N, Rao TR (1994) Influence of ontogenetic changes in prey selection on the survival and growth of rohu, *Labeo rohita* and singhi, *Heteropneustes fossilis* larvae. J Fish Biol 44:479–490

Nagdali SS, Gupta PK (2002) Impact of mass mortality of a mosquito fish, *Gambusia affinis* on the ecology of a freshwater eutrophic lake (Lake Nainital, India). Hydrobiologia 468:45–52

Oaten A, Murdoch WW (1975) Predator switching, functional response, and stability. Am Nat 109:299–318

Olden JD, Kennard MJ, Leprieur F, Tedesco PA, Winemiller KO, García-Berthou E (2010) Conservation biogeography of freshwater fishes: recent progress and future challenges. Divers Distrib 16:496–513

Paradis AR, Pepin P, Brown JA (1996) Vulnerability of fish eggs and larvae to predation: review of the influence of the relative size of prey and predator. Can J Fish Aquat Sci 53:1226–1235

Peck GW, Walton WE (2008) Effect of mosquitofish (*Gambusia affinis*) and sestonic food abundance on the invertebrate community within a constructed treatment wetland. Freshw Biol 53:2220–2233

Pyke GH (1984) Optimal foraging theory: a critical review. Annu Rev Ecol Syst 15:523–575

Rao TR (2003) Ecological and ethological perspectives in larval fish feeding. J Appl Aquac 13:145–178

Rao TR, Kumar R (2002) Patterns of prey selectivity in the cyclopoid copepod *Mesocyclops thermocyclopoides*. Aquat Ecol 36:411–424

Rehage JS, Sih A (2004) Dispersal behavior, boldness, and the link to invasiveness: a comparison of four *Gambusia* species. Biol Invasions 6:379–391

Rehage JS, Barnett BK, Sih A (2005) Foraging behaviour and invasiveness: do invasive *Gambusia* exhibit higher feeding rates and broader diets than their noninvasive relatives? Ecol Freshw Fish 14:352–360

Rice JA, Crowder LB, Binkowski FP (1987) Evaluating potential sources of mortality for larval bloater (*Coregonus hoyi*): Starvation and vulnerability to predation. Can J Fish Aquat Sci 44:467–472

Scheiner SM (2001) MANOVA: multiple response variables and multispecies interactions. In: Scheiner SM, Gurevitch J (eds) *Design and analysis of ecological experiments*, 2nd edn. Oxford University Press, Oxford, UK, pp 99–115

Schoener TW (1987) A brief history of optimal foraging theory. In: Kamil AC, Krebs JR, Pulliam HR (eds) Foraging behavior. Plenum Press, New York, pp 5–67

Smith GR, Terlecky AJ, Dayer CB, Boyd A, Ogle ME, Dibble CJ (2011) Effects of mosquitofish and ammonium nitrate on activity of green frog (*Lithobates clamitans*) tadpoles: a mesocosm experiment. J Freshw Ecol 26:59–63

Sokal RR, Rohlf FJ (2012) Biometry: the principles and practice of statistics in biological research, 4th edn. W. H. Freeman and Co., New York, 937 pp
Sornsupharp B, Lomthaisong K, Dahms HU, Sanoamuang L (2013a) Effects of dried fairy shrimp *Streptocephalus sirindhornae* meal on pigmentation and carotenoid deposition in flowerhorn cichlid; *Amphilophus citrinellus* (Günther, 1864) X *Cichlasoma trimaculatum* (Günther, 1867). Aquaculture Research: 1–12
Sornsupharp S, Dahms H-U, Sanoamuang L (2013b) Nutrient composition of fairy shrimp *Streptocephalus sirindhornae* nauplii as live food and growth performance of giant freshwater prawn postlarvae. Aquac Nutr 19:349–359
Specziár A (2004) Life history pattern and feeding ecology of the introduced eastern mosquito fish, *Gambusia holbrooki* in a thermal spa under temperate climate of Lake Hévíz, Hungary. Hydrobiologia 522:249–260
Swanson C, Cech JJJ, Pietrahita RH (1996) Mosquitofish, biology, culture and use in mosquito control. Mosquito and Vector Control Association of California and University of California, Sacramento, CA
Tseng L-C, Kumar R, Dahms H-U, Chen Q-C, Hwang J-S (2008) Copepod gut contents, ingestion rates and feeding impact in relation to their size structure in the southeastern Taiwan Strait. Zool Stud 47(4):402–416
Vargas MJ, de Sostoa A (1996) Life history of *Gambusia holbrooki* (Pisces: Poeciliidae) in the Ebro Delta (NE Iberian Peninsula). Hydrobiologia 341:215–224
Vidal O, García-Berthou E, Tedesco PA, García-Marín L (2010) Origin and genetic diversity of mosquitofish (*Gambusia holbrooki*) introduced to Europe. Biol Invasions 12:841–851
Webb C, Joss J (1997) Does predation by the fish *Gambusia holbrooki* (Atheriniformes: Poeciliidae) contribute to declining frog populations? Aust J Zool 30:316–326
Werner EE, Hall DJ (1974) Optimal foraging and the size selection of prey by the bluegill sunfish (*Lepomis macrochirus*). Ecology 55:1042–1052
Werner EE, Gilliam JF, Hall DJ, Mittelbach GG (1983) An experimental test of the effects of predation risk on habitat use in fish. Ecology 64:1540–1548
Williamson CE (1983) Behavioural interaction between a cyclopoid predator and its prey. J Plankton Res 5:701–711
Wurtsbaugh W, Cech JJ, Compton J (1980) Effect of fish size on prey size selection in *Gambusia affinis*. In: Proceedings and Papers of the Forty-eighth Annual Conference of the California Mosquito and Vector Control Association., pp 48–51

13

Spatial heterogeneity of benthic copepods: a comparative aspect on composition, abundance, and correlation

R Jayabarathi*, G Padmavati and I Anandavelu

Abstract

Background: Comparing meiofaunal assemblages in the seagrass zone with bare sediment will provide information on the structuring factors and phytal preferences of meiobenthic invertebrates since differences in density and diversity of meiofauna are to be expected between vegetated and bare zones.

Results: A total of 11 groups of meiofauna, with harpacticoids dominating (51 %) and comprising 48 species within 14 families, have been identified. At all localities, the following harpacticoids were found to be relatively abundant, contributing 30.9 % of all harpacticoids: *Longipedia weberi*, *Canuellina nicobaris*, *Scottolana longipes*, and *Parastenhelia hornelli*. A highly significant correlation ($r = 0.987$, $r^2 = 0.974$, $F_{(1,9)} = 337.3$, $P < 0.001$) of meiofaunal assemblage was found between seagrass leaf blades and the canopy sediment compared to bare sediment which was found to have a moderate correlation ($r = 0.543$, $r^2 = 0.294$, $F_{(1,9)} = 3.756$, $P = 0.085$). In addition, the abundance of harpacticoids was significantly higher (ANOVA, $F_{(2,144)} = 19.53$, $P < 0.001$) in seagrass sediments and differed markedly from blades and bare sediments, and the composition was unique in the different zones of the present study.

Conclusions: Productive seagrass ecosystems are as yet inadequately studied in the Andaman Islands. This study provides a first step to characterize a faunal group from the seagrass community.

Keywords: Andaman; Harpacticoids; Meiofauna; Seagrass; Sediment

Background

Meiofauna has been considered as a dynamic element of the marine environment and an integral part of the benthic trophic food chain (Watzin 1983; Higgins and Thiel 1988; Mascart 2010). They inhabit a wide diversity of habitats from the littoral zone to great depths, as associates or parasites of several multicellular animals and plants, and also in sea ice (Coull et al. 1982; Grainger 1991). The patchiness of meiobenthos has been influenced by availability of food and oxygen and other environmental factors (Hicks and Coull 1983; Mantha et al. 2012a) and even by macrofaunal bioturbation (Ólafsson and Moore 1992; Ólafsson 2003). Although sediment granulometry is a major structuring factor of meiofaunal population (Ólafsson 1995; Boucher 1997; Ndaro and Ólafsson 1999; Semprucci et al. 2010), benthic primary producers such as micro- and macroalgae and higher plants have great influence on the assemblages of benthic fauna which continue to live an epifaunal transition mode of life (Jayabarathi et al. 2012; Anandavelu et al. 2013). The physical and biological characteristics of macrophytes form a dynamic community of animals which subsist as epifauna on them. Seagrass beds are some of the most productive parts of coastal ecosystems with estimated annual primary production comparable to that of tropical rain forests (Duarte and Chiscano 1999; Hemminga and Duarte 2000). They provide energy through detritus decomposition and direct herbivory pathways. The rhizomal root system stabilizes the sediment, while densely growing leaves reduce current velocity and encourage the settling of further particles from suspension.

Shallow water vegetation of seagrass patches is also widespread and has an important role in the coastal ecosystem of the Andaman archipelago (Jagtap 1991; Thangaradjou et al. 2010). The coastal waters of

* Correspondence: arjayabarathi@gmail.com
Department of Ocean Studies and Marine Biology, Pondicherry University, Brookshabad Campus, P. Bag. No. 01, Chakkargoan, Port Blair 744112, Andaman and Nicobar Islands, India

Andaman Islands are recognized as an important stronghold for marine biodiversity and contain some of the least impact marine habitats such as seagrass. They are among the world's most productive marine environments, hosting a wide variety of associated macro- and meiofauna (Stoner et al. 1995; Nakaoka 2005). The reviews of seagrass associates reveal that harpacticoid copepods were the key taxon among the macrophytal communities (Arunachalam and Nair 1988; De Troch et al. 2001; 2003). In addition to that, copepods are often considered as the most important taxon as they contribute to the total biomass in terms of faunal abundance, both in pelagic (Padmavati et al. 1998) and benthic environs (Wells 1976; Wells and Rao 1987; Mantha et al. 2012a, b). Harpacticoida are primarily benthic and a key component of the coastal benthos which contributes significantly to the transfer of energy to higher trophic levels (Coull 1990; De Troch et al. 1998; Turner 2004; Danovaro et al. 2007).

The archipelago system has been playing an important role for the protection of marine biodiversity and comprised of nine species of seagrass (Jagtap 1991). Earlier studies have dealt mainly with the taxonomic and distributional aspects and environmental assessment of Indian seagrass (Jagtap and Inamdar 1991; Thangaradjou et al. 2007, 2010; Umamaheswari et al. 2009; Nobi et al. 2011; Sulochanan et al. 2011). However, with a sole reference work of Das (1996), the structure and function of the seagrass of Andaman and Nicobar Islands remain poorly understood. The importance of such fragile communities as a major ecosystem that improves the fishery productivity of the marine systems (De Troch et al. 1998; Unsworth et al. 2014) remains largely unevaluated. The studies relating meiofaunal assemblage with the seagrass ecosystem around the Indian waters are meager (Arunachalam and Nair 1988; Jayabarathi et al. 2012). Further, the knowledge of abundance and composition of faunal assemblages in seagrass beds compared to unvegetated sediment will provide information on the structuring factors and the phytal preferences of meiobenthic invertebrates. The epiphytic load and organic nutrients are often associated with macrophyte production in the benthic region, and thus, differences in density and diversity of meiobenthos may be expected between vegetative and un-vegetative areas. In the present study, the meiofaunal assemblage was investigated by comparing sediments of seagrass vegetation and the adjacent bare region.

Fig. 1 Study area

Methods

Study area

The unique and protected areas of Andaman and Nicobar Islands are prominent biodiversity hotspots. This Archipelago system partitions the Andaman Sea from the Bay of Bengal (Varkey et al. 1996). All samples examined in this study were collected from Kodiyaghat (11° 31′ N, 092° 43′ E), South Andaman (Fig. 1). The locality is a rocky shore with intermittent sediment supporting macrovegetation of seagrass *Thalassia hemprichii* patches throughout the area. The prevailing sediment types in the area are sandy sediment mixed with gravel or boulder.

Sampling

A small patch of seagrass was selected for our study near Rajiv Gandhi Centre for Aquaculture (RGCA, MPEDA) aquaculture station. Samples were collected for a 3-month period from December 2010 to February 2011 at low tide. We used a 32 cm × 32 cm quadrat for collection of seagrass blades. Grass blades inside the quadrat were gently clipped at the base without uprooting the plant and transferred immediately into a plastic bag. Care was taken to avoid the contamination of blades by sediment-dwelling species. Subsequently, collection of vegetated sediment and bare sediment was carried out with the help of PVC cores of 3-cm internal diameter (surface area 7.065 cm^2) inserted into the sediment down to a depth of 7 cm. The core was removed and the content was placed into a plastic bag. The collection of bare sediment did not exceed about 5 m from the seagrass patch. Three replicates were collected from vegetated and bare sediment zones, whereas in the case of seagrass blades, only two replicates were collected because of the intention to condense overexploitation of sparsely distributed seagrass patches in the sampling site. In parallel to all sampling events, environmental parameters such as temperature, salinity, and pH (7.4 to 8.7) were measured in the adjacent regions from the study area. Temperature in the study area was decreased from December 2010 to February 2011 such as 31, 28, and 26 °C, whereas the salinity was increased by 21, 29, and 34 PSU, respectively. Variations in salinity and temperature generally resulted from local precipitation and the time of sampling (low tide) during December to February in the study area. In the laboratory, all samples were stained with Rose Bengal solution (0.5 g L^{-1}). The content of core samples was sieved through 500- and 63-μm standard test sieves and placed into plastic vials. The same strategy was followed for extruding the phytal fauna after rinsing the leaf blades with filtered seawater. All samples were preserved with 4 % buffered formaldehyde. In the present study, meiobenthos other than harpacticoids was identified up to a possible level. Harpacticoid copepods were sorted, enumerated, and identified to species level (Lang 1948; Lang 1965; Wells 1976; Wells and Rao 1987; Giere 1993; Higgins and Thiel 1988) by using a stereoscope and inverted microscope. Animal abundance was expressed as the number of individuals per 10 cm^2 of the bottom surface.

Statistical analysis

The variation in the abundance of Harpacticoida between the sampled zones was accessed by one-way ANOVA. The statistical analysis of correlation and regression was carried out using Microsoft Excel (ver. MS Office 2007) to access the relationship between the meiofaunal abundance in seagrass compared to bare sediment. Diversity indices such as the Shannon-Wiener diversity (H') based on $\log_e$ and Pielou's evenness (J') of Harpacticoida between the seagrass leaf blade (SLB), seagrass canopy sediment (SSD), and unvegetated bare sediment (BSD) were conducted. The Bray-Curtis similarity index was applied on square-root transformed abundance data to access the relationship between the sampling zones and between the harpacticoid species of various zones of collection. Plymouth Routines in Multivariate Ecological Research (PRIMER, Clarke and Gorley 2006) computer package was used for diversity indices and cluster analysis.

Results

Totally, 11 major groups of meiofaunal size organisms were identified by this study. The dominant groups were Harpacticoida (51 %) followed by Nematoda (12.6 %) and Foraminifera (12 %). Other groups consist of Ostracoda and Polychaeta which are 7 and 6.3 %, respectively. The remaining groups such as Nemertea, Decapoda, Mollusca, Amphipoda, Isopoda, and Calanoida were less than 2 %. Harpacticoid copepods (SSD: $X \pm SD = 21.38 \pm 5.34$, $n = 3$; SLB: $X \pm SD = 5.76 \pm 4.34$, $n = 3$) were the most dominating meiofaunal component in the seagrass zone (Fig. 2a, b). In addition to harpacticoids ($X \pm SD = 4.31 \pm 5.08$, $n = 3$), other groups such as foraminiferans ($X \pm SD = 4.26 \pm 4.31$, $n = 3$) and nematodes ($X \pm SD = 4.05 \pm 4.54$, $n = 3$) were also dominant in the unvegetated region (Fig. 2c). A significantly very high correlation ($r = 0.987$, $r^2 = 0.974$, $F_{(1,9)} = 337.3$, $P < 0.001$) of meiofaunal assemblage was found between SLB and SSD compared to the bare sediment which was found to have only a moderate correlation ($r = 0.543$, $r^2 = 0.294$, $F_{(1,9)} = 3.756$, $P = 0.085$).

A total of 14 families comprising 48 species of harpacticoids have been recorded. All families were found in the seagrass canopy sediment and leaf blades except Paramesochridae in the latter zone. Eight of the families were common in all zones (Canuellidae, Cylindropsyllidae, Diosaccidae, Ectinosomatidae, Laophontidae, Longipediidae, Parastenheliidae, and Tetragonicipitidae). Families such as Porcellidiidae, Peltidiidae, Thalestridae, Ameiridae, and Cletodidae were not found in the bare sediment (Table 1).

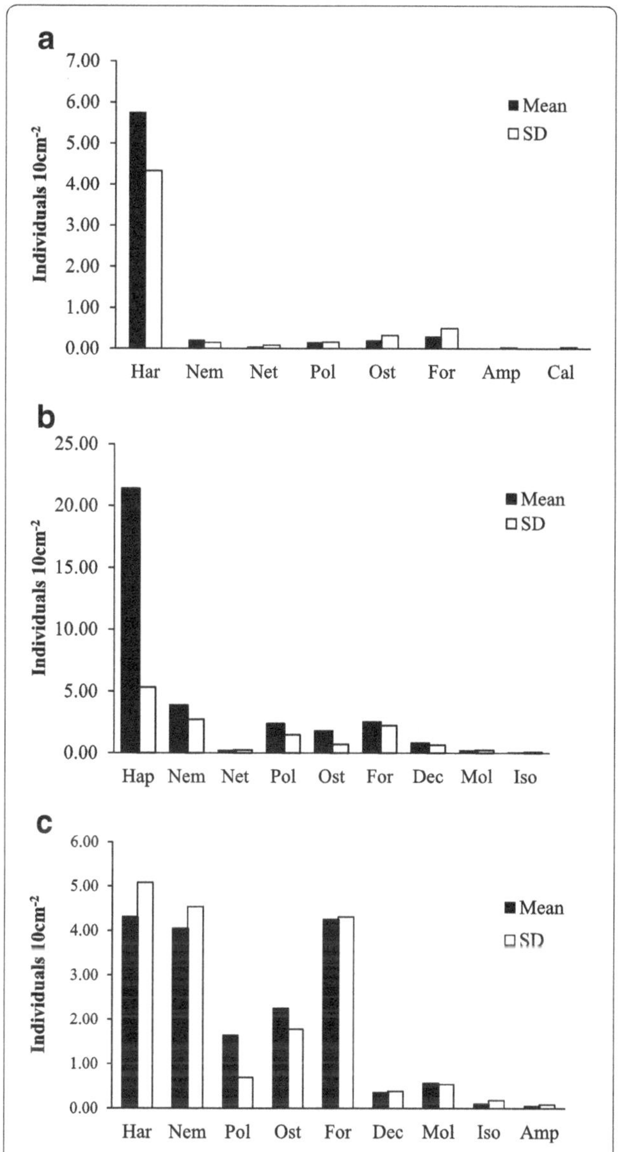

Fig. 2 Meiofaunal abundance. **a** Seagrass leaf blade. **b** Seagrass sediment. **c** Bare sediment. Harpacticoida (*Har*), Nematoda (*Nem*), Polychaeta (*Pol*), Ostracoda (*Ost*), Foraminifera (*For*), Decapoda (*Dec*), Mollusca (*Mol*), Isopoda (*Iso*), Amphipoda (*Amp*), Nemertea (*Net*), Calanoida (*Cal*)

Diosaccidae was the most highly represented family with 12 species in the SLB. Canuellidae with six species was found to be common in both zones of seagrass. Other families such as Thalestridae and Ectinosomatidae were represented by five and four species, respectively, in the blades. The dominant species of each family in the zones were marked with an asterisk (Additional file 1: Table S1). The dendrogram resulting from the cluster analysis of the average abundance shows the similarity of the families of harpacticoid copepods (Fig. 3). The presence of two major clusters of families became apparent with one of the clusters comprising families such as Porcellidiidae, Ameiridae, and Peltidiidae which were found to be the numerically dominant harpacticoid copepods of the seagrass leaf blade and sediment zones. The second major cluster was further classified with two families Paramesochridae and Cylindropsyllidae forming one cluster, with rare individuals found in the bare sediments. Cluster analysis for the harpacticoid species showed three distinct assemblages of species based on the zone of sample collection. Species harbored on SLB appeared as a separate group, while the species found in SSD and BSD were clustered, respectively, in discrete groups (Fig. 4 and Additional file 1: Table S1).

The Harpacticoida were numerically dominant in the canopy sediment, whereas assemblage in the leaf blade and canopy sediment showed high species richness (Fig. 5a, b) and was numerically dominated by *Canuellina nicobaris* and *Scottolana longipes*. Overall, 17 species were found to be regular in all regions. Species, namely, *Laophonte spinicauda, Helmutkunzia variabilis, Metamphiascopsis nicobaricus, Diosaccus monardi, Metamphiascopsis hirsutus, Neodatylopus trichodes*, and *Eupelte aurulenta*, were found only on the blades, whereas *Longipedia andamanica, Peltidium ovale, Apolaophonte hispida*, and *Echinolaophonte mirabilis* were observed in the sediment zones.

In all three zones, *Longipedia weberi, Canuellina nicobaris, Scottolana longipes*, and *Parastenhelia hornelli* were found to be relatively abundant, contributing 30.9 % of all harpacticoids. Other species such as *Noodtiella ornamentalis* (6.4 %) were the next dominant in blades. *Ectinosoma reductum* and *Halectinosoma tenuireme* were common in relative abundances of 5.4 and 4.3 %, respectively, in the seagrass zone. In addition, *Halophytophilus simplex* (3.0 %), *Diathrodes cystoecus* (4.1 %), *Idomene maldivae* (3.0 %), *Cletodes dentatus* (3.2 %), and *Euterpina acutifrons* (3.2 %) were found on seagrass leaf blades. In seagrass sediment, *Typhlamphiascus ovale, Phyllopodopsyllus crenulatus, Scottolana rostrata, Apolaophonte hispida*, and *Macrosetella gracilis* occurred with a relative abundance between 3.6 and 4.6 %.

The abundance of harpacticoids was significantly higher (ANOVA, $F_{(2,144)} = 19.53$, $P < 0.001$) in vegetated sediments and differs markedly from seagrass blades and bare sediments. Among the three zones, higher congregation of copepods was found to be in the seagrass canopy sediment. Regarding the species richness, leaf blades were higher than in the remaining regions (Fig. 5). The Shannon index shows higher diversity of copepods in the vegetated region than in the bare sediment where as the equitability of species was more or less equal in all zones during the period of collection (Fig. 5c, d).

Discussion and conclusions

Associated meiofauna of major ecosystems like coral reef and seagrass habitats have been investigated from Indian

Table 1 Density (mean ± SE) and number of species of harpacticoid families

Families	Density (individuals 10 cm^{-2})			Number of species		
	SLB	SSD	BSD	SLB	SSD	BSD
Ameiridae	0.22 ± 0.06	0.92 ± 0.18	–	1	1	–
Canuellidae	3.98 ± 0.78	19.85 ± 0.65	0.77 ± 0.12	6	6	2
Cletodidae	0.68 ± 0.23	2.92 ± 0.24	–	2	2	–
Cylindropsyllidae	0.66 ± 0.14	0.62 ± 0.05	0.36 ± 0.10	2	3	1
Diosaccidae	1.96 ± 0.25	5.54 ± 0.19	4.06 ± 0.20	12	7	7
Ectinosomatidae	2.98 ± 0.6	11.54 ± 0.45	0.69 ± 0.07	4	4	2
Laophontidae	0.28 ± 0.02	3.69 ± 0.55	0.71 ± 0.06	2	3	3
Longipediidae	1.38 ± 0.16	3.23 ± 0.37	3.37 ± 0.91	1	2	1
Paramesochridae	–	0.92 ± 0.09	1.42 ± 0.38	0	1	1
Parastenheliidae	0.94 ± 0.11	3.23 ± 0.18	0.49 ± 0.08	1	1	1
Peltidiidae	0.04 ± 0.01	0.31 ± 0.1	–	1	1	–
Porcellidiidae	0.28 ± 0.09	0.15 ± 0.05	–	1	1	–
Tetragonicipitidae	0.8 ± 0.12	4.00 ± 0.35	0.84 ± 0.08	3	3	3
Thalestridae	2.08 ± 0.2	2.62 ± 0.18	–	5	4	–

SLB seagrass leaf blade, *SSD* seagrass sediment, *BSD* bare sediment

waters (Arunachalam and Nair 1988; Ansari and Parulekar 1994) and elsewhere worldwide (Mascart et al. 2013, 2015; De Troch et al. 2008; Gheerardyn et al. 2008, 2009, 2010; Callens et al. 2011). As for the meiofauna, nematodes generally dominate benthic sediments followed by harpacticoid copepods (Ólafsson 1995), whereas occasionally other taxa may also dominate marine sediment because of the patchiness of meiofauna in distribution (Thistle 1978; McLachlan 1978; Findlay 1981; Mantha et al. 2012a, b). Reviews of the meiobenthos of marine sediments have showed that lower copepod abundances are encountered in all depth zones (Ólafsson 1995). In the present report, out of the 11 recorded meiofaunal taxa, harpacticoid copepods were the dominant component of sediments in the seagrass zone (64 %) which differed markedly from bare sediment and leaf blades (ANOVA, $F_{(2,144)} = 19.53$, $P < 0.001$) (Fig. 2). A similar pattern of harpacticoid dominance was reported in the Bay of Bengal; however, the samples were collected from the intertidal region of the Chennai coast in comparison with the present analysis of the bare sediment zone (Mantha et al. 2012a, b). It may be convenient to consider seagrass and canopy sediment as separate environments, while the correlation between meiofaunal taxa of SLB and SSD was significantly very high ($r = 0.987$, $r^2 = 0.974$, $F_{(1,9)} = 337.3$, $P < 0.001$) which provided a conclusive evidence of interaction within the vegetative environment. The results of the present study are comparable

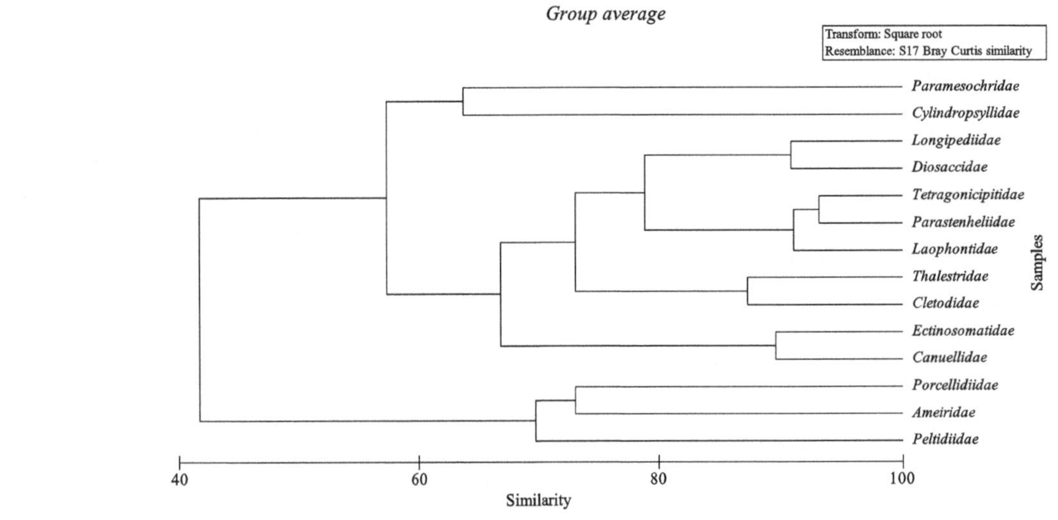

Fig. 3 Classification of families using the Bray-Curtis similarity based on abundance data from the sampled zones

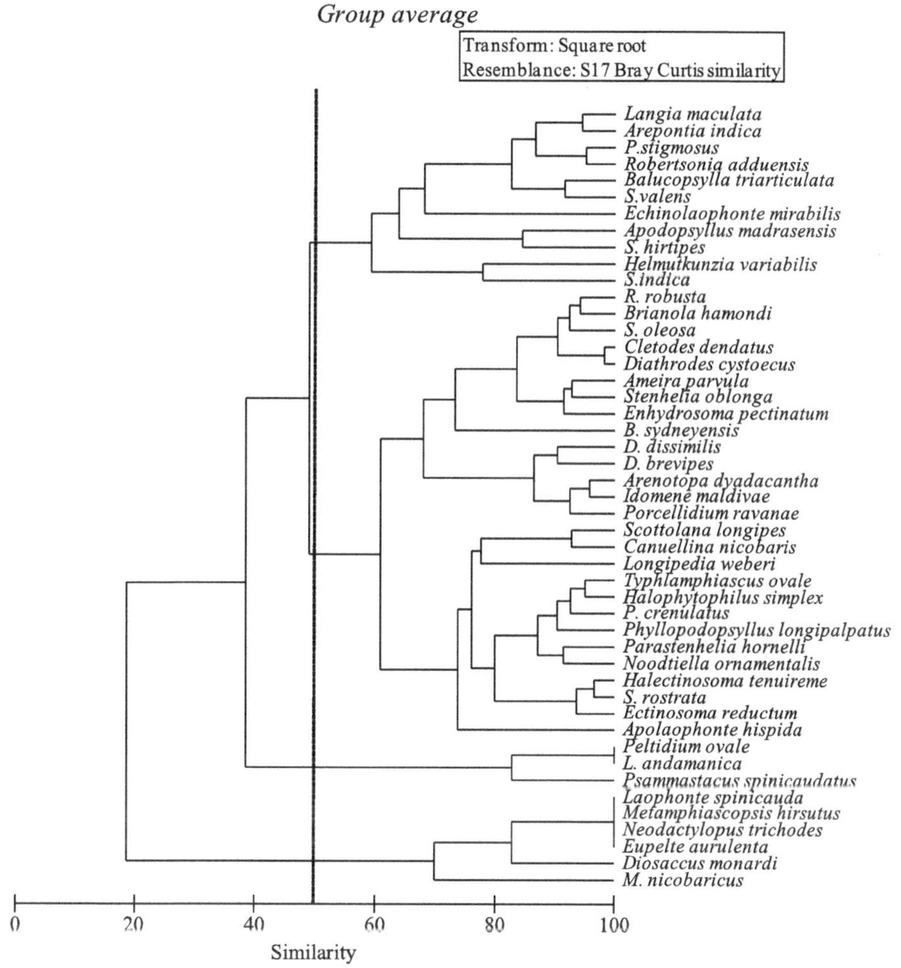

Fig. 4 Dendrogram of cluster analysis. The Bray-Curtis similarity index of the harpacticoid copepods of the studied areas in South Andaman

with the classification of harpacticoids as rare species, migrators, and non-migrators as reported by Walters and Bell (1986) and other investigations of a similar habitat (Mascart et al. 2013, 2015).

A total of 14 families of Harpacticoida were observed, with 8 families (Canuellidae, Ectinosomatidae, Longipediidae, Parastenheliidae, Diosaccidae, Tetragonicipitidae, Cylindropsyllidae, and Laophontidae) found in all zones and 5 families recorded only in the vegetated zone (Porcellidiidae, Peltidiidae, Thalestridae, Ameiridae, and Cletodidae), whereas the family Paramesochridae was confined to the sediments of both zones. True phytal-dwelling harpacticoids belong to seven families, namely, Harpacticidae, Tisbidae, Porcellidiidae, Tegastidae, Thalestridae, Diosaccidae, and Peltidiidae (Hicks and Coull 1983); however, in this study, Porcellidiidae and Peltidiidae were least encountered, whereas families such as Harpacticidae, Tisbidae, and Tegastidae were found to be absent. The families Diosaccidae and Canuellidae were rich in species number (37.5 %), contributing 12 and 6 species, respectively, whereas most of the other families contributed 2 to 4 species. Some families like Porcellidiidae, Peltidiidae, Parastenheliidae, Ameiridae, and Paramesochridae contributed only with a single representative throughout the study. The seagrass *Thalassia hemprichii* preferred for the present study is a subtidal species, and the association of families of Harpacticoida is primarily based on the tidal position (De Troch et al. 2003) and sediment characterization (Gheerardyn et al. 2008; Semprucci et al. 2010). Families like Thalestridae, Laophontidae, Diosaccidae, and Cletodidae were supported by tidal position and showed a strong preference for the subtidal zone (De Troch et al. 2003). The families Porcellidiidae, Peltidiidae, and Thalestridae have been regarded as typical phytal dwellers (Arroyo et al. 2006) and were found only in the vegetated samples. Most species represented in such families typically have flattened bodies and mouth parts adapted to facilitate adhesions to flat leaf or algal surfaces and cope with turbulence and current (Hicks 1985). The family Parastenheliidae was found in all samples of leaf blades, canopy sediment, and unvegetated sediments throughout the period of

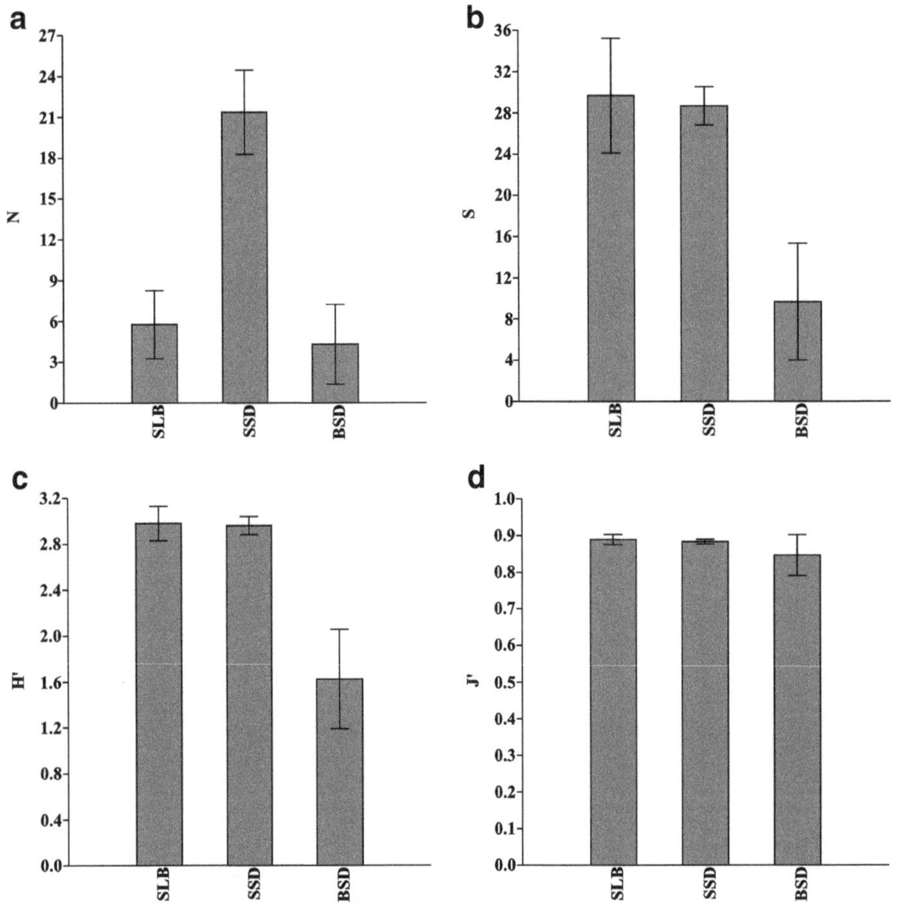

Fig. 5 Diversity indices of harpacticoid copepods. **a** Abundance (*N*). **b** Species richness. **c** Shannon diversity (H'). **d** Equitability (J') in seagrass leaf blade (*SLB*), seagrass sediment (*SSD*), and bare sediment (*BSD*)

sampling. This is explained by the non-specificity of its representatives with respect to substrate.

Most harpacticoid copepods species were found to be common and regular in the vegetated region with higher density in sediments (Fig. 5a). From this, it may be assumed that benthic copepods migrate from sediments to phytal structures. The dendrogram based on the average abundance of species provided five clusters. In the vegetative zone, sediment under seagrass contributed to the species abundance with a total of 13 species with the dominating species *Longipedia weberi*, *Canuellina nicobaris*, and *Scottolana longipes*. Certain species such as *Diathrodes cystoecus*, *Cletodes dentatus*, and *Scottolana oleosa* were found to be abundant and occurred only in the SSD zone congregated as a single cluster (Fig. 4). The local concentrations of harpacticoids could arise in nature from the patchiness in the distribution of microbes (Gray 1968) and dissolved organic carbon from seagrass exudates (Penhale and Smith 1977). Among the benthic harpacticoids inhabiting shallow waters, some species have been recognized as temporary plankton because they migrate from the benthic to the pelagic environment (Walters and Bell 1986; Schimode and Shirayama 2006). Compared to the vegetated environment, organisms of the unvegetated environment are commonly exposed to high levels of environmental stress, gravimetric shifts. Food like organic matter are washed away due to water movement of different kinds. Seagrass environments commonly provide a wider range of niche availability for food, shelter, and refuge from predation. The organisms comprising the seagrass communities are not restricted to seagrass habitats but are simply more abundant than in the surrounding environment (Hogarth 2007).

Abundance of crustaceans in seagrass meadows contribute to reduction in the epiphytic biomass, due to grooming which benefits the seagrass plant (Hemminga and Duarte 2000; Hogarth 2007). Niche diversity, providing shelter from predation and physical stress, and available food resources through sediment trapping (Hicks 1985) create a seagrass ecosystem as a highly complex habitat that has been found to increase faunal abundance and diversity relative to unvegetated habitats

(Connolly 1994; Jenkins et al. 1997; Semprucci et al. 2010). A decline of seagrasses worldwide might be due to direct anthropogenic activities, such as eutrophication and pollution with hazardous agents, invasive species, climate change, and shoreline development. Another reason might be herbivores associated with seagrasses causing secondary damage by allowing the water or pathogens to enter weakened leaves and increase the chance of leaf loss due to wave action (Gambi et al. 2003). Recent studies on meiobenthic assemblages around the coast of islands and mainland of the Indian Ocean continued to be restricted on the relationships that exist between sediment and meiofauna (Semprucci et al. 2010; Mantha et al. 2012a, b). Despite reports on species lists and the distribution of seagrasses, their importance as productive ecosystems remains inadequately evaluated from Andaman Islands as yet.

Competing interests
The authors declare that they have no competing interests.

Authors' contributions
RJ designed the study; carried out the collection, analysis, and identification of the meiofauna and copepod species; and performed data analysis. GP designed and coordinated the study and helped to draft the manuscript. IA participated in sample collection and performed the statistical analysis. All authors read and approved the final manuscript.

Acknowledgements
We thank the authorities of Pondicherry University for providing necessary facilities. The authors are thankful to the reviewers for their critical comments that improved the quality of this paper.

References
Anandavelu I, Jayabarathi R, Padmavati G, Jayaraj KA (2013) Epifaunal assemblage on morphologically distinct intertidal seaweeds of Kodiyaghat (South Andaman). India Proc Int Acad Ecol Environ Sci 3(3):229–237
Ansari ZA, Parulekar AH (1994) Meiobenthos in the sediments of seagrass meadows of Lakshadweep atolls, Arabian Sea. Vie Milieu 44(3/4):185–190
Arroyo NL, Maldonado M, Walters K (2006) Within and between plant distribution of harpacticoid copepods in a North Atlantic bed of *Laminaria ochroleuca*. J Mar Biol Assoc UK 86:309–316
Arunachalam M, Nair NB (1988) Harpacticoid copepods associated with the seagrass *Halophila ovalis* in the Ashtamudi Estuary, south-west coast of India. Hydrobiologia 167(168):515–522
Boucher G (1997) Structure and biodiversity of nematode assemblages in the SW lagoon of New Caledonia. Coral Reefs 16:177–186
Callens M, Gheerardyn H, Ndaro SGM, De Troch M, Vanreusel A (2011) Harpacticoid copepod colonization of coral fragments in a tropical reef lagoon (Zanzibar, Tanzania). J Mar Biol Assoc UK 92:1–11
Clarke KR, Gorley RN (2006) PRIMER v6: user manual/tutorial. PRIMER-E, Plymouth, UK
Connolly RM (1994) Removal of seagrass canopy: effects on small fish and their prey. J Exp Mar Biol Ecol 184:99–110
Coull BC (1990) Are members of the meiofauna food for higher trophic levels? T Am Microsc Soc 109:233–246
Coull BC, Zo Z, Tietjen JH, Williams BS (1982) Meiofauna of the southeastern United States continental shelf. Bull Mar Sci 32:233–240
Danovaro R, Scopa M, Gambi C, Franschetti S (2007) Trophic importance of subtidal metazoan meiofauna: evidence from in situ exclusion experiments on soft and rocky substrates. Mar Biol 152:339–350
Das HS (1996) Status of seagrass habitats of Andaman and Nicobar coast. SACON Technical Report No.4, Coimbatore, p 32
De Troch M, Mees J, Wakwabi E (1998) Diets of abundant fishes from beach seine catches in seagrass beds of a tropical bay (Gazi Bay, Kenya). Belg J Zool 128:135–154
De Troch M, Friers F, Vincx M (2001) Alpha and beta diversity of harpacticoid copepods in a tropical seagrass bed: the relation between diversity and species' range size distribution. Mar Ecol Prog Ser 215:225–236
De Troch M, Fiers F, Vincx M (2003) Niche segregation and habitat specialisation of harpacticoid copepods in a tropical seagrass bed. Mar Biol 142:345–355
De Troch M, Raes M, Muthumbi A, Gheerardyn H, Vanreusel A (2008) Spatial diversity of nematode and copepod genera of the coral degradation zone along the Kenyan coast, including a test for the use of higher-taxon surrogacy. Afr J Mar Sci 30(1):25–33
Duarte CM, Chiscano CL (1999) Seagrass biomass and production: a reassessment. Aquat Bot 65:159–174
Findlay SEG (1981) Small scale spatial distribution of meiofauna on a mud and sandflat. Estuar Coast Shelf Sci 12:471–484
Gambi MC, van Tussenbroek BI, Brearley A (2003) Mesofaunal borers in seagrasses: world-wide occurrence and a new record of boring polychaetes in the Mexican Caribbean. Aquat Bot 76:65–77
Gheerardyn H, De Troch M, Ndaro SGM, Raes M, Vincx M, Vanreusel A (2008) Community structure and microhabitat preferences of harpacticoid copepods in a tropical reef lagoon (Zanzibar Island, Tanzania). J Mar Biol Assoc UK 88(4):747–758
Gheerardyn H, De Troch M, Vincx M, Vanreusel A (2009) Harpacticoida (Crustacea: Copepoda) associated with cold-water coral substrates in the Porcupine Seabight (NE Atlantic): species composition, diversity and reflections on the origin of the fauna. Sci Mar 73(4):747–760
Gheerardyn H, De Troch M, Vincx M, Vanreusel A (2010) Diversity and community structure of harpacticoid copepods associated with cold-water coral substrates in the Porcupine Seabight (North-East Atlantic). Helgol Mar Res 64:53–62
Giere O (1993) Meiobenthology. The microscopic fauna in aquatic sediments. Springer, New York
Grainger EH (1991) Exploitation of Arctic sea ice by epibenthic copepods. Mar Ecol Prog Ser 7:119–124
Gray JS (1968) An experimental approach to the ecology of the harpacticoid *Leptastacus constrictus* Lang. J Exp Mar Biol Ecol 2:278–292
Hemminga MA, Duarte CM (2000) Seagrass ecology. Cambridge University Press, Cambridge
Hicks GRF (1985) Meiofauna associated with rocky shore algae. In: Moore PG, Seed R (eds) The ecology of rocky coasts. Hodder and Stoughton, London, pp 36–56
Hicks GRF, Coull BC (1983) The ecology of marine meiobenthic harpacticoid copepods. Oceanogr Mar Biol Annu Rev 21:67–175
Higgins RP, Thiel H (1988) Introduction to the study of meiofauna. Smithsonian Institution Press, London
Hogarth PJ (2007) The biology of mangroves and seagrasses, 2nd edn. Oxford University Press, New York, p 273
Jagtap TG (1991) Distribution of seagrasses along the Indian coast. Aquat Bot 40(4):379–386
Jagtap TG, Inamdar SN (1991) Mapping of seagrass meadows from the Lakshadweep Islands (India), using aerial photographs. J Indian Soc Remote 19:77–81
Jayabarathi R, Padmavati G, Anandavelu I (2012) Abundance and species composition of harpacticoid copepods from a seagrass patch of South Andaman, India. Curr Res J Biol Sci 4(6):717–724
Jenkins GP, May HMA, Wheatley MJ, Holloway MG (1997) Comparison of fish assemblages associated with seagrass and adjacent unvegetated habitats of Port Phillip Bay and Corner Inlet, Victoria, Australia, with emphasis on commercial species. Estuar Coast Shelf Sci 44:569–588
Lang K (1948) Monographie der Harpacticiden, 2 vols. Hakan Ohlsson, Lund
Lang K (1965) Copepoda Harpacticoidea from the Californian Pacific Coast. Almqvist and Wiksell, Stockholm, Sweden
Mantha G, Moorthy MSN, Altaff K, Dahms H, Sivakumar K, Hwang JS (2012a) Seasonal shifts of meiofauna community structures on sandy beaches along the Chennai coast, India. Crustaceana 85(1):27–53

Mantha G, Moorthy MSN, Altaff K, Dahms H, Sivakumar K, Hwang JS (2012b) Community structure of the Harpacticoida (Crustacea: Copepoda) on the Coast of Chennai, India. Zool Stud 51(4):463–475
Mascart T (2010) The role of meiofauna in the energy transfer in a Mediterranean seagrass bed (Calvi, Corsica), France. Master thesis, Ghent University
Mascart T, Lepoint G, De Troch M (2013) Meiofauna and harpacticoid copepods in different habitats of a Mediterranean seagrass meadow. J Mar Biol Assoc UK 93:1–10
Mascart T, Lepoint G, Deschoemaeker S, Binard M (2015) Seasonal variability of meiofauna, especially harpacticoid copepods, in *Posidonia oceanica* macrophytodetritus accumulations. J Sea Res 95:149–160
McLachlan A (1978) A quantitative analysis of the meiofauna and the chemistry of the redox potential discontinuity zone in a sheltered sandy beach. Estuar Coast Mar Sci 7:275–290
Nakaoka M (2005) Plant-animal interactions in seagrass beds: ongoing and future challenges for understanding population and community dynamics. Popul Ecol 47:167–177
Ndaro S, Ólafsson E (1999) Soft-bottom fauna with emphasis on nematode assemblage structure in a tropical lagoon in Zanzibar, eastern Africa: I. Spatial variability. Hydrobiologia 405:133–148
Nobi EP, Dilipan E, Sivakumar K, Thangaradjou T (2011) Distribution and biology of seagrass resources of Lakshdweep group of Islands, India. Indian J Mar Sci 40(5):624–634
Ólafsson E (1995) Meiobenthos in mangrove areas in eastern Africa with emphasis on assemblage structure of free-living marine nematodes. Hydrobiologia 312:47–57
Ólafsson E (2003) Do macrofauna structure meiofauna assemblages in marine soft-bottoms? A review of experimental studies. Vie Milieu 53:249–265
Ólafsson E, Moore CG (1992) Effects of macroepifauna on developing nematode and harpacticoid assemblages in a subtidal muddy habitat. Mar Ecol Prog Ser 84:161–171
Padmavati G, Haridas P, Nair KKC, Gopalakrishnan TC, Shiney P, Madhupratap M (1998) Vertical distribution of mesozooplankton in the central and eastern Arabian Sea during the winter monsoon. J Plankton Res 20:343–354
Penhale PA, Smith WO (1977) Excretion of dissolved organic carbon by eelgrass (*Zostera marina*) and its epiphytes. Limnol Oceanogr 22:400–407
Schimode S, Shirayama Y (2006) Diel vertical migration and life strategies of two phytal-dwelling harpacticoids *Ambunguipes rufocincta* and *Eudactylops spectabilis*. Plankton Benthos Res 1(1):42–53
Semprucci F, Colantoni P, Baldelli G, Rocchi M, Balsamo M (2010) The distribution of meiofauna on back-reef sandy platforms in the Maldives (Indian Ocean). Mar Ecol 31:592–607
Stoner AW, Ray M, Waite JM (1995) Effects of a large herbivorous gastropod on macrofauna communities in tropical seagrass meadows. Mar Ecol Prog Ser 121:125–137
Sulochanan B, Kumaraguru AK, Korabu LS (2011) Hydrological conditions in seagrass beds in Plak Bay and Gulf of Mannar, southeast coast of India. J Mar Biol Assoc India 53(1):108–115
Thangaradjou T, Sridhar R, Senthilkumar S, Kannan S (2007) Seagrass resource assessment in the Mandapam coast of the Gulf of Mannar Biosphere Reserve, India. Appl Ecol Env Res 6(1):139–146
Thangaradjou T, Nobi EP, Dilipan E, Sivakumar K, Susila S (2010) Heavy metal enrichment in seagrasses of Andaman Islands and its implication to the health of the coastal ecosystem. Indian J Mar Sci 39(1):85–91
Thistle D (1978) Harpacticoid dispersion patterns: implications for deep-sea diversity maintenance. J Mar Res 36:377–397
Turner JT (2004) The importance of small planktonic copepods and their roles in pelagic marine food webs. Zool Stud 43:255–266
Umamaheswari R, Ramachandran S, Nobi EP (2009) Mapping the extend of seagrass meadows of Gulf of Mannar Biosphere Reserve, India using IRS ID satellite imagery. Int J Biodivers Conserv 1(5):187–193
Unsworth RKF, Hinder SL, Bodger OG, Unsworth LCC (2014) Food supply depends on seagrass meadows in the coral triangle. Environ Res Lett 9:1–9
Varkey MJ, Murty VSN, Suryanarayana A (1996) Physical oceanography of the Bay of Bengal and Andaman Sea. In: Ansell AD, Gibson RN, Barnes M (eds) Oceanography and marine biology: an annual review. University college London press, London, pp 1–70
Walters K, Bell SS (1986) Diel patterns of active vertical migration in seagrass meiofauna. Mar Ecol Prog Ser 34:95–103
Watzin MC (1983) The effects of meiofauna on setting macrofauna: meiofauna may structure macrofaunal communities. Oecologia (Berl) 59:163–166
Wells JBJ (1976) Key to aid in the identification of marine harpacticoid copepods. University of Aberdeen, UK, Department of Zoology
Wells JBJ, Rao GC (1987) Littoral Harpacticoida (Crustacea: Copepoda) from Andaman and Nicobar Islands. Memoirs of the Zoological Survey of India 16:1–385

14

Stock assessment and management implications of an endemic fish, *Oxygymnocypris stewartii*, in the Yarlung Zangbo River in Tibet, China

Bin Huo[1], Bao Shan Ma[2], Cong Xin Xie[1*], You Jian Duan[1], Xue Feng Yang[1] and Hai Ping Huang[1]

Abstract

Background: *Oxygymnocypris stewartii* is an endemic species only inhabiting the middle and upper reaches of the Yarlung Zangbo River and its tributaries at altitudes above 3600 m in the Qinghai-Tibet Plateau and has suffered a continuous decline of population size owing to anthropogenic disturbance. However, little is known about the stock status of this species and how to protect it. In this study, we evaluated the population status of *O. stewartii* and investigated two management policies including the minimum age limits and seasonal closure in the Yarlung Zangbo River.

Results: Estimates of the total mortality rate were 0.35 $year^{-1}$ for females and 0.65 $year^{-1}$ for males. The natural mortality rates were estimated as 0.10 and 0.22 $year^{-1}$ for females and 0.12 and 0.29 $year^{-1}$ for males using two empirical methods. The *O. stewartii* population may be in near full exploitation under the current harvesting strategy in the Yarlung Zangbo River, with a high chance of recruitment failure in the future. Although both minimum age limits and seasonal closure could be effective measures to conserve the *O. stewartii* stock, it was proposed that a seasonal closure was implemented each year from September to December according to the complexity of the application and the effects on the income of local fishermen and on the biological invasion.

Conclusions: The stock of *O. stewartii* may be in near full exploitation under the current harvesting strategy, with a high chance of recruitment failure in the future. Our results indicated that the imposition of the seasonal closure from September to December might be the effective measure to conserve the *O. stewartii* population.

Keywords: *Oxygymnocypris stewartii*; Per-recruit; Seasonal closure; Mortality; Conservation

Background

Due to anthropogenic disturbance, many natural aquatic resources have suffered severe decline in abundance and yields (Dichmont et al. 2000; Hutchings and Lamberth 2002; Pauly et al. 2002; Myers and Worm 2003) and some have even been driven to the point of collapse (Hauck and Sweijd 1999; Griffiths 2000; Cowley et al. 2002). Therefore, considerable measures, including imposition of minimum captured age, protected areas, as well as closed seasons and areas in which harvesting may not take place, were implemented to better manage natural aquatic resources (Ye 1998; Turpie et al. 2000; Pradervand and Baird 2002; Sun et al. 2005; Govender et al. 2006).

With lack of long-term catch data and little knowledge on the spawner biomass recruit relationship, per-recruit analysis provides an alternative method to assess fish stocks for which there is some information on growth and mortality (Beverton and Holt 1957; Ricker 1975; Quinn and Deriso 1999). Among the per-recruit models, spawning stock biomass per recruit (SSBR) analyses are often employed to establish long-term management strategies to keep or recover the reproductive capacity of the population (Gabriel et al. 1989; Goodyear 1993).

Management strategies of fisheries may be established based on biological, sociological, or economic objectives. To achieve the objectives including the maximum long-

* Correspondence: xiecongxin@mail.hzau.edu.cn
[1]College of Fisheries, Huazhong Agricultural University, Wuhan, Hubei 430070, China
Full list of author information is available at the end of the article

term catch and socio-economic considerations (e.g., maximizing jobs), the biological approach will often lead to higher fishing levels and corresponding lower yields. Therefore, fisheries managers require estimates of harvest levels that provide maximum long-term yield as well as those at which the risk of stock depletion is low (Griffiths 1997). For many fish populations, there is insufficient information available to determine optimal harvesting policies. Much attention has therefore been given in the literature to determining biological reference points (BRPs) that can be readily calculated from available information (Quinn and Deriso 1999). These BRPs are usually fishing mortalities or abundance levels (thresholds) and may be specified either as a target for optimal harvesting or a danger zone to be avoided (Leaman 1993).

Oxygymnocypris stewartii (family Cyprinidae, subfamily Schizothoracinae) is one of the large predatory freshwater cyprinids distributed in the middle and upper reaches of the Yarlung Zangbo River and its tributaries at altitudes above 3600 m in the Qinghai-Tibet Plateau (Wu and Wu 1992; Chen and Cao 2000). Not only are *O. stewartii* important to local commercial fisheries, they are also one of the most ecologically important species structuring their aquatic ecosystems, mainly owing to their relatively large size and predatory nature. The field investigation found that this species was preferentially captured from September to December and local fishing communities targeted this species primarily using gill nets and traps. In winter, the preferred fishing methods are set gill nets, while drift gill nets are prone to being used for other seasons, and more rarely the traps are used to capture the fry. This species was characterized by slow growth, a long life-span, late maturation, and once-a-year spawning during February to April (Huo et al. 2012, 2013), which made it particularly sensitive to anthropogenic disturbance. Recently, overexploitation and biological invasions, including those of *Carassius auratus* and *Pseudorasbora parva*, have caused the *O. stewartii* population to rapidly decline. This species is listed in the IUCN Red List of Threatened Species as a near threatened fish (Ng 2010); nevertheless, currently the *O. stewartii* fishery is unregulated. Scarce stock information hinders the management for this species population. Thus, to collect critical information for this stock, our specific objectives were to (1) estimate the sex-specific mortalities, (2) evaluate the population status of *O. stewartii* in the Yarlung Zangbo River, and (3) investigate two management policies including the minimum age limits and seasonal closure for the *O. stewartii* stock using the per-recruit model.

Methods

Ethics statement

This study was conducted according to "Instructive notions with respect to caring for laboratory animals," issued by the Ministry of Science and Technology of the People's Republic of China. All researchers had received appropriate training before performing animal studies.

Sample collection

A total of 579 individuals were monthly collected using gill nets (mesh sizes 4, 6, 8, 12, and 18 cm) between August 2008 and August 2009 in the Yarlung Zangbo River and its tributary (Xiang Qu) (Fig. 1). Fish were euthanized with MS-222, and standard length (*SL*) and body weight (*BW*) were measured to the nearest 1 mm and 0.1 g, respectively. Lapillus otoliths were extracted from each fish and were used for age estimation. Gender was determined by examining the gonad morphology.

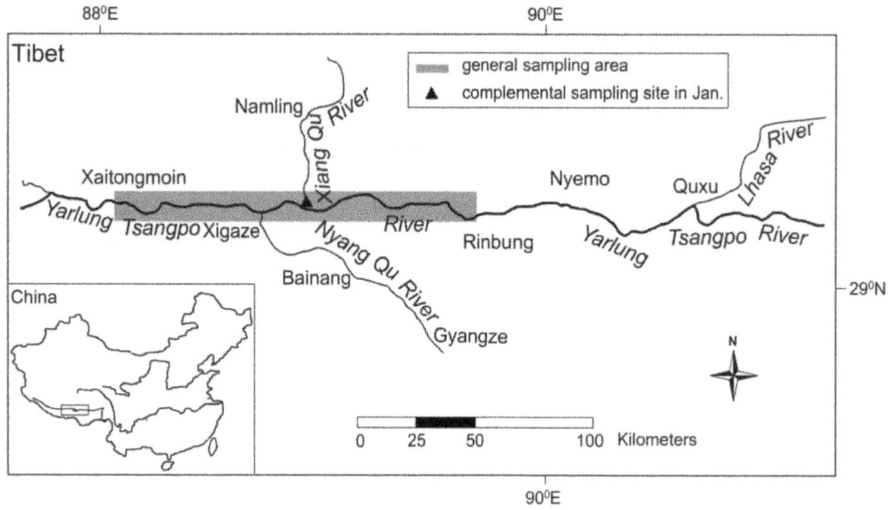

Fig. 1 The sampling locations of *O. stewartii* in the Yarlung Zangbo River

Growth

Growth in length with age was modeled using the von Bertalanffy growth equation of Huo et al. (2012):

$$L_t = 526.8\{1-\exp[-0.141(t-0.491)]\} \text{ for males} \quad (1)$$

$$L_t = 618.2\{1-\exp[-0.106(t-0.315)]\} \text{ for females} \quad (2)$$

where L_t is the standard length in millimeters at age t. Standard length was transformed to weight in kilograms using the length-weight relationship reported by Huo et al. (2012):

$$W_t = 9.872 \times 10^{-6} L_t^{3.052} \text{ for males} \quad (3)$$

$$W_t = 6.108 \times 10^{-6} L_t^{3.126} \text{ for females} \quad (4)$$

Reproduction

O. stewartii spawns just once each year from February to April. The fraction of mature fish at age (G_t) was modeled by the logistic function published by Huo et al. (2013):

$$G_t = \frac{1}{1+e^{-1.545(t-5.1)}} \text{ for males} \quad (5)$$

$$G_t = \frac{1}{1+e^{-1.376(t-7.3)}} \text{ for females} \quad (6)$$

Mortality

Instantaneous total annual mortality (Z), for each sex, was estimated from age-based catch curves (Ricker 1975; Quinn and Deriso 1999; Smith et al. 2012). The instantaneous natural mortality (M) was estimated using two empirical equations (Then et al. 2014):

$$M = 4.899 t_{max}^{-0.916} \quad (7)$$

$$M = 4.118 k^{0.73} L_{\infty}^{-0.33} \quad (8)$$

where t_{max} is the longevity of *O. stewartii*, estimated at 29 years for females and 22 years for males using Eq. (9); L_∞ and k are derived from Eqs. (1) and (2). The current instantaneous fishing mortality rate ($F_{current}$) was calculated as $Z - M$. The estimates of M and F were assumed to be constant and independent of age.

Per-recruit analysis

Two variables, the spawning potential ratio (SPR) and yield per recruit (YPR), were calculated to assess the status of the *O. stewartii* stock for various fishing mortalities ranging from 0 to large values (Goodyear 1993; Quinn and Deriso 1999). The traditional per-recruit model was modified to evaluate the effects of closed seasons on SPR and YPR, which was achieved by assuming a time step of 1 month in the per-recruit model. SPR and YPR models were developed separately for each sex.

The maximum life-span (t_{max}) of *O. stewartii* was calculated using the empirical relationship of Taylor (1958):

$$t_{max} = t_0 + \frac{2.996}{k} \quad (9)$$

where t_0 and k are derived from Eqs. (1) and (2).
The SPR was calculated using the following equations:

$$\text{SPR} = \frac{\text{SSBR}_F}{\text{SSBR}_{F=0}} \quad (10)$$

$$\text{SSBR} = \frac{\text{SSB}}{R} = \sum_{t=t_r}^{t_{max}} \exp((-FS_tA_t-M)(t-t_c)) \times \exp(-M(t_c-t_r)) a L_t^b G_t \quad (11)$$

where SSB is the total spawning stock biomass; SSBR is the spawning stock biomass per recruit; R is the number of recruits at age t_r and set to 1; SSBR_F is the spawning stock biomass per recruit with fishing; $\text{SSBR}_{F=0}$ is the spawning stock biomass per recruit without fishing; F and M are the fishing and natural mortality rates, respectively; a and b are the coefficients of the length-weight relationships and derived from Eqs. (3) and (4); t is the age and the unit is in months; L_t is the predicted mean length at age t and derived from Eqs. (1) and (2); t_{max} is the maximum life-span of *O. stewartii*; t_c is the age at first capture and was taken as the age corresponding to the top of the catch curve (Griffiths 1997); t_r is the age at recruitment; and A_t indicates whether a particular month corresponding to age t is open to fishing or not. If it is open to fishing, it takes on a value of 1 or 0 otherwise. The month

Table 1 Biological parameters for per recruit analysis of *O. stewartii*

Parameters	Female	Male	Source
k	0.0088 month^{-1}	0.0118 month^{-1}	Huo et al. (2012)
t_0	3.78 months	5.89 months	
L_∞	618.2 mm	526.8 mm	
a	0.000006108	0.000009872	
b	3.162	3.052	
Z	0.0292 month^{-1}	0.0542 month^{-1}	This study
M	0.0083–0.0183 month^{-1}	0.0100–0.0242 month^{-1}	
$F_{current}$	0.0108–0.0208 month^{-1}	0.0300–0.0442 month^{-1}	
t_r	12 months	12 months	
t_c	60 months	60 months	
k_m	0.1147 month^{-1}	0.1288 month^{-1}	Huo et al. (2013)
A_{50}	87.6 months	61.2 months	

a and b are the coefficients of Eqs. (3) and (4); k_m and A_{50} are the coefficients of Eqs. (5) and (6)

Table 2 The different harvesting strategies for the *O. stewartii* stock spawning from February to April

Management policies	The age at first capture (years)	Seasonal closure
Current policy	5	–
Policy 1	1	–
Policy 2	3	–
Policy 3	7	–
Policy 4	1	Feb–May
Policy 5	3	Feb–May
Policy 6	5	Feb–May
Policy 7	7	Feb–May
Policy 8	1	Sep–Dec
Policy 9	3	Sep–Dec
Policy 10	5	Sep–Dec
Policy 11	7	Sep–Dec

when $t = t_r$ was set to March; G_t is derived from Eqs. (5) and (6); and S_t is the gear selectivity at age t and is assumed to be knife-edged selectivity:

$$S_t = \begin{cases} 0, & t < t_c \\ 1, & t \geq t_c \end{cases} \quad (12)$$

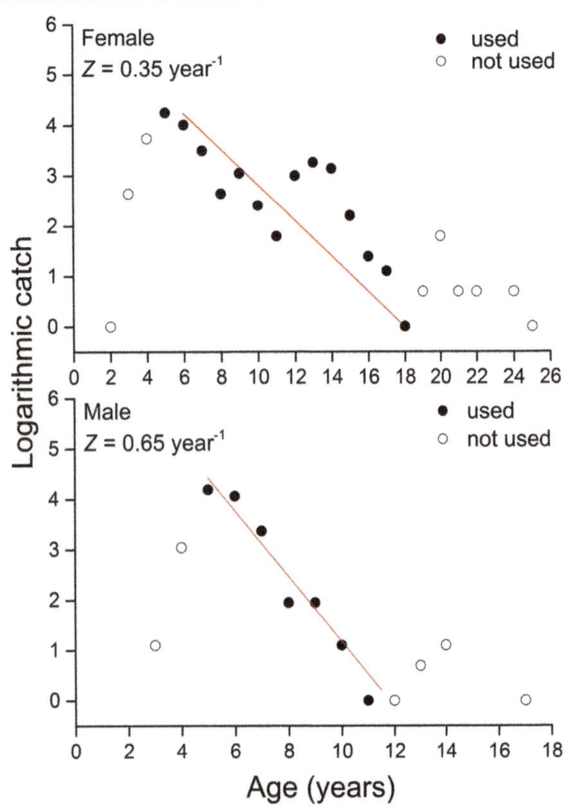

Fig. 2 Catch curves of *O. stewartii* in the Yarlung Zangbo River. The procedure recommended by Smith et al. (2012) was used to fit this regression line to the log-transformed catch data

The YPR was calculated from the following formula:

$$\mathrm{YPR} = \frac{Y}{R} = \sum_{t=t_r}^{t_{\max}} \frac{FS_tA_t}{FS_tA_t+M}\exp((-F_sS_tA_t-M)(t-t_c)) \times \exp(-M(t_c-t_r))(1-\exp(-FS_tA_t-M))aL_t{}^b \quad (13)$$

where Y is the total yield for a cohort throughout its life-span.

The parameters used in this study are given in Table 1. Given the difficulties in obtaining accurate estimates of natural mortality for exploited populations, the sensitivity of the per-recruit models to various estimates of M was tested. Moreover, we explored the changes in YPR and SPR under different combinations of the age at first capture and fishing mortality using the isopleth plot in order to evaluate the best compromise between the conservation of *O. stewartii* population and maximizing YPR. We also simulated 11 different harvesting strategies that were

Table 3 The age frequency of *O. stewartii* in the Yarlung Zangbo River

Age group (year)	Catch (number)	
	Female	Male
1	0	0
2	1	0
3	14	3
4	42	21
5	70	66
6	56	58
7	31	29
8	15	7
9	21	7
10	11	3
11	6	1
12	20	1
13	26	2
14	23	3
15	9	0
16	4	0
17	3	1
18	1	0
19	2	0
20	7	0
21	2	0
22	2	0
23	0	0
24	2	0
25	1	0

different from the current state to evaluate whether the closure season is an effective conservation for the *O. stewartii* population. These harvesting strategies were termed policies 1–11 (Table 2).

Biological reference point

The following reference points were used to determine the current status of the *O. stewartii* stock as well as to evaluate the effectiveness of different regulations: $F_{25\%}$ and $F_{40\%}$ are defined as fishing mortality rates that reduce SPR to 25 and 40 %, respectively, when compared to an unfished state where SPR is assumed to be at a level of 100 % (Gabriel et al. 1989; Goodyear 1993; Sun et al. 2005); F_{max} is defined as the fishing mortality rate that produces the maximum yield per recruit.

Results

Mortality

The estimates of Z for female and male *O. stewartii* from the slope of the descending limb of the catch curve were 0.35 and 0.65 year^{-1}, respectively (Fig. 2, Table 3). The values of M estimated by the empirical Eqs. (7) and (8) were 0.10 and 0.22 year^{-1} for females and 0.12 and 0.29 year^{-1} for males, respectively. The instantaneous natural mortalities of *O. stewartii* were therefore assumed to be between 0.10 and 0.22 year^{-1} for females and between 0.12 and 0.29 year^{-1} for males, respectively. The corresponding estimates of $F_{current}$ were 0.13 and 0.25 year^{-1} for females and 0.36 and 0.53 year^{-1} for males, respectively.

Per-recruit analysis

Per-recruit analysis was conducted at three alternative values of natural mortality (0.10, 0.17, and 0.22 year^{-1} for females; 0.12, 0.21, and 0.29 year^{-1} for males) within the predicted range for each gender. At the current harvesting strategy and three corresponding alternative values of M, the range of SPR for females was 10.93–38.14 % while that for males was 9.17–28.78 %, when compared to the state of no fishing (Fig. 3, Tables 4 and 5). In addition, the estimates of $F_{25\%}$ and $F_{40\%}$ fluctuated widely with M. The estimates of $F_{25\%}$ were 0.14–0.20 year^{-1} for females and 0.22–0.45 year^{-1} for males, respectively, and the estimates of $F_{40\%}$ were 0.08–0.12 year^{-1} for females and 0.13–0.23 year^{-1} for males, respectively (Fig. 3, Tables 4 and 5).

The effects of varying the age at first capture on the YPR and SPR under three corresponding values of M for each gender are demonstrated in Figs. 4 and 5. The YPR increased rapidly at low levels of F for most of the range of t_c. Increasing the t_c to between 6 and 13 years for females and between 5 and 9 years for males would have relatively little impact on the YPR under almost any level of fishing, when compared to the current YPR (Fig. 4). Moreover, the SPR would be maintained above 25 % under almost all ranges of fishing mortality, when enhancing the

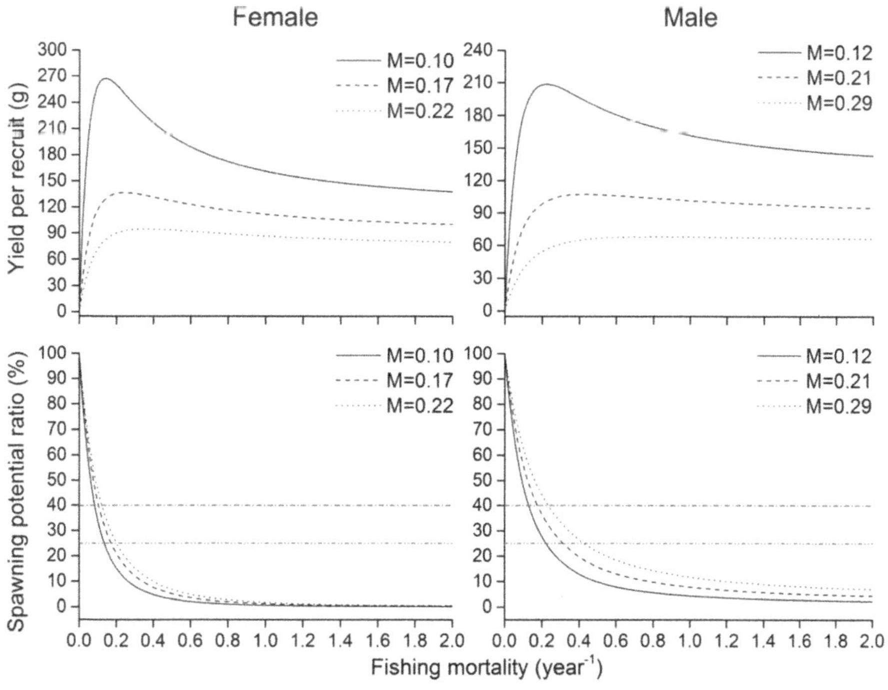

Fig. 3 Yield per recruit and spawning potential ratio curves for *O. stewartii* by sex. Yield per recruit and spawning potential ratio curves for *O. stewartii* by sex under a range of fishing mortality rates and the current harvesting strategy of the first age at capture at 5 years and no closed season. The *dashed-dotted-dotted blue and dashed-dotted red lines* represent SPR at 25 and 40 %, respectively

Table 4 Estimates of YPR, SPR, and reference points under different harvesting strategies for female *O. stewartii* stock

Management policy	M (year^{-1})	F_{max} (year^{-1})	$F_{25\%}$ (year^{-1})	$F_{40\%}$ (year^{-1})	YPR$_{current}$ (g)	YPR$_{max}$ (g)	YPR$_{25\%}$ (g)	YPR$_{40\%}$ (g)	SPR$_{current}$ (%)
Current policy	0.10	0.14	0.14	0.08	248.33	267.24	267.24	245.84	10.93
	0.17	0.24	0.17	0.11	134.65	136.77	133.71	121.18	23.89
	0.22	0.36	0.20	0.12	79.90	94.57	89.85	77.56	38.14
Policy 1	0.10	0.08	0.09	0.06	111.41	188.93	188.27	182.41	4.02
	0.17	0.11	0.11	0.07	79.97	88.79	88.76	83.62	11.63
	0.22	0.13	0.12	0.08	57.95	57.95	57.89	54.04	22.68
Policy 2	0.10	0.10	0.11	0.07	176.81	224.78	224.59	213.60	6.63
	0.17	0.15	0.13	0.08	110.31	111.21	110.48	99.46	16.67
	0.22	0.20	0.15	0.09	72.32	75.49	74.13	64.80	29.41
Policy 3	0.10	0.20	0.18	0.11	306.80	309.17	308.56	288.25	17.83
	0.17	0.44	0.26	0.14	144.54	158.42	154.04	134.79	33.98
	0.22	0.85	0.33	0.18	77.65	108.27	102.07	88.19	49.18
Policy 4	0.10	0.08	0.09	0.06	81.08	171.45	169.03	168.10	4.16
	0.17	0.09	0.11	0.07	58.16	72.95	71.74	71.13	11.87
	0.22	0.10	0.12	0.08	42.16	43.39	42.82	42.49	22.98
Policy 5	0.10	0.09	0.11	0.07	132.28	202.00	199.60	196.51	6.72
	0.17	0.12	0.13	0.08	82.48	89.01	88.59	84.76	16.82
	0.22	0.13	0.15	0.09	54.12	54.12	53.81	51.30	29.59
Policy 6	0.10	0.13	0.14	0.08	218.09	250.13	249.32	235.28	10.95
	0.17	0.18	0.17	0.11	118.22	118.22	118.12	110.78	23.92
	0.22	0.22	0.20	0.12	70.17	75.59	75.40	68.53	38.18
Policy 7	0.10	0.20	0.18	0.11	306.80	309.17	308.56	288.25	17.83
	0.17	0.44	0.26	0.14	144.54	158.42	154.04	134.79	33.98
	0.22	0.85	0.33	0.18	77.65	108.27	102.07	88.19	49.18
Policy 8	0.10	0.07	0.16	0.08	77.07	139.49	108.13	138.98	20.75
	0.17	0.10	–	0.13	55.32	60.90	–	59.85	34.53
	0.22	0.14	–	0.22	40.08	40.13	–	37.93	47.04
Policy 9	0.10	0.09	0.19	0.09	120.18	161.73	135.40	161.73	22.08
	0.17	0.16	–	0.15	74.99	75.15	–	75.10	37.09
	0.22	0.36	–	0.28	49.16	55.26	–	54.89	50.47
Policy 10	0.10	0.11	0.23	0.11	154.97	181.70	158.46	181.70	24.26
	0.17	1.09	–	0.19	84.02	103.63	–	84.43	40.75
	0.22	1.21	–	0.41	49.86	81.39	–	68.31	54.90
Policy 11	0.10	5.01	0.32	0.13	152.87	218.67	137.21	185.53	27.69
	0.17	5.62	–	0.26	71.98	141.44	–	69.85	45.75
	0.22	6.30	–	2.72	38.69	104.15	–	98.54	60.38

–: no data

t_c to 12 years for females and 9 years for males (Fig. 5). Thus, setting the t_c at 12 years for females and 9 years for males could ensure that the SPR was above the threshold at almost any level of fishing with relatively little impact on YPR regardless of M.

It is apparent that an introduction of the seasonal closure helps to conserve the *O. stewartii* population at the expense of decreasing the YPR. The introduction of the seasonal closure caused the YPR to significantly decline, regardless of when the closure season is set (Figs. 6 and 7, Tables 4 and 5). However, the SPR is related to when the closure season is set (Figs. 8 and 9, Tables 4 and 5). Based on a scenario of closed season from February to May, significantly increasing the SPR is dependent on both the age at first capture and the level of fishing mortality. On the other hand, the SPR can be dramatically increased

Table 5 Estimates of YPR, SPR, and reference points under different harvesting strategies for male *O. stewartii* stock

Management policy	M (year^{-1})	F_{max} (year^{-1})	$F_{25\%}$ (year^{-1})	$F_{40\%}$ (year^{-1})	$YPR_{current}$ (g)	YPR_{max} (g)	$YPR_{25\%}$ (g)	$YPR_{40\%}$ (g)	$SPR_{current}$ (%)
Current policy	0.12	0.22	0.22	0.13	185.50	208.42	208.42	195.64	9.17
	0.21	0.43	0.31	0.18	107.44	107.46	106.07	96.16	17.92
	0.29	0.82	0.45	0.23	63.84	67.94	65.92	57.15	28.78
Policy 1	0.12	0.11	0.13	0.08	44.95	134.78	132.59	130.90	1.15
	0.21	0.14	0.15	0.09	37.31	63.80	63.61	60.06	3.19
	0.29	0.17	0.17	0.11	30.53	37.78	37.78	35.50	7.03
Policy 2	0.12	0.15	0.16	0.10	110.23	168.97	168.58	161.56	3.31
	0.21	0.22	0.20	0.12	76.44	85.11	84.91	77.55	7.69
	0.29	0.32	0.24	0.14	53.29	53.48	52.61	46.15	14.43
Policy 3	0.12	0.36	0.39	0.20	239.59	242.55	242.41	230.99	20.28
	0.21	1.24	0.86	0.32	115.91	121.65	121.18	110.15	34.21
	0.29	–	3.36	0.58	58.68	74.03	72.77	65.25	48.42
Policy 4	0.12	0.09	0.13	0.08	16.82	112.58	106.44	111.67	4.18
	0.21	0.11	0.16	0.10	13.96	45.42	42.06	45.34	8.63
	0.29	0.12	0.21	0.12	11.42	22.41	19.17	22.41	14.92
Policy 5	0.12	0.12	0.16	0.10	44.09	137.97	132.60	136.60	5.61
	0.21	0.14	0.21	0.13	30.58	57.47	53.85	57.30	11.62
	0.29	0.16	0.28	0.16	21.32	28.95	25.29	28.95	19.84
Policy 6	0.12	0.18	0.22	0.13	127.27	183.02	180.24	178.21	9.76
	0.21	0.24	0.32	0.18	73.72	82.36	80.35	80.48	18.83
	0.29	0.30	0.46	0.24	43.80	44.32	41.67	43.76	29.95
Policy 7	0.12	0.36	0.39	0.20	239.59	242.55	242.41	230.99	20.28
	0.21	1.24	0.87	0.32	115.91	121.65	121.22	110.15	34.21
	0.29	–	3.36	0.58	58.68	74.03	72.77	65.25	48.42
Policy 8	0.12	0.10	–	0.16	39.45	91.88	–	85.69	29.36
	0.21	0.16	–	0.33	32.75	45.03	–	38.52	37.85
	0.29	0.22	–	0.61	26.79	29.40	–	19.38	43.74
Policy 9	0.12	0.15	–	0.20	94.35	113.46	–	111.48	30.98
	0.21	0.40	–	0.51	65.42	65.51	–	64.87	41.22
	0.29	0.52	–	1.13	45.61	46.98	–	41.90	49.28
Policy 10	0.12	1.06	–	0.29	139.67	149.43	–	131.09	35.28
	0.21	1.27	–	1.87	80.90	96.46	–	95.05	48.72
	0.29	1.57	–	–	48.06	66.16	–	–	59.82
Policy 11	0.12	5.46	–	1.11	107.30	208.46	–	145.79	41.86
	0.21	–	–	–	51.91	119.64	–	–	58.44
	0.29	–	–	–	26.28	74.03	–	–	71.60

–: no data

regardless of the level of fishing mortality and the age at first capture when assuming a closed season from September to December.

Discussion

Biological reference points such as F_{max}, $F_{25\%}$, and $F_{40\%}$ have often been used to develop fishery management strategies. Several authors have advocated designating $F_{40\%}$ as a target reference point and $F_{25\%}$ as a threshold reference point in order to obtain near optimal yields while insuring against stock collapse (Hildén 1993; Leaman 1993; Mace and Sissenwine 1993; Rivard and Maguire 1993; Mace 1994; Griffiths 1997; Kirchner 2001; Sun et al. 2005). Our results based on per-recruit analyses suggest that the current fishing mortalities for each sex population are constantly higher than the corresponding

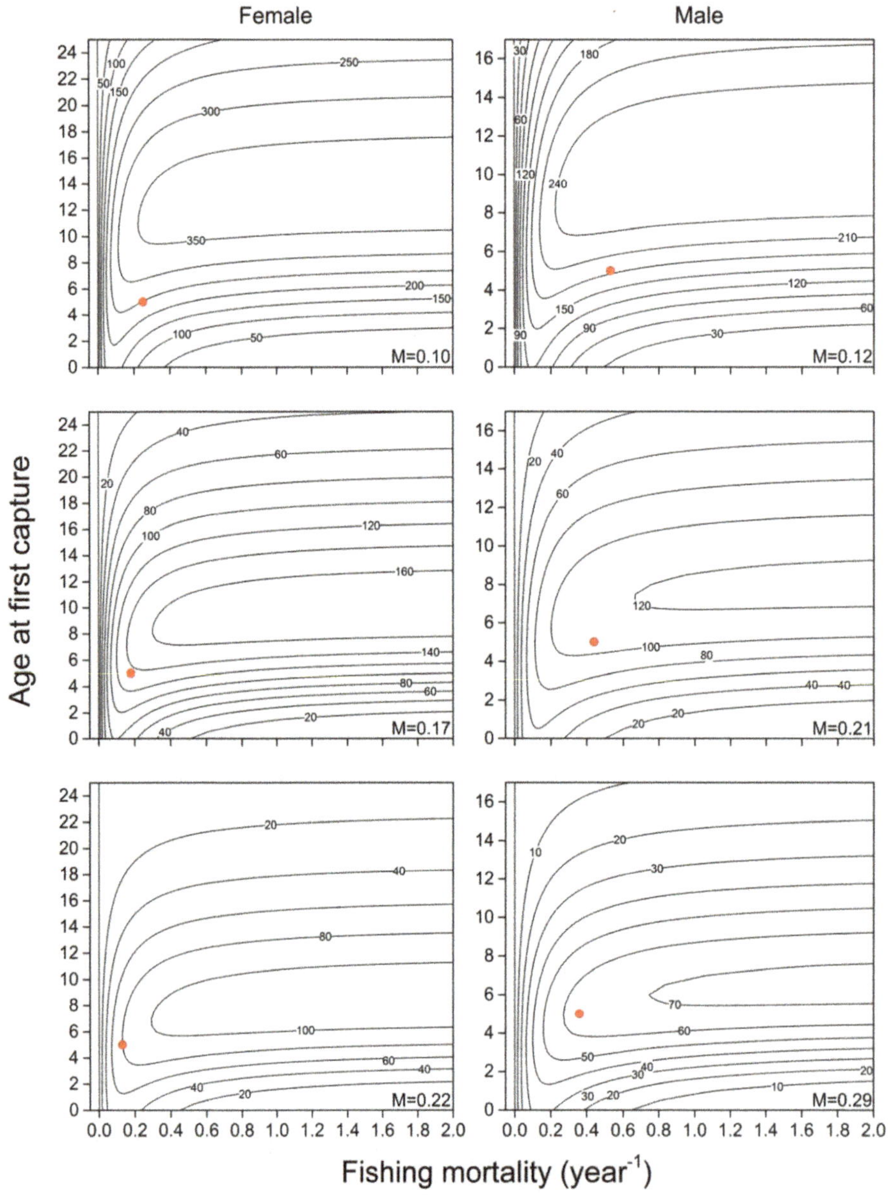

Fig. 4 Isopleths of yield per recruit for *O. stewartii* in the Yarlung Zangbo River. *Red circles* represent the current estimated yield per recruit

threshold reference point $F_{25\%}$ at the lower and intermedia values of *M*, while the current fishing mortalities are lower than the corresponding threshold reference point $F_{25\%}$ but higher than the corresponding target reference point $F_{40\%}$ at the higher values of *M*, which indicates that the stock status of *O. stewartii* in the Yarlung Zangbo River is sensitive to the estimated values of *M*. However, the evaluation of various empirical estimators, implemented by Then et al. (2014), shows clearly that Eq. (7) is both the best and a sufficient predictor of *M*, indicating that the estimated higher value of *M* may be more reliable than the others. In addition, Smith et al. (2012) reported that the weighted regression estimator of fishing mortality (derived from *Z*) was always negatively biased, indicating that the estimated $F_{current}$ might be underestimated. Therefore, the *O. stewartii* population may be in near full exploitation in the Yarlung Zangbo River.

Imposition of a closed season during the breeding period is only effective under three circumstances: (1) if fishing disturbs and reduces reproduction of individuals that are not captured, (2) if the target species aggregates to breed and becomes vulnerable to capture, and (3) if the closed season achieves a reduction of annual fishing effort (Arendse et al. 2007). Our results indicated that imposition of the closed season during the spawning season (February to May) was not effective, while that during

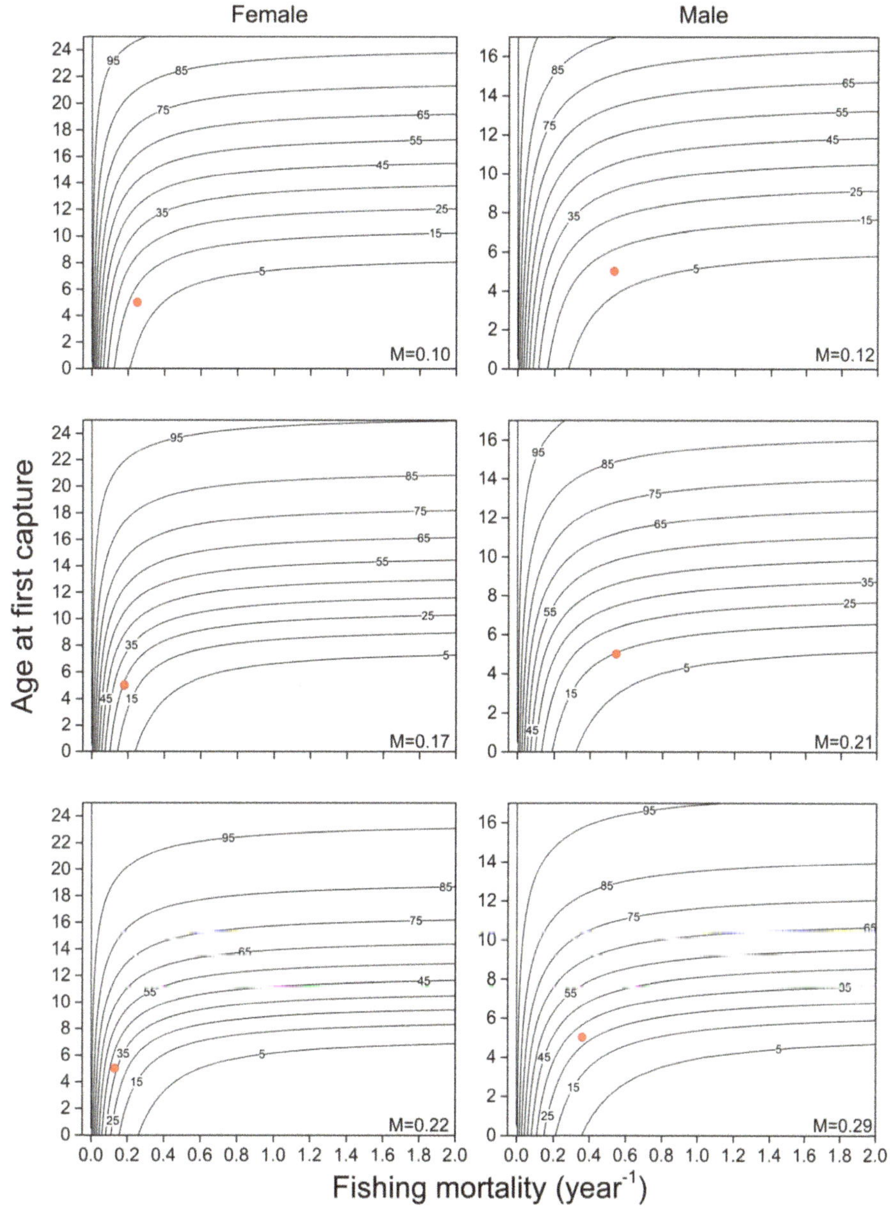

Fig. 5 Isopleths of spawning potential ratio for *O. stewartii* in the Yarlung Zangbo River. *Red circles* represent the current estimated SPR

September to December was effective. Our field survey found that *O. stewartii* aggregated to lay eggs and became vulnerable to capture during the spawning season. However, there were no spawning sites in our sampling locations. Moreover, *O. stewartii* is intensively captured during September to December every year, which corresponded to the gonad development period. Thus, the imposition of the closed season during September to December could reduce the annual fishing effort and help to conserve the *O. stewartii* stock.

Although increasing the age at first capture and introducing the closed season could be effective measures to conserve the *O. stewartii* stock, we prefer the imposition of the closed season to manage its population based on three reasons: (1) Compared with increasing the age at first capture, the closed season could be easily implemented and monitored, especially for the vast territory of Tibet. (2) The field survey found that the local people had an important custom that they would buy a lot of all kinds of small fish, including juveniles of *O. stewartii*, to release back to the river. Thus, the income of local fishermen is derived from both the adult fish and small fish. If the measure of increasing the age at first capture was carried out, part of the income from the sales of small fish would be reduced. If the closed season was implemented, the income from the sales of adult fish

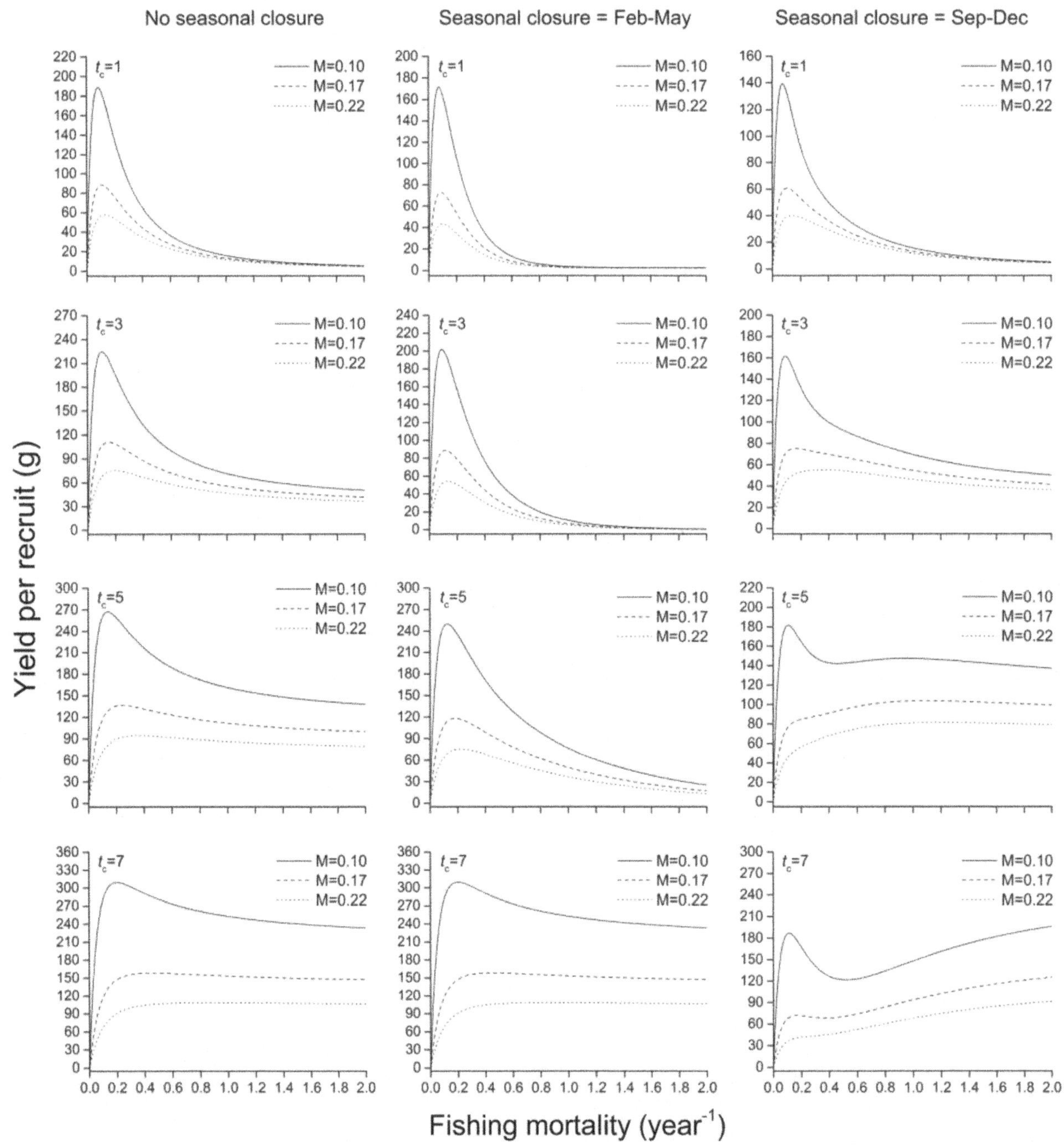

Fig. 6 Yield per recruit curves for female *O. stewartii* under different harvesting strategies. Please refer to Table 2 for the corresponding harvesting strategies

would be decreased. Therefore, regardless of which measures are taken, they have similar impact on the income of local fishermen. (3) By increasing the age at first capture, the catch of endemic small fish would likely decline dramatically, which likely will cause local people to switch to releasing the alien small fish back to the river. Invasive species usually threaten the survival of the native species due to competition for habitat and dietary resources.

Our results based on the per-recruit analysis can be misleading if the assumptions of the analysis and its various inputs, namely growth parameters, sampling bias, fishing, and natural mortality, are not made explicit. First of all, the per-recruit analysis is heavily affected by von Bertalanffy growth parameters. A meaningful estimate of growth parameters is related to age and size inputs. Compared with other age materials, pre-analyses showed that lapillus, which provided the age input for

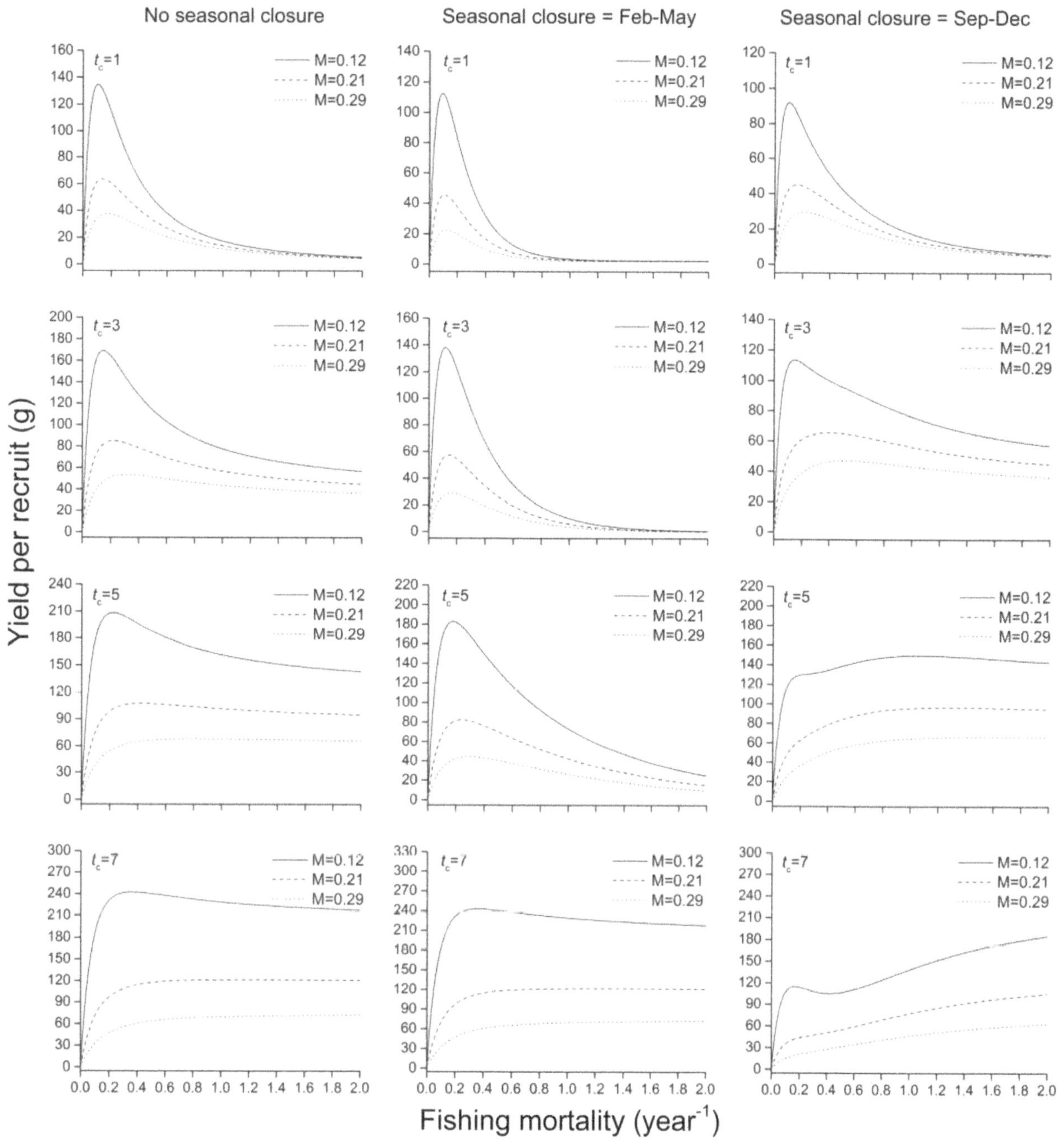

Fig. 7 Yield per recruit curves for male *O. stewartii* under different harvesting strategies. Please refer to Table 2 for the corresponding harvesting strategies

the von Bertalanffy growth model, is the most precise and accurate structure for age estimation. In addition, growth model estimates are greatly influenced by the lack of very young or old individuals (Cailliet and Goldman 2004). Huo et al. (2012) reported that the size composition including enough juveniles (<100 mm *SL*) and old individuals (>500 mm *SL*) provided a reliable estimate of growth parameters. Secondly, we acknowledge that the gill net has its own sampling bias. However, to reduce the sampling bias, the gill nets with multiple mesh sizes were deployed to collect *O. stewartii*, and the ranges in standard length and age extended broadly indicating that sampling bias did not drastically appear. Thirdly, our results are only valid if natural and fishing mortality rates used in our analysis are representative of the long-term situation. Our field investigations find that the catches of this species fluctuate little in recent years and the possibility for this trend may be continuous in

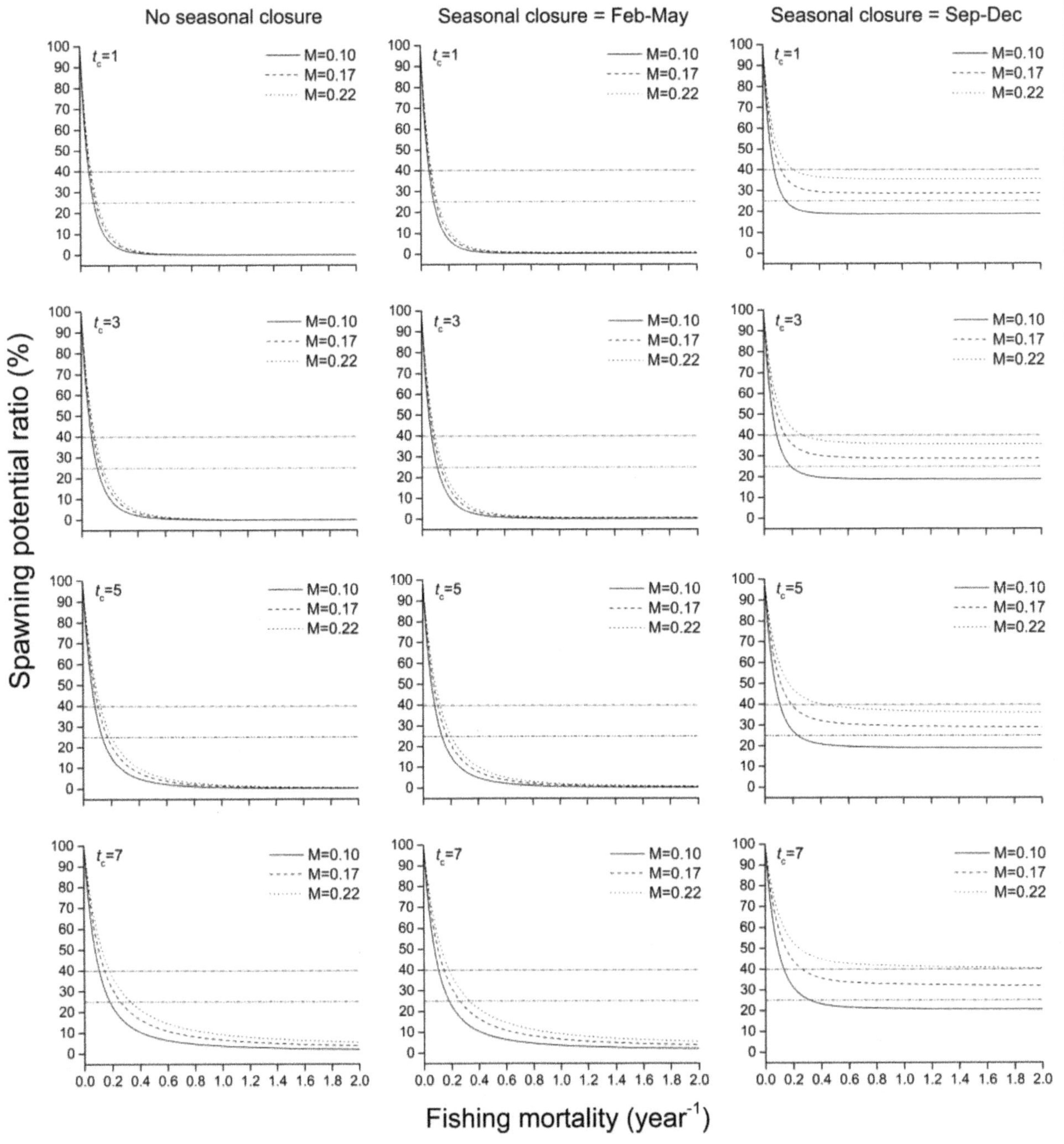

Fig. 8 Spawning potential ratio curves for female *O. stewartii* under different harvesting strategies. Please refer to Table 2 for the corresponding harvesting strategies. The *dashed-dotted-dotted blue and dashed-dotted red lines* represent SPR at 25 and 40 %, respectively

the future. Thus, it is possible that the age structure of the population may remain the same in the future as we have documented here. Fourthly, the natural mortality is generally difficult to be estimated reliably and directly. By direct, the natural mortality prefers to be estimated using information strictly pertaining to the stock of interest. However, direct estimation methods of *M* are often data intensive, thus limiting their application to relatively data-rich stocks (Then et al. 2014). For data-poor stocks, indirect or empirical equations are often used to estimate *M*. Although these empirical methods are less reliable than their data-rich counterparts, a consensus is that empirical methods are useful and very important particularly for the data-poor stocks. In our study, we evaluated the sensitivity of the per-recruit models to various estimates of *M* within the range of plausible values for *M* for the *O. stewartii* stock by applying two empirical methods, which represent both the best and sufficient predictors of *M*.

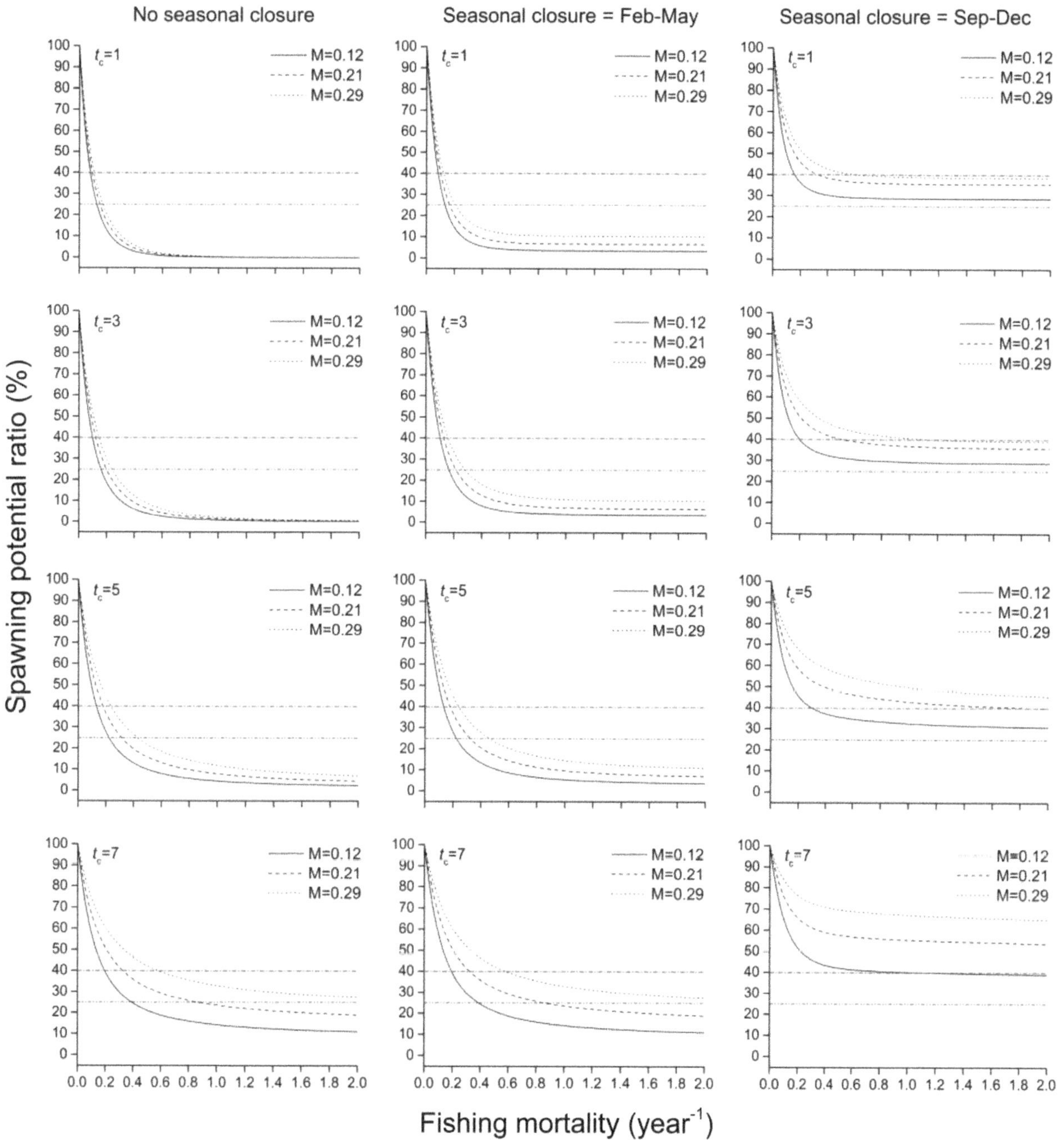

Fig. 9 Spawning potential ratio curves for male *O. stewartii* under different harvesting strategies. Please refer to Table 2 for the corresponding harvesting strategies. The *dashed-dotted-dotted blue and dashed-dotted red lines* represent SPR at 25 and 40 %, respectively

Fifthly, catch curve analysis can be used to estimate the current fishing mortality (derived from Z) under the assumption that all fish are equally vulnerable to the fishing gear above some determined age. This assumption is often not satisfied in many fisheries, particularly gill net fisheries, due to size selectivity effects of the mesh size. Due to lack of fishing gear selectivity data, the gear selectivity is assumed to be a knife-edge selection in our per-recruit analysis. However, it is clear from the catch curve analysis that size selectivity does exist in the old age groups of *O. stewartii* fishery, indicating that the estimated F_{current} may be negatively biased. Moreover, the estimated F_{current} may also be negatively biased using the weighted regression method proposed by Smith et al. (2012). Overall, our estimate of F_{current} may therefore be underestimated and should be viewed with caution. Finally, although the stock status of *O. stewartii* is sensitive to the estimated values of M, the effects of increasing the age at first capture and introducing the closed season are insensitive to the values of M, indicating that both of measures can effectively prevent the *O. stewartii* population collapse.

Our results showed that the estimated $F_{current}$ for males was two times higher than that for females. Generally, females may prefer to be captured, given that females actually grow to a larger size. However, for the fisheries using gill nets, small fish can pass through the meshes as is the case for trawl nets, while large fish may also avoid being caught in a gill net, because their heads are so large that they cannot be gilled (Sparre and Venema 1998). Thus, the $F_{current}$ for males was estimated using higher age groups than that for females, indicating that the phenomenon that our estimation of $F_{current}$ for males was higher than that for females may be reasonable.

O. stewartii belongs to the subfamily Schizothoracinae, which originated from primitive barbine fishes distributed in Tibet during the late Tertiary and subsequently evolved into the Qinghai-Tibet plateau fish fauna owing to the vicariance induced by the uplift of Tibet (Cao et al. 1981; Wu and Tan 1991; He and Chen 2006; He and Chen 2007). The isolation made this specialized fish fauna, including *O. stewartii*, more vulnerable to anthropogenic activities. Moreover, *O. stewartii* is the top predator inhabiting the Yarlung Zangbo River (Huo et al. 2014) and changes in its abundance can alter ecosystem processes and community structure in aquatic ecosystems (McQueen et al. 1989; Carpenter and Kitchell 1993; Schindler et al. 1997). Thus, knowledge of the population dynamics of this species is of fundamental importance to conserving this stock. Our findings indicate that the stock of *O. stewartii* is likely to be in the near full exploitation under the current harvesting strategy, and both measures of increasing the age at first capture and introducing the closed season can be effective to protect the *O. stewartii* population. Our study represents an important investigation in better understanding the population dynamics of this species and its possible management policies within the Yarlung Zangbo aquatic ecosystem. Although our study characterized the likely stock status of *O. stewartii* over a wide range of harvesting strategies using the per-recruit model, we were unable to estimate the age-specific abundance and biomass of the *O. stewartii* population. More research work is needed to fill this information gap for *O. stewartii*.

Conclusions

We have, for the first time, quantitatively evaluated the population dynamics of *O. stewartii*. Our results revealed that the stock of *O. stewartii* may be in near full exploitation under the current harvesting strategy, with a high chance of recruitment failure in the future. Although both minimum age limits and seasonal closure could be effective measures to conserve the *O. stewartii* stock, we proposed a seasonal closure to be implemented each year from September to December. Our study represents a first step in better understanding the population dynamics of this species as well as its possible management policies in the Yarlung Zangbo aquatic ecosystem.

Competing interests

The authors declare that they have no competing interests.

Authors' contributions

BH and CXX conceived and designed the study. BH, BSM, YJD, XFY, and HPH performed the experiments and acquired the data. BH analyzed the data and performed the statistical analysis. BH prepared the manuscript. BSM and YJD edited the manuscript. CXX reviewed the manuscript and guaranteed the integrity of the entire study. All authors read and approved the final manuscript.

Acknowledgements

We are grateful to Xin Yang for his assistance and cooperation in carrying out the present study. This study is funded by the special project of the Ministry of Agriculture of the People's Republic of China (201203086).

Author details

[1]College of Fisheries, Huazhong Agricultural University, Wuhan, Hubei 430070, China. [2]Yangtze River Fisheries Research Institute, Chinese Academy of Fishery Sciences, Wuhan, Hubei 430223, China.

References

Arendse CJ, Govender A, Branch GM (2007) Are closed fishing seasons an effective means of increasing reproductive output? A per-recruit simulation using the limpet *Cymbula granatina* as a case history. Fish Res 85:93–100

Beverton RJH, Holt SJ (1957) On the dynamics of exploited fish populations. In: Fishery Investigations Ministry of Agriculture and Food (Great Britain), vol 2., p 533

Cailliet GM, Goldman KJ (2004) Age determination and validation in chondrichthyan fishes. In: Carrier J, Musick JA, Heithaus M (eds) The biology of sharks and their relatives. CRC Press, New York, pp 399–447

Cao WX, Chen YY, Wu YF, Zhu SQ (1981) Origin and evolution of schizothoracine fishes in relation to the upheaval of the Xizang Plateau. In: Tibetan Expedition Team of the Chinese Academy of Science (ed) Studies on the period, amplitude and type of the uplift of the Qinghai-Xizang Plateau. Science Press, Beijing, pp 118–130, in Chinese

Carpenter SR, Kitchell JF (1993) The trophic cascade in lakes. Cambridge University Press, Cambridge

Chen YF, Cao WX (2000) Schizothoracinae. In: Yue PQ (ed) Fauna sinica osteicthtyes cypriniformes III. Scince Press, Beijing, pp 273–388 (In Chinese)

Cowley PD, Brouwer SL, Tilney RL (2002) The role of the Tsitsikamma National Park in the management of four shore-angling fish along the south-eastern Cape Coast of South Africa. S Afr J Mar Sci 24:27–35

Dichmont CM, Butterworth DS, Cochrane KL (2000) Towards adaptive approaches to management of the South African abalone *Haliotis midae* fishery. S Afr J Mar Sci 22:33–42

Gabriel WL, Sissenwine MP, Overholtz WJ (1989) Analysis of spawning stock biomass per recruit: an example for Georges Bank haddock. N Am J Fish Manage 9:383–391

Goodyear CP (1993) Spawning stock biomass per recruit in fisheries management: foundation and current use. In: Smith SJ, Hunt JJ, Rivard D (eds) Risk evaluation and biological reference points for fisheries management, vol 120, Can Spec Pub Fish Aquat Sci., pp 67–81

Govender A, Al-Oufi H, McIlwain JL, Claereboudt MC (2006) A per-recruit assessment of the kingfish (*Scomberomorus commerson*) resource of Oman with an evaluation of the effectiveness of some management regulations. Fish Res 77:239–247

Griffiths MH (1997) The application of per-recruit models to *Argyrosomus inodorus*, an important South African sciaenid fish. Fish Res 30:103–115

Griffiths MH (2000) Long-term trends in catch and effort of commercial linefish off South Africa's Cape province: snapshots of the 20th century. S Afr J Mar Sci 22:81–110

Hauck M, Sweijd MA (1999) A case study of abalone poaching in South Africa and its impact on fisheries management. ICES J Mar Sci 56:1024–1032
He DK, Chen YY (2006) Biogeography and molecular phylogeny of the genus *Schizothorax* (teleostei: Cyprinidae) in China inferred from cytochrome *b* sequence. J Biogeogr 33:1448–1460
He DK, Chen YY (2007) Molecular phylogeny and biogeography of the highly specialized schizothoracine fishes. Chin Sci Bull 52:303–312 (in Chinese)
Hildén M (1993) Reference points for fisheries management: the ICES experience. Can Spec Publ Fish Aquat Sci 120:59–66
Huo B, Xie CX, Ma BS, Yang XF, Huang HP (2012) Age and growth of *Oxygymnocypris stewartii* in the Yarlung Zangbo River, Tibet, China. Zool Stud 51:185–194
Huo B, Xie CX, Ma BS, Yang XF, Huang HP (2013) Reproductive biology of *Oxygymnocypris stewartii* in the Yarlung Zangbo River in Tibet, China. Environ Biol Fish 96:481–493
Huo B, Xie CX, Madenjian CP, Ma BS, Yang XF, Huang HP (2014) Feeding habits of an endemic fish, *Oxygymnocypris stewartii*, in the Yarlung Zangbo River in Tibet, China. Environ Biol Fish 97:1279–1293
Hutchings K, Lamberth SJ (2002) Catch-and-effort estimates for the Gillnet and Beach-seine fisheries in the Western Cape, South Africa. S Afr J Mar Sci 24:205–225
Kirchner CH (2001) Fisheries regulations based on yield-per-recruit analysis for the linefish silver kob *Argyrosomus inodorus* in Namibian waters. Fish Res 52:155–167
Leaman BM (1993) Reference points for fisheries management: the western Canadian experience. Can Spec Publ Fish Aquat Sci 120:15–30
Mace PM (1994) Relationships between common biological reference points used as thresholds and target of fisheries management strategies. Can J Fish Aquat Sci 51:110–122
Mace PM, Sissenwine MP (1993) How much spawning per recruit is enough? Can Spec Publ Fish Aquat Sci 120:101–118
McQueen DJ, Johannes MRS, Post JR, Stewart TJ, Lean DRS (1989) Bottom-up and top-down impacts on freshwater pelagic community structure. Ecol Monogr 59:289–309
Myers RA, Worm B (2003) Rapid worldwide depletion of predatory fish communities. Nature 423:280283
Ng HH (2010) *Oxygymnocypris stewartii*. IUCN Publishing Red List of Threatened Species., http://www.iucnredlist.org/details/168408/0. Accessed Apr 2010
Pauly D, Christensen V, Guénette S, Pitcher TJ, Sumaila UR, Walters CJ, Watson R, Zeller D (2002) Towards sustainability in world fisheries. Nature 418:689–695
Pradervand P, Baird D (2002) Assessment of the recreational linefishery in selected Eastern Cape estuaries: trends in catches and effort. S Afr J Mar Sci 24:87–101
Quinn TJ II, Deriso RB (1999) Quantitative fish dynamics. Oxford University Press, New York, pp 1–542
Ricker WE (1975) Computation and interpretation of biological statistics of fish populations. Bull Fish Res Bd Can 191:1–333
Rivard D, Maguire JJ (1993) Reference points for fisheries management: the eastern Canadian experience. Can Spec Publ Fish Aquat Sci 120:31–58
Schindler DE, Carpenter SR, Cole JJ, Kitchell JF, Pace ML (1997) Influence of food web structure on carbon exchange between lakes and the atmosphere. Sci 277:248–251
Smith MW, Then AY, Wor C, Ralph G, Pollock KH, Hoenig JM (2012) Recommendations for catch-curve analysis. N Am J Fish Manag 32:956–967
Sparre P, Venema SC (1998) Introduction to tropical fish stock assessment. Part 1. Manual. FAO fisheries technical paper. No. 306. 1, Rev. 2. FAO, Rome
Sun CL, Wang SP, Porch CE, Yeh SZ (2005) Sex-specific yield per recruit and spawning stock biomass per recruit for the swordfish, *Xiphias gladius*, in the waters around Taiwan. Fish Res 71:61–69
Taylor CC (1958) Cod growth and temperature. J Cons Int Explor Mer 23:366–370
Then AY, Hoenig JM, Hall NG, Hewitt DA (2014) Evaluating the predictive performance of empirical estimators of natural mortality rate using information on over 200 fish species. ICES J Mar Sci 72(1):1–6
Turpie JK, Beckley LE, Katua SM (2000) Biogeography and the selection of priority areas for conservation of South African coastal fishes. Biol Conserv 92:59–72
Wu YF, Tan QJ (1991) Characteristics of the fish-fauna of the characteristics of Qinghai-Xizang plateau and its geological distribution and formation. Acta Zool Sin 37:135–152
Wu YF, Wu CZ (1992) The fishes of the Qinghai-Xizang Plateau. Sichuan Publishing House of Science & Technology, Chengdu (In Chinese)
Ye Y (1998) Assessing effects of closed seasons in tropical and subtropical penaeid shrimp fisheries using a length-based yield-per-recruit model. ICES J Mar Sci 55:1112–1124

15

Seasonal dynamics of the copepod community in a tropical monsoonal estuary and the role of sex ratio in their abundance pattern

G. Vineetha*, N. V. Madhu, K. K. Kusum and P. M. Sooria

Abstract

Background: As salinity is considered the prime "ecological master factor" governing the zooplankton distribution and abundance in estuaries, a spatio-temporal interactive approach was followed to assess whether the responses of the estuarine copepod community towards the salinity are always direct with a strong positive affinity or whether there exist any complexities in their interrelationship. The study, also for the first time, addressed the role of sex ratio in governing the abundance and the population structure of copepods in the tropical monsoonal estuarine system.

Results: The ecological scenario in the Cochin estuary revealed that irrespective of the season, higher zooplankton abundance occurred in the mesohaline zone (MSZ; salinity 5–18) of the estuary, despite the pronounced spatial shift of the MSZ from the lower reaches of the estuary to upstream locations, in conjunction with the varying seasonal fluvial influx and marine water intrusion. In the case of the sex ratio of copepods, the seasonal scenario revealed that the dominant copepods had a lower sex ratio during the period of higher abundance, and a negative relation was observed between the abundance and the sex ratio of copepod species during most of the seasons.

Conclusions: The preponderance of the mesohaline and euryhaline species of the copepod community formed the key contributing factor towards the maintenance of higher abundance in the MSZ of the estuary. The bias in the sex ratio towards females favored higher reproductive output, resulting in the higher abundance during the favorable pre-monsoon period.

Keywords: Zooplankton; Monsoonal estuary; Mesohaline; Copepod; Sex ratio

Background

Estuaries, the confluence of marine and limnetic ecosystems, are considered one of the most productive natural habitats on earth (Alongi 1998). Though representing only 6 % of the total coastal zone area, these aquatic ecosystems play a pivotal role in the global carbon balance (Smith and Hollibaugh 1993) and nutrient recycling (Fisher et al. 1982) and serve as a conducive breeding and nursing environment for many commercially important shell and fin fishes (Haedrich 1983).The high physicochemical variability associated with estuarine ecosystems often affects the abundance and distribution of the estuarine organisms immensely, leading to prominent spatial and temporal heterogeneity in their abundance and community structure (Collins and Williams 1981; Madhupratap 1987; Laprise and Dodson 1994). Of the multitude of environmental variables, salinity is often considered the prime "ecological master" (Kinne 1966), evoking pronounced spatio-temporal variability in the estuarine zooplankton dynamics, as salinity experiences wide gradients in the spatial distribution, with high-salinity zones in the inlet locations to limnetic environments in the upstream regions, which in turn undergo prominent temporal variability associated with seasonal fluvial influxes. Hence, the interrelation between salinity and zooplankton and also the mode of salinity regulation on zooplankton dynamics have been central aims for many plankton ecological studies in estuaries worldwide (Baretta and Malschaert 1988; Kimmerer 2002).

* Correspondence: vineetha.varada@gmail.com
CSIR-National Institute of Oceanography, Regional Centre, Dr. Salim Ali Road, Kochi, India

The Cochin estuary (CE), heavily influenced by the Indian summer monsoon, differs from all other estuaries by its unsteadiness in salinity characteristics and also in its total runoff, which is many times higher than the estuarine volume (Vijith et al. 2009; Revichandran et al. 2012) (June–September), resulting in a possible washout during the monsoon period. This uniqueness in the estuarine features of the CE thus demands a comprehensive study for identifying the regulatory mechanisms underlying the seasonal variability in the abundance, distribution, and population maintenance of the endemic copepoda community of this estuarine ecosystem. In the CE, as salinity is taken as the major limiting factor influencing the zooplankton standing stock and abundance (Madhupratap 1987; Madhu et al. 2007), we aimed to evaluate whether the responses of the estuarine copepods are always direct with a strong affinity towards salinity or whether the estuarine copepods exhibit wide complexities in their responses to the variability in salinity characteristics.

As copepods form the predominant zooplankton taxon in terms of both abundance and diversity and since they form a critical component of the estuarine trophic ecology (Baretta and Malschaert 1988; Soetaert and Rijswijk 1993; Froneman 2004), the variability in their population dynamics, in conjunction with the prevailing environmental conditions, is very much decisive in the better understanding of the seasonal dynamics of the zooplankton community in the estuarine ecosystem. Copepods, being dioecious organisms with sexually dimorphic males and females, each exhibiting prominent sex-specific differences in their physiological longevity (Rodríguez-Graña et al. 2010), growth, and mortality rates (Ohman and Hsieh 2008), the responses of individuals belonging to varied sexes to environmental variability are expected to play a crucial role in the population dynamics of the estuarine copepod community. Though there are ample studies regarding biasness from the Fischerian 1:1 sex ratio in the natural population of various copepod species (Moraitou-Apostolopaulou 1972; Hirst et al. 2010) and also on the factors responsible for the biasness in sex ratio (Gusmão et al. 2013), the evaluation of the influence that sex ratio has on the abundance pattern and population dynamics of the estuarine copepod community is mostly overlooked. In this context, we aimed to study the variability in sex ratio of the dominant species among the copepod community in the CE and to assess the determining role that sex ratio has on their seasonal community structure and population maintenance in a tropical monsoonal estuarine system.

Methods

Study area

CE is an oxbow-shaped monsoonal estuarine system located along the southwest coast of India (Gupta et al. 2009) (Fig. 1). This microtidal estuary (~1 m) (Qasim 2003), with a surface area of 231 km^2, makes perennial connections to the Arabian Sea through two inlets: one at Cochin (width 450 m) and the other at Azhikode (width 250 m). Heavily influenced by the Indian summer monsoon, the annual precipitation of the CE is around 320 cm, of which around 75 % occurs during the summer monsoon period (June–September) (Qasim 2003). The estuary receives an influx of about 22,000 × 10^6 m^3 of freshets annually (Revichandran et al. 2012) from seven perennial rivers—Periyar, Chalakudy, Pamba, Muvattupuzha, Manimala, Meenachil, and Achancoil—and their tributaries. The seasonal asymmetry in the balance between the fluvial runoff associated with the Indian summer monsoon and the tidal incursion from the Arabian Sea often evokes intense heterogeneity in the spatio-temporal characteristics of the CE.

Sampling

Monthly sampling was carried out at eight locations along the salinity gradient, covering the entire CE (Fig. 1), from June 2008 to May 2009. Stations 1 and 2 were the two inlets (Azhikode and Cochin, respectively) at the lower reaches of the estuary. Stations 3, 4, and 5 were the near-inlet locations to the middle reaches of the estuary, and stations 6, 7, and 8 were in the southern upper reaches of the estuary. An SBE Seabird 19 plus CTD was deployed at every station location to measure the surface water temperature and salinity. Water samples were collected from the surface (0.5 m) using Niskin samplers for the estimation of salinity, pH, dissolved oxygen (DO), suspended particulate matter (SPM), and chlorophyll *a*. Surface salinity values obtained through the CTD were further counterchecked by using a digital salinometer for better precision of the data (DIGI-AUTO, 3G Tsurumi Seiki, Japan, accuracy ±0.001), and pH was estimated using a pH meter (ELICO LI610, accuracy ±0.01). The estimation of DO was carried out following Winkler's method (Grasshoff 1983). SPM was determined by filtering 250 ml of water sample onto pre-weighed Millipore filter paper (0.45-μm pore size) and drying the residue at 80 °C. The weight of the particles collected on the filter paper was estimated gravimetrically following the standard procedure (APHA 2005). Chlorophyll *a* was measured using a flourometer (Trilogy, Turner Designs, USA) after filtering 500 ml water on GF/F filter papers (pore size 0.7 μm), followed by overnight extraction in 10 ml 90 % acetone (Strickland and Parsons 1972). For phytoplankton species identification, 500 ml of water sample from the surface (0.5 m) was fixed in 1 % Lugol's iodine and enumerated under an inverted microscope (OLYMPUS, CK-30, magnification ×100) in a Sedgewick–Rafter counting chamber following standard protocols (Tomas 1997). Zooplankton

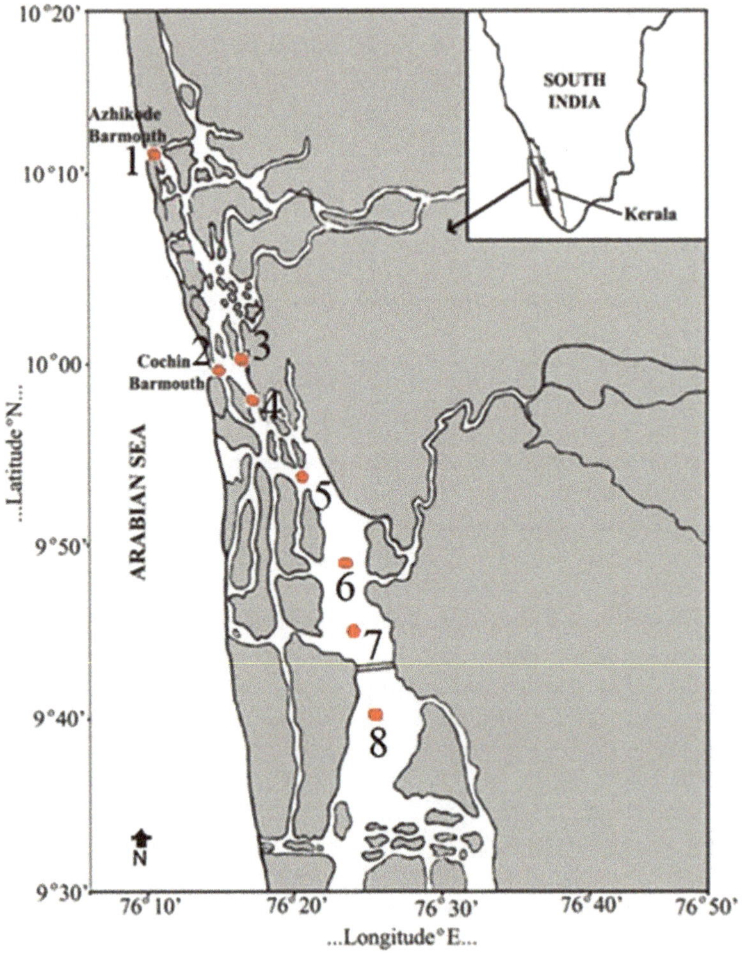

Fig. 1 Sampling locations in Cochin estuary

samples were collected using a Working Party net-2 (mesh size 200 μm) with a mouth area of 0.28 m^2. The net was hauled horizontally just below the surface for 10 min with an approximate speed of 1 knot to minimize net avoidance by the larger zooplankton. A digital flow meter (Hydro Bios, Model-438110) was attached across the mouth of the net to estimate the amount of water filtered. Zooplankton biomass was estimated by the displacement volume method after removing large detrital particles and was preserved in 4 % formaldehyde solution (Harris et al. 2000). Zooplankton groups were sorted, counted, and identified to the lowest possible taxon, and the abundance was expressed as individuals per cubic meter (ind. m^{-3}). Copepod species level identification was carried out following the identification keys of Sewell (1999), Kasturirangan (1963), and Wellershaus (1969). For calculating the sex ratio (male:female) of the dominant copepod species, the identification of their sex was carried out mostly on the C5 and adult stages, since sexes in copepod can only be morphologically distinguished in the late stages of development (Gusmão et al. 2013).

Data analysis

To check whether a significant variation exists between the seasonal values of both abiotic and biotic parameters, one-way analysis of variance (ANOVA) was performed. Before the analysis, the D'Agostino and Pearson omnibus normality test was carried out to check their normality in distribution, and based on the result, parametric or non-parametric ANOVA was performed for the variables with two-tailed *P* values and 95 % confidence intervals.

The dominant species of both the phytoplankton and copepod communities are expected to play a vital role in the ecological process of the estuarine system because of their higher numerical abundance. Hence, for identifying the dominant species in the estuarine system, along with their abundance, the spatial frequency of occurrence was also considered and calculated according to Yang et al. (1999) as:

$$Y_i = (N_i/N) \times f_i$$

where Y_i is the dominance of species i
N_i is the number of individuals of species
N is the number of individuals of all species at all locations
f_i is the frequency of locations at which species i occur.

The species with a Y value of ≥0.02 were considered the dominant species (Yang et al. 1999; Lee et al. 2009).

While assessing the seasonal dynamics of the copepod community, the characterizing species of each season were identified following the similarity percentage (SIMPER) routine implemented in PRIMER (Clarke and Gorley 2006) to get the best-suited representative of a particular season. These species deal with good intragroup similarity and less standard deviation. The characterizing species among the phytoplankton for each season were also identified following the same procedure.

Redundancy analysis (RDA) was performed on the physicochemical variables and chlorophyll *a* and for both the characterizing and dominant species of copepods during each season for a better elucidation of the relationship existing between them (Van den Wollenberg 1977). The advantage of RDA is that it allows one to obtain a simultaneous representation of the observations of the response variables (species) and the explanatory variables (physicochemical variables and chlorophyll *a*) in two or three dimensions, and the position of the variables in the triplot helps in visualizing their interrelationships. Further, for a detailed understanding of the interrelation existing among the phytoplankton and the copepod community of the CE, RDA was carried out between the characterizing and dominant species along different seasons using the statistical software, CANOCO 4.5.

To check the similarity in the distribution of copepod species, agglomerative hierarchical cluster analysis was performed based on the Bray-Curtis similarity during each season, and a similarity profile (SIMPROF) test was done further to identify the significantly correlated species. Further, to find out the similarity between station locations, non-metric multidimensional scaling (NMDS) was performed based on the similarity in the distribution of different copepod species, using PRIMER (Clarke and Gorley 2006).

To analyze the seasonal variability in the assemblages of the copepod community, the diversity indices, such as species diversity (H') and species evenness (J'), were calculated using PRIMER (Clarke and Warwick 2001).To identify the interrelation between the sex ratio and the abundance of the dominant copepod species, Pearson's correlation analysis was performed followed by a linear regression analysis based on the monthly data of the sex ratio and abundance during each season. Considering the wide variations in the abundance, $\log_e$-transformed abundance value was used for the analysis. As the dominant copepod species were not observed at few station locations during some months, the number of samples (n) varied along seasons. To check whether the regression lines (slopes) are significantly different from zero, a one-sample t test was also carried out. In estuaries, along with the sex ratio, as the freshwater gradient along stations also may influence the zooplankton distribution and abundance, the effect of the freshwater content (FWC) on distribution of the dominant species was also assessed by plotting it against the residuals extracted from the regression plot of the sex ratio and abundance. The amount of FWC along each location was calculated as follows:

$$\text{FWC at station } i\ (\%) = \frac{\text{Salinity of the open ocean} - \text{Salinity of the station } i}{\text{Salinity of the open ocean}} \times 100$$

Results

Physicochemical variables

Among the environmental variables, the spatio-temporal distribution in the water temperature did not show much variation during the sampling periods (Fig. 2a), whereas salinity exhibited marked seasonal variation in the estuary ($P < 0.05$) (Fig. 2b). Seasonal averages showed relatively higher salinity during the pre-monsoon (average (av.) 19.2 ± 6.8) when compared to the monsoon (av. 5.4 ± 9.7) and post-monsoon seasons (av. 16.1 ± 11.8). During the pre-monsoon season, salinity was relatively high in all station locations, and a mesohaline zone with salinity 5–18 was observed in the stations located at the upper reaches of the estuary (stations 6, 7, and 8) (Fig. 2b). With the onset of the monsoon season, a sudden drop in salinity was observed throughout the estuary, and it was <1 at the stations in the southern upper reaches of the estuary (stations 6, 7, and 8). During this period, relatively higher salinity was observed, mainly in the inlet stations, among which the maximum value was recorded at station 1 (Fig. 2b). With the post-monsoon season, salinity progressively increased in the estuary, and the mesohaline zone prevailed at the middle reaches of the estuary, mostly at station 5. Stations 1–4 exhibited relatively higher salinity, whereas a low-salinity environment (salinity <5) prevailed at the upstream locations of the estuarine region (stations 6, 7, and 8). Though the dissolved oxygen (DO) showed higher values during the monsoon season compared to both the pre- and post-

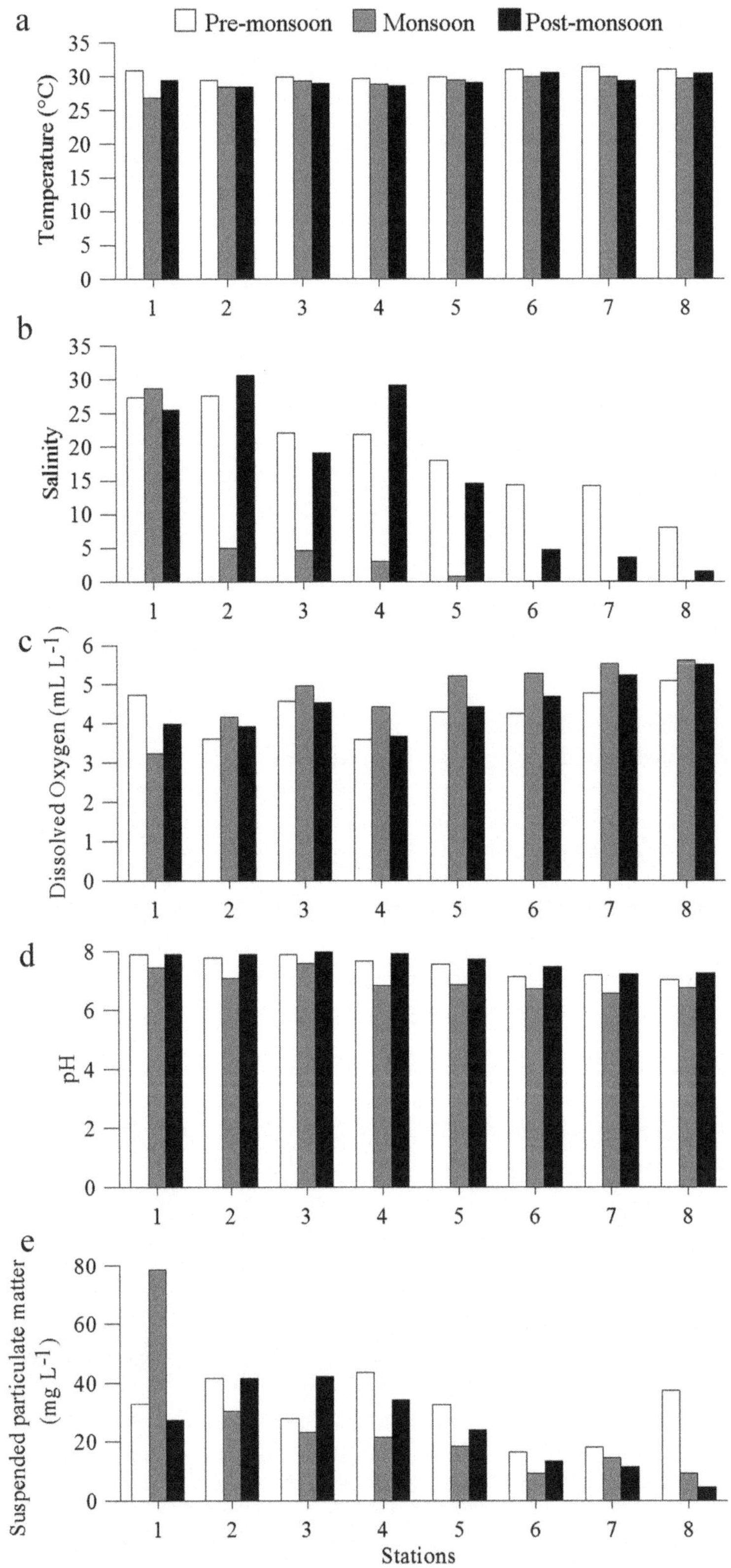

Fig. 2 Distribution of the physicochemical variables **a** temperature, **b** salinity, **c** dissolved oxygen, **d** pH, and **e** suspended particulate matter along the sampling locations in each season

monsoon periods at all station locations except station 1 (Fig. 2c), the difference was not statistically significant ($P > 0.05$). The pH values at all station locations were relatively lower during the monsoon period compared to the other two seasons (Fig. 2d) and exhibited a prominent seasonal variation ($P < 0.05$), which was opposite to the trend observed in the DO distribution. Though SPM varied seasonally throughout the station locations (Fig. 2e), the difference was statistically insignificant ($P > 0.05$).

Chlorophyll *a* and phytoplankton composition

The seasonal scenario in the chlorophyll *a* distribution during the study period revealed relatively higher values during the pre-monsoon (5.6–31.2 mg m^{-3}, av. 10.9 ± 8.4 mg $^{-3}$) compared to the monsoon (4.8–17.6 mg m^{-3}, av. 8.7 ± 4.1 mg m^{-3}) and post-monsoon seasons (2.7–16.8 mg m^{-3}, av. 6.9 ± 4.5 mg m^{-3}) (Fig. 3a), but this difference was statistically insignificant ($P > 0.05$). As the phytoplankton community composition and their relative contribution exhibited both spatial and temporal variabilities, to assess

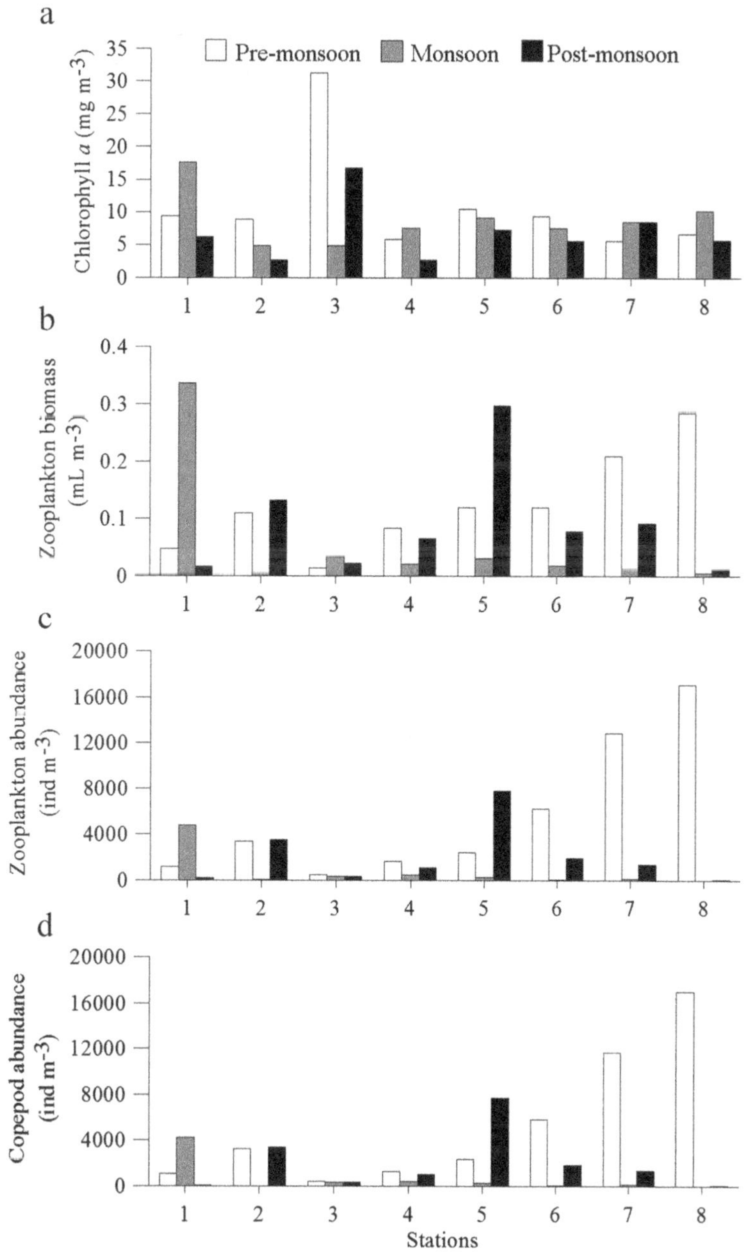

Fig. 3 Distribution of the biotic parameters **a** chlorophyll *a*, **b** zooplankton biomass, **c** zooplankton abundance, and **d** copepod abundance along the sampling locations in each season

their influence on the copepod species, the dominant and characterizing species of phytoplankton were identified (for details, see Additional file 1). In general, diatoms and dinoflagellates contributed a relatively major share among the phytoplankton community during each season.

Zooplankton biomass, abundance, and composition

Pronounced spatio-temporal variation in the zooplankton biomass and abundance was observed in the estuary during the study period (Fig. 3b, c). Both biomass and abundance were higher during the pre-monsoon when compared to the monsoon and post-monsoon periods (Fig. 3b, c), and the seasonal variation was significantly different ($P < 0.05$). During the pre-monsoon season, relatively higher biomass and abundance were observed at the mesohaline zone located in the upper reaches of the estuary (stations 6, 7, and 8) than in other locations. During the monsoon season, a sharp decline in zooplankton biomass and abundance (<0.1 ml m^{-3}) occurred in all stations throughout the estuary except at station 1 (0.34 ml m^{-3}). With the post-monsoon season, a slight increase in the zooplankton biomass and abundance was observed in most of the stations, and the maximum value was observed at the mesohaline location at station 5 (0.30 ml m^{-3}) (Fig. 3b, c).

The zooplankton community of the CE was represented by 16 groups of holoplankton and seven groups of meroplankton (Table 1). Irrespective of seasons, copepoda formed the most dominant zooplankton taxon, and though there was a significant variation in their seasonal abundance ($P < 0.05$), similar to zooplankton distribution, relatively higher abundance of copepoda was observed in the mesohaline zones of the estuary (Fig. 3d). During the pre-monsoon season, copepoda contributed to 94.9 % of the total population, and with the onset of the monsoon season and a subsequent drop in the salinity, the holoplankton composition changed drastically throughout the estuary. Though copepods formed the dominant representative, during the monsoon season, their percentage contribution (87.8 %) decreased considerably, and cladocera (9.8 %) formed the next abundant group. Though the holoplanktonic taxa, like mysida and amphipoda, were present throughout the seasons, their abundance varied greatly between the seasons. The holoplanktonic groups, like thaliacea and pteropoda, with comparatively high-salinity preferences, were observed only during the pre-monsoon and post-monsoon periods, respectively (Table 1). Among the meroplankton, decapod larvae, gastropod larvae, and lamellibranch larvae dominated, with relatively higher abundance during the pre-monsoon period (Table 1).

Table 1 Zooplankton composition and abundance (ind. m^{-3}) during each season

Group	PRM	MS	PM
Holoplankton			
Hydromedusae	5	0.034	0.004
Siphonophore	0.001	–	0.1
Ctenophore	0.018	0.474	–
Polycheata	0.084	0.031	0.025
Chaetognatha	2	7	4
Copepoda	5369	673	2152
Ostracoda	–	0.004	–
Cladocera	0.039	75	0.287
Mysid	–	0.014	0.0157
Isopoda	0.001	0.002	–
Amphipoda	0.157	0.054	0.528
Stomatopoda	0.001	–	0.005
Luciferiidae	15	2	1
Thaliacea	–	0.011	–
Appendicularia	–	0.069	0.14
Pteropoda	–	–	0.003
Copepoda nauplii	13	–	12
Meroplankton			
Polychaete larvae	0.015	0.132	–
Decapod larvae	57	3.5	6
Cirriped larvae	2	–	0.112
Gastropod larvae	162	0.438	6
Lamellibranch larvae	24	0.648	1.469
Fish eggs	5	0.186	7.5
Fish larvae	1	4	1

PRM pre-monsoon, *MS* monsoon, *PM* post-monsoon

Copepod composition and abundance

The copepod abundance during the pre-monsoon period (av. 5369 ± 5954 ind. m^{-3}) was markedly higher than in the monsoon (av. 673 ± 1433 ind. m^{-3}) and post-monsoon periods (av. 2152 ± 2566 ind. m^{-3}) (Fig. 3d). Copepod, the dominant taxon among the zooplankton community in the CE, was represented by 31 species of calanoids, 3 species of cyclopoids, 2 species of poecilostomatoids, and 1 species of harpacticoid (Table 2). Among the calanoids, the family Acartiidae and Paracalanidae dominated throughout the study period, of which the most abundant species were *Bestiolina similis*, *Acartia plumosa*, *Acartiella gravelyi*, *Acartiella keralensis*, and *Acartia centrura*. High-saline copepods, like *Labidocera pectinata*, *Acartia erythraea*, *Acartia spinicauda*, and *Subeucalanus subcrassus*, were observed in varying densities, mostly in or near the two inlets. *Oithona brevicornis*, belonging to the family Oithonidae, showed higher abundance among the cyclopoids. The

Table 2 List of the copepod species along different seasons in Cochin estuary

Copepoda species	Pre-monsoon	Monsoon	Post-monsoon
Acartia centrura Giesbrecht, 1889	****	***	***
Acartia spinicauda Giesbrecht, 1889	***	**	***
Acartia erythraea Giesbrecht, 1889	*	–	*
Acartia plumosa Scott T., 1894	*****	***	*****
Acartia bowmani Abraham, 1976	**	**	–
Acartiella gravelyi Sewell, 1919	***	***	**
Acartiella keralensis Wellershaus 1969	****	**	***
Paracalanus aculeatus Giesbrecht, 1888	***	–	***
Bestiolina similis Sewell, 1914	*****	****	****
Acrocalanus monachus Giesbrecht, 1888	–	–	**
Pseudodiaptomus annandalei Sewell, 1919	***	**	***
Pseudodiaptomus serricaudatus Scott T., 1894	**	*	***
Pseudodiaptomus malayalus Wellershaus 1969	*	**	***
Pseudodiaptomus mertoni Früchtl, 1924	–	**	–
Heliodiaptomus cinctus Gurney, 1907	–	**	–
Allodiaptomus mirabilipes Kiefer, 1936	–	**	–
Subeucalanus subcrassus Giesbrecht, 1888	***	*	***
Subeucalanus crassus Giesbrecht, 1888	*	**	**
Undinula vulgaris Dana, 1849	**	*	***
Canthocalanus pauper Giesbrecht, 1888	*	*	**
Centropages alcocki Sewell, 1912	**	***	***
Centropages furcatus Dana, 1849	***	*	***
Centropages tenuiremis Thompson & Scott, 1903	**	–	–
Centropages trispinosus Sewell, 1914	**	***	**
Centropages calaninus Dana, 1849	**	*	**
Centropages orsiini Giesbrecht, 1889	–	–	**
Candacia bradyi Scott A., 1902	–	–	**
Labidocera pectinata Thompson & Scott, 1903	**	–	**
Labidocera acuta Dana, 1849	**	–	–
Temora turbinata Dana, 1849	–	–	**
Tortanus forcipatus Giesbrecht, 1889	–	–	*
Oithona brevicornis Giesbrecht, 1891	***	***	***
Oithona rigida Giesbrecht, 1896	–	*	–
Oncaea venusta Philippi, 1843	*	–	*
Mesocyclops sps	*	**	**
Euterpina acutifrons Dana, 1847	*	–	**
Onychocorycaeus catus Dahl, 1894	*	–	**

*indicates abundance <1 ind. m^{-3}; **indicates 1–10 ind. m^{-3}; ***indicates 10–100 ind. m^{-3}; ****indicates 100–1000 ind. m^{-3}; *****indicates >1000 ind. m^{-3}

poecilostomatoids, *Oncaea venusta* and *Onychocorycaeus catus*, and the harpacticoid, *Euterpina acutifrons*, were also observed in varying densities during the pre- and post-monsoon periods (Table 2).

During the pre-monsoon season, a total of 28 species were observed throughout the estuary, and the higher copepod abundance at the mesohaline zone at stations 6, 7, and 8 was mostly contributed by the medium-saline (salinity preference ~10–20) *A. plumosa*, *Ac. keralensis*, and euryhaline *B. similis*. Based on the dominancy analysis, these three were found as the dominant species of this period. The SIMPER analysis revealed a total of 8 species as the characterizing species of the pre-monsoon period, which are listed in Table 3. With the onset of the

Table 3 Result of the SIMPER analysis exhibiting the characteristic species of each season

Species	Av. Abund	Av. Sim	Sim/SD	Contrib %	Cum. %
Pre-monsoon					
Bestiolina similis	5.85	16.66	4.25	36.59	36.59
Acartia plumosa	4.61	11.17	1.52	24.53	61.11
Acartiella keralensis	2.94	3.76	0.48	8.26	69.37
Acartia centrura	1.87	3.14	0.71	6.91	76.28
Oithona brevicornis	1.54	2.86	0.7	6.29	82.56
Pseudodiaptomus annandalei	1.07	1.54	0.67	3.38	85.94
Labidocera pectinata	0.99	1.45	0.71	3.19	89.13
Centropages alcocki	0.88	1.01	0.5	2.22	91.35
Monsoon					
Acartia plumosa	2.2	7.92	2.93	16.44	16.44
Bestiolina similis	2.39	7.43	3.29	15.43	31.87
Acartiella gravelyi	1.66	6.78	2.01	14.07	45.94
Mesocyclops sps	1.08	4.92	1.44	10.22	56.16
Allodiaptomus mirabilipes	1.09	4.19	1.41	8.69	64.85
Oithona brevicornis	1.45	3.74	2.47	7.76	72.61
Pseudodiaptomus malayalus	0.9	2.98	0.94	6.18	78.79
Heliodiaptomus cinctus	0.99	2.85	1.26	5.91	84.71
Pseudodiaptomus annandalei	1.01	2.6	0.92	5.41	90.11
Post-monsoon					
Bestiolina similis	4.23	10.2	3.99	24.1	24.1
Acartia plumosa	3.48	6.45	1.42	15.24	39.34
Oithona brevicornis	2.37	5.01	1.63	11.85	51.19
Acartia centrura	1.95	2.98	1.02	7.03	58.23
Pseudodiaptomus annandalei	1.26	2.49	1.5	5.89	64.12
Acartiella keralensis	1.45	1.86	0.58	4.39	68.51
Pseudodiaptomus serricaudatus	1.3	1.78	1.02	4.21	72.72
Centropages alcocki	1.17	1.23	0.71	2.9	75.62
Pseudodiaptomus malayalus	1.07	1.12	0.5	2.66	78.28
Centropages furcatus	1.3	1.11	0.51	2.61	80.89
Subeucalanus subcrassus	1.07	0.95	0.51	2.25	83.14
Centropages calaninus	0.75	0.95	0.51	2.25	85.38
Acartiella gravelyi	0.83	0.94	0.46	2.23	87.61
Undinula vulgaris	1.09	0.91	0.51	2.15	89.76
Centropages trispinosus	0.95	0.84	0.51	1.99	91.75

The analysis was performed based on the fourth root transformed abundance data of copepod species. As for a good charactering species, the average similarity for a particular taxa within a group should be greater than its intragroup standard deviation (SD) by a factor of at least one, they were marked with *bold letters Av* average, *Abund* abundance, *Sim* similarity, *Contrib* contribution, *Cum.* cumulative

monsoon season, copepod diversity dropped throughout the estuary, and the community was represented by about 24 species of copepods. During this period, maximum density (4198 ind. m^{-3}) and species diversity were observed at station 1, the Azhikode inlet. *A. plumosa*, *A. centrura*, and *B. similis* and the low-saline (salinity preference <10) *Ac. gravelyi* were found as the dominant species during this period. Another low-saline species, *Heliodiaptomus cinctus* (av. 5 ind. m^{-3}), was also observed but with varying density. A total of 9 species were found as the characterizing species, and among them, 4 species were observed to be common during both the pre-monsoon and monsoon periods (Table 3). During the post-monsoon season, species diversity among copepods was high and was represented by a total of 30 species and was dominated by *A. plumosa*, *B. similis*, and *A.*

centrura. The maximum abundance of copepods was observed at the mesohaline zone at station 5 (7700 ind. m^{-3}). Among the 15 species found as the characterizing species during this period, *A. plumosa*, *B. similis*, *Pseudodiaptomus annandalei*, and *O. brevicornis*, were found as common during all three seasons (Table 3).

Multivariate analysis

The RDA triplot helped in unfolding the preferred abiotic environment for both the characterizing and dominant species of the copepods in each season and also indicated the influence of chlorophyll *a* on them (Fig. 4). During the pre-monsoon season, when a high-salinity environment prevailed in a major part of the estuary, the dominant species exhibited a negative relation with salinity and a positive relation with temperature and DO, whereas several characterizing species showed close affinity to SPM but had no proper trend with salinity (Fig. 4a). During the low-salinity period of the monsoon season, few dominant and characterizing species showed positive affinity with salinity, SPM, and pH (Fig. 4b). During the post-monsoon period, the dominant species showed neither positive nor negative affinity with salinity (Fig. 4c). During most of the study period (except monsoon), the majority of the dominant and characterizing species did not show any definite relationship with chlorophyll *a* (Fig. 4). Hence, a detailed study on the interrelation among the species of the copepods and phytoplankton (both characterizing and dominant species) was performed for each season, and the triplot in the RDA was helpful in both visualizing all the data points plotted in the coordinate system and identifying the interrelationship among them (Fig. 5). Irrespective of season, *A. plumosa* and *B. similis*, which were dominant throughout the study period, showed a preference for the autotrophic dinoflagellate species

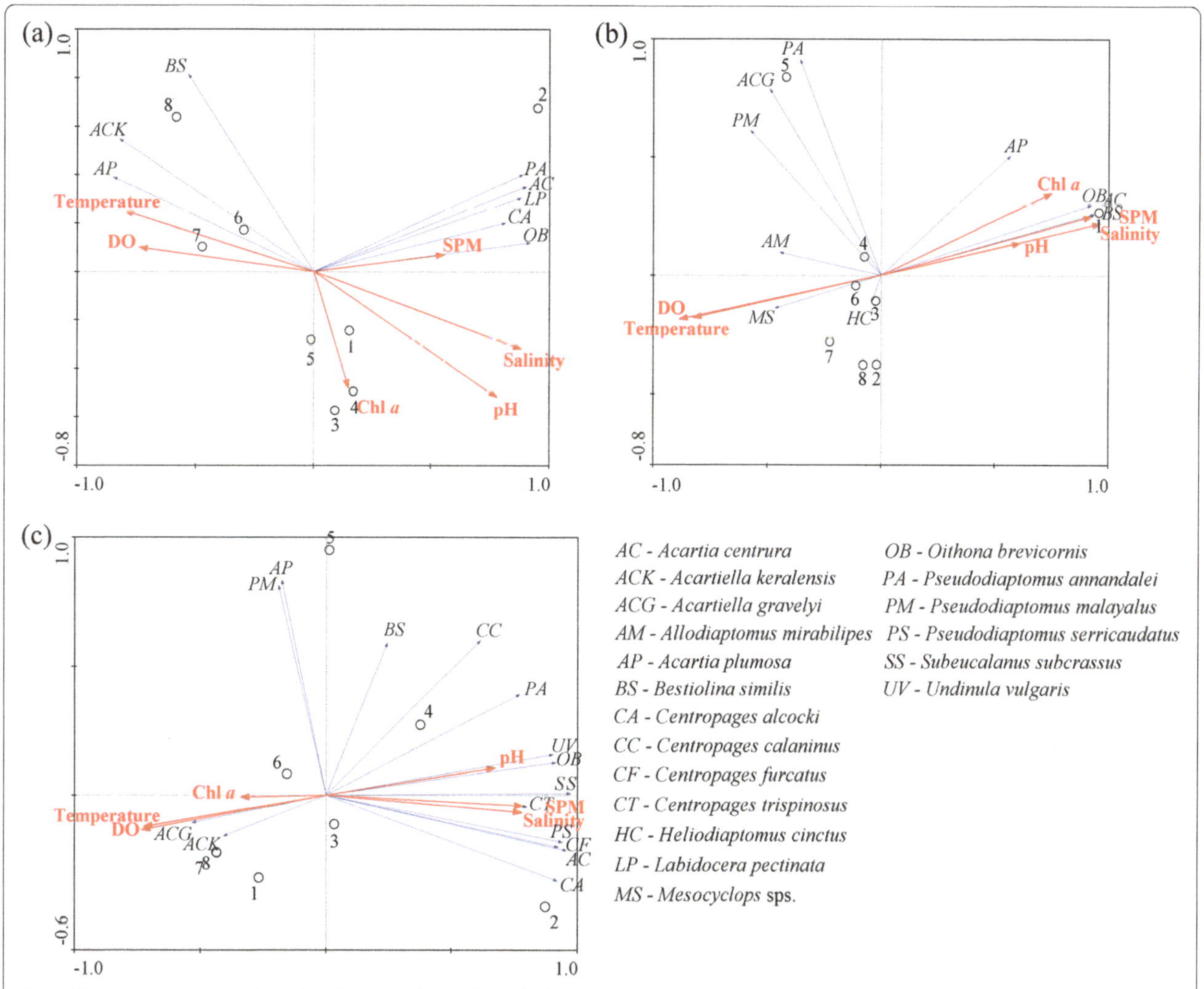

Fig. 4 The redundancy analysis triplot displaying the ecological relationship between physicochemical variables, chlorophyll *a*, and dominant and characterizing copepod species during the **a** pre-monsoon, **b** monsoon, and **c** post-monsoon periods. The direction of an *arrow* dictates the direction of increase of that particular variable on spatially assembled sampling locations

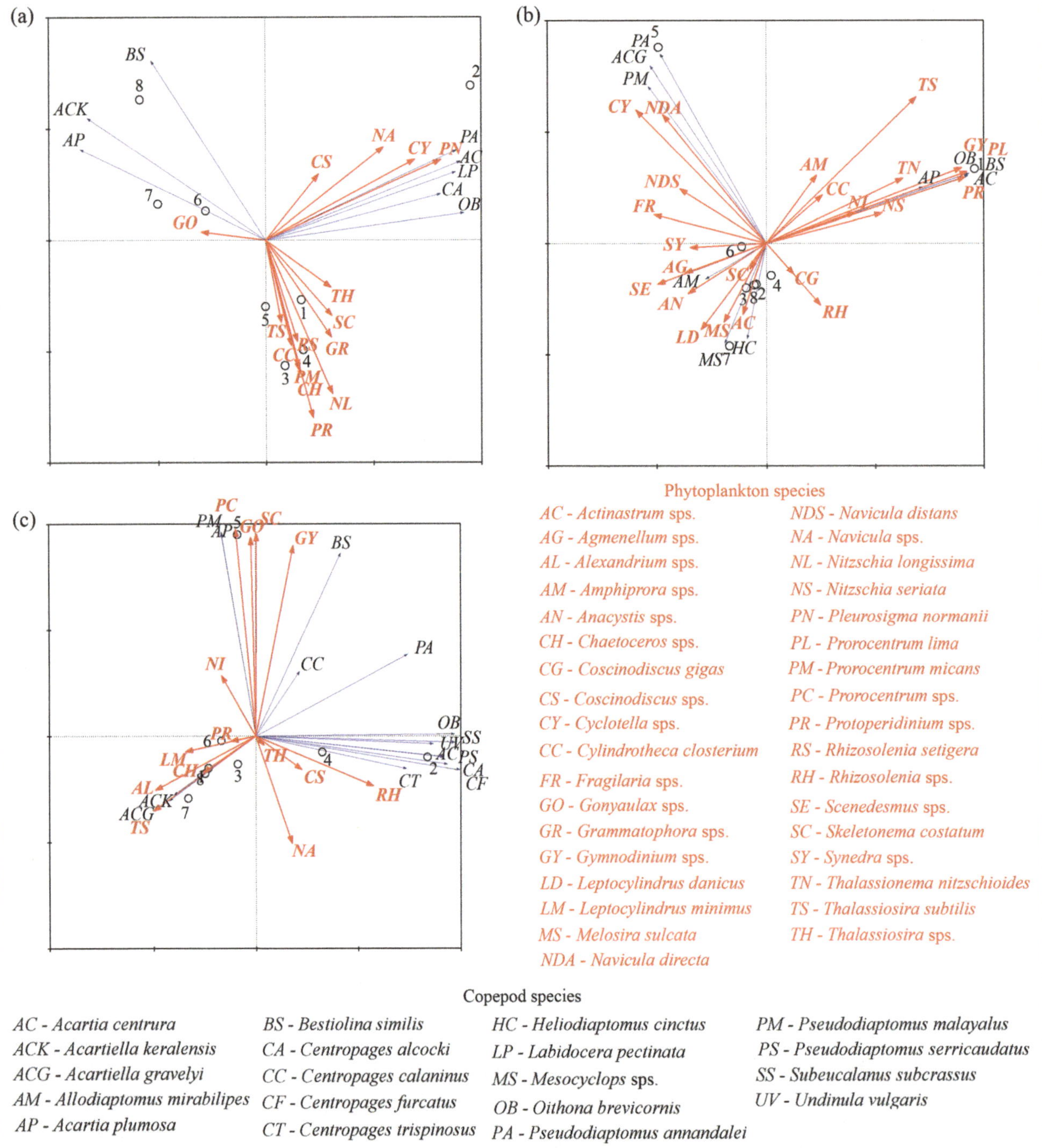

Fig. 5 The RDA triplot showing interrelation between copepod and phytoplankton species (dominant and characterizing species) during the **a** pre-monsoon, **b** monsoon, and **c** post-monsoon periods. The *red color* indicates the phytoplankton species, and the *black color* indicates the copepod species

compared to the diatoms. The preference of the low-saline species of copepods (*H. cinctus*, *Allodiaptomus mirabilipes*, *Mesocyclops* sps) for the phytoplankton species of the low-saline environment (*Anacystis* sps, *Actinastrum* sps, *Synedra* sps, *Scenedesmus* sps, *Melosira sulcata*) was evident during the monsoon season (Fig. 5).

Cluster analysis and non-metric multidimensional scaling

The dendogram of the copepod species, based on their abundance pattern along different seasons, helped considerably in understanding the similarity in their distribution in the estuary. During the pre-monsoon season, the three species that formed the dominant as well as the characterizing species were found in a single cluster, and most of the other characterizing species were found in another cluster (Fig. 6a). During the monsoon season, all the dominant and characterizing species were found in a single large cluster (Fig. 6b). During the post-monsoon season, the two dominant species (*A. plumosa* and *B. similis*) were found in a separate cluster, whereas most of the characterizing species were observed in a large cluster. The characterizing species that had preferences for a low-salinity environment (*Pseudodiaptomus malayalus*, *Ac. gravelyi*, and *Ac. keralensis*) were also found in a separate cluster (Fig. 6c).

Figure 7a describes the similarity in the station locations along different seasons derived through the NMDS analysis, based on the Bray-Curtis similarity indices, using the abundance of different copepod species present along each season. The variability of the spatial abundance of the two common dominant as well as characterizing species of each season (*A. plumosa*, *B. similis*), overlaid on the MDS plot, also indicated their affinity towards the medium-salinity environment during each season (Fig. 7b, c).

Diversity indices

Species diversity (H') of the copepods varied both spatially and temporally (Fig. 8a). As the inlet locations are relatively more influenced by the nearby marine environment compared to the other locations, the seasonal scenario of H' was checked separately for the inlet and the non-inlet locations and also for all the station locations together. In both cases, though copepod abundance was higher during the pre-monsoon season, H' was higher during the post-monsoon period. Species evenness (J') did not show much variation with the seasons (0.6 ± 0.1, 0.6 ± 0.2, and 0.5 ± 0.1 for the pre-monsoon, monsoon, and post-monsoon seasons, respectively) (Fig. 8b). The maximum evenness value was observed at station 8 (0.9) during the monsoon season.

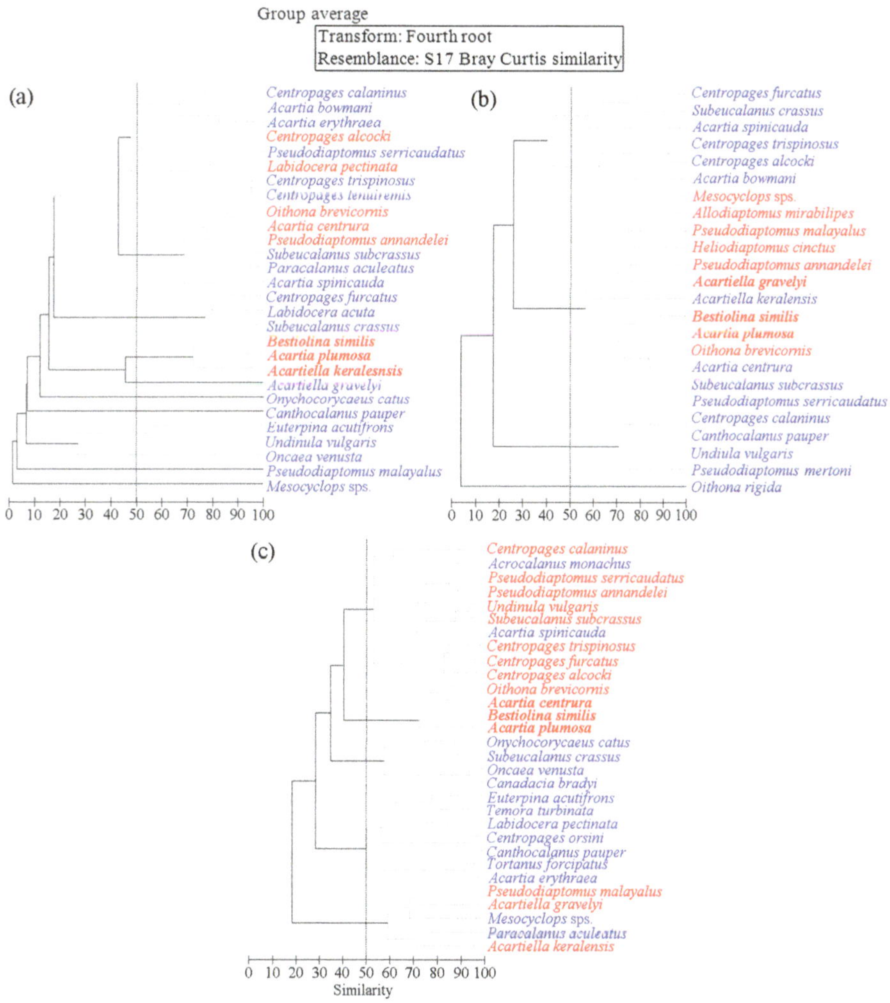

Fig. 6 Bray-Curtis similarity-based hierarchical clustering of copepod species during the **a** pre-monsoon, **b** monsoon, and **c** post-monsoon periods. The SIMPROF analysis was overlaid with the cluster, and the species connected by *gray lines* could not be significantly differentiated. The *red color* indicates the characterizing species, and the *bold letter* indicates the dominating species

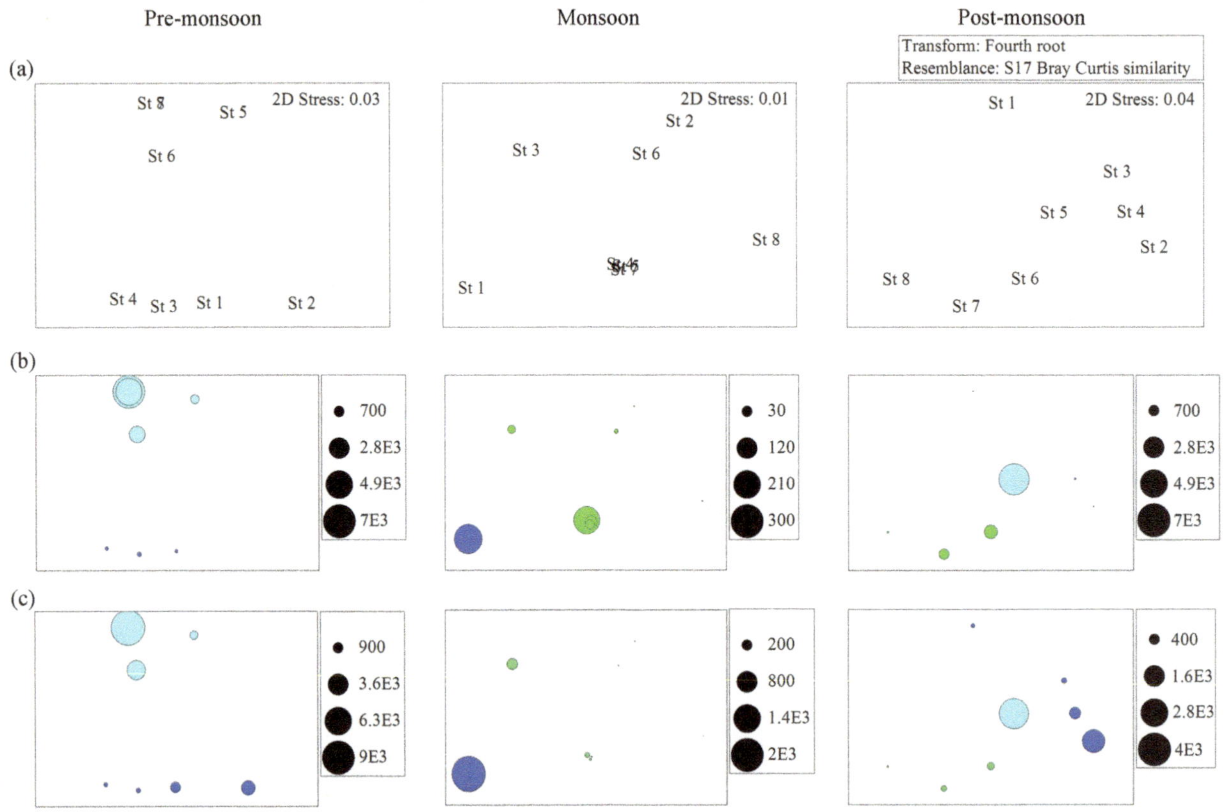

Fig. 7 **a** Bray-Curtis similarity-based non-metric multidimensional scaling (NMDS) plot of sampling locations based on the abundance of different copepod species. **b** The NMDS overlaid with the bubble plots of *Acartia plumosa* showing their abundance and distribution. **c** The NMDS overlaid with the bubble plots of *Bestiolina similis* showing their abundance and distribution. The bubble positions in these plots correspond to the positions of the sampling locations in Fig. 7a. The bubble size in each plot is proportional to the abundance (ind. m^{-3}). Among the bubble colors, *green* indicates low-saline, *cyan* indicates mesohaline, and *blue* indicates high-saline characteristics of the sampling locations

Sex ratio and abundance pattern

Since *A. plumosa* and *B. similis* were found as the only two species that remained the dominant as well as characterizing species during all three seasons, a detailed investigation of their sex ratio was carried out to understand whether sex-specific differences had any determining role in the maintenance of their population in the estuarine system. In the case of both *A. plumosa* and *B. similis*, the seasonal scenario revealed relatively higher sex ratio values during the monsoon season and lower values during the pre-monsoon period (Fig. 9). A prominent skewness towards female dominance was evident in the sex ratio of both the species, and in *A. plumosa*, it varied between 0.02 and 0.98 (av. 0.36 ± 0.2). Except for one higher value (1.99) during the monsoon season, the sex ratio of *B. similis* was also limited between 0.05 and 0.8 (av. 0.31 ± 0.26). The interrelation between sex ratio and the abundance of *A. plumosa* showed a significant negative relation ($P < 0.05$) during both the pre-monsoon and monsoon periods, whereas during the post-monsoon season, it did not show any particular trend ($P > 0.05$) (Fig. 10a). The result of the one-sample t test also showed that the regression lines are significantly different from zero during the pre-monsoon and monsoon periods (Fig. 10a). The analysis carried between the residuals extracted from the regression plot of *A. plumosa* and the freshwater content did not show any significant relationship ($P > 0.05$) in any season (Fig. 10b). *B. similis* exhibited a negative relation with abundance in all three seasons, and it was significant during the pre-monsoon and post-monsoon periods ($P < 0.05$) (Fig. 10c), indicating an influence of the biased sex ratio in the variability of their abundance. In case of *B. similis*, the result of the one-sample t test revealed that the regression lines are significantly different from zero during the pre-monsoon and post-monsoon periods (Fig. 10c) but the interrelation between the residuals extracted from the regression plot and the freshwater content showed a significant negative relation ($P < 0.05$) only during the monsoon period (Fig. 10d).

Discussion

While assessing the seasonal variability of the abiotic parameters and their control on copepod dynamics, the

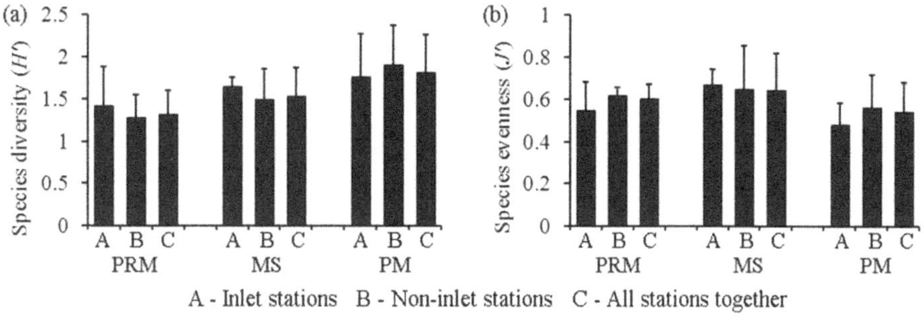

Fig. 8 Diversity indices **a** species diversity and **b** species evenness, along the three seasons. The *vertical line* on the bar diagram indicates the standard deviation. *PRM* pre-monsoon, *MS* monsoon, *PM* post-monsoon)

spatial scale gradients in the biotic and abiotic variables and their subsequent effects on the estuarine ecology were also given importance for a better elucidation of the spatio-temporal heterogeneity occurring in the zooplankton community of this dynamic estuarine system. Marked seasonality and unsteadiness in the salinity values, regulated by the monsoonal rainfall, were prominent in the system, supporting its categorization as a typical monsoonal estuary. During the monsoon season, due to the heavy rainfall (~300 cm) and associated high river influx, the entire estuarine system behaved as a limnetic region with low-salinity values. Revichandran et al. (2012) have shown that in the CE, nearly 60 % of the total annual rainfall occurs during the summer monsoon period, and this, along with the subsequent river runoff (60–70 % of the total), contributes greatly in the conversion of the estuary into a freshwater-dominated region. Therefore, a marked salinity range (0.1–28.5) prevailed between the upper and lower reaches of the estuary, and the salinity value could not reach >5, except at one inlet location. With the post-monsoon season, due to the gradual intrusion of the Arabian Sea water (Menon et al. 2000), though the salinity of the entire study area progressively increased, a salinity gradient prevailed between the upper and lower reaches, with a mesohaline zone in the middle reaches. During the pre-monsoon season, with markedly less river influx (1.69 %, Srinivas et al. 2003) and high incursion of the Arabian Sea water, the stations located at the lower reaches of the CE behaved as an extended part of the Arabian Sea (Madhupratap 1987). The low-salinity gradient (20.1 ± 6.4) during this period was the result of the progression of high-salinity water up to the middle reaches of the estuary, resulting in the development of a mesohaline zone in the upstream locations of the estuary (stations (St.) 6–8). The limited seasonal variation in the temperature distribution throughout the estuary might be due to its tropical nature, whereas the less conspicuous seasonal variation in the DO could be due to the existence of higher phytoplankton standing stock, irrespective of the season (Madhu et al. 2007). The significant seasonal variation observed in the pH distribution might have happened by the difference occurring in the seasonal fluvial runoff characteristics and saline water intrusion (Saraladevi et al. 1983).

The higher zooplankton abundance observed during the pre-monsoon season and the lower abundance during the monsoon season, in accordance with the average seasonal salinity pattern of the estuarine system, generally put forth the question of whether an increase in the salinity always leads to higher zooplankton abundance. The earlier studies depicting the zooplankton distribution, based on the average seasonal salinity and abundance values, support the concept of salinity-influenced zooplankton abundance (Rao et al. 1975; Jyothibabu et al. 2006; Madhu et al. 2007) in the CE.

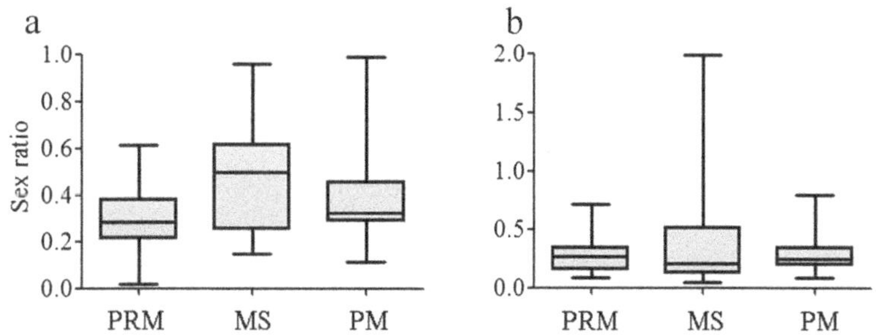

Fig. 9 The seasonal variability of the sex ratio of the dominant species **a** *Acartia plumosa* and **b** *Bestiolina similis*

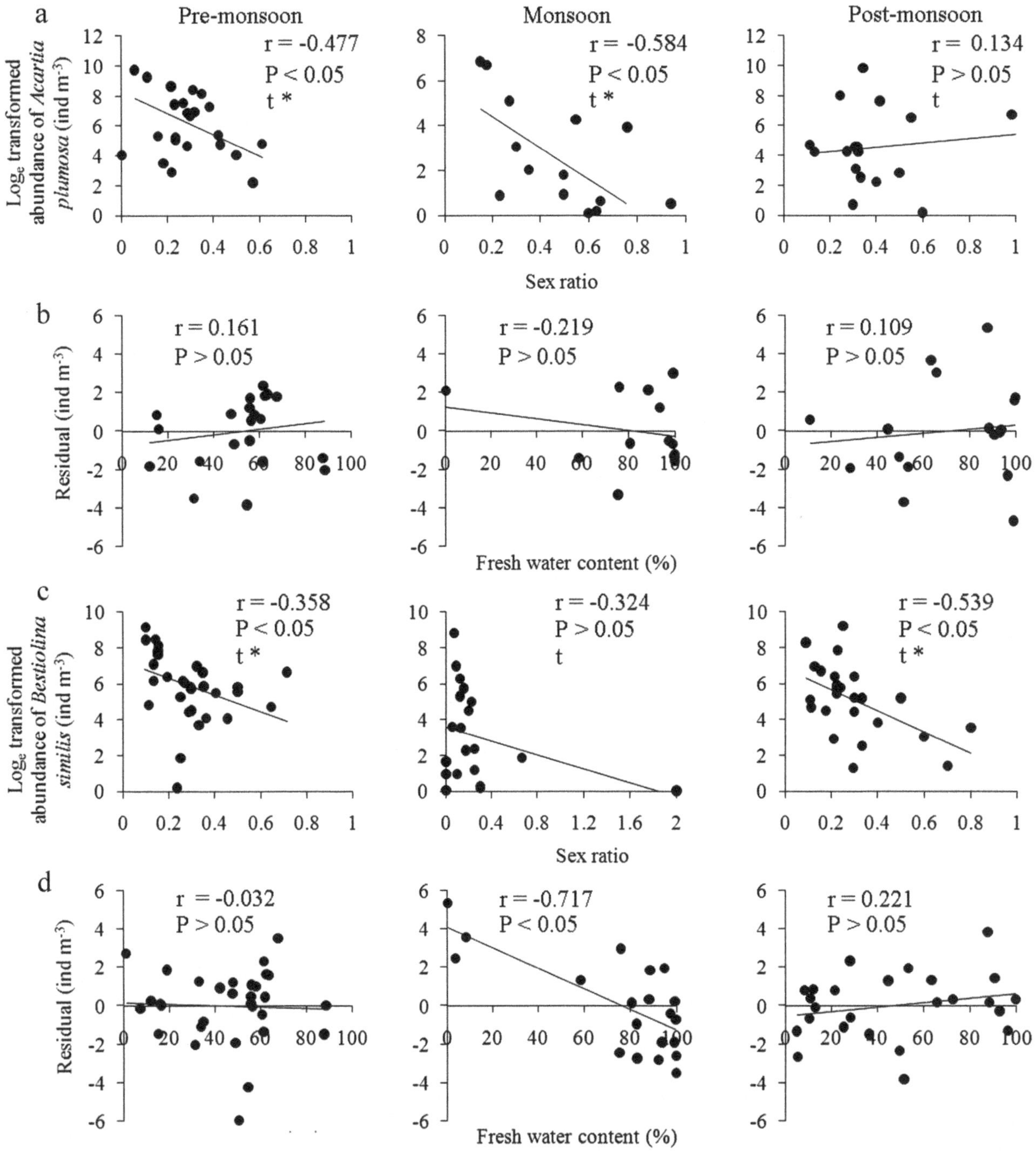

Fig. 10 The relation between the abundance and sex ratio of the dominant copepods **a** *Acartia plumosa* and **c** *Bestiolina similis* along different seasons and the interrelation between the residuals extracted from the regression plot and the freshwater content (**b**, **d**). The *t* value with an *asterisk* indicates that the regression line is significantly different from zero. The residual represents the difference between the observed and the estimated value in the abundance of the respective species

However, the detailed spatio-temporal analysis during the present study evidenced a varied view in the interrelation between salinity and the zooplankton community in this monsoonal estuary. During the pre-monsoon season, though a high-salinity environment (salinity >18) prevailed in the lower reaches of the estuarine zone, higher zooplankton abundance was observed, mostly at the mesohaline zones (salinity 5–18) at the upstream locations, which indicates a positive affinity of the estuarine organisms towards the mesohaline salinity ranges. During the monsoon season, among all the sampling locations in the CE, only the inlet location, station 1, had salinity >5, and concurrently, higher zooplankton abundance was observed at that station. The higher abundance of zooplankton, mainly at the mesohaline zone of the estuary, during the post-monsoon period also (Figs. 2 and 3) further validates our concept of the preference of mesohaline salinity ranges by the estuarine zooplankton community compared to the very high-salinity environments or very low-salinity environments. Though the result of our study differs from the observation of Kimmerer et al. (1998) on the distribution of a major population of estuarine zooplankton in the low-salinity zone of the temperate San Francisco estuary, the findings of Cronin et al. (1962) in the Delaware estuary clearly support our view of the affinity of estuarine organisms towards the mesohaline zone.

Since copepods formed the predominant zooplankton taxon, contributing >87 % of the total abundance, and are in accordance with earlier observations (Madhupratap 1987; Madhu et al. 2007), the present study, depicting the detailed spatio-temporal distribution, community composition, and species-specific abundance pattern, helped considerably in a better understanding of the ecology of the zooplankton community of this estuarine system. Irrespective of the season, the higher preponderances of the copepod community in the mesohaline zones of the estuary clearly validate their affinity towards the mesohaline environment. With the evident intense seasonality in the abiotic variables and in copepod abundance, a prominent change is also expected in the copepod community structure and in the relative contribution of different species towards the total copepod population. But, the higher contribution of the mesohaline species *A. plumosa* and *B. similis*, a euryhaline opportunistic colonizing paracalanoid, towards the higher abundance in the mesohaline zones of the CE, irrespective of season, points towards a significant role they play in the sustenance of the copepod community in this estuarine system. Prior reports on the higher abundance of *B. similis* and *A. plumosa* in different tropical estuaries along the Indian coast (Goswami 1982; Ramaiah et al. 1996; Dalal and Goswami 2001) are in agreement with our findings in the CE. The predominantly omnivorous diet of both these copepod species, along with their short generation time of 7–10 days (VanderLugt and Lenz 2009), might have favored their contribution towards higher abundance in the mesohaline zones of the estuarine region. Hence, the observation of the higher abundance and dominance of these copepods and their preferences for the mesohaline zones of the estuary throughout the study period further substantiate the concept of considering them the most suitable estuarine endemic among the copepod community for this tropical monsoonal estuary. The evident pattern of distribution and abundance of the copepods *P. annandalei* and *O. brevicornis* as the characterizing species of all three seasons, along with *A. plumosa and B. similis*, suggest their supporting role in the sustenance of the estuarine copepod community. Hence, the view on the vital role of the allochthonous population in the maintenance of the estuarine zooplankton abundance compared to the autochthonous populations, as observed in the temperate EMS estuary by Baretta and Malschaert (Baretta and Malschaert 1988), seems inapplicable in the case of this typical tropical monsoonal estuarine system.

The interrelation between the copepod species abundance and the distribution, based on their preferred environment, was examined in detail to have a better understanding on the determining role of the estuarine characteristics on the population dynamics and also on the complexities in relation between copepod species and the seasonal salinity patterns. As observed in the triplot of the RDA, during the pre-monsoon season, when a major part of the estuary behaves as an extension of the neighboring Arabian Sea due to the high marine water incursion (Madhupratap 1987; Srinivas et al. 2003), *A. plumosa* and *B. similis* showed negative affinity with salinity, and in contrast, during the monsoon season, when the estuary gets transformed as a freshwater-dominated system due to the higher precipitation and fluvial runoff, their abundance exhibited positive affinity with salinity, whereas in the transitional phase during the post-monsoon season, no particular trend was found, further affirming their preferences towards the mesohaline environment. Among the characterizing species, the observed negative affinity of the low-saline species *A. gravelyi*, *P. malayalus*, and *Al. mirabilipes* and the positive affinity of the high-saline species *Undinula vulgaris*, *S. subcrassus*, *A. centrura*, *Pseudodiaptomus serricaudatus*, *Centropages calaninus*, and *Centropages furcatus* towards salinity along different seasons truly characterize their preferences for the specific saline zones in the estuarine system. Similar observations on salinity-specific habitat preferences have been reported in the zooplankton community of temperate estuaries also (Collins and Williams 1981) as the occurrence in the habitat of preferred salinity ranges aids in the higher survival rate and also boosts their toleration capacity (Lance 1963). Among the copepod species, the assemblages observed among the dominant and characterizing species of the copepod (Fig. 6) further point towards their similarity in habitat preferences. The unique seasonal copepod community structure in the CE might have been regulated by the distinct species-specific salinity preferences and tolerance of the copepod population to the variability in the abiotic factors generated by the estuarine seasonality. Among these species, the ones with more affinity and adaptability to the seasonal estuarine characteristics might have gotten an advantage to be the dominant and characterizing species of the respective season.

Though in CE relatively low copepod diversity ($H' < 2$) was observed across the seasons, the seasonal variability in the H' value observed during the study period helped further in a better understanding of the variability occurring in the seasonal composition of the copepod community. The low species diversity of copepods during the monsoon period might have resulted due to the absence of high-saline marine migrants with a salinity preference >20, and a similar feature has also been observed in other monsoonal estuarine systems along the Indian coast (Goswami 1982) and also in the coastal waters of this region (Stephen 1977). The interesting feature evident in the seasonal diversity and abundance pattern with high abundance during the pre-monsoon season and high H' during the post-monsoon season augmented the need to understand the role of community composition during periods of varying abundance and diversity. During the post-monsoon season, though salinity progressively increased, a wide salinity range prevailed in the estuary (salinity—St. 1–4 > 18, St. 5 > 5 and <18, and St. 6, 7, and 8, <5), which in turn triggered different copepod species to inhabit niches of preferred salinity ranges in the estuary with maximum abundance at the mesohaline middle reaches. This was evident from the restricted occurrence of the low-saline copepod species *P. malayalus* and *H. cinctus* (salinity preference <10) in the southern limb of the estuary and several high-saline species (salinity preference >20) in the polyhaline region (salinity >18) in or near the two inlets. Most of the medium-saline species had wide distributions in the estuary, based on their specific salinity preferences. In contrast, during the pre-monsoon season, though the abundance was higher, it was mostly contributed by the mesohaline and euryhaline copepods, with maximum abundance in the mesohaline upstream locations. Though several high-saline species were present in varying abundance during this period, the species that had preferences to low salinity were mostly absent in the estuarine zone, resulting in the lower H' during this period.

As a fundamental trait of the sexually reproducing population, studies on the sex ratio, its variability from the Fischerian 1:1 ratio in the natural population, and the factors responsible for the skewness have received attention from the scientific community (Gusmão et al. 2013). Both laboratory studies (Irigoien et al. 2000; Dur et al. 2012) as well as field observations (Moraitou-Apostolopaulou 1972) imply the role of variability of sex ratio in governing the population dynamics of the copepod community, and hence, in the present study, it became an important tool to unfold the critical role they play in sustaining the population of the two dominant copepod species, *A. plumosa* and *B. similis*, in this dynamic estuarine ecosystem. In the seasonal scenario, the observation of the lower sex ratio during the period of higher abundance and vice versa clearly indicates the leaning of sex ratio towards female dominance during higher abundance. This observation of skewness in sex ratio towards females, in conjunction with the abundance pattern, is against the view of Moraitou-Apostolopaulou (1972) in the temperate Aegean Sea but is in agreement with the observations made by Saraswathy and Santhakumari (1982) on the copepod population along the coastal waters of India. The incident high female skewness in the copepod population might be an advantage towards easier mate finding for the male copepods and in turn facilitates higher reproductive output during the stable (with the lowest riverine influx) and food-rich environment (maximum chlorophyll *a*) in the pre-monsoon period, leading to the higher abundance. In the CE, the abundance of microzooplankton, another major diet of the copepod (Calbet and Saiz 2005), has also been recorded at maximum levels during the pre-monsoon period (Jyothibabu et al. 2006). Hence, along with higher chlorophyll *a*, this might also have played a role in supporting the biased sex ratio and, in turn, the higher abundance of the dominant omnivorous copepod species, *A. plumosa* and *B. similis*, during the pre-monsoon season. As the female copepods have comparatively higher feeding efficiency and are reported to be voracious feeders than their male counter parts, they might have utilized the food-rich environment more efficiently than the males, resulting in their dominance. The negative relation observed between the abundance and the sex ratio of both copepod species during most of the seasons (Fig. 10) further supports the inclination towards female skewness during the period of higher abundance. The insignificant relation observed between the freshwater content and the residuals extracted from the regression plots of both species also clearly substantiates the role of sex ratio in their abundance. The preference for mesohaline environment over the low- and high-saline environments might be the reason for the insignificant relation observed between the freshwater content and the residuals extracted from the regression plots (Fig. 10). The interrelation between the dominant and characterizing species of the copepod and phytoplankton community further helped to understand the influence that the autotrophic community has on the sex ratio-abundance pattern of the dominant copepod species. Irrespective of the season, *A. plumosa* and *B. similis*, which remained dominant throughout the study period, exhibited a preference for the autotrophic dinoflagellate species than the diatoms (Fig. 5). Compared to the diatoms, the dinoflagellates have been reported to have potential influence on egg production, hatching viability, and, in turn, the reproductive success of copepod species (Ceballos and Ianora 2003), as they provide more protein, carbohydrate, and lipid than diatoms of equivalent volume

(Kleppel 1993). Thus, the female-biased sex ratio and the feeding preference towards dinoflagellates over diatoms of both dominant estuarine endemic copepod species (Fig. 5) might have helped them in maintaining the dominance and also in their population sustenance during each season.

The NMDS plot expressing the similarity between stations along different seasons, based on the abundance of all copepod species, and addressing the variability in the spatial abundance of the two dominant species further helped us to get a better inference of their spatio-temporal distribution. The distinct groups of the station locations, based on the abundance of different copepod species, showed a close affinity towards the different salinity ranges, signifying the role of salinity in their abundance and distribution (Fig. 7a). The higher abundance of both *A. plumosa* and *B. similis* in the mesohaline region (Fig. 7b, c) and the negative relation of sex ratio with their abundance values (Fig. 10) indicate towards the role of sex ratio in maintaining their high population in the mesohaline environment of the estuary.

Conclusions

The evident heterogeneity in the spatial and temporal abundance, community composition, and diversity indices of the copepod community in the CE is governed mainly by the species-specific sensitivity and subsequent responses of the organisms towards the prime limiting abiotic variable, salinity. In spite of the prominent unsteadiness in the salinity characteristics induced by the seasonal variability in the precipitation regime and fluvial fluxes influenced by the Indian summer monsoon (Revichandran et al. 2012), the maximum zooplankton abundance always occurred at the mesohaline zone of the estuary, which experienced a seasonal shift from the lower reaches of the estuary during the monsoon season to the middle reaches during the post-monsoon season and ultimately towards the upstream locations of the estuary during the pre-monsoon period, in contrast to the expected high-salinity zones of the estuary. The dominant zooplankton taxon, copepods (>87 %), played a crucial role in the maintenance of the higher abundance of the zooplankton community in the mesohaline zones of the estuary. Depending on the seasonal variability in salinity, though a few high-saline and low-saline species flourished in the lower and upper reaches of the estuary, respectively, the estuarine copepod community was dominated by the preponderance of the mesohaline and euryhaline copepods *A. plumosa* and *B. similis*, along with *O. brevicornis* and *P. annandalei*, as the characterizing species, irrespective of the season. The role of sex ratio in the maintenance of the higher abundance of the dominant estuarine endemic species in the mesohaline region was another important finding in the present study, and in the seasonal scenario, the bias in the sex ratio towards female skewness might have favored higher reproductive output, resulting in the higher abundance during the favorable pre-monsoon period. Hence, the species-specific sensitivity to the seasonal changes in the salinity distribution of the copepod community played a prominent role in determining their abundance pattern in the CE, and the trend in the sex ratio of the dominant copepods towards female dominance further aided in sustaining the higher abundances in the mesohaline zone. Considering the limited study on tropical monsoonal estuarine zooplankton dynamics, the results of the present study will be relevant in the better understanding of the unique population sustenance of the estuarine endemic zooplankton community. Further studies can be focused towards understanding the mechanisms that lead to the biasness in the sex ratio of the estuarine endemic copepod species.

Competing interests

The authors declare that they have no competing interests.

Authors' contributions

VG and NVM designed the study. VG analyzed the sample. VG, NVM, and KKK analyzed and interpreted the data. All four authors helped in drafting the manuscript. All authors read and approved the final manuscript.

Acknowledgements

We express our sincere gratitude to the Director, CSIR-National Institute of Oceanography, for the facilities provided. This work forms a part of the Supra Institutional Project-1302—"Eco biogeography of the estuarine and coastal waters of south-west coast of India," of NIO. We thank all the scholars of the project SIP 1302 for their help rendered during sampling. We also convey our sincere thanks to both the anonymous reviewers for their valuable suggestions to improve the manuscript. The first author acknowledges Dr. M. Rakhesh, University of Bergen, Norway, and Dr. K.V. Jayalakshmi, CSIR-NIO, for the constructive suggestions which helped to improve the manuscript. The first author is also thankful to CSIR, New Delhi for the Senior Research Fellowship. This is NIO contribution number 5777.

References

Alongi DM (1998) Coastal ecosystem processes. CRC Press, Boca Raton, Florida

APHA (2005) Standard methods for the examination of water and wastewater, 21st edn. American Public Health Association/American Water Works Association / Water Environment Federation, Washington DC

Baretta JW, Malschaert JFP (1988) Distribution and abundance of the zooplankton of the EMS estuary (North Sea). Neth J Sea Res 22:69–81

Calbet A, Saiz E (2005) The ciliate-copepod link in marine ecosystems. Aquat Microb Ecol 38:157–167

Ceballos S, Ianora A (2003) Different diatoms induce contrasting effects on the reproductive success of the copepod *Temora stylifera*. J Exp Mar Biol Ecol 294:189–202

Clarke KR, Gorley RN (2006) PRIMER v6: user manual/tutorial. PRIMER-E, Plymouth

Clarke KR, Warwick RM (2001) Change in marine communities: an approach to statistical analysis and interpretation, 2nd edn. PRIMER-E, Plymouth

Collins NR, Williams R (1981) Zooplankton of the Bristol Channel and Severn estuary. The distribution of four copepods in relation to salinity. Mar Biol 64:273–283
Cronin LE, Daiber JC, Hulbert EM (1962) Quantitative seasonal aspects of zooplankton in the Delaware River estuary. Chesapeake Sci 3:63–93
Dalal SG, Goswami SC (2001) Temporal and ephemeral variations in copepod community in the estuaries of Mandovi and Zuari-west coast of India. J Plankton Res 23:19–26
Dur G, Souissi S, Schmitt FG, Cheng SH, Hwang JS (2012) Sex ratio and mating behaviour in the calanoid copepod *Pseudodiaptomus annandalei*. Zool Stud 51:589–597
Fisher TR, Carlson PR, Barber RT (1982) Sediment nutrient regeneration in three North Carolina estuaries. Estuar Coast Shelf Sci 14:101–116
Froneman PW (2004) Zooplankton community structure and biomass in a southern African temporarily open/closed estuary. Estuar Coast Shelf Sci 60:125–132
Goswami SC (1982) Distribution and diversity of copepods in the Mandovi-Zuari estuarine system, Goa. Indian J Mar Sci 11:292–295
Grasshoff K (1983) Determination of oxygen. In: Grasshoff K, Ehrhardt M, Kremling K (eds) Methods of sea water analysis. Verlag Chemie, Weinheim, pp 61–72
Gupta GVM, Thottathil SD, Balachandran KK, Madhu NV, Madeswaran P, Nair S (2009) CO_2 supersaturation and net heterotrophy in a tropical estuary (Cochin, India): influence of anthropogenic effect. Ecosystems 12:1145–1157
Gusmão LFM, McKinnon AD, Richardson AJ (2013) No evidence of predation causing female-biased sex ratios in marine pelagic copepods. Mar Ecol Prog Ser 482:279–298
Haedrich R (1983) Estuarine fishes. In: Ketchum BH (ed) Estuaries and enclosed seas. Elsevier, Amsterdam, pp 183–208
Harris R, Wiebe P, Lenz J, Skjoldal HR, Huntley ME (eds) (2000) ICES zooplankton methodology manual. Academic Press, London
Hirst AG, Bonnet D, Conway DVP, Kiørboe T (2010) Does predation control adult sex ratios and longevities in marine pelagic copepods? Limnol Oceanogr 55:2193–2206
Irigoien X, Obermüller B, Head RN, Harris RP, Rey C, Hansen BW, Hygum BH, Heath MR, Durbin EG (2000) The effect of food on the determination of sex ratio in *Calanus* spp.: evidence from experimental studies and field data. ICES J Mar Sci 57:1752–1763
Jyothibabu R, Madhu NV, Jayalakshmi KV, Balachandran KK, Shiyas CA, Martin GD, Nair KKC (2006) Impact of fresh water influx on micro zooplankton mediated food web in a tropical estuary (Cochin backwaters), India. Estuar Coast Shelf Sci 69:505–515
Kasturirangan LR (1963) A key for the identification of the more common planktonic copepod of Indian coastal waters. INCOR (CSIR), New Delhi
Kimmerer WJ (2002) Physical, biological, and management responses to variable freshwater flow into the San Francisco estuary. Estuaries 25:1275–1290
Kimmerer WJ, Burau JR, Bennett WA (1998) Tidally oriented vertical migration and position maintenance of zooplankton in a temperate estuary. Limnol Oceanogr 43:1697–1709
Kinne O (1966) Physiological aspects of animal life in estuaries with special reference to salinity. Netherlands J Sea Res 3:222–244
Kleppel GS (1993) On the diets of calanoid copepods. Mar Ecol Prog Ser 99:183–195
Lance J (1963) The salinity tolerance of some estuarine planktonic copepods. Limnol Oceanogr 8:440–449
Laprise R, Dodson JJ (1994) Environmental variability as a factor controlling spatial patterns in distribution and species diversity of zooplankton in the St. Lawrence Estuary. Mar Ecol Prog Ser 107:67–81
Lee CY, Liu DC, Su WC (2009) Seasonal and spatial variations in the planktonic copepod community of Ilan Bay and adjacent Kuroshio waters off northeastern Taiwan. Zool Stud 48:151–161
Madhu NV, Jyothibabu R, Balachandran KK, Honey UK, Martin GD, Vijay JG, Shiyas CA, Gupta GVM, Achuthankutty CT (2007) Monsoonal impact on planktonic standing stock and abundance in a tropical estuary (Cochin Backwaters—India). Estuar Coast Shelf Sci 73:54–64
Madhupratap M (1987) Status and strategy of zooplankton of tropical Indian estuaries: a review. Bull Plankton Soc Japan 34:65–81
Menon NN, Balchand AN, Menon NR (2000) Hydrobiology of the Cochin backwater system—a review. Hydrobiologia 430:149–183
Moraitou-Apostolopaulou M (1972) Sex ratio in the pelagic copepods, *Temora stylifera* Dana and *Centropages typicus* Kroyer. J Exp Mar Biol Ecol 8:83–87
Ohman MD, Hsieh CH (2008) Spatial differences in mortality of *Calanus pacificus* within the California current system. J Plankton Res 30:359–366
Qasim SZ (2003) Indian estuaries. Allied Publication, Mumbai
Ramaiah N, Chatterji A, Madhupratap M (1996) A study on the zooplankton of the Burhabalanga estuary, Orissa coast. Proc Indian Natn Sci Acad B62:1–4
Rao TSS, Madhupratap M, Haridas P (1975) Distribution of zooplankton in space and time in a tropical estuary. Bull Dept Mar Sci Univ Cochin 7:695–704
Revichandran C, Srinivas K, Muraleedharan KR, Rafeeq M, Shivaprasad A, Vijayakumar K, Jayalakshmy KV (2012) Environmental set-up and tidal propagation in a tropical estuary with dual connection to the sea (SW coast of India). Environ Earth Sci 66:1031–1042
Rodríguez-Graña L, Calliari D, Tiselius P, Hansen BW, Sköld HN (2010) Gender-specific ageing and non-Mendelian inheritance of oxidative damage in marine copepods. Mar Ecol Prog Ser 401:1–13
Saraladevi K, Venugopal P, Remani KN, Zacharias D, Unnithan RV (1983) Nutrients in some estuaries of Kerala. Mahasagar 16:161–173
Saraswathy M, Santhakumari V (1982) Sex ratio of five species of pelagic copepods from Indian Ocean. Mahasagar 15:37–42
Sewell RBS (1999) The copepod of Indian seas. Daya Books, India
Smith SV, Hollibaugh JT (1993) Coastal metabolism and the oceanic organic carbon balance. Rev Geophys 31:75–89
Soetaert K, Rijswijk PV (1993) Spatial and temporal pattern of the zooplankton in the Westerschelde estuary. Mar Ecol Prog Ser 97:47–59
Srinivas K, Revichandran C, Maheswaran PA, Mohamed ATT, Murukesh N (2003) Propagation of tides in the Cochin estuarine system, southwest coast of India. Indian J Mar Sci 32:14–24
Stephen R (1977) Calanoid copepods from the shelf and slope waters off Cochin: distribution, biomass and species diversity, Proc Symp Warmwater Zooplankton. Unesco/NIO 21-27
Strickland JDH, Parsons TR (1972) A practical handbook of seawater analysis. Bulletin of Fisheries Research Board, Canada
Tomas CR (1997) Identifying marine phytoplankton. Academic, USA
Van den Wollenberg AL (1977) Redundancy analysis. An alternative for canonical correlation analysis. Psychometrika 42:207–219
VanderLugt K, Lenz PH (2009) Handbook for the Cultivation of two Hawaiian paracalanid copepods. University of Hawaii, Honolulu
Vijith V, Sundar D, Shetye SR (2009) Time-dependence of salinity in monsoonal estuaries. Estuar Coast Shelf Sci 85:601–608
Wellershaus S (1969) On the taxonomy of planktonic copepod in the Cochin backwater (a south Indian estuary). Veröff Inst Meeresforsch Bremerh 11:245–286
Yang GM, He DH, Wang CS, Miao YT, Yu HH (1999) Study on the biological oceanography characteristics of planktonic copepods in the waters north of Taiwan, II. Community characteristics. Acta Oceanol Sin 21:72–80

16

The function of body coloration of the hai coral snake *Sinomicrurus japonicus boettgeri*

Koji Mochida[1*], Wan-Yu Zhang[2] and Mamoru Toda[1]

Abstract

Background: Prey animals often protect themselves from visual hunting predators via their body coloration, which encompasses various visual effects. When a prey animal displays a certain color pattern on its body surface, its protective function and effect are largely dependent on how a predator would encounter and perceive the prey animal. Asian coral snakes of the genus *Sinomicrurus*, which are venomous, display black bands and stripes on their orange body coloration. The banded pattern has been characterized as an aposematic signal in the New World coral snakes, but the stripes generally occur in cryptic snakes. We investigated the function of this complex color pattern, which might be interpreted as aposematic and cryptic, in *Sinomicrurus japonicus boettgeri*.

Results: First, plasticine replica experiments were conducted to assess whether natural avian predators avoid the color pattern of *S. japonicus boettgeri*; the results showed that they attacked the coral snake replicas and the control replicas with coloration similar to another prey snake, suggesting that the body coloration of *S. japonicus boettgeri* did not function aposematically in the wild. Second, we evaluated the chromatic contrast of the snake coloration with backgrounds from their natural habitats based on the avian predator visual systems. The body coloration of *S. japonicus boettgeri* showed the same, or lower, contrast levels with natural backgrounds than those of sympatric cryptic snakes, suggesting that the coloration was ineffective as an aposematic signal.

Conclusions: These results imply that the body coloration of *S. japonicus boettgeri* would work as crypsis through background matching or disruptive camouflage rather than aposematism.

Keywords: Aposematism; Disruptive coloration; Background matching; Coral snake mimicry

Background

Animals can protect themselves from visual hunting predators using their body coloration. They display a wide range of coloration and patterns, bearing several visual effects to reduce the risk of being detected by predators (e.g., background matching, disruptive camouflage, and optical illusion), to advertise their unprofitability for predators (aposematism) or to deceive predators into misidentifying it as something else (e.g., mimicry and masquerade) (Cott 1940; Edmunds 1974; Ruxton et al. 2004). More than one visual effect might coexist on the body surface of a prey animal, as the function of protective coloration often changes depending on the circumstances of when and where prey animals encounter predators (Hailman 1977; Tullberg et al. 2005; Bohlin et al. 2008). When prey animals use different visual effects, such as crypsis and aposematism, these often exploit similar features of their predator's visual perception (Stevens 2007). The physiological abilities of the predators to detect and identify objects might also enhance or suppress the potency of the defense (Exnerovà et al. 2003; Mochida 2011; Valkonen et al. 2012), and considering the perception of the predator in quantitative color analysis of any protective coloration is therefore critical (Endler and Mielke 2005). Thus, various theoretical frameworks and techniques have been developed in the study of visual signals (Endler and Mielke 2005; Stevens 2007).

Snakes are one of the most useful groups for understanding the processes involved in visual signal design in predator-prey relationships, as they would often display the protective coloration on their body surface, and the coloration is classified broadly into three types generated

* Correspondence: kj.mochida@gmail.com
[1]Tropical Biosphere Research Center, University of Ryukyus, Nishihara, Okinawa 903-0213, Japan
Full list of author information is available at the end of the article

by a relatively simple number of pattern elements: longitudinal stripes, spots (blotches or unicolored-speckled), and transverse bands (Murray and Myerscough 1991; Allen et al. 2013). Although most body coloration works cryptically through background matching, disruptive camouflage, or optical illusion (Brattstrom 1955; Jackson et al. 1976; Allen et al. 2013), some are used as aposematic signals (Green and McDiamond 1981; Pough 1988; Mappes et al. 2005). The New World coral snakes of the genera *Micrurus* and *Micruroides*, which are venomous, display red, yellow, white, and black transverse bands and are good examples of aposematism, as supported by laboratory and field experiments (Smith 1975; Brodie 1993; Brodie and Janzen 1995). Brodie (1993) conducted the field experiments using plasticine replicas of snakes to assess the function of the color pattern in coral snakes and showed that snake replicas with transverse bands, such as coral snakes, were rarely attacked by avian predators compared to plain control replicas, suggesting that the coral snake banded patterns functioned aposematically. Another study using plasticine snake replicas with a zigzag pattern, generally perceived as cryptic coloration, showed that the pattern had an aposematic function (Valkonen et al. 2011). Also, while the red or yellow coloration of the tropical snake is apparently conspicuous to the human eye, it would work as camouflage in their natural habitats from an avian predator's perspective (Wilson et al. 2007). Thus, if we are to understand how a particular coloration functions in nature, we need to consider how predators encounter and perceive their prey animals. This will provide new insights into the functional study of protective coloration, especially in cases in which prey animals display mismatched coloration in the wild and when that coloration is not effectively employed for defense (Stevens 2007).

The Asian coral snakes of the genus *Sinomicrurus*, the closest sister taxon of *Micrurus* and *Micruroides*, might have been expected to be aposematic, as they are venomous and display similar transverse bands to the New World coral snakes (Honma et al. 1967; Slowinski et al. 2001; Castoe et al. 2007). Notably, several species of *Sinomicrurus* exhibit not only transverse bands, but also longitudinal stripes on their body surfaces (Figure 1). The latter might be linked to cryptic defense, contrary to an aposematic strategy that makes the snake obvious against its background (Jackson et al. 1976). In this study, we investigated the function of the body coloration of *Sinomicrurus japonicus boettgeri*, which is distributed on several islands of the Ryukyu Archipelago, Japan, and whose body coloration consists of transverse bands and longitudinal stripes, which might be interpreted as both aposematism and crypsis. First, we conducted a plasticine replica experiment using plasticine snake replicas to assess whether natural avian predators avoided the color pattern of *S. japonicus boettgeri* as a prey item. Second, we measured the reflectance spectra of the body coloration of the snakes and constructed a tetrahedral color space to evaluate the chromatic contrasts with background objects, based on the visual systems of their avian predators (Endler and Mielke 2005; Stoddard and Prum 2008); these chromatic contrasts can produce several visual effects, such as background matching, disruptive camouflage, optical illusion, and aposematic signals on the body surface (Cott 1940; Gamberale-Stille 2001; Cuthill et al. 2005).

Methods

Plasticine replica experiment

We conducted plasticine replica experiments using natural avian predators in the mountain forests of Okinawajima Island, Ryukyu Archipelago, Japan. At the field site, *S. japonicus boettgeri* is active during the day and at night on the ground (Obara 2013; Mochida and Toda, personal observation). We constructed two types of snake

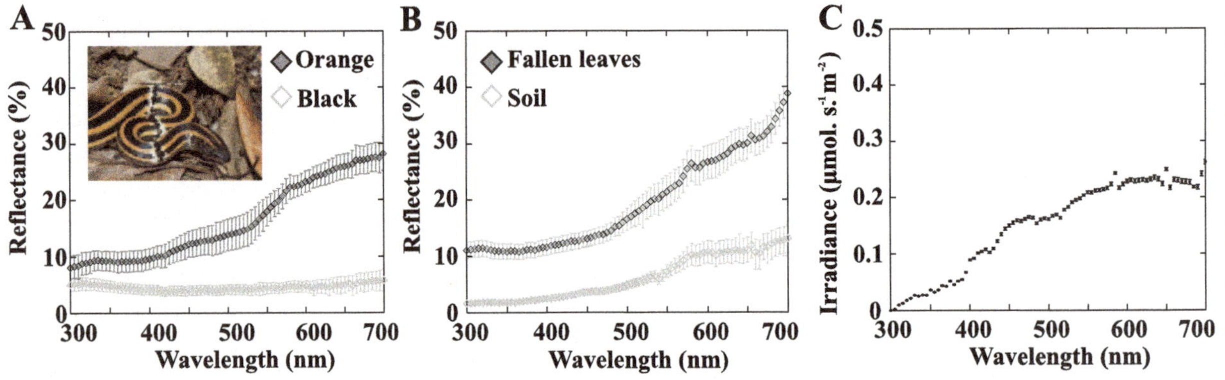

Figure 1 Body coloration of *Sinomicrurus japonicus boettgeri* and its color environments. Average reflectance spectra of two color elements of the body surface of *S. japonicus boettgeri* **(A)** and background subjects **(B)**, and average irradiance of their habitats **(C)**. Error bars indicate the standard error (SE).

replicas using pre-colored plasticine (Sculpy-III; Polyform Products Co., Elk Grove Village, IL, USA) threaded onto an S-shaped wire frame (Brodie 1993). Coral snake replicas were painted black with narrow white transverse bands and black stripes on an orange surface similar to the body color pattern of *S. japonicus boettgeri*, and the control replicas were plain green, similar to *Cyclophiops semicarinatus*, which is another prey item of avian predators at the field site (Tanaka and Mori 2000). We verified that the coloration of the replicas did not exhibit appreciable difference in reflectance spectra to that of target snakes using a USB2000+ spectrometer (Ocean Optics, Dunedin, FL, USA) and that there were no snake species showing coloration and pattern similar to replicas except for target snakes. The size of both replicas was 10 mm diameter and 180 mm length. Both types of replica were placed into one location side by side, enabling the avian predators to choose a profitable prey item (Figure S1 in Additional file 1). If the body color pattern of *S. japonicus boettgeri* functions aposematically, avian predators would avoid predation attempts on replicas of a similar color pattern. The replica pairs were placed on forest floor 0.5 to 1.0 m from edge of mountain roads. We prepared two types of background, an artificial white board and natural backgrounds, as predators might not find them on natural backgrounds if both replicas were too cryptic for them to locate. In one experiment, 25 pairs of replicas were set on each background in an alternating arrangement. Fifty pairs of replicas were divided into three blocks at intervals of 2 km, with each pair set at over 10-m intervals within a block. The number of pairs was from 12 to 20 in each block. After 72 h, the replicas were collected, and the U-shaped or V-shaped bill marks left by avian predators on the replicas were recorded as attacking attempts on them (Brodie 1993; Brodie and Janzen 1995; Valkonen et al. 2011). Multiple marks on one replica were scored as a single predation attempt. The experiments were conducted in November 2012 and June 2013, using a total of 100 pairs of snake replicas. November is the season when the snake-eating buzzard *Butastur indicus* is temporally abundant at the field site because of its migration from northern areas. Some of them remain there during winter, and then the numbers increase again in early spring until they disappear from the field site by May. Besides the buzzards, there are also a few avian species, which are potential predators on snakes, such as crows and kingfishers, throughout the year. Because some of the replicas were carried away from their original locations in the first experiment (see Results), both sides of the replicas were fixed to the ground in the second experiment.

Color analysis

The chromatic contrasts between the color elements of the snake body and background objects, and between the color elements within a snake body in *S. japonicus boettgeri*, were analyzed. These chromatic contrasts were compared to those of the two nonvenomous colubrid species, *C. semicarinatus* and *Dinodon semicarinatum*, and an ambush venomous pit viper, *Ovophis okinavensis*, all found on Okinawajima Island and living sympatrically with *S. japonicus boettgeri*. *D. semicarinatum* has large black blotches and a reddish white color, and the body coloration of *O. okinavensis* consists of small dark brown blotches on a light brown color surface (Figure S2 in Additional file 1). The main habitat of these four snake species is the forest floor, which is covered with red clay and fallen leaves. The reflectance spectra were measured of the orange and black color elements of *S. japonicus boettgeri* ($N = 5$), a green color element of *C. semicarinatus* ($N = 5$), the black and reddish-white color elements of *D. semicarinatum* ($N = 8$), and the light and dark brown color elements of *O. okinavensis* ($N = 11$). The spectra of the soil and fallen leaves in their habitats were analyzed as background color elements. For each background type, 20 subjects were measured. In these analyses, a USB2000+ spectrometer was used with a DT-Mini-2-GS deuterium and halogen light source (Ocean Optics, Dunedin, FL, USA). The average reflectance of each color element was calculated by computing the means from three replicate spectra measured for each subject for each data point between 300 and 700 nm, expressed relative to a 99% white reflectance standard. Irradiance of their habitats was also measured using a USB2000+ spectrometer with a cosine adapter head. We used LS-1 CAL as the standard reflectance light. In April 2013, we measured the irradiance at three landscape locations in their habitats: the ridge, valley, and the gentle slope of the mountain forest, under full sun conditions between 10:00 and 14:00 h. The average irradiance of the habitat was calculated by computing the means from 10 irradiance data measures for each location for each data point between 300 and 700 nm. To evaluate the chromatic contrasts of the snake body coloration, all color spectra data were mapped in a tetrahedral space, based on environmental irradiance and ultraviolet-sensitive (UVS) or violet-sensitive (VS) visual systems, as birds have two types of color vision (Endler and Mieke 2005). To calculate the chromatic contrast in the tetrahedral space, the coloration, represented by the color covering the largest surface area of each species, was set as the base color element (*S. japonicus boettgeri*; orange, *C. semicarinatus*; green, *D. semicarinatum*; black, and *O. okinavensis*; light brown). Based on these color elements, chromatic contrasts were calculated using formula (22) in Endler and Mieke (2005). We also calculated these contrasts based on a black color element that covers the second largest area of its body surface in *S. japonicus boettgeri*. *C. semicarinatus* shows plain green coloration;

therefore, we did not calculate intrabody contrast in this species. Animals were handled and experiments were conducted in accord with the Guideline for Animal Experiments in University of the Ryukyus.

Results

In 25 of 100 replica pairs, replicas were either attacked by birds (16 replicas) or missing (26 replicas). In 17 of these 25 pairs, both replicas were attacked or missing. We found a difference in the number of avian predation attempts on the pairs of replicas between November (20/50 pairs) and June (5/50 pairs). No significant differences were observed in the number of replicas remaining after predation attempts between the coral snakes and controls on natural (McNemar pairwise test; χ^2 = 36.54, $P > 0.05$) and white backgrounds ($\chi^2 = 40.02$, $P > 0.05$) when the 'missing' events were treated as non-predation attempts. Small pieces of the replicas were often found at such 'missing' locations, suggesting that avian predation events were also assumed. The statistical results did not change when the 'missing' events were treated as predation attempts by birds (natural background; $\chi^2 = 25.52$, $P > 0.05$, white background; χ^2 = 10.02, $P > 0.05$) (Figure 2). Although we know that small mammals, which are distributed on the field sites, such as rats and mice, bite or shave the surface of plasticine replicas regardless of their shape (Mochida and Toda, unpublished data), there were few replicas showing such signs on their surface in this experiment.

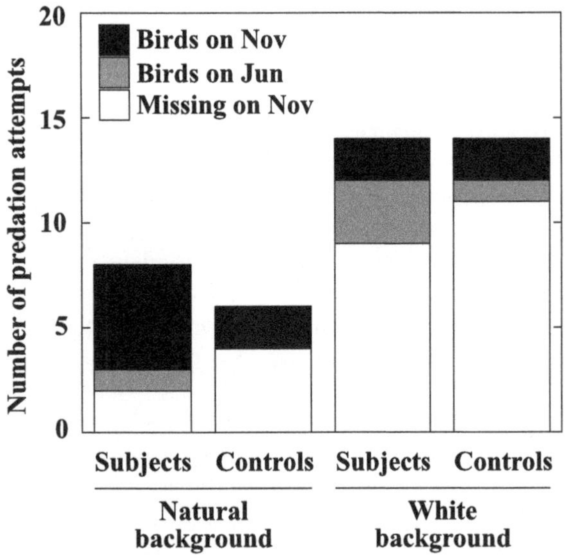

Figure 2 The number of avian attacks on snake replicas. Black and gray columns denote the number of attacking attempts on snake replicas in November and June, respectively. The white column indicates the number of missing replicas.

Color analyses based on environmental irradiance and two types of avian visual systems showed interspecific differences in intrabody contrasts (ANOVA: UVS system, $F_{2,21} = 16.9272$, $P < 0.0001$; VS system, $F_{2,21} = 11.0245$, $P = 0.0005$) and in background contrasts with soils (UVS system, $F_{3,25} = 26.8817$, $P < 0.0001$; VS system, $F_{3,25}$ = 13.1676, $P < 0.0001$) but not in fallen leaves (UVS system, $F_{3,25} = 2.0281$, $P = 0.1356$; VS system, $F_{3,25} = 1.8053$, P = 0.1720) (Figures 1 and 3; Table S1 in Additional file 1). The orange coloration of *S. japonicus boettgeri* had the same level or lower contrast with two types of backgrounds compared to that of the sympatric three species (Soil-UV type; 0.09 ± 0.01, Soil-V type; 0.08 ± 0.01, Fallen leaf-UV type; 0.06 ± 0.01. Fallen leaf-V type; 0.06 ± 0.02), and its black coloration of transverse bands and stripes also showed similar reflectance curve, which produces same level background contrasts (Soil-UV type; 0.22 ± 0.02, Soil-V type; 0.18 ± 0.02, Fallen leaf-UV type; 0.14 ± 0.02. Fallen leaf-V type; 0.11 ± 0.02) to black elements of *D. semicarinatum* (Figure 1; Table S1 in Additional file 1).

Discussion

The plasticine replica experiments showed that the avian predators did not avoid attacking the coral snake replicas or the controls. Both replicas were identified as prey items by the predators, as *C. semicarinatus* has been known as a prey item of birds (Tanaka and Mori 2000). A previous study suggested that predators that specialized in feeding on snakes might disrupt plasticine replica experiments used for testing aposematic signals of prey, as these animals might be capable of coping with the secondary defenses of the snakes (Valkonen et al. 2012). In November, a snake-eating buzzard is temporally abundant. However, it appeared that another avian predator also attacked the coral snake replicas ($N = 4$) over the controls ($N = 1$) in June when the buzzards were not present, although the small sample size prevented any statistical significance in the differences of the number of replicas attacked by predators between the two types. Thus, our result suggested that the body color pattern of *S. japonicus boettgeri* did not function aposematically in the wild.

Our color analyses supported the above interpretation of the plasticine replica experiments. The body coloration of *S. japonicus boettgeri* had the same level or lower contrast with two types of backgrounds compared to that of the sympatric species, including the two cryptic species *C. semicarinatus* and *O. okinavensis*. The two background subjects had a similar curve of reflectance spectra to that of the orange coloration of *S. japonicus boettgeri*, leading to a decrease in contrast. Background matching and disruptive camouflage would work best when large sections of body coloration blended into the background (Cott 1940; Stevens et al. 2006; Stevens and Merilaita 2009),

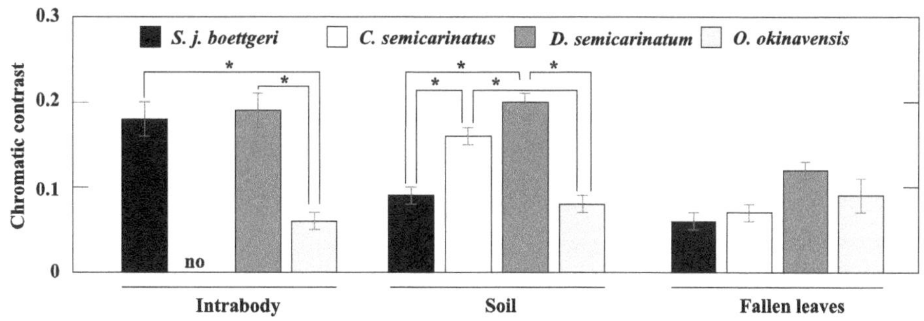

Figure 3 Chromatic contrasts of the body coloration of snakes based on the UV type of the avian visual system. Asterisks indicate significant differences in each chromatic contrast between species (Tukey-Kramer test: $P < 0.05$). Error bars indicate the standard error (SE).

contrary to the aposematic signals that should show high contrasts with backgrounds (Gamberale-Stille 2001). In cryptic defense, low contrast with a background is not always a prerequisite, as a high contrast will still often induce the optical illusion when striped or banded snakes can move faster; this is known as the flicker-fusion effect (reviewed by Stevens 2007; Allen et al. 2013), but it might not applicable to *S. japonicus boettgeri*, which moves relatively slowly (Authors, personal observation). The roles played by intrabody contrast on disruptive camouflage and background matching through longitudinal stripes and transverse bands are controversial and not clear in snakes (Allen et al. 2013). Studies on butterflies showed that a high contrast within a body could enhance the disruptive effects to conceal the body outline compared to a low contrast with the background (Cuthill et al. 2005); however, too much contrast within a body might increase predation risks due to conspicuousness (Stobbe and Schaefer 2008). We did not test the effects of intrabody contrast of *S. japonicus boettgeri* on their defenses. A direct comparison of the chromatic contrasts of body coloration with New World coral snakes that have been confirmed as aposematic is needed, but our results suggested that the orange coloration of *S. japonicus boettgeri* with a black stripe and band patterns may work cryptically through background matching or disruptive camouflage rather than aposematism.

In the Ryukyu Archipelago, there are two subspecies, *Sinomicrurus japonicus japonicus* and *Sinomicrurus japonicus takarai*, and kin species, *Sinomicrurus macclellandi iwasakii*. *S. japonicus japonicus* and *S. japonicus takarai* display black longitudinal stripes on orange body coloration similar to *S. japonicus boettgeri*, and also *S. japonicus takarai* does not show transverse bands, implying that their body coloration may not function as an aposematic signal, as in the case of *S. japonicus boettgeri*. On the other hand, *S. macclellandi iwasakii* displays wide transverse bands throughout its entire body and has no longitudinal stripes, and therefore, its body color pattern is more similar to that of the New World coral snakes than to other species of *Sinomicrurus*. *S. macclellandi iwasakii* is the most southerly distributed of the *Sinomicrurus* species in the Ryukyu Archipelago, and its distribution range overlaps with that of the resident raptor *Spilornis cheela*, a well-known snake-eating predator. This coral snake might encounter avian predators at a higher frequency than other *Sinomicrurus* throughout the year in this region. Establishment of aposematism is advantageous for prey animals in cases when the predators can learn about the unprofitability of prey carrying aposematic signals (Mappes et al. 2005). Empirical studies have revealed that an abundance of visually oriented predators would drive the acquisition of aposematic defenses in prey animals (Ratcliffe and Nydam 2008; Mochida 2011). Further work is required to clarify the relationships between interspecific variation in body coloration and differences in local predator composition in *Sinomicrurus*, as well as considering the influences of their phylogenetic relationships and other defensive traits such as toxicity.

Conclusion

Our result implies that the body coloration of *S. japonicus boettgeri* would work as crypsis through background matching or disruptive camouflage rather than aposematism. This would give us an understanding as to when and where some Asian coral snakes lose and acquire their aposematic transverse bands and longitudinal stripes, which could be linked with crypsis.

Competing interests

The authors declare that they have no competing interests.

Authors' contributions

KM participated in the design of the study, conducted all experiments, analyzed data, and wrote the manuscript. WYZ also carried out the plasticine replica experiments. MT participated in the design of the study and wrote the manuscript. All authors read and approved the final manuscript.

Acknowledgements

We thank the Environmental Forest Yona Field, University of Ryukyus for support of this research. We are grateful for helps to collect the reflectance spectral data of snakes to T. Jono, Y. Kadota, T. Kaito, T. Otani, K. Terada, and A. Mori. T. Jono and T. Kaito also gave us good photographs. This study was financially supported by a JSPS Research Fellowship for Young Scientists and a Grant-in-Aid for Research Activity Start-up (228700159) to KM.

Author details

[1]Tropical Biosphere Research Center, University of Ryukyus, Nishihara, Okinawa 903-0213, Japan. [2]Graduate School of Engineering and Science, University of Ryukyus, Nishihara, Okinawa 903-0213, Japan.

References

Allen WL, Baddeley R, Scott-Samuel NE, Cuthill IC (2013) The evolution and function of pattern diversity in snakes. Behav Ecol 24:1237–1250

Bohlin T, Tullberg BS, Merilaita S (2008) The effect of signal appearance and distance on detection risk in an aposematic butterfly larva (*Parnassius apollo*). Anim Behav 76:577–584

Brattstrom BH (1955) The coral snake 'mimic' problem and protective coloration. Evolution 9:217–219

Brodie ED III (1993) Differential avoidance of coral snake banded patterns by free-ranging avian predators in Costa Rica. Evolution 47:227–235

Brodie ED III, Janzen FJ (1995) Experimental studies of coral snake mimicry: generalized avoidance of ringed snake patterns by free-ranging avian predators. Funct Ecol 9:186–190

Castoe TA, Smith EN, Brown RM, Parkinson CL (2007) Higher-level phylogeny of Asian and American coralsnakes, their placement within the Elapidae (Squamata), and the systematic affinities of the enigmatic Asian coralsnake *Hemibungarus calligaster* (Wiegmann, 1834). Zool J Linn Soc 151:809–831

Cott HB (1940) Adaptive Coloration in Animals. Methuen, London

Cuthill IC, Stevens M, Sheppard J, Maddocks T, Párraga CA, Troscianko TS (2005) Disruptive coloration and background pattern matching. Nature 434:72–74

Edmunds M (1974) Defence in Animals: A Survey of Antipredator Defences. Longman, Essex

Endler JA, Mielke PW (2005) Comparing entire colour patterns as birds see them. Biol J Linn Soc 86:405–431

Exnerovà A, Landovà E, Stys P, Fuchs R, Prokopovà M, Cehlàrikovà P (2003) Reactions of passerine birds to aposematic and nonaposematic firebugs (*Pyrrhocoris apterus*; Heteroptera). Biol J Linn Soc 78:517–525

Gamberale-Stille G (2001) Benefit by contrast: an experiment with live aposematic prey. Behav Ecol 12:768–772

Green HW, McDiamond RW (1981) Coral snake mimicry: does it occur? Science 213:1207–1212

Hailman JP (1977) Optical Signals. Indiana University Press, Bloomington

Honma M, Okonogi T, Kosuge T, Mishima S (1967) Studies on the toxicity of striped coral snake ("Hyan", *Calliophis japanicus japanicus*) venom. Nippon nettai igaku zasshi 8:65–69 (in Japanese)

Jackson JF, Ingram W III, Campbell HW (1976) The dorsal pigmentation pattern of snakes as an antipredator strategy: a multivariate approach. Am Nat 110:1029–1053

Mappes J, Marples N, Endler JA (2005) The complex business of survival by aposematism. Trens Ecol Evol 20:598–603

Mochida K (2011) Combination of local selection pressures drives diversity in aposematic signals. Evol Ecol 25:1017–1028

Murray JD, Myerscough MR (1991) Pigmentation pattern formation on snakes. J Theor Biol 149:339–360

Obara Y (2013) A case of predation on *Achalinus werneri* by *Sinomicrurus japonicus boettgeri*. AKAMATA 24:1-2 (in Japanese)

Pough FH (1988) Mimicry of vertebrates: are the rules different? Am Nat 131:S67–S102

Ratcliffe JM, Nydam ML (2008) Multimodal warning signals for a multiple predator world. Nature 455:96–99

Ruxton GD, Sherratt TN, Speed MP (2004) Avoiding Attack: The Evolutionary Ecology of Crypsis, Warning Signals & Mimicry. Oxford University Press, New York

Slowinski JB, Boundy J, Lawson R (2001) The phylogenetic relationships of Asian coral snakes (Elapidae: *Calliophis* and *Maticora*) based on morphological and molecular characters. Herpetologica 57:233–245

Smith SM (1975) Innate recognition of coral snake pattern by a possible avian predator. Science 187:759–760

Stevens M (2007) Predator perception and the interrelation between different forms of protective coloration. Proc R Soc B 274:1457–1464

Stevens M, Merilaita S (2009) Defining disruptive coloration and distinguishing its functions. Philo Trans R Soc B 364:481–488

Stevens M, Cuthill IC, Windsor AMM, Walker HJ (2006) Disruptive contrast in animal camouflage. Proc R Soc B 273:2433–2438

Stobbe N, Schaefer HM (2008) Enhancement of chromatic contrast increases predation risk for striped butterflies. Proc R Soc B 275:1535–1541

Stoddard MC, Prum RO (2008) Evolution of avian plumage color in tetrahedral color space: a phylogenetic analysis of New World buntings. Am Nat 171:755–776

Tanaka K, Mori A (2000) Literature survey on predators of snakes in Japan. Current Herp 19:97–111

Tullberg BS, Merilaita S, Wiklund C (2005) Aposematism and crypsis combined as a result of distance dependence: Functional versatility of the colour pattern in the swallowtail butterfly larva. Proc R Soc B 272:1315–1321

Valkonen J, Niskanen M, Björklund M, Mappes J (2011) Disruption or aposematism? Significance of dorsal zigzag pattern of European vipers. Evol Ecol 25:1047–1063

Valkonen JK, Nokelainen O, Niskanen M, Kilpimaa J, Björklund M, Mappes J (2012) Variation in predator species abundance can cause variable selection pressures on warning signaling. Ecol Evol 2:1971–1976

Wilson D, Heinsohn R, Endler JA (2007) The adaptive significance of ontogenetic colour change in a tropical python. Biol Lett 3:40–43

17

Abundance and diversity of amphipods (Crustacea: Peracarida) on shallow algae and seagrass in lagoonal ecosystem of the Mediterranean Tunisian coast

Wahiba Zaabar, Rym Zakhama-Sraieb*, Faouzia Charfi-Cheikhrouha and Mohamed Sghaïer Achouri

Abstract

Background: Bizerte lagoon is a semi-enclosed marine ecosystem, where various types of human activities have been developed. To investigate the distribution and temporal variation of amphipod assemblage, monthly sampling was conducted at the Menzel Jemil site from October 2009 to September 2010.

Results: A total number of 3,620 specimens were collected from floating algae and seagrass allowing the identification of 10 amphipod species. Moreover, several indices, species richness, dominance, mean density, and diversity index were estimated to characterize the amphipode assemblage. *Gammarus aequicauda* was the most dominant species in all seasons. In addition, the minimum and maximum values of species richness of amphipod were observed in January (3 species) and April (8 species), respectively. The mean density and species richness were negatively correlated with plant biomass. Mean Shannon index (H') and evenness (J') were 1.62 ± 0.34 and 0.67 ± 0.16, respectively. Non-metric multidimensional scaling (MDS) analysis based on the mean species density showed three seasonal groups of samples. Therefore, canonical correspondence analysis (CCA) made it possible to summarize the overall situation for the species, monthly sampling, and environmental parameters on a single graph.

Conclusions: Thus, the temperature, turbidity, and chlorophyll *a* content are the most often reported factors for the distribution and structure of amphipods in the Bizerte lagoon.

Keywords: Coastal lagoon; Temporal distribution; Amphipod assemblage; Tunisia

Background

Wetlands are ecologically very important and extremely productive ecosystems, however very sensitive essentially in transitional locations subject to environmental and anthropogenic constraints (Mouillot et al. 2005; Rossi et al. 2006; Blanchet et al. 2008). The Tunisian lagoons, like several Mediterranean wetlands, are subject to an increasing pressure in the anthropogenic activities (urbanization, industry, pollution, aquaculture, tourism, and overfishing). The consequences can be detected on the general state of ecosystems, mainly in macrofauna that is more sensitive and more exposed (Ben Mustapha et al. 1999; Ayari and Afli 2003). In the last decades, Bizerte lagoon in the northeast of Tunisia is the best example that illustrated the disturbances that were mainly caused by the worn water rejections coming from the bordering cities (Dellali et al. 2001), the naval port and the metallurgic factory of Menzel Bourguiba, and some other industries (iron and steel plant, cement factory, refinery) established on its shoreline (Essid and Aissa 2002).

Most studies are carried out on invertebrates in the northern Mediterranean lagoon (Migné and Davoult 1997; Danesi et al. 1999; Mizzan 1999; Kevrekidis et al. 2000). In Tunisia, information on lagoonal invertebrate biodiversity is relatively scarce (Diawara et al. 2008; Tlig-Zouari and Maamouri-Mokhtar 2008; Tlig-Zouari et al. 2009; Afli et al. 2009).

* Correspondence: zakhamarym@yahoo.fr
S11UR11 Bio-Ecologie et Systématique Evolutive, Département de Biologie Animale, Faculté des Sciences de Tunis, Université Tunis El Manar, 2092 Manar II, Tunis, Tunisia

Amphipods represent one of the most important groups of invertebrates associated with algae and seagrass, playing an important link in trophic webs from producers to higher consumers such as fish populations (Sanchez-Jerez et al. 1999; Zakhama-Sraieb et al. 2006; Fernandez-Gonzalez and Sanchez-Jerez 2014). Moreover, this fauna is sensitive to environmental conditions and therefore constitutes a good bioindicator of pollution (Bellan-Santini 1980; Virnstein 1987; Conradi et al. 1997; Guerra-Garcia and Garcia-Gomez 2001). Amphipods have proven to be a difficult group to identify due to their small size and morphology; they are however scantily sampled and studied in the lagoon systems (Diviacco and Bianchi 1987; Procaccini and Scipione 1992).

The present study focuses on the dynamics during an annual cycle, with the aim to assess the influence of plant biomass and abiotic factors on the dynamics of amphipod assemblages in the Bizerte lagoon. The selected location appears suitable to analyze the temporal distribution of amphipods and the possible influence of some environmental factors on them. Results will be compared with similar studies on other Mediterranean lagoons. Our hypotheses are that algae and seagrass of the Bizerte lagoon function as a refuge for the amphipod assemblages and that they may change during the year in relation to environmental factors.

Methods

Area of study

The Bizerte lagoon is located in the northeast part of Tunisia, between latitudes: 37°8′ N and 37°14′ N, and longitudes: 9°46′ E and 9°56′ E (Figure 1). It covers an area of 130 km^2 and has a mean depth of 7 m. It is known for its geostrategic position since it is connected to the Mediterranean Sea through a 6 km long inlet and to the Ichkeul Lake through the Tinja channel, which is approximately 5 km long and a few meters in depth (3 m in the overflow period). The tide undergoes changes in the water level of the Mediterranean, while that own Bizerte lagoon is negligible (Chebbi 2010). The lagoon lies in the vicinity of several cities (Bizerte, Zarzouna, Menzel Abderrahmen, Menzel Jemil, and Menzel Bourguiba) and industrial units.

Sampling was performed in Menzel Jemil (37°13′2″ N 9°55′8″ E) in the northeast of Bizerte lagoon from October 2009 to September 2010 where algae and seagrass were developed during all the years. In this station, the bottom is sandy to sandy-mud.

Sampling procedure

The benthic macrofauna was sampled monthly during 1 year (from October 2009 to September 2010) using a metal quadrat of 25 × 25 cm with 12 replicates between 20 and 80 cm depth according to the tide. Those replicates were taken from one sampling site in Menzel Jemil and separated from a minimum of 2 m. Specimens were collected from a mix of algae *Gracilariopsis longissima* (S.G. Gmelin) M. Steentoft, L.M. Irvine & W. F. Farnham 1995, *Gracilaria bursa-pastoris* (S.G. Gmelin) P. C. Silva 1952, *Cladophora* sp., *Ulva lactuca* (Linnaeus, 1753), and seagrass *Cymodocea nodosa* (Ucria) Ascherson, 1870. Animals were removed by washing the vegetation in a big tray and recovered on a sieve of a 1 mm mesh, which retained all individuals including amphipods. Retained specimens were sorted, fixed in 70% alcohol, and then identified to species and counted. After that, the plant biomass was estimated by weighting algae and seagrass after being dried at 70°C for 48 h.

A trophic guild analysis was done attributing the identified species to trophic categories, according to the literature (Guerra-Garcia et al. 2014) as follows: S, suspension feeders; DS, deposit-suspension feeders; He, herbivores; De, plant detritus feeders; and O, omnivores.

Experimental observation

The following parameters were measured monthly *in situ*: temperature (*T*), salinity (*S*), pH, dissolved oxygen (O_2), and turbidity (Tr). Temperature, salinity, pH, and dissolved oxygen were measured at approximately 10 cm below the surface using a salinometer (WTW Cond 315i, SUNTEX, Weilheim, Germany), a pH meter (pH 330i/SET, SUNTEX, Weilheim, Germany) and oximeter (WTW Oxi315i/SET, SUNTEX, Weilheim, Germany) calibrated beforehand. The laboratory analysis of the surface water samples (at 10 cm) was performed for nitrites (NO_2^-), nitrates ($NO^-{}_3$), and phosphorous ($PO_4{}^{3-}$). The chlorophyll *a* (Chl *a*) content was determined using the spectrophotometric method of Lorenzen (1967) and following the procedure given by Parsons et al. (1984) after 24 h extractions in 90% acetone at 5°C in the dark.

Data analysis

To evaluate the importance of the different species, (i) the total abundance (over the study period) (Ni) and monthly mean abundances, (ii) the total dominances (Di%), and (iii) the frequencies (Ci) were estimated.

The mean density (individuals.m^{-2}) was calculated for each month. Collections in April to September were grouped and considered as samples from the dry season, whereas collections in December to March were considered as samples from the rainy season.

Total abundance of amphipod (*N*), number of species (*S*), Shannon-Wiener diversity index (*H*′) (Shannon and Weaver 1963), and Pielou's evenness index (*J*′) (Pielou 1966) were monthly calculated. Non-metric multidimensional scaling (MDS) (Kruskal and Wish 1978) were performed based on mean abundance of species at each sample.

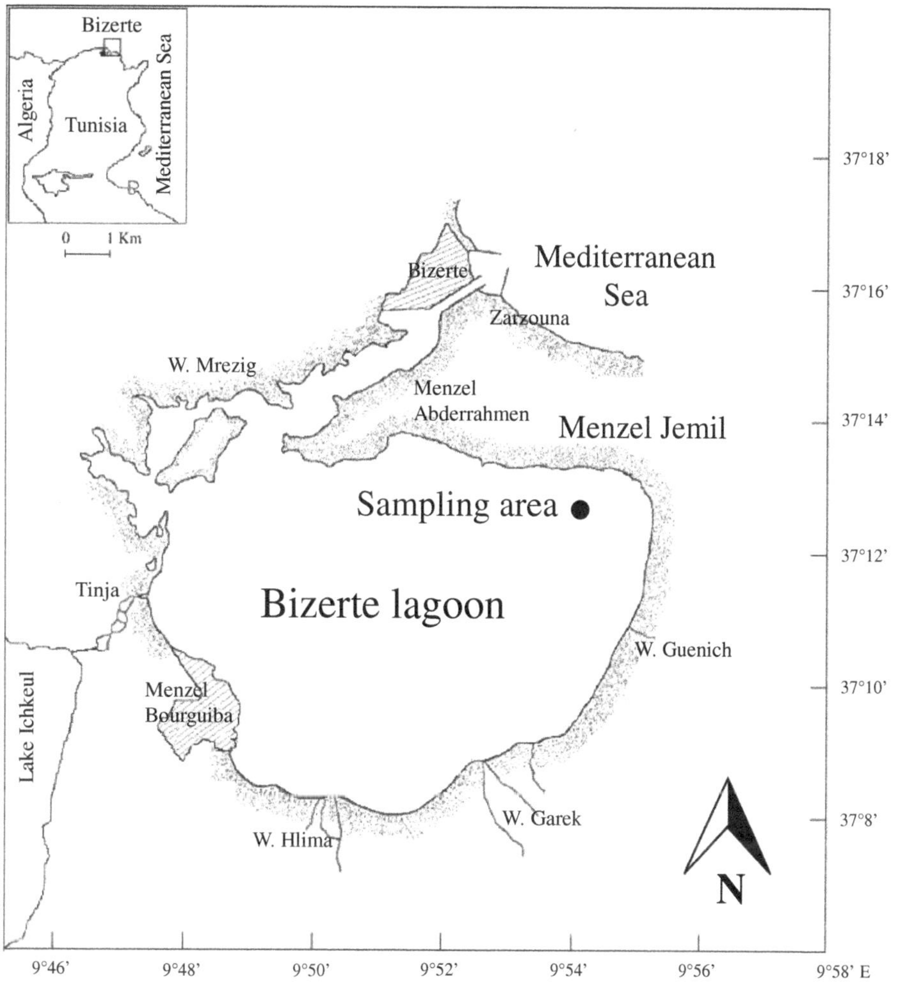

Figure 1 Map of the Bizerte lagoon area showing the location of the study site (Menzel Jemil) and wadis (W).

To investigate the relationship between the environmental variables and the species abundance data, the canonical correlation analysis (CCA) (Anderson and Willis 2003), extremely informative, was applied to the data to show which was the most contributed parameter in the differences reported among samples. The analysis was performed using PRIMER v5.2 (Plymouth Routines of the Multivariate Ecological Research software package) (Clarke and Warwick 1994) and XLSTAT 2013.

Results

Vegetal composition

The biomass of each plant species (Table 1) collected during an annual cycle shows a seasonal variation. In fact, the biomass of *G. bursa-pastoris* is maximized in the spring while that of *U. lactuca* peaked in summer (68.6 g.m^{-2} in August). Regarding *Cladophora* sp., it appears in winter, when the three algae biomass decreases and reappears in summer with the proliferation of *U. lactuca*. The seagrass *C. nodosa* is present all year, and its biomass reaches a maximum in autumn (38.48 g.m^{-2} in November).

The highest total plant biomass was recorded in early summer reaching 127.12 g.m^{-2} in June, followed by spring and autumn, whereas the lowest one was reported in winter (23.11 g.m^{-2} in January), showing a clear seasonal trend (Table 1).

Abiotic factors

Due to the connection of the Bizerte lagoon to the Mediterranean Sea from the north and to the Ichkeul Lake from the south, the exchange of seawater and the freshwater impact from wadi runoff defines the brackish characteristics and the presence of seasonal spatial gradients in the distribution of the abiotic and biotic variables inside the lagoon. In the present study, the term *wadi* means a temporary or permanent watercourse, whose flow depends on rains. The principal *wadis* which feed the Bizerte lagoon with freshwater are the *wadis* of Mrezig, Hlima, Garek, Guenich, and the channel of

Table 1 Monthly variation of total plant biomass (gDW.m^{-2}) and amphipod abundance and values of Shannon-Wiener (*H'*) and evenness indexes (*J'*)

	October	November	December	January	February	March	April	May	June	July	August	September	Ni	Di	Ci	Ci × Di	TG
Macroalgae and seagrass																	
Ulva lactuca	4	7	5.15	2.11	3	3	8	10.73	60	66.10	68.60	30					
Cladophora sp.	0	0	7	2	3	3.73	0	0	8	11	0	0					
Gracilariopsis longissima	1	4	0	3	14	20	27	30	20.10	11.01	8	9.08					
Gracilaria bursa-pastoris	1	4	0	4	15	20	30.18	32	23	11.10	8.01	9					
Cymodocea nodosa	38	38.48	30	12	8	7	8	8	16.2	17	17	37.08					
Monthly total biomass	44	61.48	41.15	23.11	44	53.73	73.18	80.73	127.12	116.21	96.7	85.16					
Amphipods																	
Gammarus aequicauda (Martynov, 1931)	99	199	87	111	128	132	320	165	172	116	41	233	1,803	49.80	1	49.80	De
Gammarus insensibilis Stock, 1966	89	19	38	113	92	11	89	12	18	13	6	94	594	16.40	1	16.40	De
Gammarella fucicola (Laech, 1814)					15		49	32	18	34	87	81	325	8.97	0.66	5.92	He
Elasmopus rapax Costa, 1853	9	1				5							15	0.41	0.25	0.10	S
Monocorophium acherusicum (Costa, 1853)	7	3					1					4	15	0.41	0.33	0.13	Ds
Ericthonius difformis (Milne-Edwards, 1830)	21	20	28	67	11	48	8	14	8	32	29		286	7.90	0.91	7.18	Ds
Cymadusa filosa Savigny, 1816	13	19	12	6		15	5	14		153	159	84	480	13.26	0.66	8.75	He
Microdeutopus gryllotalpa Costa, 1853							5	10	1	9		9	34	0.94	0.41	0.38	He
Dexamine spinosa (Montagu, 1813)							11	10	4	42			67	1.85	0.33	0.61	De
Caprella sp. (Laech, 1814)		1											1	0.02	0.08	0.00	O
N	238	262	165	297	246	220	488	257	221	399	322	505	3,620	100			
H'	2.19	1.53	1.70	1.54	1.37	1.05	1.74	1.74	1.03	1.92	1.68	1.98					
J'	0.78	0.54	0.85	0.97	0.68	0.45	0.58	0.62	0.4	0.74	0.72	0.76					

Tinja (Figure 1). In rainy seasons, autumn, winter, and early spring, the winds induce a vertical mixture of the water column, the rains are strong, and the freshwater flow coming from the Ichkeul Lake is important. On the other hand, in summer, the dry season, the influence of seawater is important and the water column warming can induce its stratification.

Physical properties, temperature, and salinity of the Bizerte lagoon varied, depending on rainy season and atmospheric forcing (Table 2). Water temperature and salinity ranged from 9°C in January to 30.86°C in August and from 30.8 psu in February to 39.2 psu in August, respectively, indicating a seasonal gradient of temperature and salinity in this lagoon. However, dissolved oxygen content showed an obvious seasonal variation from 4.8 $mg.l^{-1}$ in October (autumn) to 7.3 $mg.l^{-1}$ in May (spring). The turbidity varied from 2.1 NTU in April (spring) to 15.44 NTU in October (autumn). The high nitrogen anion content (NO_3^- and NO_2^-), measured during the sampling period, was about 3.7 $\mu mol.l^{-1}$ in winter. The maximum phosphorous content was about 0.38 $\mu mol.l^{-1}$ in August. The Chl *a* content ranged between 3.3 $\mu g.l^{-1}$ in April and 6.3 $\mu g.l^{-1}$ in August.

Except for the pH, homogeneous during the sampling period, the temperature, salinity, dissolved oxygen, and turbidity fluctuated with season (Table 2). In fact, the winter cold and wet conditions decrease the water temperature and salinity at 9°C and 30.86 psu, respectively. In summer, the hot and dry conditions lead to the opposite situation because of evaporation (30.86°C and 39.2 psu).

Amphipod assemblages and species affinities

Results of the frequencies (Ci) are divided into several categories in ascending order of constant according to the Lopez De La Rosa et al. (2006) classification. Therefore, *Caprella* sp. (Ci <12%) is considered as a scarce species; *Elasmopus rapax* (13% ≤ Ci ≤ 25%) a restricted species; *Monocorophium acherusicum*, *Dexamine spinosa*, and *Microdeutopus gryllotalpa* (26% ≤ Ci ≤ 50%) are common species; and *Gammarus aequicauda*, *Gammarus insensibilis*, *Gammarella fucicola*, *Cymadusa filosa*, and *Ericthonius difformis* are constant species (51% ≤ Ci ≤ 100%). According to the Ci% × Di% values (>5), the most important species in the amphipod assemblage are *G. aequicauda*, *G. insensibilis*, *G. fucicola*, *E. difformis*, and *C. filosa*, representing 88.074% of the total number of specimens.

Our results showed that seasons have a noticeable effect on the amphipod assemblage structure, at the Menzel Jemil station, affecting species richness, mean density, and diversity. *Caprella* sp. was absent in the samples except in November, where the presence of one specimen was probably due to the frequent tides in autumn. Although they are typical lagoon species, *M. acherusicum* and *E. rapax* were rarely observed in Menzel Jemil. In fact, the presence of the two species *D. spinosa* and *M. gryllotalpa* is limited to the spring period during which the *Gracilaria* are dominant. However, *E. difformis*, *C. filosa*, *G. insensibilis*, and *G. aequicauda* were collected throughout the sampling period and whatever the season, with varied abundance. The presence of these species coincides with that of the seagrass *C. nodosa* and the macrophyte *U. lactuca*.

The monthly species richness fluctuated between 3 and 8 in January and April, respectively (Figure 2). The lowest species richness was recorded in winter.

In fact, during the most part of the study period, the mean density according to the four depths (20, 40, 60, and 80 cm) exhibited a similar evolution to the species richness (Figure 2). In December, the mean density showed the lowest value and then increased from spring to summer. Shannon-Wiener diversity ranged between

Table 2 Monthly variation of physicochemical parameters of water in Menzel Jemil during the sampling period

Months	*T* (°C)	*S* (psu)	pH	O_2 ($mg.l^{-1}$)	Tr (NTU)	NO_2^- ($\mu M.l^{-1}$)	NO^-_3 ($\mu M.l^{-1}$)	PO^3_4 ($\mu M.l^{-1}$)	Chl *a* ($\mu g.l^{-1}$)
October	19.73	35.40	7.90	*(4.80)*	*(15.44)*	0.50	1.60	0.13	5.00
November	13.25	34.60	8.30	4.90	14.06	0.69	1.80	0.14	4.40
December	10.00	31.80	8.39	5.90	4.70	0.49	2.00	0.16	4.20
January	*(9.00)*	32.00	8.62	6.70	4.17	0.43	2.10	0.2	4.00
February	11.20	*(30.80)*	8.65	6.50	3.01	0.50	*(3.20)*	0.18	*(3.80)*
March	15.10	31.10	8.62	6.00	2.45	0.57	3.00	0.24	4.50
April	26.13	36.80	8.20	5.50	*(2.10)*	0.30	1.80	0.23	*5.30*
May	25.63	37.20	8.30	*(7.30)*	5.12	0.28	1.70	0.26	4.70
June	27.73	37.90	8.10	6.70	4.76	*(0.10)*	1.00	0.27	5.00
July	28.46	39.00	8.20	5.70	6.42	0.70	0.81	0.28	*5.70*
August	*(30.86)*	*(39.20)*	8.10	5.90	3.20	*(1.00)*	(0.73)	0.38	*(6.30)*
September	26.70	38.60	8.38	5.60	6.18	0.49	1.40	0.12	5.40

For each parameter, maximum and minimum values were indicated in italic and in parentheses.

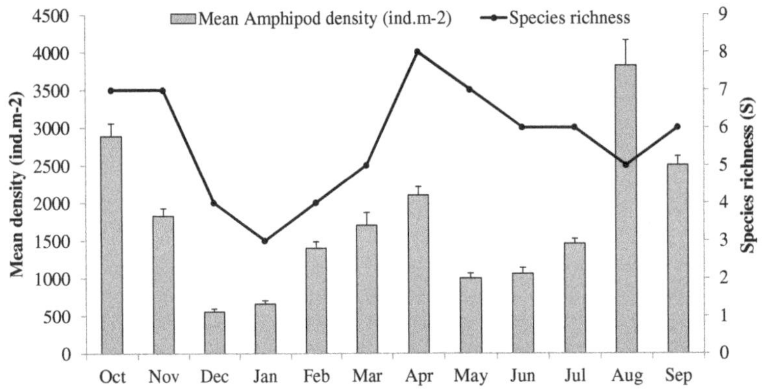

Figure 2 Monthly mean density and species richness of amphipods between October 2009 and September 2010.

1.03 bits in June and 2.19 bits in October. However, eveness index (*J'*) showed low values in June (0.4) and peaked in January (0.97) (Table 1).

Relationships between amphipod fauna and environmental variables

MDS ordination plot based on the mean species density revealed three distinct groups of samples corresponding to different periods of the year (Figure 3). Group I included the samples of October and November (autumn), in which species richness and Shannon index were higher. *G. aequicauda* and *G. insensibilis* exhibited the most important mean density in this group. Group II was made up of months from December to March (winter) in which greater mean densities of *G. aequicauda* and *E. difformis* were found with decrease of species richness and increase of mean evenness index (*J'*). Group III, composed of months from April to September (spring and summer), has a greater mean density of *C. filosa*. The number of species increased and peaked in April.

According to the results of canonical correspondence analysis between amphipods species and environmental variables, the first two axes accounted for 77.95% of the variance of species-environment relation (Figure 4). Forward selection in this analysis selected water chlorophyll *a* content, turbidity, and temperature as the variables explaining most of the variance in the species data ($p < 0.01$). Nitrate concentration, pH, and salinity had less influence on the system, while nitrite and phosphate concentrations and dissolved oxygen had a minimal influence. The ordination showed that temperature, dissolved oxygen, and nitrite concentration were related to axis F1 and that pH, salinity, and chlorophyll *a* content, phosphate and nitrate concentrations, and turbidity were related to axis F2 (Table 3).

G. fucicola and *M. gryllotalpa* were associated with chlorophyll *a* content and pH. *C. filosa* was related to high values of temperature and salinity. *G. aequicauda* was associated with turbidity and nitrate concentration.

Dominant feeding guilds in all studied period were herbivores and plant detritus feeders (three species) followed by deposit-suspension feeders (two species), but suspensivores and omnivores were only represented with one species each (Table 1). Herbivores reached their highest values in summer, represented by *G. fucicola* and *C. filosa*; however, plant detritus feeders were abundant in autumn and represented by the genus *Gammarus*.

Discussion

The vegetation in Menzel Jemil consists essentially of five species with a seasonal variation in their distribution. Indeed, *G. longissima* and *G. bursa-pastoris* dominate other species in the spring, *U. lactuca* whose biomass increases from the spring peaked in summer, *Cladophora* sp. appears in the winter, while *C. nodosa*, all year round, dominates in autumn. The temporal distribution of *G. bursa-pastoris* and *U. lacuta* in Menzel Jemil station is similar to that described by Sahli-Hazami (2004). In the lagoon of Venice, Italy, Fava et al. (1992) describe another seasonal succession of vegetation cover which consists mainly of *Ulva rigida*; the *G. bursa-pastoris* is present during the cold season, *Cladophora* sp. is occasional, and marine phanerogam *Zostera noltii* is found in limited fields.

Plant species richness in the Menzel Jemil station is lower than that described by Zaouali (1980) and Frisoni et al. (1986) in the lagoon of Bizerte; this could be due to eutrophication resulting in the presence of *Ulva lacuta* (Anonyme 2000), or the increase in salinity caused a decrease in species diversity (Zaouali 1980).

Monthly plant biomass fluctuates with an increase in early spring, peaking in June, and a decrease in summer. Indeed, the rise in water temperature in the spring causes the increase in plant biomass, but when the temperature reaches its maximum in August, plant biomass starts to decrease to reach the lowest values in winter. Highlighting the close relationship between temperature and plant biomass confirms the work of Antit-Ben Rejeb (2012) which showed the same effect of

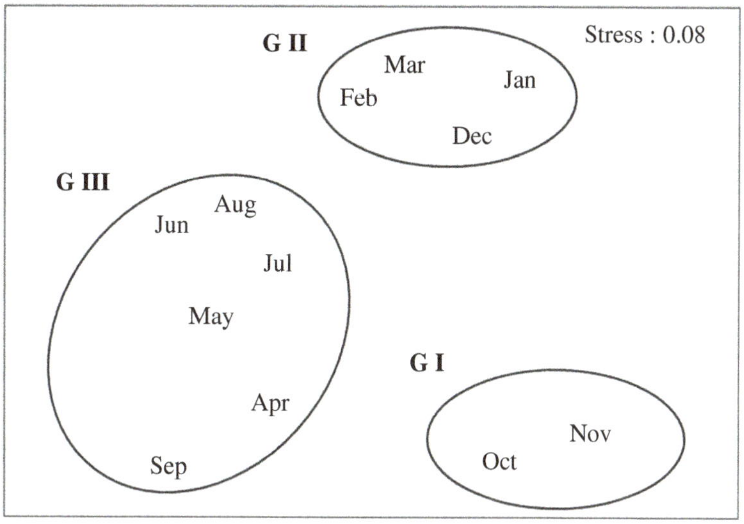

Figure 3 MDS ordination plots based on amphipod abundance of monthly sample.

temperature on the variation of biomass photophilic algae in the bay of Tunis.

In the Bizerte lagoon, water temperature ranges between 9°C in January to 30.86°C in August. Harzallah (2002) and Béjaoui et al. (2008) reported however a seasonal cycle in the lagoons, ranging between 10°C in winter and 28°C in summer, and from 11.5°C in January to 29.5°C in August, respectively. The water salinity ranges from 30.8 in February to 39.2 in August, which is higher than the values reported by ANPE (1990), 30 in winter to 38 in summer (Béjaoui et al. 2005). Thus, seasonal gradients of temperature and water salinity in the Bizerte lagoon are relatively large. Moreover, during the last decade, the temperature and salinity have increased. The evaporation in the lagoon is very important particularly in summer with 166 million $m^3.year^{-1}$. Precipitation does not exceed 146 million $m^3.year^{-1}$ in winter while it is weak and scarce in summer (74 million $m^3.year^{-1}$). The water input from the Ichkeul Lake has dramatically decreased during the last decades as an effect of the construction of dams discharging into this lake (MAERH 2003). This has led to a significant alteration of the water characteristics of the lagoon, particularly a dramatic increase in salinity (Béjaoui et al. 2005).

Dissolved oxygen content shows an obvious seasonal variation. Therefore, low water oxygenation logically

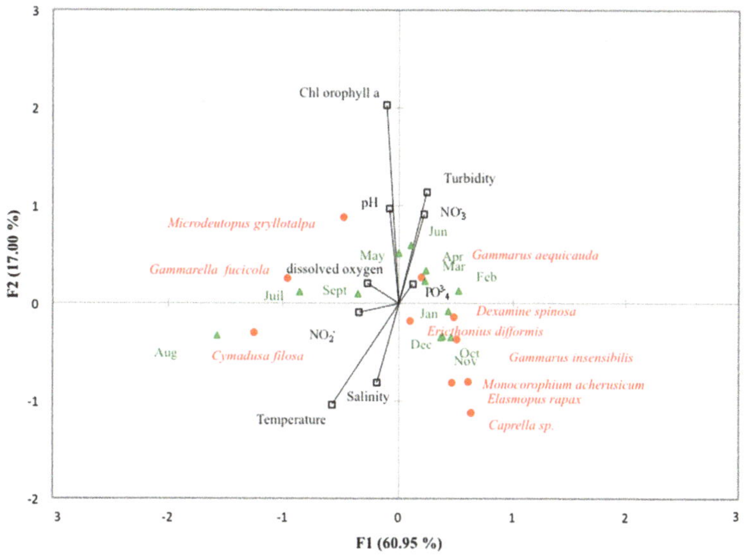

Figure 4 Canonical correspondence analysis ordination of environmental variables and sampled time. Red spot: species; green triangle: month; black square: physicochemical parameter.

Table 3 CCA: correlation matrix of environmental variables with the first two axes

	F1	F2
Temperature (°C)	−0.344	−0.096
Salinity (psu)	−0.188	−0.816
Dissolved oxygen ($mg.l^{-1}$)	−0.273	0.209
pH	−0.079	0.964
Turbidity (NTU)	0.244	1.140
NO^{-}_{2} ($\mu mol.l^{-1}$)	−0.574	−1.038
NO^{-}_{3} ($\mu mol.l^{-1}$)	0.223	0.905
PO^{3-}_{4} ($\mu mol.l^{-1}$)	0.119	0.195
Chl *a* ($\mu g.l^{-1}$)	−0.106	2.022

characterizes the dry period. These results agree with those of Béjaoui et al. (2008). Nutrient variability appears to be related to two factors, by seawater coming from the ship canal and also by freshwater coming from the Ichkeul Lake and its large catchment area. The high nitrogen anion concentrations (NO^{-}_{3} and NO^{-}_{2}) and phosphorous content can be related to the domestic sewage input from towns around the lagoon and specially Menzel Abderrahmen city or to the freshwater from Oued Guenich located on the eastern side of the Bizerte lagoon which is widely open to an agricultural zone where fertilizers are constantly applied. These factors may provide the explanation of the various content of chlorophyll *a* during the sampling period. However, the turbidity appears to be related to two factors, (i) hydrodynamic conditions and (ii) rainy season, which create water movements. Indeed, studies performed by Harzallah (2003) on the currents of the Bizerte lagoon reveal that the flow of water in both the surface and the bottom follows the wind direction, which is often northwest with low amplitude. Nevertheless, the interpretation of these results must take into account the special conditions of the Bizerte lagoon, which is a transitional area currently affected by many environmental variables (Afli et al. 2008). Moreover, nutrients and chlorophyll *a* concentrations recorded during the sampling period are lower than those observed before the connection of the urban and industrial waste to the ONAS network (National Sanitation Utility) (MAERH 2003; Béjaoui et al. 2008). Besides, coastal lagoons are considered to be highly productive ecosystems but they are vulnerable to human disturbance as a result of their semi-enclosed situation and their proximity to the sources of terrestrial effluents (Bazairi et al. 2005).

In Menzel Jemil (Bizerte lagoon, Tunisia), 3,620 specimens of amphipods were collected from October 2009 to September 2010 belonging to 8 families and 9 genera; the species are common in Tunisian lagoons (Diawara et al. 2008; Zakhama-Sraieb et al. 2009). Differences in the abundance of the ten species were observed. While *G. aequicauda* was the most abundant, *G. fucicola*, *C. filosa*, and *Caprella* sp. were recorded for the first time in the Bizerte lagoon. *M. gryllotalpa* was typically a lagoonal species. Compared to Diawara et al. (2008), the number of amphipod species herein recorded is lower than that found in the Tunis north lagoon; however, a relatively high number of amphipod species are found in the coastal lagoon of Smir (northwest of Morocco) by Chaouti and Bayed (2011), by Cherkoui et al. (2003) in the Bou Regreg estuary Moroccan Atlantic coast, and by Mogias and Kevrekidis (2005) in the Laki lagoon (Northern Aegean).

Our results show that the distribution of the amphipod fauna in the Bizerte lagoon is primarily linked to temperature and salinity. Oscillations in these environmental variables are related to the climatic season. An increase in temperature and salinity is characteristic of the dry season (summer), when the seawater penetrates the lagoon in a greater volume as a result of reduced rainfall, whereas in the wet season (winter), freshwater comes from rivers and the Ichkeul Lake and thereby reduces salinity. These results agree with studies on distribution patterns for other crustaceans in relation to the temperature and salinity of the water, which are considered extremely important for a better understanding of the dynamics of these species (Pinheiro 1991). Variations in temperature and salinity have been cited by a number of authors as important factors for the occurrence of particular amphipod species (Nevis et al. 2009; Mogias and Kevrekidis 2005). In fact, many ecological studies have shown salinity to be the main factor controlling zonation in the structure of the benthic communities (Cacabelos et al. 2010). However, Bazairi et al. (2003) have emphasized that this is not the main factor in brackish environments. Some other authors have demonstrated that modification in habitat complexity affects crustacean assemblages (Sanchez-Jerez et al. 1999; Ayala and Martin 2003; Vazquez-Luis et al. 2008). Indeed, the importance of temperature should be interpreted carefully, since Ysebaert and Herman (2002) pointed out that long-term averages of environmental variables are more important than values obtained during samplings (Cacabelos et al. 2010). Nevertheless, the difference in mean density between months can be explained by the ecological preferences of each species. Amphipod assemblages at the Bizerte lagoon are dominated in terms of density and species richness by the genus *Gammarus*. The temporal variability of the total density is mainly due to the numerically dominant species, such as *G. aequicauda*, *G. insensibilis*, and *C. filosa*. The seasonal variation in mean density recorded from our study area was similar to that described in other Mediterranean lagoons (Mogias and Kevrekidis 2005; Procaccini and

Scipione 1992). With respect to changes in algae and seagrass, which create refuge and food resource to diverse amphipods, an important reduction in species richness was recorded in January in our study contrary to Mogias and Kevrekidis (2005) in July.

Concerning the diversity of the communities, the diversity index (H') and evenness (J') values are low in the different seasons. Moreover, abundance and species richness values showed a sharp decrease during December, January, and February when both temperature and salinity diminish, whereas equitability and Shannon-Wiener indexes in the studied amphipod assemblage are more stable over time than those found in other areas.

The temporal distribution of amphipod assemblage studied by analyses of similarity between the samples, based on the abundance of species, exhibited three major assemblages whose distribution is strongly dependent on the water temperature and shows a seasonal pattern. Similar distribution patterns for amphipods and other crustaceans have been reported in several studies (Bazairi et al. 2003; Mogias and Kevrekidis 2005; Lopez De La Rosa et al. 2006; Vazquez-Luis et al. 2008; Nevis et al. 2009; Cacabelos et al. 2010; Zakhama-Sraieb et al. 2010). In fact, canonical correspondence analysis (CCA) made it possible to summarize the overall situation for the species, sampling months, and environmental parameters on a single graph. In accordance with our results, the temperature, turbidity, and chlorophyll *a* content are the most often reported factors in determining the distribution and composition of amphipods in Menzel Jemil. However, phosphate and nitrite concentrations do not show large variability. Thus, amphipod assemblages are probably not very influenced by their modest variations. On the other hand, these changes in environmental factors in this study may affect algae and seagrass which are support of the amphipod assemblages.

Amphipods play an important role in structuring benthic assemblages (Duffy and Hay 2000) as secondary and tertiary producers in marine communities (Beare and Moore 1996). Amphipods are important source of food for benthic fauna of commercial interest and also very ecologically sensitive organisms and good indicators of natural or disturbed environmental conditions (Conradi et al. 1997). Moreover, benthic databases are essential for comparisons valuable for impact studies or monitoring programs, in order to preserve the environment and the species of commercial importance that they support (Desroy et al. 2002).

Conclusions

A total of 3,620 amphipod individuals were collected in Menzel Jemil (Bizerte lagoon, Tunisia) belonging to ten species. *G. aequicauda* was the most abundant species, and *G. fucicola*, *C. filosa*, and *Caprella* sp. were recorded for the first time in the Bizerte lagoon. The diversity index (H') and evenness (J') values are low in the different seasons, while abundance and species richness values showed a sharp decrease during December, January, and February when both temperature and salinity diminish. According to canonical correspondence analysis, the temperature, turbidity, and chlorophyll *a* content are the most often reported factors in determining the distribution and composition of amphipods in Menzel Jemil.

Competing interests

The authors declare that they have no competing interests.

Authors' contributions

All the authors performed the field sampling, participated and coordinated in the design and analysis of the study, and drafted the manuscript. All authors read and approved the final manuscript.

Acknowledgements

The present study was funded by the Research Unit of Bio-ecology and Evolutionary Systematics (UR11ES11), Faculty of Science of Tunis, University of Tunis El Manar. We would like to thank Miranda Lowe (Natural History Museum of London) for improving the English language.

References

Afli A, Ayari R, Zaabi S (2008) Ecological quality of some Tunisian coast and lagoon locations, by using benthic community parameters and biotic indices. Estuar Coast Shelf Sci 80:269–280, doi:10.1016/j.ecss.2008.08.010

Afli A, Boufahja F, Sadraoui S, Ben Mustapha K, Aissa P, Mrabet R (2009) Functional organization of the benthic macrofauna in the Bizerte lagoon (SW Mediterranean Sea), semi-enclosed area subject to strong environmental/anthropogenic variations. Cah Biol Mar 50:105–117

Anderson MJ, Willis TJ (2003) Canonical analysis of principal coordinates: a useful method of constrained ordination for ecology. Ecol 84:511–525

Anonyme (2000) Evaluation accélérée du niveau de la mer en Tunisie, vulnérabilité et adaptation. Rapport définitif. TUN/95/G31: Projet National sur les changements climatiques. République Tunisienne, Ministère de l'Environnement et de l'Aménagement du territoire, 119 p

ANPE (1990) Etude préliminaire de l'écologie du lac de Bizerte. Rapport, Agence Nationale de Protection de l'Environnement, Tunis, 100 pp

Antit-Ben Rejeb M (2012) *Caractérisation des communautés de mollusques dans des milieux littoraux de la baie de Tunis*. Thèse de doctorat en sciences. Biologiques, Faculté des Sciences de Tunis, 384 p

Ayala Y, Martin A (2003) Relaciones entre la comunidad de anfipodos y las macroalgas a las que estàn asociados, en una plataforma rocosa del litoral central de Venezuela. Bol Inst Esp Oceanogr 19(1–4):171–183

Ayari R, Afli A (2003) Bionomie benthique du petit golfe de Tunis. Bulletin de l'Institut National des Sciences et Technologies de la Mer de Salammbô, Tunisia 30:79–90

Bazairi H, Bayed A, Glémarec M, Hily C (2003) Spatial organisation of macrozoobenthic communities in response to environmental factors in a coastal lagoon of the NW African coast (Merja Zerga, Morocco), Oceanol Acta 26: 457–471. doi:10.1016/S0399-1784(03)00041-0

Bazairi H, Bayed A, Hily C (2005) Structure et bioévaluation de l'état écologique des communautés benthiques d'un écosystème lagunaire de la côte atlantique marocaine. C R Biologies 328:977–990, doi:10.1016/j.crvi.2005.09.006

Beare DJ, Moore PG (1996) The distribution, growth and reproduction of *Pontocrates arenarius* and *P. altamarinus* (Crustacea: Amphipoda) at Millport, Scotland. J Mar Biol Ass UK 76:931–950, doi:10.1017/S0025315400040893

Béjaoui B, Ben Charrada R, Moussa M, Ben Hamadou R, Harzallah A, Chapelle A (2005) Caractérisation hivernale de la lagune de Bizerte. Bull Inst Natn Scien Tech Mer de Salammbô 32:79–91

Béjaoui B, Harzallah A, Moussa M, Chapelle A, Solidoro C (2008) Analysis of hydrobiological pattern in the Bizerte lagoon (Tunisia). Estuar Coast Shelf Sci 80:121–129, doi:10.1016/j.ecss.2008.07.011

Bellan-Santini D (1980) Relationship between populations of amphipods and pollution. Mar Poll Bull 11:224–227

Ben Mustapha K, Hattour A, Mhetli M, El Abed A, Tritar B (1999) Bionomie des étages infra et circalittoral du golfe de Gabès. Bull Inst Natn Scien Tech Mer de Salammbô, Tunisia 26:5–48

Blanchet H, Lavesque N, Ruellet T, Dauvin JC, Sauriau PG, Desroy N, Desclaux C, Leconte M, Bachelet G, Janson AL, Bessineton C, Duhamel S, Jourde J, Mayot S, Simon S, de Montaudouin X (2008) Use of biotic indices in semi-enclosed coastal ecosystems and transitional waters habitats - implications for the implementation of the European Water Framework Directive. Ecol Indic 8:360–372, doi:10.1016/j.ecolind.2007.04.003

Cacabelos E, Lourido A, Troncoso JS (2010) Composition and distribution of subtidal and intertidal crustacean assemblages in soft-bottoms of the Ria de Vigo (NW Spain). Scient Mar 74(3):455–464

Chaouti A, Bayed A (2011) Structure et organisation trophique du peuplement macrobenthique de la lagune méditerranéenne de Smir (Maroc). Bull Inst Sci Rabat 33(1):1–12

Chebbi N (2010) Etude systématique, bio-écologique et chimique des ascidies de Tunisie. Thèse de Doctorat, Institut National Agronomique de Tunisie, 301 p

Cherkoui E, Bayed A, Hilly C (2003) Organisation spatiale des peuplements macrozoobenthiques subtidaux, d'un estuaire de la côte atlantique marocaine. Cah Biol Mar 44:339–352

Clarke KR, Warwick RM (1994) Similarity-based testing for community pattern: the 2-way layout with no replication. Mar Biol 118:167–176, doi:10.1007/BF00699231

Conradi M, Lopez-Gonzalez PJ, Garcia-Gomez JC (1997) The amphipod community as a bioindicator in Algeciras Bay (Southern Iberian Peninsula) based on a spatio-temporal distribution. Mar Ecol 18(2):97–111, doi:10.1111/j.1439-0485.1997.tb00430.x

Danesi P, Gola G, Tagnin L (1999) Structure e dinamiche del macrobenthos ai confini dei partiacque nel bacino centrale della laguna di Venezia. In: Bon M., Sburlino G. And Zcarello V. (eds), Aspetti Ecologici et Naturalistici dei sistemi lagunairi e Costieri. Atti13. Conv. Gruppo Ecologia di Base "G. Gardio" Venezia 267–269

Dellali M, Elbour M, Aïssa P (2001) Evaluation de la pollution bactérienne dans la lagune de Bizerte : Résultats préliminaires. J Rech Oceanogr 26:18–28

Desroy N, Warembourg C, Dewarumez JM, Dauvin JC (2002) Macrobenthic resources of the shallow soft-bottom sediments in the eastern English Channel and southern North Sea. ICES J Mar Sci 60: 120–131. doi:org/10.1006/jmsc.2002.1333

Diawara M, Tlig-Zouari S, Rabaoui L, Ben Hassine OK (2008) Impact of management on the diversity of macrobenthic communities in Tunis north lagoon: systematics. Cah Biol Mar 49:1–16

Diviacco G, Bianchi CN (1987) Fauna interrelationships between lagoonal and marine amphipod crustacean communities of the Po River Delta (Northern Adriatic). Ann Biol (Biologia Ambiental) 12(3):67–77

Duffy JE, Hay ME (2000) Strong impacts of grazing amphipods on the organization of a benthic community. Ecol Monogr 70:237–263

Essid N, Aissa P (2002) Etude quantitative des nématodes libres des secteurs Nord et Est de la lagune de Bizerte (Tunisie). Bull Inst Natn Scien Tech Mer de Salammbô 29:53–63

Fava G, Zangaglia A, Cervelli M (1992) Ecology of *Idotea balthica* (Pallas) populations in the lagoon of Venice. Oceanol Acta 15(6):651–660

Fernandez-Gonzalez J, Sanchez-Jerez P (2014) First occurrence of *Caprella scaura* Templeton, 1836 (Crustacea: Amphipoda) on off-coast fish farm cages in the Mediterranean Sea. Helgol Mar Res 68:187–191, doi:10.1007/s 10152-013-0375-y

Frisoni G F, Guelorget O, Pertuisot J P, Fresi E (1986) Diagnose écologique et zonation biologique du lac de Bizerte. Applications aquacoles. Rapport du projet MEDRAP : regional Méditerranean developpement of aquaculture. FAO, 41 p

Guerra-Garcia JM, Garcia-Gomez JC (2001) The spatial distribution of Caprellidea (Crustacea: Amphipoda): a stress bioindicator in Ceuta (North Africa, Gibraltar area). Mar Ecol PSZN 22:357–367

Guerra-Garcia JM, Tierno de Figueroa JM, Navarro-Barranco C, Ros M, Sanchez-Moyano JE, Moreira J (2014) Dietary analysis of the marine Amphipoda (Crustacea: Peracarida) from the Iberian Peninsula. J Sea Res 85: 508–517. doi: org/10.1016/j.seares.2013.08.006

Harzallah A (2002) Etat actuel et évolution de l'exploitation halieutique et aquacole des lagunes - cas de la lagune de Bizerte. Rap Inst Natn Scien Tech Mer de Salammbô

Harzallah A (2003) Transport des polluants dans la lagune de Bizerte simulé par un modèle de circulation de l'eau. Bull Inst Natn Scien Tech Mer de Salammbô 30:115–133

Kevrekidis T, Mogias A, Gouvis N (2000) Interannual changes in the composition of the macrobenthic fauna of Drana lagoon (Evros Delta, N. Aegean Sea): preliminary note. Belg J Zool 130(Suppl 1):101–107

Kruskal JB, Wish M (1978) Multidimensional Scaling. Sage Publications, Beverly Hills, California, 93 p

Lopez De La Rosa I, Rodriguez A, Garcia Raso JE (2006) Seasonal variation and structure of a decapod (Crustacea) assemblage living in a *Caulerpa prolifera* meadow in Càdiz Bay (SW Spain). Estuar Coast Shelf Sci 66:624–633, doi:10.1016/j.ecss.2005.11.008

Lorenzen CJ (1967) Determination of chlorophyll and pheopigments by spectrophotometric equations. Limnol Oceanogr 12:343–346

MAERH (2003) Etude sur la dépollution industrielle dans le bassin versant du lac de Bizerte. Rapport, Ministère de l'Agriculture, de l'Environnement et des Ressources Hydrauliques, Tunis, 182 pp

Migné A, Davoult D (1997) Distribution quantitative de la macrofaune benthique du peuplement des cailloutis dans le détroit du Pas de Calais (Manche orientale, France). Oceanol Acta 20(2):453–460

Mizzan L (1999) Le specie alloctone del Macrozoobenthos della laguna di Venezia: Il punto della situazione. Boll Mus civ St nat Venezia 49(1998):145–177

Mogias A, Kevrekidis T (2005) Macrozoobenthic community structure in a poikilohaline Mediterranean lagoon (Laki Lagoon, northern Aegean). Helgol Mar Res 59: 167–176. doi:10.007/s10152-004-0215-1

Mouillot D, Laune J, Jean-Antoine Tomasini JA, Aliaume C, Brehmer P, Dutrieux E, Do Chi T (2005) Assessment of coastal lagoon quality with taxonomic diversity indices of fish, zoobenthos and macrophyte communities. Hydrobiologia 550:121–130, doi:10.1007/s10750-005-4368-y

Nevis AB, Martinelli M, Carvalho ASS, Nahum VJI (2009) Abundance and spatial-temporal distribution of the family Portunidae (Crustacea, Decapoda) in the Curuça estuary on the northern coast of Brazil. Braz J Aquat Sci Technol 13(1):71–79

Parsons TR, Maita Y, Lalli CM (1984) A manual of chemical and biological methods for seawater analysis. Pergamon Press, Oxford, 173 p

Pielou EC (1966) The measurement of diversity in different types of biological collections. J Theor Biol 13:131–144

Pinheiro MAA (1991) Distribuição e biologia populacional de *Arenaeus cribrarius* (Lamarck, 1818) (Crustacea, Brachyura, Portunidae), na enseada da Fortaleza. Ubatuba, SP. Dissertação de Mestrado. Instituto de Biociências – UNESP, 192 p

Procaccini G, Scipione MB (1992) Observations on the spatio-temporal distribution of crustacean amphipods in the Fusaro coastal lagoon (central Tyrrhenian Sea, Italy) and some notes on their presence in Mediterranean lagoons. Mar Ecol 13:203–224, doi:10.1111/j.1439-0485.1992.tb00351.x

Rossi F, Castelli A, Lardicci C (2006) Distribution of macrobenthic assemblages along a marine gradient in Mediterranean eutrophic coastal lagoons. Mar Ecol 27:66–75

Sahli-Hazami K (2004) Evaluation des potentialités de régénération de l'algue rouge Gracilaria sp. du lac de Bizerte et détermination de sa faune associée. Diplôme d'Etude Approfondes en Ecologie Générale. Faculté des Sciences de Tunis, 108 p

Sanchez-Jerez P, Barberà-Cebriàn C, Ramos-Espla AA (1999) Comparison of the epifauna spatial distribution in Posidonia oceanica, Cymodocea nodosa and unvegetated bottoms: importance of meadow edges. Acta Oecolog 20: 391–405. doi:org/10.1016/S1146-609X(99)00128-9

Shannon CE, Weaver W (1963) The mathematical theory of communication. Illinois University Press, Urbana, 117 p

Tlig-Zouari S, Maamouri-Mokhtar F (2008) Macrozoobenthic species composition and distribution in the northern lagoon of Tunis. Transit Waters Bull 2:1–15

Tlig-Zouari S, Mami T, Maamouri F (2009) Structure of benthic macroinvertebrates and dynamics in the northern lagoon of Tunis. J Mar Biol Ass UK 89(7):1305–1318, doi:10.1017/S0025315409000721

Vazquez-Luis M, Sanchez-Jerez P, Bayle-Sempere JT (2008) Changes in amphipod (Crustacea) assemblages associated with shallow-water algal habitats invaded by *Caulerpa racemosa var. cylindracea* in the western Mediterranean Sea. Mar Environ Res 65:416–426, doi:10.1016/j.marenvres.2008.01.006

Virnstein RW (1987) Seagrass-associated invertebrate communities of the southeastern USA: a review. Fla Mar Res Publs 42:89–116

Ysebaert T, Herman PMJ (2002) Spatial and temporal variation in benthic macrofauna and relationships with environmental variables in an estuarine, intertidal soft-sediment environment. Mar Ecol Prog Ser 244:105–124, doi:10.3354/meps244105

Zakhama-Sraieb R, Sghaier YR, Charfi-Cheikhrouha F (2006) Is amphipod diversity related to the quality of *Posidonia oceanica* beds? Biol Mar Medit 13:174–180

Zakhama-Sraieb R, Sghaier YR, Charfi-Cheikhrouha F (2009) Amphipod biodiversity of the Tunisian coasts: update and distributional ecology. JMBA2 Biodiversity Records, 2, e155 doi:10.1017/S1755267209990820 (published online)
Zakhama-Sraieb R, Sghaier YR, Charfi-Cheikhrouha F (2010) Community structure of amphipods on shallow *Posidonia oceanica* meadows off Tunisian coasts. Helgol Mar Res 65(2):203–212, doi:10.1007/s10152-010-0216-1
Zaouali J (1980) Flore et faune benthiques de deux lagunes tunisiennes: le lac de Bizerte (Tunisie septentrionale) et la mer de Bougrara (Tunisie méridionale). Bull Off natn Pêch Tunisie. Tunisie 4(1):196–200

18

Habitat suitability models of mountain ungulates: identifying potential areas for conservation

Prakash Kumar Paudel[1,2,4*], Martin Hais[2] and Pavel Kindlmann[1,3]

Abstract

Background: Determining the distribution of species and of suitable habitats is a fundamental part of conservation planning. We used slope and ruggedness of the terrain, forest type and distance to the nearest village to construct habitat suitability maps for three mountain ungulates (barking deer (*Muntiacus muntjak*), Himalayan goral (*Naemorhedus goral*) and Himalayan serow (*Capricornis thar*)) in the midhills of western Nepal. We used locations of sightings and signs of presence of these mountain ungulates collected during surveys along transect to derive a suitability value for each variable using Jacob's index. A multiplication approach was used to combine environmental variables and produce a habitat suitability map for each of the three species. An independent dataset was used to evaluate the maps using Boyce's index. This approach provides an overview of the probable distributions of the species in question.

Results: We predict that of the total area studied, 57% is suitable for *M. muntjak*, 67% for *N. goral* and 41% for *C. thar*. Although there are suitable habitats for all three species throughout the study area, the availability of high-quality habitats for these species varied considerably.

Conclusions: Suitable habitats for *N. goral* and *C. thar* were fragmented and mostly confined to the southern and northern parts of the study area. This study provides important baseline information for conservation biologists concerned with maintaining biodiversity in the midhills of Nepal.

Keywords: *Capricornis thar*; Habitat model; Midhills; *Muntiacus muntjak*; *Naemorhedus goral*; Nepal

Background

Human interference in the last remaining wilderness areas has resulted in a drastic decline in population size and distribution range of many species of wildlife (Mills 2009; Paudel and Kindlmann 2012a; Morrison et al. 2012). However, actions aimed to minimize this effect are often launched too late, usually after species and their habitats have already been seriously affected (Mittermeier et al. 1998; Myers et al. 2000; Brooks et al. 2002). Hence, it is important to maintain critical wildlife habitats (Poiani et al. 2000; Sala et al. 2000) and establish what determines the present distribution of species, which is especially difficult in mountainous areas and when species population density is low (Gibson et al. 2004). Generally, large-scale species distribution maps based on small-scale species-habitat association data were developed with the aid of Geographical Information Systems (GIS) (Osborne et al. 2001; Boitani et al. 2007). Such maps provide an important baseline for designing reserves (Abbitt et al. 2000; Ferrier 2002), species reintroduction (Pearce and Lindenmayer 1998) and rare species assessment (Engler et al. 2004).

The midhills in Nepal are an intermediary landscape between low-lying Tarai in the south and Himalayan region in the north. This region harbours the highest species diversity in the country (Paudel et al. 2012; Primack et al. 2013). However, ecosystems in the midhills were poorly studied in the past; therefore, there is almost no information on their biodiversity and consequently they are very poorly represented in the protected area network (Hunter and Yonzon 1993; Paudel et al. 2012; Paudel and Heinen 2015). The midhills are densely populated and the forest areas there are highly fragmented due to human exploitation for firewood, fodder and

* Correspondence: pk.paudel@gmail.com
[1]Department of Biodiversity Research, Global Change Research Centre AS CR, Bělidla 4a, 603 00 Brno, Czech Republic
[2]Faculty of Sciences, University of South Bohemia, Branišovská 31, 370 05 České Budějovice, Czech Republic
Full list of author information is available at the end of the article

timber including agricultural lands (Paudel and Šipoš 2014). Subsistence hunting, detrimental to wildlife, is also widespread here (Jackson 1979; Paudel 2012; Paudel and Kindlmann 2012a). Such activities in fragmented forests adversely affect the community structure of wildlife (Andrén 1994; Peres 2001; Fahrig 2003; Paudel and Kindlmann 2012a). For example, Himalayan tahr (*Hemitragus jemlahicus*), which was once common in the midhills of Nepal, is now extinct there because of habitat loss and hunting (Green 1979), and Himalayan serow (*Capricornis thar*) is confined to remote forest areas on the slopes of mountains (Wegge and Oli 1997; Paudel and Kindlmann 2012b). The same is true for many other wildlife species living in the midhills. Because of the remoteness and inaccessibility, the distributions of most species in this region are unknown.

Two broad approaches - (1) expert opinion integrated into GIS and (2) mathematical and machine learning algorithms (e.g. tree-based classification, neural networks and random forest, ecological niche factor analysis) - are used to produce habitat suitability maps (Pearce and Ferrier 2000; Guisan and Zimmermann 2000; Store and Kangas Store and Kangas 2001; Breiman 2001; Yamada et al. 2003). Expert opinion has been used with GIS modelling techniques to provide a basis for developing habitat suitability model when adequate empirical data are not available (Store and Kangas Store and Kangas 2001; Yamada et al. 2003; see details in Thuiller and Münkemüller 2010). There is some uncertainty regarding the reliability of such methods mainly because of potential errors in the expert judgments (Maddock and Samways 2000). Similarly, it is possible to use robust statistical approaches to produce habitat suitability maps (e.g. MADIFA, ENFA and BIOMOD) (Guisan and Zimmermann 2000; Hirzel et al. 2002; Rushton et al. 2004; Thuiller and Münkemüller 2010). Such approaches require advanced statistical and computing knowledge and a large dataset on environment collected using a stratified survey (Austin and Meyers 1996). Prior knowledge of environmental determinants of species distribution is crucial to improve the quality and reliability of predictions (Hirzel and Le Lay 2008). There is therefore a need for an alternative modelling approach that does not require sophisticated tools and can be implemented using limited environmental data based on documented species-habitat associations (Gavashelishvili and Lukarevskiy 2008). A multiplicative operation for combining important environmental variables is one of the widely used techniques of model construction (e.g. Liu et al. 2001; Dayton and Fitzgerald 2006).

This study aimed to produce habitat suitability maps for three mountain ungulates, still relatively common in the Nepalese midhills (barking deer (*Muntiacus muntjak*), Himalayan goral (*Naemorhedus goral*) and Himalayan serow (*Capricornis thar*)) and verify their accuracy. All of these ungulates are widely distributed across the Himalayan landscapes, but they exhibit distinct habitat affinities (see the 'Methods' section for details of the species studied). These maps have important applications not only for the conservation of these species, but also because these species serve as indicator species of healthy mountain ecosystem and form a part of the diet of endangered predators in the region: common and clouded leopard at low and snow leopard at high altitudes.

Methods

Study area

The study area (81°17′30.216″ to 82°42′58.158″ E, 28°28′ 57.261″ to 29°14′11.418″ N) comprises approximately 2,844 km^2 in the midhills of Western Nepal, situated between the Bardia National Park in the lowland Tarai and the Shey Phoksundo National Park in the mountain region (Figure 1).

The habitats vary from dense subtropical deciduous sal forest (*Shorea robusta*) in the south to alpine pastures in the north. The study area spans over five ecoregions: Himalayan subtropical broadleaf forest, endangered western Himalayan broadleaf forest (Olson et al. 2001), Himalayan subtropical pine forest, western Himalayan subalpine coniferous forest and western Himalayan alpine shrub and meadow (Wikramanayake 2002). The areas below 1,000 m, with a tropical climate, are dominated by sal (*Shorea robusta*) forest. Areas at altitudes of 1,000 to 2,000 m, with a subtropical climate, are dominated by chir pine (*Pinus roxburghii*) with an admixture of *Quercus incana*, *Quercus lamellosa*, *Rhododendron arboreum*, *Alnus nepalensis* etc. Areas at altitudes of 2,000 to 3,000 m, with a temperate climate, are dominated by oak, rhododendron and laurel stands along with conifers such as blue pine, fir and hemlock. In the subalpine zone (3,000 to 4,000 m), several conifers (*Pinus wallichiana*, *Abies spectabilis*, *Abies pindrow*, *Tsuga dumosa*) and broad-leaved species (*Betula utilis*, rhododendrons and *Quercus semecarpifolia)* form diverse types of vegetation (Paudel et al. 2012). In the alpine zone, above 4,000 m, there are some shrubs, but grassland is more common. The region also supports a high faunal (especially mammal) diversity, including 7% of the mammals endemic to this region (Paudel et al. 2012; Primack et al. 2013). Summer monsoon (June-September) accounts for about 80% of the precipitation and winter rain is infrequent (Das 1968).

Subsistence agriculture, supporting not much more than the people's basic survival, is the main occupation. Because of rapid population growth and expansion of the road network, human activities are putting an ever increasing pressure on the forests and the species living there (Paudel 2012).

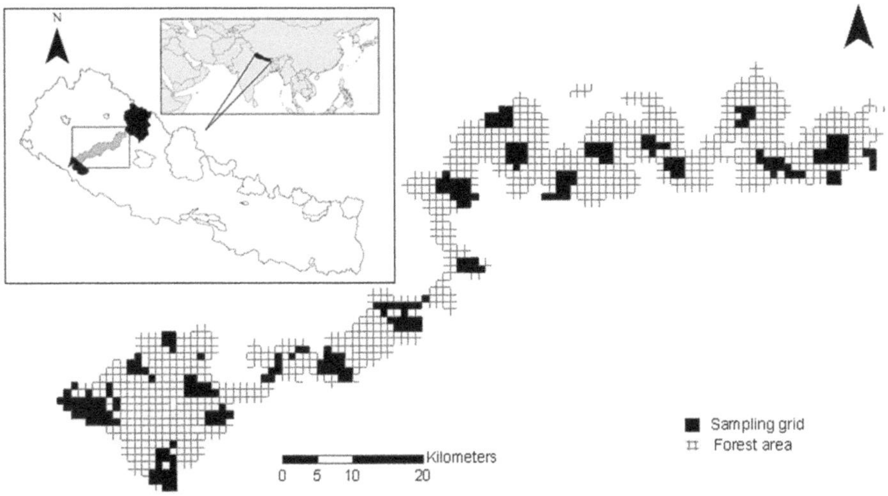

Figure 1 Map of a 1 km × 1 km grid superimposed over the study area. The black regions are the grids that were sampled. The inset shows a map of the study area (in grey) showing the protected areas Bardia National Park in the south and Shey Phoksundo National Park in the north (in black) of Nepal. Location of the study area in the box (in black).

Species studied

Barking deer is a relatively small and solitary cervid, which is widely distributed throughout large parts of Southeast Asia (Ohtaishi and Gao 1990; Prater 1990; Roberts 1999). It prefers dense forest with a high understory (Teng et al. 2004; Odden and Wegge 2007) but is also found in degraded forest areas near human settlements (Oka 1998). Because it is territorial, large concentrations of individuals of this species are rarely seen (Odden and Wegge 2007). Barking deer prefers plains and gentle slopes, but such areas have been degraded over the years. Their inability to live in the open and on rugged mountain slopes is limiting them to areas around human settlements (Paudel and Kindlmann 2012a).

Himalayan goral is associated with a wide variety of habitats throughout the mountains (Green 1987; Wegge and Oli 1997). It prefers open plant communities with good grass cover and avoids shrub-rich patches (Cavallini 1992), feeds primarily on grass in all seasons (Mishra and Johnsingh 1996; Fakhar-I-Abbas et al. 2008) and is adapted to steep and rugged mountain terrain (Green 1987; Cavallini 1992; Roberts 1999; Paudel and Kindlmann 2012a). Thus, the steep slope is an important feature preferred by goral (Cavallini 1992; Mishra and Johnsingh 1996; Paudel and Kindlmann 2012a).

Himalayan serow is an IUCN's near threatened (NT) mountain ungulate. It was widely distributed in the mountains of Himalaya (eastern part of the river Jamuna in Bangladesh, northern India, Tibet and Nepal). Now, it is very rare and confined to a few scattered, isolated populations in its former ranges (Green 1987; Grubb 2005) because of habitat loss and hunting (Prater 1990; Paudel and Kindlmann 2012a). In Nepal, the exact distribution of serow is unknown and is restricted to steep forested hills and cliffs, in areas relatively inaccessible to the humans and in protected areas (Wegge and Oli 1997; Paudel and Kindlmann 2012a). Serow inhabits rugged mountains or ridges covered with thick bushes or forests (Nowak and Wilson 1999).

Species distribution data

We collected presence data for these three mountain ungulates by means of surveys along transects in 21 'clusters' in 2008, 2009 and 2010 (December-June), which were selected as follows: First, we superimposed a 1 km × 1 km grid on the map of the study area using ArcMap 9.2© Geographic Information System. Then, we excluded pixels with less than 50% forest cover, which yielded a set of pixels shown in Figure 1. We generated random points with an 'inhibition distance' (i.e. the smallest distance allowed between any two random placed points) of 5 km. This procedure gave us 24 random points (hereafter 'cluster'). Among them, three points occurred in human habitation and therefore they were excluded. Around each of these 21 clusters, we laid out 10 to 20 transects (depending on its size) at least 100 m from the edge of a forest. We maintained at least 100-m distance between two transects based on the accessibility and availability of the forests. The transects included riversides and plains, ravines, mountain ridges and steep cliffs.

In each transect, the presence of species was assessed on the basis of the presence or absence of footprints and faeces within a circular plot with a radius of 2.5 m and by direct sighting at 100-m intervals along the transect, henceforth referred to as 'sampling points'. Two people, mostly local people acquainted with the forests, were involved in searching for signs of the presence of the species in question at each sampling point. Since sampling points were relatively small (approximately 20 m^2), we searched carefully for signs of presence (footprints and faeces), and it is likely that all such signs were detected. If the animal is observed during searching of presence signs, we counted a corresponding sampling point as present. Here, we should be cautious that the absence site could be a pseudo-absence one. This will not affect our results because we are predicting the relative probability of occurrence of the species and the absences are interpreted as unused sites. In order to avoid confusing signs of the presence of livestock (e.g. sheep and goats) with those of wild herbivores, we tested our ability to distinguish footprints and faeces of barking deer, Himalayan goral and Himalayan serow from those of livestock. This test indicated that we were able to distinguish between the signs left by the three ungulates and livestock. Surveys were carried out at 4,328 sampling points along 432.8 km of transects, which varied in length from 0.3 to 2.5 km depending on the steepness of the terrain and vegetation. These surveys revealed a total of 687 signs of presence: 229 of barking deer, 316 of Himalayan goral and 142 of Himalayan serow. At each sampling point, we determined (1) vegetation type, elevation and (2) slope (Table 1). Other explanatory data (e.g. distance to the nearest village, topographic ruggedness; see Table 1 for details) were derived from the GIS.

Environmental variables

Based on documented species-habitat associations (Green 1987; Prater 1990; Cavallini 1992; Mishra and Johnsingh 1996; Wegge and Oli 1997; Roberts 1999; Grubb 2005; Odden and Wegge 2007; Paudel and Kindlmann 2012b; Paudel and Kindlmann 2012a), our field experience and models developed for similar species (Kushwaha et al. 2000), we used various numbers of categories of four main environmental variables (Table 1): We selected environmental variables describing landscape (i.e. slope, topographic ruggedness), habitat type (i.e. vegetation) and human influences (i.e. distance to the nearest village) based on the documented habitat affinities. We tested the performance of these environmental variables for predicting species presence using logistic regressions (Menard 2001). Before performing the regression analyses, we tested for multicollinearity among the data using a Pearson correlation matrix, because environmental variables that are correlated ($r > 0.7$) can bias a model (Berry and Feldman 1985). Nagelkerke R^2 and receiver operating characteristic curves (ROC) or area under curve (AUC) were used to evaluate the association of the habitat variables with the presence of species. ROC is a plot of true positive cases (or sensitivity) on the y-axis against corresponding false positive cases (or specificity) on the x-axis for a range of threshold values (Fielding and Bell 1997). It provides a measure of discrimination ability, which varies from 0.5, when it is no better than expected on the basis of random choice, to 1.0, when it is perfectly discriminatory. Our results indicated that the environmental variables we selected had a high discriminatory ability (Table 2).

We classified environmental variables into a number of classes. Such classes are important when associations between species and environmental variables are non-linear. Class boundaries here are based on ecological requirement of species in question or types of data:

(i) slope (6 categories, spanning 15° each: 0 to 15, 15 to 30, ..., 75 to 90),
(ii) topographic ruggedness (categorized into four classes using natural breaks based upon Jenks optimization: 0 to 0.05, 0.05 to 0.19, 0.19 to 0.44 and 0.44 to 0.99 - see below for how these values were derived),
(iii) forest type (11 classes of forests and 2 classes of land type: bushy areas and alpine meadow), and
(iv) distance to the nearest village (less than 250, 500, 750, and 1,000 m and greater than 1,000 m)

The environmental data used here were the presence/absence data, collected in 2008 and 2009 along 325.8 km of transects (76% of total 432.8 km) for model development (see below about the model: 'Preparation and evaluation of the suitability map'). This dataset was used to estimate relative probability of usage of the different categories of habitat based on four environmental variables. The remaining presence records, approximately 24% of the transect data, collected along 102 km of transects (collected in 2010), were used to evaluate the above model (see 'Preparation and evaluation of the suitability map').

Two Landsat ETM scenes taken on 25 December 2001 (path 143, row 40) and on 21 February 2003 (path 144, row 40), when there was no cloud cover, were obtained from the http://glovis.usgs.gov web pages. Both scenes were geometrically corrected. The data were classified using supervised classification (maximum likelihood - Geomatica 2005) based on 146 training areas (polygons), which included 11 vegetation and two land use units (see Table 1). The geographic coordinates of the training areas were determined in the field, using GPS mapping. As some types of vegetation have identical spectra (e.g. grassland vs. alpine meadow; lower temperate broad-leaved forest vs. temperate broad-leaved forest), it was not possible to distinguish between them so they were either aggregated or when possible separated on the basis of altitude (Digital Elevation Model). We separated types of vegetation on the basis of altitude only when they do not overlap

Table 1 Description of environmental variables used to produce the suitability maps

Variable	Description
Slope	
0 to 15	*Slope* of the land surface relative to the horizontal measured in degrees
15 to 30	
30 to 45	
45 to 60	
60 to 75	
75 to 90	
Ruggedness	
0 to 0.05	*Topographic ruggedness* measured in terms of the variability in slope and aspect (Sappington et al. 2007)
0.05 to 0.19	
0.19 to 0.44	
0.44 to 0.99	
Forest type	
STSF	*Subtropical sal forest*: north tropical dry deciduous forest (Champion and Seth 1968), sub-tropical hill sal forest (Stainton 1972), hill sal forest (TISC 2002); altitudinal range: 300 to 1,000 m
SPF	*Subtropical pine forest*: chir pine forest (TISC 2002), sub-tropical pine forest (Champion and Seth 1968); altitudinal range: 1,000 to 2,000 m
STPBLF	*Subtropical pine broad-leaved forest*: chir pine broad-leaved forest (TISC 2002); altitudinal range: 1,000 to 2,000 m
UTCF	*Upper temperate blue pine forest*: (TISC 2002); altitudinal range: 2,500 to 3,000 m
RARA	*Grass and rocky areas*: barren hillsides devoid of trees because of rocks, frequently with dense forest nearby; altitudinal range: 300 to 3,200 m
TMF	*Temperate mixed forest*: fir-hemlock-oak forest (TISC 2002), fir-oak-rhododendron forest (TISC 2002), mixed blue pine-oak forest (TISC 2002); altitudinal range: 2,000 to 3,000 m
LTBLF	*Lower temperate broad-leaved forest*: lower temperate oak forest (TISC 2002), *Alnus* woods (Stainton 1972); altitudinal range: 2,000 to 2,500 m
TBLF	*Temperate broad-leaved forest*: oak-rhododendron forest (TISC 2002), rhododendron forest (TISC 2002), temperate mixed broad-leaved forest (Stainton 1972)
SABLF	*Subalpine broad-leaved forest*: birch-rhododendron forest (TISC 2002), subalpine mountain oak forest (TISC 2002), *Quercus semecarpifolia* forest (Stainton 1972; Champion and Seth 1968); altitudinal range: 3,000 to 4,000 m
SAMF	*Subalpine mixed forest*: fir-oak-rhododendron forest (TISC 2002), fir-hemlock-oak forest (TISC 2002); altitudinal range: 3,000 to 4,000 m
SACF	*Subalpine conifer forest*: fir forest (TISC 2002), *Abies spectabilis* forest (Stainton 1972), *Abies pindrow* forest (Stainton 1972); altitudinal range: 3,000 to 4,000 m
BA	*Bushy area*: forest area covered with bushes; altitudinal range: 300 to 4,000 m

Table 1 Description of environmental variables used to produce the suitability maps *(Continued)*

AM	*Alpine meadow*: altitudinal range: >3,500 m
Distance to the nearest village	
250	*Forest within a 250-m radius* of the nearest village
500	*Forest within a 500-m radius* of the nearest village
750	*Forest within a 750-m radius* of the nearest village
1,000	*Forest within a 1,000-m radius* of the nearest village
>1,000	*Forest outside a 1,000-m radius* of the nearest village

in their altitudinal ranges. The resulting data were filtered (SIEVE) in order to remove disjointed single pixels. The accuracy of this classification was evaluated using the data (783 points) that were not used in the original classification.

ASTER Global Digital Elevation Model (ASTER GDEM) data were used to derive terrain data. ASTER GDEM data were accessed from http://gdem.ersdac.jspacesystems.or.jp using a map of the study area. The data are available at a resolution of 30 m per pixel. Topographic ruggedness was calculated using the VRM (vector ruggedness measure) (Sappington et al. 2007) using an ArcGIS script in a 3×3 window size (http://arcscripts.esri.com/details.asp?dbid=15423). It is based on a geomorphological method for measuring vector dispersion that is less correlated with slope (Sappington et al. 2007). The dimensionless ruggedness number ranges from 0 (flat) to 1 (most rugged).

Using the land cover map of the study area (Department of Survey), a map of distance to the village was prepared by classifying forest area into five categories based on its proximity to the nearest village (less than 250, 500, 750, and 1,000 m and greater than 1,000 m) in ArcGIS. This map was rasterized at a resolution of 30 m.

Preparation and evaluation of the suitability map

We used approximately 76% of data (collected along 325.8 km of transects in 2008 and 2009) for evaluating species-habitat associations (model development) and remaining dataset for model verification. To see species-habitat associations, we defined the availability (p) for each category of environmental variable as the proportion of all records in all clusters, for which the corresponding value of the environmental variable was 'true' (e.g. in which the slope was between 15° and 30° if p for the environmental variable 'slope between 15° and 30°').

For each species and environmental variable, we defined species usage (r) as the proportion of records for which the corresponding value of this environmental variable was 'true' and associated with signs of this species occurrence (e.g. r value for barking deer in areas

Table 2 Results of binomial logistic regressions of the selected habitat variables

Model	Barking deer				Himalayan goral				Himalayan serow			
	Parameter estimates	SE	Wald	*P* value	Parameter estimates	SE	Wald	*P* value	Parameter estimates	SE	Wald	*P* value
Topographic variables												
Slope	−0.052	1.010	47.886	0.000	0.066	0.01	99.32	0.000	0.020	0.010	3.810	0.050
Ruggedness	−3.363	0.940	12.810	0.000	0.908	0.76	1.446	0.229	5.130	1.960	6.83	0.009
Constant	2.495	0.320	62.020	0.000	−2.492	0.25	98.34	0	−1.8	0.340	31.400	0.000
Nagelkerke R^2		0.335				0.411				0.331		
AUC		0.805				0.830				0.800		
Distance to the village	0.002	0.000	45.706	0.000	0.001	0.000	66.345	0.000	0.001	0.000	33.365	0.000
Constant	−1.299	0.209	38.553	0.000	−1.513	0.201	56.458	0.000	−1.514	0.292	26.967	0.000
Nagelkerke R^2		0.241				0.217				0.261		
AUC		0.786				0.741				0.837		

with a slope between 15° and 30° was calculated as the ratio of the number of records of signs of the presence of barking deer divided by the total number of sampling points where the slope was between 15° and 30°).

We then calculated Jacob's index (*D*) as $D = \frac{r-p}{r+p-2rp}$ (Jacobs 1974) for each species and each category of environmental variable. The value of *D* varies from 1 for maximum association to −1 for no association.

For each species and each of the four main environmental variables (slope, topographic ruggedness, forest type and distance to the village), a suitability map was constructed in ArcGIS as follows. Each pixel of the study area was categorized into one of the five classes of habitat suitability according to the *D* value corresponding to this pixel and of environmental variables and species in question (described in Figure 2 and Table 3). The final *D* value of all different categories of environmental variables is given in Table 4.

For each species, the four suitability maps (one for each of the four environmental variables) were multiplied using the map calculator in ArcGIS 9.2 to create a single suitability map for the species (Liu et al. 2001; Dayton and Fitzgerald 2006). This is a better method than a weighted summation because it eliminates optimistic prediction (e.g. 0 suitability in any of the environmental variables gets unsuitable score finally). Furthermore, environmental variable here are already weighted while assigning suitability scores based on Jacob index (*D*). The map suitability value calculated in this way ranged from 0 (no suitability) to 144 (very high suitability). Finally, we derived four categories (poorly suitable, moderately suitable, suitable and highly suitable) from this continuous scale of 0 to 144 based on the multiplication of three suitability values for each category (e.g. 4 is the value for a highly suitable category of environmental variable. The multiplicative value of the three highly suitable categories is 64. Thus, a pixel with a value greater than 64 was classified as highly suitable in the final map). Thus, we classified each pixel as follows: 0 - unsuitable, 1 to 8 - poorly suitable, 9 to 27 - moderately suitable, 28 to 64 - suitable and greater than 64 - highly suitable. Because we ranked some environmental factors as 0 and used a multiplicative approach, the final map had many areas with a value 0, indicating that they are unsuitable for the species in question regardless of the suitability of the other environmental variables.

In surveys of large mobile wildlife, absence data are in fact 'pseudo absence', because it is not possible to be certain that the particular location is not used by the target species. We, therefore, used a presence-only data for model evaluation that discriminates a model predicting presence everywhere from a more contrasted model by using predicted-to-expected (*R*) ratio of each habitat suitability category (Boyce et al. 2002) as $R = P_i/E_i$, where P_i is the proportion of presence records for evaluation points with a habitat suitability class *i* and E_i is the relative area covered by the suitability class *i*. The basic assumption of this evaluation procedure is that a good suitability map should be based on a better evaluation of presence than expected by chance, resulting in $R > 1$. Thus, highly suitable habitats should have a proportionally higher number of presence records (Hirzel et al. 2006). We calculated the Boyce index (B_b) by computing a Spearman rank correlation between *R* and class rank, which varies between −1 and 1. Positive values indicate a model, whose predictions are consistent with the distribution of presences in the dataset evaluated, negative values indicate an incorrect model and values close to zero indicate that the model's prediction is not different from random (Boyce et al. 2002; Hirzel et al. 2006). The model evaluation was carried out using an independent dataset of species presences that was collected along 104 km of transects. It included 52 records of the

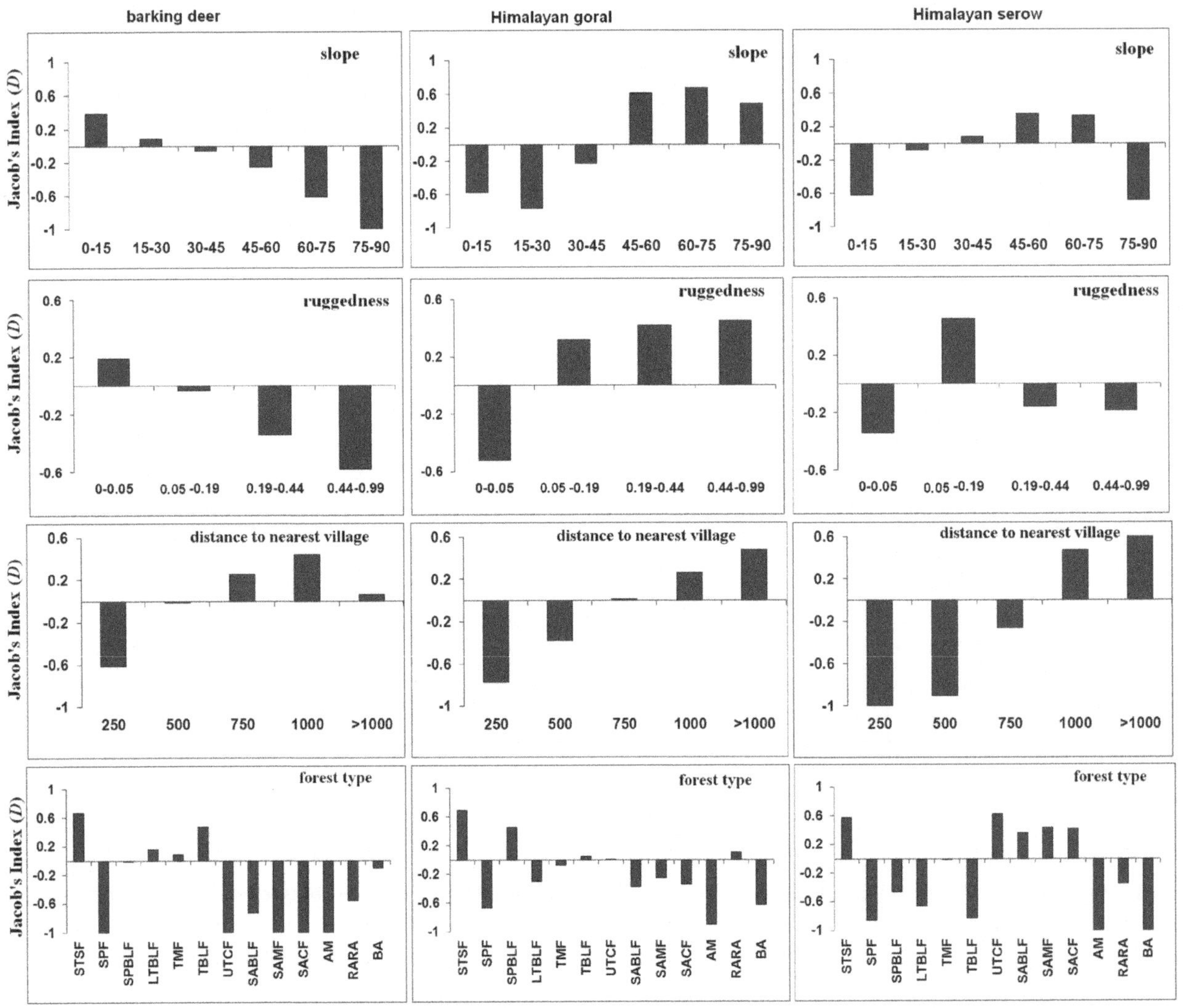

Figure 2 Habitat preferences of each species based on Jacob's index. The first, second and third columns correspond to barking deer, Himalayan goral and Himalayan serow, respectively (see Table 1 for the abbreviations).

presence of barking deer, 63 of Himalayan goral and 38 of Himalayan serow.

We also evaluated whether the observed frequencies of occurrence of the three species in each category of habitat quality were different from those expected based on the availability of habitat categories using a chi-square goodness of fit test. Bonferroni confidence interval (adjusted):

$$p_i - z_{(1-\alpha/2k)}\sqrt{p_i(1-p_i)/n} \le p_i \le p_i + z_{(1-\alpha/2k)}\sqrt{p_i(1-p_i/n}$$

was used to determine, which habitat categories were drivers of significant changes in the model's prediction. Here, p_i is the proportion of presence signs in the ith suitability category, n the sample size (number of recorded presences) and k the number of parameters used in the model.

Table 3 The conversion of Jacob's index (*D*) into corresponding suitability values

Jacob's index (*D*) range	Suitability categories	Suitability value
−1	Unsuitable	0
−1 to −0.5	Poorly suitable	1
−0.5 to 0	Moderately suitable	2
0 to 0.5	Suitable	3
0.5 to 1	Highly suitable	4

The lower bound of the class is not included in the corresponding class.

Results

Logistic regression revealed the discriminatory power of the habitat variables for indicating the presence of all the species (Table 2). *Barking deer* was not recorded in areas with a highly rugged and steep terrain and those close to

Table 4 Suitability rating of the four factors described in Table 3

Variable	Barking deer	Himalayan goral	Himalayan serow
Slope (degrees)	Suitable (<30), moderately suitable (30 to 60), poorly suitable (60 to 75), unsuitable (>75)	Highly suitable (45 to 75), suitable (75 to 90), moderately suitable (30 to 45), poorly suitable (<30)	Suitable (30 to 75), moderately suitable (15 to 30), poorly suitable (0 to 15, 75 to 90)
Ruggedness	Suitable (<0.05), moderately suitable (0.05 to 0.44), poorly suitable (0.44 to 0.99)	Suitable (>0.05), poorly suitable (<0.05)	Suitable (0.05 to 0.19), moderately suitable (>0.19), poorly suitable (<0.05)
Distance to the nearest village (metres)	Suitable (>500), moderately suitable (250 to 500), poorly suitable (>250)	Suitable (>750), moderately suitable (250 to 500), poorly suitable (<250)	Highly suitable (>1,000), suitable (750 to 1,000), moderately suitable (500 to 750), poorly suitable (250 to 500), unsuitable (<250)
Forest type	Highly suitable (STSF), suitable (LTBLF, TMF, TBLF), moderately suitable (BA, STPBLF), poorly suitable (SABLF, RARA), unsuitable (SPF, UTCF, SAMF, SACF, AM)	Highly suitable (STSF), suitable (SPBLF, TBLF, UTCF, RARA), moderately suitable (LTBLF, TMF, SABLF, SAMF, SAMF), poorly suitable (SPF, AM, BA)	Highly suitable (STSF, UTCF), suitable (SACF, SAMF, SABLF), poorly suitable (SABLF, TMF, RARA), unsuitable (AM, BA)

human settlements (Figure 2); it occurred mainly in subtropical sal forest, lower temperate broad-leaved forest, temperate mixed forest and temperate broad-leaved forest (Figure 2). Topographic variables were stronger indicators of the presence (Nagelkerke R^2 0.335, AUC 0.805) than the distance to the nearest village (Nagelkerke R^2 0.241, AUC 0.786) for this species. *Himalayan goral* occurred mainly in subtropical sal forest, subtropical pine broad-leaved forest, temperate broad-leaved forest and rocky and grassy areas (Figure 2). Topographic variables were stronger indicators of the presence (Nagelkerke R^2 0.411, AUC 0.830) than distance to the nearest village (Nagelkerke R^2 0.217, AUC 0.741) for this species (Table 2). *Himalayan serow* occurred mainly in subtropical sal forest, upper temperate coniferous forest, subalpine coniferous forest and subalpine broad-leaved forest (Figure 2). Both topography (Nagelkerke R^2 0.331, AUC 0.800) and distance to the nearest village (Nagelkerke R^2 0.261, AUC 0.837) were important indicators of the presence of this species (Table 2).

The Boyce indices (B_b) for barking deer and Himalayan goral were at their maximum theoretical limit (1), confirming excellent discrimination (Table 5). The index for Himalayan serow was also high (0.6), which indicates that the model consistently indicates a distribution for goral that accords with its distribution. Predicted-to-expected (R) ratio was also consistently high for the high suitability class and greater than one for the highly suitable habitat for all target species. According to Boyce et al. (2002) and Hirzel et al. (2006) such patterns in suitability maps are consistently close to reality.

Suitable habitat based on all the categories for barking deer covered 57% of the study area. Of all the suitable habitats for this species, 18%, 19%, 33% and 30% are poorly suitable, moderately suitable, suitable and highly suitable habitats, respectively (Figure 3). Approximately 79% of the signs of the presence of barking deer occurred in the suitable and highly suitable habitats. The records of barking deer in the predicted habitat categories were disproportionate relative to their availability ($\chi^2 = 11.95$, $df = 3$, $P < 0.007$) as this species was recorded in highly suitable habitats significantly more frequently than expected by chance (Table 6).

Suitable habitat in terms of all the categories for Himalayan goral covered 67% of the study area. Of all the suitable habitats, 38%, 26%, 32% and 4% were poorly suitable, moderately suitable, suitable and highly suitable habitats, respectively (Figure 4). Approximately 63% of the signs of the presence of Himalayan goral were recorded in suitable and highly suitable habitats. The records of goral in the predicted habitat categories were disproportionate relative to their availability ($\chi^2 = 56.86$, $df = 3$, $P = 0.001$) as this species was recorded in highly suitable habitats significantly more frequently than expected by chance (Table 6).

Suitable habitats in terms of all the categories covered 41% of the study area for Himalayan serow. Of all the suitable habitats, 13%, 23%, 24% and 40% were poorly suitable, moderately suitable, suitable and highly suitable habitats, respectively (Figure 5). Approximately 89% of the records of the presence of Himalayan serow were in suitable and highly suitable habitats. The records of serow in predicted habitat categories were disproportionate relative to their availability ($\chi^2 = 10.84$, $df = 3$, $P = 0.012$) as significantly fewer than expected were for poorly suitable and moderately suitable habitats and significantly more for suitable and highly suitable habitats (Table 6).

Discussion

When identifying areas for conservation it is important to know the spatial distribution of good-quality habitat for the target species. We provide landscape-scale maps of the distribution of three mountain ungulates based on

Table 5 Boyce index (B_b) of the habitat suitability maps for barking deer, Himalayan goral and Himalayan serow

	Suitability (*i*)	Proportion of evaluation points (P_i)	Relative area covered by a particular suitability class (E_i)	Predicted-to-expected ratio	Boyce index (Bb)
Barking deer	Poorly suitable	0.08	0.18	0.44	1.00
	Moderately suitable	0.13	0.19	0.68	
	Suitable	0.27	0.33	0.82	
	Highly suitable	0.52	0.31	1.68	
Himalayan goral	Poorly suitable	0.11	0.38	0.29	1.00
	Moderately suitable	0.25	0.26	0.96	
	Suitable	0.44	0.32	1.38	
	Highly suitable	0.19	0.04	4.75	
Himalayan serow	Poorly suitable	0.03	0.13	0.23	0.60
	Moderately suitable	0.08	0.23	0.35	
	Suitable	0.37	0.24	1.54	
	Highly suitable	0.53	0.40	1.33	

High values indicate a high consistency in the evaluation of the datasets, and a value greater than 1 indicates good suitability.

habitat suitability. The Boyce index and predicted-to-expected ratio (*R*) show that the maps accurately describe the general distribution of suitable habitats and distribution of the target species. These maps reveal that all three ungulates are present significantly more often in highly suitable habitats than expected based on the availability of this habitat. Thus, wildlife mainly occurs in highly suitable habitats, which are sparsely distributed in this region, mainly due to human modification of the landscape, which has greatly reduced the proportion of the habitats in region that are of high quality (Fischer and Lindenmayer 2007).

The map for barking deer shows that there is a large area of suitable habitat available for this species throughout the study area (Figure 3), which is made up of 66% of suitable and highly suitable habitats. This may be because barking deer is adapted to live in a wide variety of habitats (Ohtaishi and Gao 1990; Prater 1990; Roberts 1999; Paudel and Kindlmann 2012a) including degraded forest near human settlements (Oka 1998; Paudel and Kindlmann 2012a). However, although there is a large area of high-quality habitat for barking deer in the middle region, it needs to be urgently conserved, and poaching there reduced because human activities in non-protected areas are likely to increase in the absence of a conservation program. The suitable habitat for goral is scattered throughout the study area (Figure 4) but occurs mainly in the south (along the boundary of the Bardia National Park) and the north of the region (along mountain ridges). Highly suitable habitat makes up only 4% of the suitable habitat for Himalayan goral. This indicates that Himalayan goral occurs abundantly in small, highly fragmented and patchily distributed habitats, as previously reported by Hajra (2002), who found that less than 1% of the area is highly suitable for goral in the Sivalik hills in Uttaranchal, India. The steep grassy slopes, which are the most preferred habitat of goral (Cavallini 1992; Mishra and Johnsingh 1996; Fakhar-I-Abbas et al. 2008; Paudel and Kindlmann 2012a), are now almost exclusively used for agriculture and settlements. The map for Himalayan serow shows its habitat as patchily distributed in the south (along the boundary of the Bardia National Park) and north (along mountain ridges) of the study area (Figure 5). Suitable habitats of all categories cover only 40% of the study area, which indicates that the area of high-quality habitats (suitable and highly suitable) is very small for Himalayan serow, even though they comprise a relatively large proportion of all categories of suitable habitats (64%). Thus, the habitat suitability maps for serow indicate that it is important to increase habitat quality and maintain corridors connecting core habitats in order to ensure the long-term survival of this species.

We provide a simple way to identify high-quality habitat for ungulates in a mountain landscape. This approach primarily results in the production of a map of the spatial distribution of the preferred habitats of the target species. It has several advantages over other methods. Many investigators select habitat variables based on the literature and expert judgment. In our study, we first tested whether habitat variables can be used to predict the presence and absence of species using logistic regression and derived suitability indices for particular habitat variables based on Jacob's index (*D*). Furthermore, the multiplication approach used to combine habitat variables with suitability results in values of 0 (not suitable) and 1 (poorly suitable) and reduces the incidence of overoptimistic predictions (Dayton and Fitzgerald 2006; Gavashelishvili and Lukarevskiy 2008). Another possible criticism may be that too few variables are included in the model.

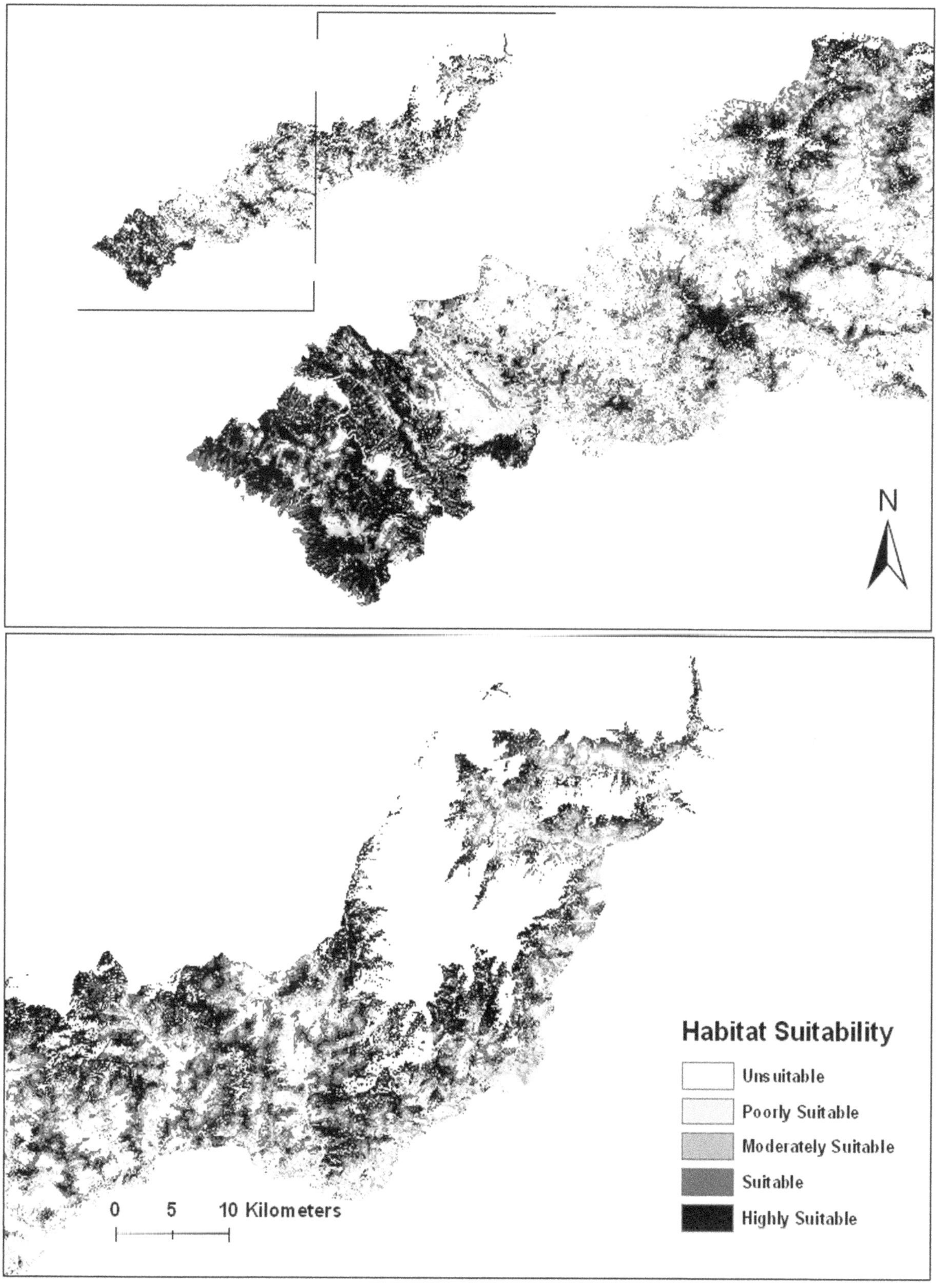

Figure 3 Map of the habitats suitable for barking deer.

Table 6 Observed and expected occurrence of three species in each of the four habitat categories

	Predicted suitability category (i)	Proportion of suitable category (p_i)	Observed frequency in the ith habitat category (O)	Expected frequency of species observation in the ith habitat category (p_iO+)	Proportional frequency in the habitat category (p_iO/p_iO+)
Barking deer	Poorly suitable	0.18	4	9.36	0.08
	Moderately suitable	0.19	7	9.75	0.13
	Suitable	0.33	14	16.91	0.27
	Highly suitable[a]	0.31	27	15.98	0.52
Himalayan goral	Poorly suitable	0.38	7	24.11	0.11
	Moderately suitable	0.26	16	16.43	0.25
	Suitable	0.32	28	20.19	0.44
	Highly suitable[a]	0.04	12	2.27	0.19
Himalayan serow	Poorly suitable[a]	0.13	1	4.89	0.03
	Moderately suitable[a]	0.23	3	8.70	0.08
	Suitable[a]	0.24	14	9.18	0.37
	Highly suitable[a]	0.40	20	15.24	0.53

O+, total number of records in all habitat categories for each species. [a]Significant at $P < 0.05$ after Bonferroni correction.

Studies on habitat suitability and distribution of wildlife show that accurate predictions can be made on the basis of a few variables (Liu et al. 2001; Dayton and Fitzgerald 2006; Gavashelishvili and Lukarevskiy 2008).

Biased sampling can result in an erroneous resource selection function. For example, transects that predominately follow a common habitat and topographic feature might not include all the important habitat features for a particular target species. The areas surveyed included significantly representative proportions of the different slopes ($r^2 = 0.97$) and topographic ruggedness ($r^2 = 0.97$), but not distances to the nearest village ($r^2 = 0.78$) and forest type ($r^2 = 0.69$). These sampling biases are a result of little sampling within 250 m of the periphery of villages (beyond 250 m from villages, $r^2 = 0.98$), which are mostly covered by highly degraded forest such as bushy areas and subtropical pine forest (for subtropical pine forest without bushy areas $r^2 = 0.91$). These are poor habitats for all species. We argue that biases introduced by not sampling these habitats do not affect the allocation of the suitability index because it is very unlikely that the target species occur in these habitats.

Generally, the prediction of those sites that are occupied is more successful than of those that are not (Osborne et al. 2001). Thus, the maps need to be evaluated in an ecological context (Fielding and Bell 1997). For example, serow locally extirpated from mid-regions although there are large fragments of suitable habitat (Paudel and Kindlmann 2012b). Hence, further research is necessary to determine whether the absence of serow in many suitable habitats is either an effect of small patch size, isolation or human influence or a combination of these factors. We argue that some external factors, such as hunting, might result in the local extinction of certain species. We found that serow in small areas of suitable habitat are more likely to be hunted than those in large areas because small areas are more often monitored by hunting groups (unpublished data). We could not use hunting data in developing the model as such data are difficult to obtain and its spatial precision is uncertain because of the mobile nature of hunting.

Habitat loss and wildlife hunting pose a formidable challenge for wildlife conservation in Nepal's mountains (Jackson 1979; Paudel 2012). Long-term survival of wildlife depends on sufficiently large areas of suitable habitat and opportunities for dispersal between such areas (Harrison 1991; Hanski 1999). Thus, we propose that areas supporting suitable habitats for all three mountain ungulates be designated as conservation areas, especially in areas adjoining the northern boundary of Bardia National Park, along the mountain tops in the middle part and forested areas in the northern part. Except in the national parks (Bardia National Park in the south and Shey Phoksundo National Park in the north), the quantity and quality of habitat is likely to decrease due to human activities such as collecting wood for fuel, harvesting timber and clearing forest for agriculture. Once forests that are easy to access or close to human settlements are exhausted, those in more remote areas at higher altitudes will be exploited. It is therefore important to conserve large patches of habitat and restore connectivity between core areas to ensure species survival. This will enable not only survival of the three target species but also others native to the midhills. The mountain ranges, where suitable habitat for serow and goral exists, in the mid and northern parts of the study area, also provide occasional habitats for a small population of blue sheep (*Pseudois nayaur*), Himalayan musk deer (*Moschus chrysogaster*)

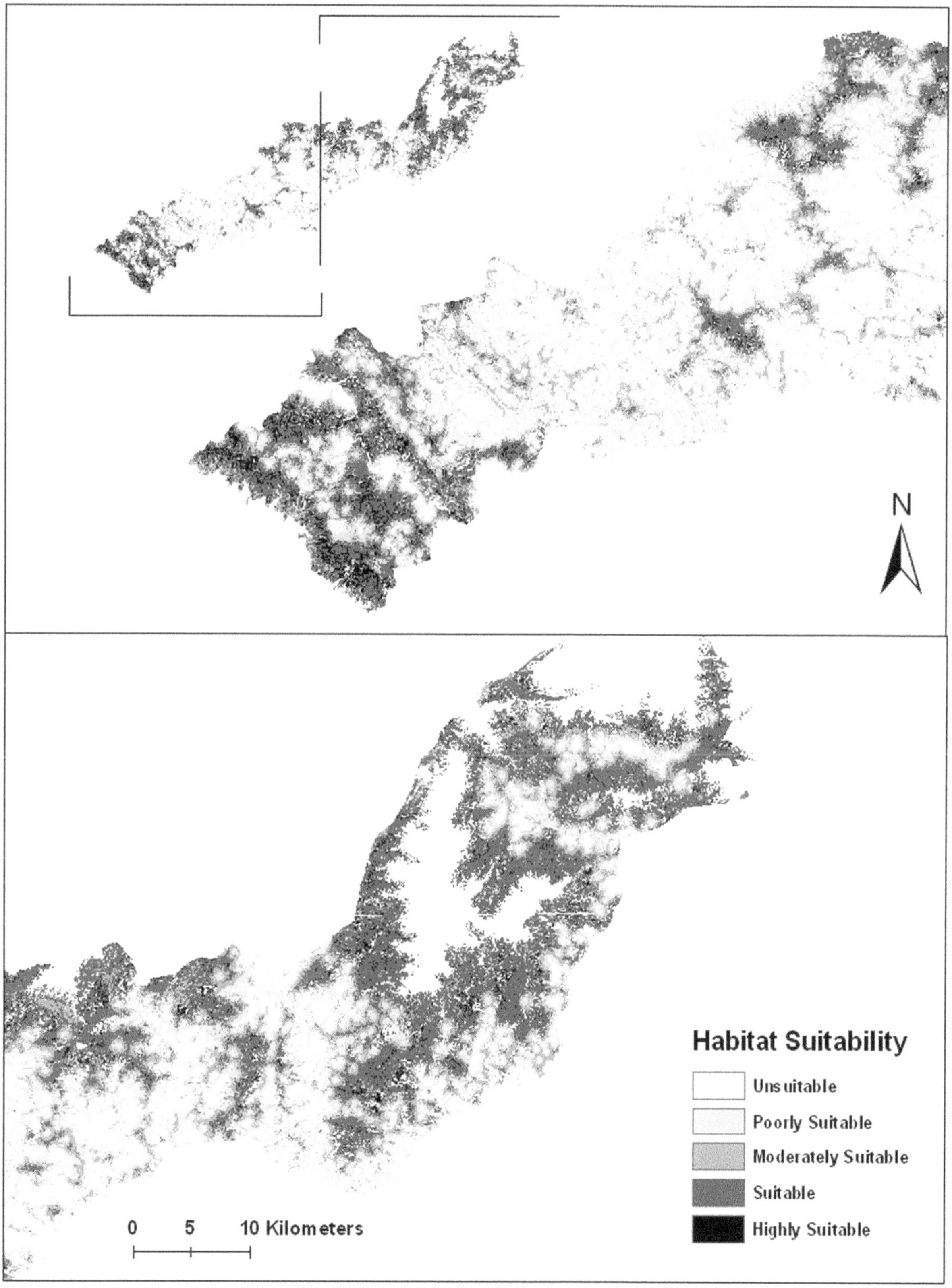

Figure 4 Map of the habitats suitable for Himalayan goral.

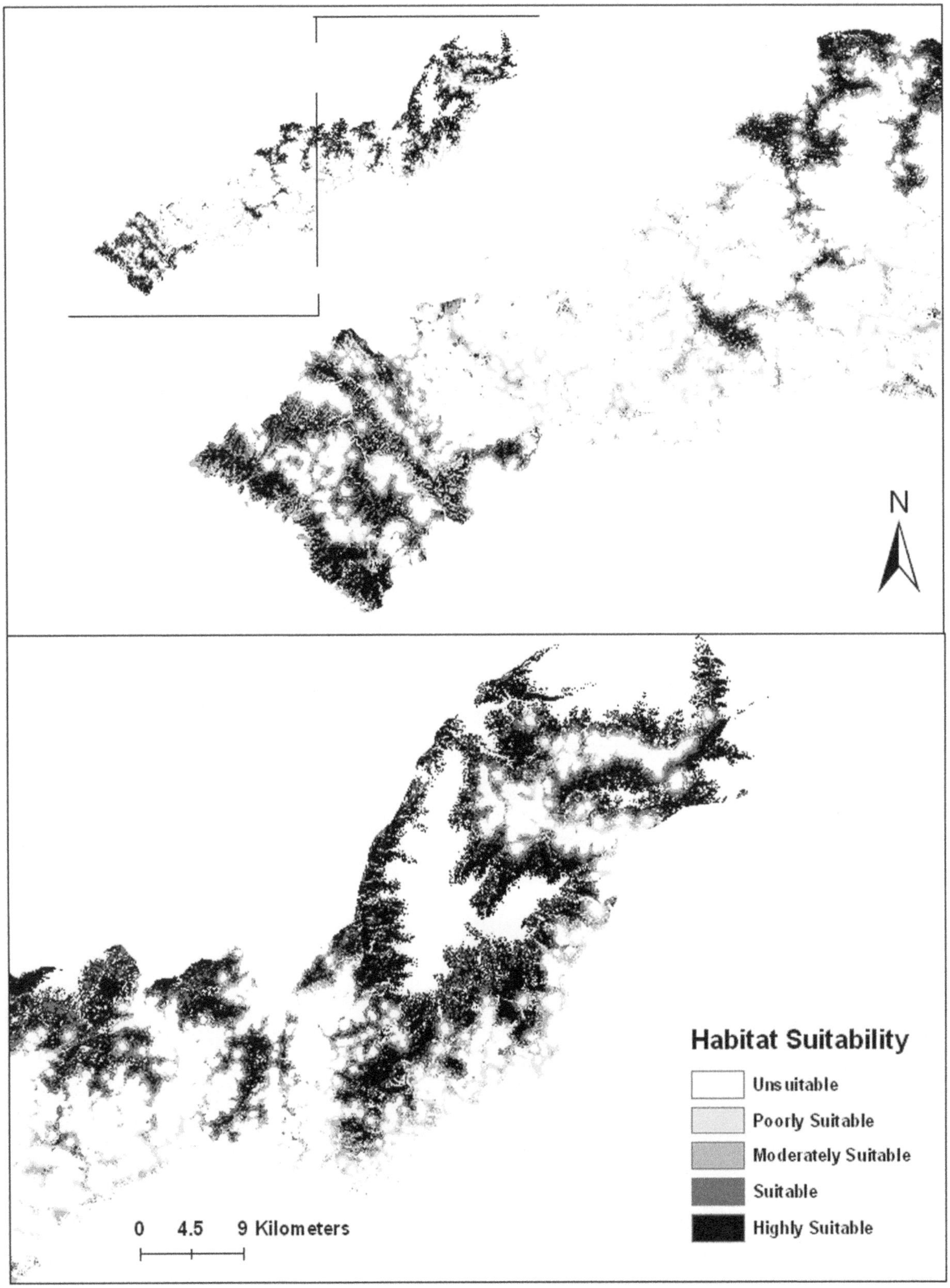

Figure 5 Map of the habitats suitable for Himalayan serow.

and Himalayan black bear (*Ursus thibetanus*) (unpublished data). These species are important prey of snow and common leopards (Prater 1990; Roberts 1999). We argue that conservation of core habitats and their connectivity leads to conservation of isolated wildlife populations and their predators such as snow and common leopards because carnivores tend to occur where their preferred prey are abundant (Carbone and Gittleman 2002).

Conclusions

The habitat suitability maps provide important baseline information for landscape-level conservation over a large altitudinal range in the western midhills of Nepal. The maps show that the majority of the suitable habitats are clustered mostly either in the northern or southern regions of the study area, whereas the mid region harbours small but highly fragmented suitable habitats. The size and spatial configuration of the suitable habitats is critical for the long-term survival of wildlife (Andrén 1994; Hanski 1999; Mech and Hallett 2001). We suggest further studies on connectivity and patch occupancy of wildlife. This is true especially for Himalayan serow that is patchily distributed in isolated habitats. The absence of serow in many large areas of suitable habitat shows the need for the restoration of corridors between such habitats (Paudel and Kindlmann 2012b). Furthermore, the map identifies probable sites for potential reintroduction of serow in areas within its historical range in the midhills of Nepal. Like serow, Himalayan goral is patchily distributed, with high-quality habitats confined to the Churia hills in the southern part of the area, which is a relatively inaccessible area with a rich cover of grass. Therefore, conservation of existing high-quality habitats should be the main focus of the conservation of Himalayan goral.

Here, we use a multi-species approach to identify the problems of conserving wildlife in Nepal. Although the use of surrogate information for certain species is questioned (Bonn and Gaston 2005), but recommended in the absence of other assessment options (Landres et al. 1988), we chose three particular species because they occupy a range of habitats that include those of many other species of wildlife. An extensive survey of the occurrence of carnivores such as leopards (*Panthera pardus*, *Uncia uncia*) and Himalayan musk deer in the study area is not possible because they are rare and the topography of the region is highly rugged. Hence, the habitat suitability maps of the species studied can be an effective tool for evaluating their conservation needs and designating areas for conservation.

Competing interests

The authors declare that they have no competing interests.

Authors' contributions

PKP conceived and executed the study, collected the field data and performed the modelling analyses. MH carried out the vegetation categorization of the study area using remote sensing. PK helped to draft the manuscript. All authors read and approved the final manuscript.

Acknowledgements

We thank our field assistants (Mr. Purna Bahadur Bishwakarma, Mr. Sahila Bishwakarma and Mr. Lil B. Khatri) for their tireless support in collecting data. The Department of National Parks and Wildlife Conservation provided permission to do this research. We also thank hundreds of local assistants and people who supported us in many ways to collect data. We thank Tony Dixon for correcting the English. This research was partially supported by the MSMT grant LC06073 and within the National Sustainability Program I (NPU I), grant number LO1415 and by grant No. 14-36098G of the GA CR.

Author details

[1]Department of Biodiversity Research, Global Change Research Centre AS CR, Bělidla 4a, 603 00 Brno, Czech Republic. [2]Faculty of Sciences, University of South Bohemia, Branišovská 31, 370 05 České Budějovice, Czech Republic. [3]Institute for Environmental Studies, Faculty of Natural Sciences, Charles University, Benátská 2, Prague, Czech Republic. [4]Current address: Center for Conservation Biology, Kathmandu Institute of Applied Sciences, PO Box 23002, Kathmandu, Nepal.

References

Abbitt RJF, Scott JM, Wilcove DS (2000) The geography of vulnerability: incorporating species geography and human development patterns into conservation planning. Biol Conserv 96:169–175, doi:10.1016/S0006-3207(00)00064-1

Andrén H (1994) Effects of habitat fragmentation on birds and mammals in landscapes with different proportions of suitable habitat: a review. Oikos 71:355–366, doi:10.2307/3545823

Austin MP, Meyers JA (1996) Current approaches to modelling the environmental niche of eucalypts: implication for management of forest biodiversity. For Ecol Manag 85:95–106, doi:10.1016/S0378-1127(96)03753-X

Berry WD, Feldman S (1985) Multiple Regression in Practice. SAGE Publications, Inc., Beverly Hills

Boitani L, Sinibaldi I, Corsi F, Biase AD, Carranza I, Ravagli M, Reggiani G, Rondinini C, Trapanese P (2007) Distribution of medium- to large-sized African mammals based on habitat suitability models. Biodivers Conserv 17:605–621, doi:10.1007/s10531-007-9285-0

Bonn A, Gaston KJ (2005) Capturing biodiversity: selecting priority areas for conservation using different criteria. Biodivers Conserv 14:1083–1100, doi:10.1007/s10531-004-8410-6

Boyce MS, Vernier PR, Nielsen SE, Schmiegelow FKA (2002) Evaluating resource selection functions. Ecol Model 157:281–300, doi:10.1016/S0304-3800(02)00200-4

Breiman L (2001) Random forests. Mach Learn 45:5–32, doi:10.1023/A:1010933404324

Brooks TM, Mittermeier RA, Mittermeier CG, Da Fonseca GAB, Rylands AB, Konstant WR, Flick P, Pilgrim J, Oldfield S, Magin G, Hilton-Taylor C (2002) Habitat loss and extinction in the hotspots of biodiversity. Conserv Biol 16:909–923, doi:10.1046/j.1523-1739.2002.00530.x

Carbone C, Gittleman JL (2002) A common rule for the scaling of carnivore density. Science 295:2273–2276, doi:10.1126/science.1067994

Cavallini P (1992) Survey of the goral Nemorhaedus goral (Hardwicke) in Himachal Pradesh. J Bombay Nat Hist Soc 89:302–307

Das P (1968) The Monsoons. St. Martin's Press, New York

Dayton GH, Fitzgerald LA (2006) Habitat suitability models for desert amphibians. Biol Conserv 132:40–49, doi:10.1016/j.biocon.2006.03.012

Engler R, Guisan A, Rechsteiner L (2004) An improved approach for predicting the distribution of rare and endangered species from occurrence and pseudo-absence data. J Appl Ecol 41:263–274, doi:10.1111/j.0021-8901.2004.00881.x

Fahrig L (2003) Effects of habitat fragmentation on biodiversity. Annu Rev Ecol Evol Syst 34:487–515, doi:10.1146/annurev.ecolsys.34.011802.132419

Fakhar-I-Abbas F, Akhtar T, Mian A (2008) Food and feeding preferences of Himalayan gray goral (Naemorhedus goral bedfordi) in Pakistan and Azad Jammu and Kashmir. Zoo Biol 27:371–380, doi:10.1002/zoo.20202

Ferrier S (2002) Mapping spatial pattern in biodiversity for regional conservation planning: where to from here? Syst Biol 51:331–363, doi:10.1080/10635150252899806

Fielding AH, Bell JF (1997) A review of methods for the assessment of prediction errors in conservation presence/absence models. Environ Conserv 24:38–49

Fischer J, Lindenmayer DB (2007) Landscape modification and habitat fragmentation: a synthesis. Glob Ecol Biogeogr 16:265–280, doi:10.1111/j.1466-8238.2007.00287.x

Gavashelishvili A, Lukarevskiy V (2008) Modelling the habitat requirements of leopard Panthera pardus in west and central Asia. J Appl Ecol 45:579–588, doi:10.1111/j.1365-2664.2007.01432.x

Geomatica (2005) Geomatica 10, Focus user guide. Geomatics Enterprises, Ontario

Gibson LA, Wilson BA, Cahill DM, Hill J (2004) Modelling habitat suitability of the swamp antechinus (Antechinus minimus maritimus) in the coastal heathlands of southern Victoria, Australia. Biol Conserv 117:143–150, doi:10.1016/S0006-3207(03)00288-X

Green MJB (1979) Tahr in a Nepal National Park. Oryx 15:140–144, doi:10.1017/S0030605300024224

Green MJB (1987) Ecological separation in Himalayan ungulates. J Zool 1:693–719, doi:10.1111/j.1096-3642.1987.tb00751.x

Grubb P (2005) Artiodactyla. In: Wilson DE, Reeder DAM (eds) Mammal Species of the World: A Taxonomic and Geographic Reference, 2-volume set, 3rd edn. Johns Hopkins University Press, Baltimore, pp 637–722

Guisan A, Zimmermann NE (2000) Predictive habitat distribution models in ecology. Ecol Model 135:147–186, doi:10.1016/S0304-3800(00)00354-9

Hajra A (2002) An ecological study of the vegetation and wildlife habitats in and around Rajaji-Corbett corridor area. PhD thesis, Wildlife Institute of India

Hanski I (1999) Habitat connectivity, habitat continuity, and metapopulations in dynamic landscapes. 87:209–219. doi:10.2307/3546736

Harrison S (1991) Local extinction in a metapopulation context: an empirical evaluation. Biol J Linn Soc 42:73–88, doi:10.1111/j.1095-8312.1991.tb00552.x

Hirzel AH, Le Lay G (2008) Habitat suitability modelling and niche theory. J Appl Ecol 45:1372–1381, doi:10.1111/j.1365-2664.2008.01524.x

Hirzel AH, Hausser J, Chessel D, Perrin N (2002) Ecological-niche factor analysis: how to compute habitat-suitability maps without absence data? Ecology 83:2027–2036, doi:10.1890/0012-9658(2002)083[2027:ENFAHT]2.0.CO;2

Hirzel AH, Le Lay G, Helfer V, Randin C, Guisan A (2006) Evaluating the ability of habitat suitability models to predict species presences. Ecol Model 199:142–152, doi:10.1016/j.ecolmodel.2006.05.017

Hunter ML Jr, Yonzon P (1993) Altitudinal distributions of birds, mammals, people, forests, and parks in Nepal. Conserv Biol 7:420–423

Jackson R (1979) Aboriginal hunting in West Nepal with reference to musk deer Moschus moschiferus moschiferus and snow leopard Panthera uncia. Biol Conserv 16:63–72, doi:10.1016/0006-3207(79)90008-9

Jacobs J (1974) Quantitative measurement of food selection: a modification of the forage ratio and Ivlev's electivity index. Oecologia 14:413–417

Kushwaha SPS, Munkhtuya S, Roy PS (2000) Geospatial modelling for goral habitat evaluation. J Indian Soc Remote Sens 28:293–303, doi:10.1007/BF02990819

Landres PB, Verner J, Thomas JW (1988) Ecological uses of vertebrate indicator species: a critique. Conserv Biol 2:316–328, doi:10.1111/j.1523-1739.1988.tb00195.x

Liu J, Linderman M, Ouyang Z, An L, Yang J, Zhang H (2001) Ecological degradation in protected areas: the case of Wolong Nature Reserve for giant pandas. Science 292:98–101, doi:10.1126/science.1058104

Maddock AH, Samways MJ (2000) Planning for biodiversity conservation based on the knowledge of biologists. Biodivers Conserv 9:1153–1169, doi:10.1023/A:1008992602151

Mech SG, Hallett JG (2001) Evaluating the effectiveness of corridors: a genetic approach. Conserv Biol 15:467–474

Menard S (2001) Applied Logistic Regression Analysis, 2nd edn. SAGE Publications, Inc., Thousand Oaks, CA

Mills LS (2009) Conservation of wildlife populations: demography, genetics and management. John Wiley & Sons

Mishra C, Johnsingh AJT (1996) On habitat selection by the goral Nemorhaedus goral bedfordi (Bovidae, Artiodactyla). J Zool 240:573–580, doi:10.1111/j.1469-7998.1996.tb05307.x

Mittermeier RA, Myers N, Thomsen JB, Da Fonseca GAB, Olivieri S (1998) Biodiversity hotspots and major tropical wilderness areas: approaches to setting conservation priorities. Conserv Biol 12:516–520, doi:10.1046/j.1523-1739.1998.012003516.x

Morrison ML, Marcot B, Mannan W (2012) Wildlife-Habitat Relationships: Concepts and Applications. Island Press, Washington

Myers N, Mittermeier RA, Mittermeier CG, da Fonseca GAB, Kent J (2000) Biodiversity hotspots for conservation priorities. Nature 403:853–858, doi:10.1038/35002501

Nowak RM, Wilson DE (1999) Walker's Mammals of the World, 6th edn. Johns Hopkins University Press, Baltimore

Odden M, Wegge P (2007) Predicting spacing behavior and mating systems of solitary cervids: a study of hog deer and Indian muntjac. Zoology 110:261–270, doi:10.1016/j.zool.2007.03.003

Ohtaishi N, Gao Y (1990) A review of the distribution of all species of deer (Tragulidae, Moschidae and Cervidae) in China. Mammal Rev 20:125–144, doi:10.1111/j.1365-2907.1990.tb00108.x

Oka GM (1998) Factors Affecting the Management of Muntjac Deer (Muntiacus muntjak) in Bali Barat National Park, Phd Thesis. University of Western Sydney, Indonesia

Olson DM, Dinerstein E, Wikramanayake ED, Burgess ND, Powell GVN, Underwood EC, D'amico JA, Itoua I, Strand HE, Morrison JC, Loucks CJ, Allnutt TF, Ricketts TH, Kura Y, Lamoreux JF, Wettengel WW, Hedao P, Kassem KR (2001) Terrestrial ecoregions of the world: a new map of life on earth. A new global map of terrestrial ecoregions provides an innovative tool for conserving biodiversity. Bioscience 51:933–938, doi:10.1641/0006-3568(2001)051[0933:TEOTWA]2.0.CO;2

Osborne PE, Alonso JC, Bryant RG (2001) Modelling landscape-scale habitat use using GIS and remote sensing: a case study with great bustards. J Appl Ecol 38:458–471

Paudel PK (2012) Challenges to wildlife conservation posed by hunting in non-protected areas north of the Bardia National Park. In: Kindlmann P (ed) Himal. Biodivers. Chang. World. Springer, Netherlands, pp 177–195

Paudel PK, Heinen JT (2015) Conservation planning in the Nepal Himalayas: effectively (re)designing reserves for heterogeneous landscapes. Appl Geogr 56:127–134, doi:10.1016/j.apgeog.2014.11.018

Paudel PK, Kindlmann P (2012a) Human disturbance is a major determinant of wildlife distribution in Himalayan midhill landscapes of Nepal. Anim Conserv 15:283–293, doi:10.1111/j.1469-1795.2011.00514.x

Paudel PK, Kindlmann P (2012b) Distribution pattern of the threatened Himalayan serow (Capricornis thar) in western midhills of Nepal: an insight for conservation along an altitudinal gradient. J Nat Conserv 20:177–180, doi:10.1016/j.jnc.2012.01.004

Paudel PK, Šipoš J (2014) Conservation status affects elevational gradient in bird diversity in the Himalaya: a new perspective. Glob Ecol Conserv 2:338–348, doi:10.1016/j.gecco.2014.10.012

Paudel PK, Bhattarai BP, Kindlmann P (2012) An overview of the biodiversity in Nepal. In: Kindlmann P (ed) Himal. Biodivers. Chang. World. Springer, Netherlands, pp 1–40

Pearce J, Ferrier S (2000) Evaluating the predictive performance of habitat models developed using logistic regression. Ecol Model 133:225–245, doi:10.1016/S0304-3800(00)00322-7

Pearce J, Lindenmayer D (1998) Bioclimatic analysis to enhance reintroduction biology of the endangered helmeted honeyeater (Lichenostomus melanops cassidix) in Southeastern Australia. Restor Ecol 6:238–243, doi:10.1046/j.1526-100X.1998.00636.x

Peres CA (2001) Synergistic effects of subsistence hunting and habitat fragmentation on Amazonian forest vertebrates. Conserv Biol 15:1490–1505, doi:10.1046/j.1523-1739.2001.01089.x

Poiani KA, Richter BD, Anderson MG, Richter HE (2000) Biodiversity conservation at multiple scales: functional sites, landscapes, and networks. Bioscience 50:133–146, doi:10.1641/0006-3568(2000)050[0133:BCAMSF]2.3.CO;2

Prater SH (1990) The Book of Indian Animals, 3rd edn. OUP, India

Primack RB, Paudel PK, Bhattarai BP (2013) Conservation Biology: A Primer for Nepal. Dreamland Publication, Kathmandu, Nepal

Roberts TJ (1999) Mammals of Pakistan, 2nd edn. OUP Pakistan, Karachi

Rushton SP, Ormerod SJ, Kerby G (2004) New paradigms for modelling species distributions? J Appl Ecol 41:193–200, doi:10.1111/j.0021-8901.2004.00903.x

Sala OE, Chapin FS 3rd, Armesto JJ, Berlow E, Bloomfield J, Dirzo R, Huber-Sanwald E, Huenneke LF, Jackson RB, Kinzig A, Leemans R, Lodge DM, Mooney HA, Oesterheld M, Poff NL, Sykes MT, Walker BH, Walker M, Wall DH (2000) Global biodiversity scenarios for the year 2100. Science 287:1770–1774, doi:10.1126/science.287.5459.1770

Sappington JM, Longshore KM, Thompson DB (2007) Quantifying landscape ruggedness for animal habitat analysis: a case study using bighorn sheep in the Mojave Desert. J Wildl Manag 71:1419–1426, doi:10.2193/2005-723
Store R, Kangas J (2001) Integrating spatial multi-criteria evaluation and expert knowledge for GIS-based habitat suitability modelling. Landsc Urban Plan 55:79–93, doi:10.1016/S0169-2046(01)00120-7
Teng L, Liu Z, Song Y-L, Zeng Z (2004) Forage and bed sites characteristics of Indian muntjac (Muntiacus muntjak) in Hainan Island, China. Ecol Res 19:675–681, doi:10.1111/j.1440-1703.2004.00683.x
Thuiller W, Münkemüller T (2010) Habitat suitability modeling. In: Fielder W, Berthold P, Moeller AP (eds) Effects of Climate Change on Birds. Oxford University Press, Oxford, pp 77–85
Wegge P, Oli M (1997) Nepal. In: Shackleton DM (ed) Wild sheep and goat and their relatives: status survey and conservation action plan for caprice. IUCN, Gland, pp 231–239
Wikramanayake ED (2002) Terrestrial ecoregions of the Indo-Pacific: a conservation assessment. Island Press
Yamada K, Elith J, McCarthy M, Zerger A (2003) Eliciting and integrating expert knowledge for wildlife habitat modelling. Ecol Model 165:251–264, doi:10.1016/S0304-3800(03)00077-2

Recolonizing wolves influence the realized niche of resident cougars

L Mark Elbroch[1*], Patrick E Lendrum[1], Jesse Newby[2], Howard Quigley[1] and Daniel J Thompson[3]

Abstract

Background: Niche differentiation may betray current, ongoing competition between two sympatric species or reflect evolutionary responses to historic competition that drove species apart. The best opportunity to test whether ongoing competition contributes to niche differentiation is to test for behavioral shifts by the subordinate competitor in controlled experiments in which the abundance of the dominant competitor is manipulated. Because these circumstances are difficult to coordinate in natural settings for wide-ranging species, researchers seize opportunities presented by species reintroductions. We tested for new competition between reintroduced wolves and resident cougars in the Southern Yellowstone Ecosystem to assess whether wolves might be impacting the realized niche of sympatric cougars.

Results: Between 2002 and 2012, a period during which wolves increased from 15 to as high as 91 in the study area, cougars significantly increased the percentage of deer and decreased the percentage of elk in their diet in summer. Our top models explaining these changes identified elk availability, defined as the number of elk per wolf each year, as the strongest predictor of changing cougar prey selection. Both elk and deer were simultaneously declining in the system, though deer more quickly than elk, and wolf numbers increased exponentially during the same time frame. Therefore, we concluded that prey availability did not explain prey switching and that competition with wolves at least partially explained cougar prey switching from elk to deer. We also recorded 5 marked cougar kittens killed by wolves and 2 more that were killed by an undetermined predator. In addition, between 2005 and 2012, 9 adult cougars and 10 cougar kittens died of starvation, which may also be in part explained by competition with wolves.

Conclusions: Direct interspecific predation and shifting cougar prey selection as wolves increased in the system provided evidence for competition between recolonizing wolves and resident cougars. Through competition, recolonizing wolves have impacted the realized niche of resident cougars in the Southern Yellowstone Ecosystem (SYE), and current resident cougars may now exhibit a realized niche more reflective of an era when these species were previously sympatric in the Yellowstone Ecosystem.

Keywords: *Canis lupus*; Competition; Prey switching; *Puma concolor*; Realized niche; Wyoming; Yellowstone

Background

Resource competition describes a diverse suite of intra- and interspecific interactions that shape ecological communities (Birch 1957; Schoener 1983; Petren and Case 1996). One way in which competition structures communities is through its influence on species ecological niches. The larger fundamental niche describes the ecological conditions in which a species may thrive on its own, whereas the smaller realized niche describes the ecological conditions in which the same species can survive while under the additional constraints imposed by predators and competitors (Hutchinson 1957). Competition, therefore, drives evolutionary niche differentiation within and among species, as well as shapes the realized niches of sympatric species in ecological communities.

Consider wolves (*Canis lupus*) and cougars (*Puma concolor*). Wolves are the dominant competitors in most physical competition over shared resources (Ruth and Murphy 2010a). Wolves are coursing, social predators that select disadvantaged prey in areas of limited structural complexity where they can test their prey's condition (Husseman et al. 2003; Peterson et al. 2003). Cougars are solitary stalk-and-pounce predators that select prey

* Correspondence: melbroch@panthera.org
[1]Panthera, 8 West 40th Street, 18th Floor, New York, NY 10018, USA
Full list of author information is available at the end of the article

opportunistically (i.e., of any health) in areas where structural complexity (e.g., slope, trees, boulders) provide them an advantage (Husseman et al. 2003; Atwood et al. 2007; Ruth and Murphy 2010b). Thus, we expect wolves and cougars to inhabit and utilize different ecological niches, allowing them to spatially and temporally coexist (Husseman et al. 2003; Atwood et al. 2007); however, in the absence of wolves, cougars utilize areas traditionally assumed to be the dominion of coursing predators (Riley et al. 2004; Elbroch and Wittmer 2012). This suggests that where wolves are sympatric with cougars, wolves may impact the realized niche of cougars and that ongoing competition between the two species plays a role in structuring ecological systems.

Our best opportunity to test whether ongoing competition contributes to niche differentiation between species is to test for behavioral shifts by the subordinate competitor in controlled experiments in which the abundance of the dominant competitor is manipulated (Connell 1983). Could we, for example, detect prey switching or changes in habitat use by cougars as we increase or decrease the number of wolves in a controlled system? It is, however, logistically and perhaps ethically difficult to manipulate large carnivore abundances in natural systems. Instead, researchers must be quick to seize rare opportunities presented by species perturbations (e.g., reintroductions or removals) to test whether there is a change in the realized niche of the subordinate competitor (Kortello et al. 2007; Seddon et al. 2007).

The introduction of wolves in Northern Yellowstone National Park and central Idaho, USA, beginning in 1995, where wolves have been absent since 1926 (Haines 1996), provided such an opportunity. Here we provide a brief summary of research findings on competition between the two species to date: Kunkel et al. (1999) did not detect separation in prey selection or habit use in newly sympatric wolves and cougars in Montana and did not detect any form of competition in their formal analyses. They did, however, report kleptoparasitism of cougar kills by wolves and interspecific killing of cougars by wolves as evidence of interference competition. Ruth (2004b) and Kortello et al. (2007) reported increased cougar starvation as wolves increased in two different systems, and both studies attributed the change largely to interference competition with wolves and declining prey. Kortello et al. (2007) also detected cougar prey switching from elk to deer as wolf numbers increased (and elk numbers decreased) in Banff National Park, Alberta, as well as spatial avoidance between the two species at fine scales; they attributed these changes to exploitation competition because they believed wolves were diminishing elk numbers to levels where cougars needed to prey switch to survive. Kortello et al. also reported the intraspecific killing of one cougar by wolves as evidence of interference competition. Atwood et al. (2007) reported that cougars increased predation on elk (*Cervus elaphus*) and reduced predation on mule deer (*Odocoileus hemionus*) with the recolonization of wolves in southwest Montana; in contrast to Kortello et al. (2007), they attributed the change to elk switching habitats from open grasslands that mitigate predation risk from cougars to more structured habitats that mitigate predation risk from wolves, suggesting facilitation rather than any form of competition between top predators. Ruth et al. (2011, 2015) found that cougars increased their use of structurally complex habitats after the reintroduction of wolves in Northern Yellowstone National Park, implying a reduction in the spatial attributes of the cougar's realized niche. Further, wolves killed three adult cougars and five kittens in Northern Yellowstone over the course of their study (Ruth 2004a, 2015). Bartnick et al. (2013) found the probability of finding mule deer relative to elk at cougar kill sites increased during summer months in the Southern Yellowstone Ecosystem (SYE). The authors attributed the increased probability of detecting a mule deer as evidence of prey switching and interference competition, rather than exploitation competition as reported in Kortello et al. (2007). Bartnick et al. also reported an increase in the number of wolf tracks at cougar kills over time. Finally, Lendrum et al. (2014) observed that cougars in the SYE selected home ranges with higher "hunting opportunity" but further from wolves than expected, suggesting spatial displacement of cougars by wolves in high prey areas and current competition between the two species. When considered together, the results of these research projects strongly suggest some form of ongoing competition is contributing to niche separation between currently sympatric wolves and cougars.

In this study, we tested for changes in prey selection exhibited by a resident cougar population during wolf recolonization of the SYE as further evidence of ongoing competition between the two species. The SYE is a large undeveloped landscape where ungulate prey exhibit seasonal migrations (Sawyer et al. 2005; Smith 2007; Nelson et al. 2012; Elbroch et al. 2013) and winter snows force wolves, cougars, and their prey to share lower elevations in late winter (Kortello et al. 2007; Ruth et al. 2011; Elbroch et al. 2013). Evidence for competition between the two species implies that ongoing competition with wolves limits the realized niche of cougars. Following Bartnick et al.'s (2013) hypothesis that cougars in the SYE are increasingly selecting deer instead of elk in the presence of wolves, we tested for changes in actual cougar prey selection between 2002 and 2012, a period during which wolves increased from 15 to as high as 91 in the system (a 600 % increase). We hypothesized that we would detect a shift from elk to deer only in summer, because in winter deer migrate almost entirely out of the study system, thus limiting the opportunities for prey switching (Elbroch et al. 2013). We predicted cougar prey selection might switch from elk to deer because of increased encounter rates with deer related to preferential hunting, an increase in cougar use of more complex habitats where deer are more abundant (Kortello et al. 2007), or declining elk availability in the study area. In addition, we expected cougar predation on elk calves to decrease over the course of the study due to exploitation competition with wolves and other predators over a potentially limited resource (Griffin et al. 2011; Ruth et al. 2015).

We also hypothesized that we would detect interference competition between wolves and cougars, as evidenced by increased cougar kill rates (in animals killed per week) between 2005 and 2012, a period during which wolves increased from 16 to as high as 91. We expected summer cougar kill rates to increase because of a shift from larger (elk) to smaller (mule deer) prey (Knopff et al. 2010). We expected winter cougar kill rates to increase because of greater wolf presence at cougar kills (Bartnick et al. 2013), forcing cougars to abandon their kills more quickly and therefore to hunt more frequently (sensu Elbroch and Wittmer 2013a). Finally, we summarized cause-specific cougar mortality data over the full course of the study. Documenting wolves killing cougars would be the strongest evidence for current interference competition between the two species (Kortello et al. 2007) and that wolves are limiting the size of the cougar's realized niche in the Southern Yellowstone Ecosystem.

Methods

Study area

Our study area encompassed approximately 2300 km^2 of the SYE, inclusive of portions of Grand Teton National Park, the National Elk Refuge, and the Bridger-Teton National Forest north of the town of Jackson, Wyoming (Fig. 1). Elevations in the study area ranged from 1800 m in the valleys to >3600 m in the mountains. Plant communities included cottonwood (*Populus angustifolia*) riparian zones interspersed with sagebrush (*Artemisia* spp.) uplands at lower elevations. At intermediate elevations, aspen (*Populus tremuloides*), Douglas fir (*Pseudotsuga menziesii*), and lodgepole pine (*Pinus contorta*) were the dominant species. Engelmann spruce (*Picea engelmannii*) and subalpine fir (*Abies lasiocarpa*) were the primary tree species at the higher elevations (Knight 1996). The area was characterized by short, cool summers and long winters with frequent snowstorms. Precipitation occurred mostly as snow, and maximum snow depths ranged from 100 cm at lower elevations to >245 cm at intermediate and higher elevations (2000+ m).

The study area supported a diverse community of large mammals. Carnivores included wolves, black bears (*Ursus americanus*), grizzly bears (*Ursus arctos*), coyotes (*Canis latrans*), and red foxes (*Vulpes vulpes*). Ungulates included elk, mule deer, moose (*Alces alces*), bison (*Bison bison*), pronghorn (*Antilocapra americana*), bighorn sheep (*Ovis canadensis*), and a small number of white-tailed deer (*Odocoileus virginianus*). Deer, elk, bighorn sheep, and pronghorn exhibited seasonal migrations (Sawyer et al. 2005; Smith 2007; Elbroch et al. 2013); elk were heavily harvested (10–12 % annually) in efforts to reduce their population, and deer and moose supported limited, male-only harvest rates. Bighorn sheep too supported a limited harvest of either sex (http://wgfd.wyo.gov/web2011/HUNTING-1000184.aspx).

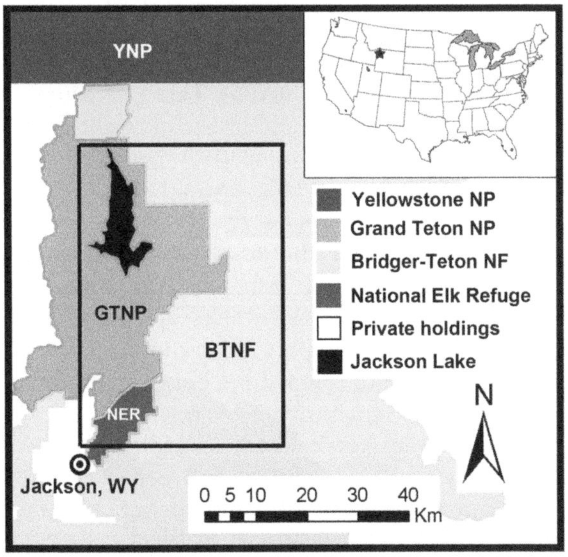

Fig. 1 Location of the study area in northwest Wyoming, USA, and a close-up of land ownership within the area of focus. The *smaller rectangle delineated by a black line* was the area in which we focused capture efforts and marked individual cougars

Cougar capture and collar programming

Beginning in 2001, we captured cougars during winters, from late November through March of the following year, when determining the presence of a cougar was facilitated by snow. We used trailing hounds to force cougars to retreat to a tree or rocky outcrop where we could safely approach them. Cougars were immobilized with ketamine (4.0 mg/kg) and medetomidine (0.07 mg/kg), and then their temperature, heart rate, and respiration were monitored at 5-min intervals while they were processed, sampled, and fitted with either a VHF (Telonics, Mesa, AZ) or GPS collar (Telonics, Mesa, AZ; Televilt, Bandygatan, Sweden; Vectronics, Berlin, Germany). Once the animal was completely processed, the effects of the capture drugs were reversed with atipamezole (0.375 mg/kg), and cougars departed the capture sites on their own. Our capture protocols for cougars followed those outlined in Quigley (2000), adhered to the guidelines outlined by the American Society of Mammalogists (Sikes et al. 2011), and were approved by the Jackson Institutional Animal Care and Use Committee (Protocol 027-10EGDBS-060210).

Documenting cause-specific mortalities

Cougar collars were equipped with mortality sensors, which alerted researchers as to when the collars had not moved for prolonged periods (8+ h). We conducted investigations of sites where cougars died as soon as was possible and used signs of struggle or other animals to aid in determining the cause of death and whether species present at the site acted as predators or scavengers (Elbroch 2003). We also conducted field or lab gross necropsies, generally led by a veterinarian, to assess interior and exterior clues to determine cause-specific mortality. Blood and tissue samples were sent to the Wyoming State Veterinary Laboratory for additional disease analyses.

Determining cougar prey selection and hunting habitats

We conducted site searches of areas where triangulation of cougars wearing VHF collars revealed that they had not moved for 24+ h, or spatially aggregated GPS points, called GPS clusters (Anderson and Lindzey 2003), indicated a cougar had remained in place for 8+ h. Prey remains, including hair, skin, rumen (stomach), and bone fragments, were used to identify prey species, and the state of prey remains, including the location of bite marks and feeding patterns, were used to determine whether the cougar had killed the animal or was scavenging. We determined the age of the prey by examining tooth eruption and wear patterns (Heffelfinger 2010) and separated prey into four age categories: calves and fawns <1 year old, yearlings 1–2 years, subadults 2–3 years, and adults 3+ years. Additionally, we recorded the general habitat type at the kill site: forest, edge, meadow, riparian, or rocky/barren. We later condensed habitat type into two categories: (1) forested, which included forest and edge, or (2) open habitat, which encompassed all remaining habitat types. To determine if cougars were hunting in forests more often over time, we employed a generalized linear model to assess changes in the percentage (%) of kills found in forests in summer and winter from 2002 to 2012. *p* values for generalized linear models were created based on asymptotic chi-square distributions (JMP 11, SAS Institute Inc., Cary, NC, USA).

Defining the seasons and testing for differences in cougar prey selection

Following well-established elk migration dates in the study area, we defined winter as December 1 of one year through May 31 of the next year, and summer as June 1 through November 30 of the same year (Elbroch et al. 2013). We examined seasonal prey selection from 2002 to 2012; however, because of limited sample sizes, we did not include 2002 or 2005 in the analysis of summer prey selection or 2003 in the analysis of winter prey selection.

To account for variable number of prey killed by individual cougars, we quantified each individual cougar's prey selection as the proportions of prey killed by each cougar, before conducting any population level analyses. We excluded non-ungulate prey for our analyses and divided the remainder into proportions of elk, deer, and "other ungulates," which included bighorn sheep, moose, and pronghorn (Fig. 2). Our analyses focused on the proportions of elk and mule deer killed by cougars because these two species composed >80 % of cougar diets in the study population (Elbroch et al. 2012). We employed generalized linear models to test whether the proportions of elk killed by cougars changed over time

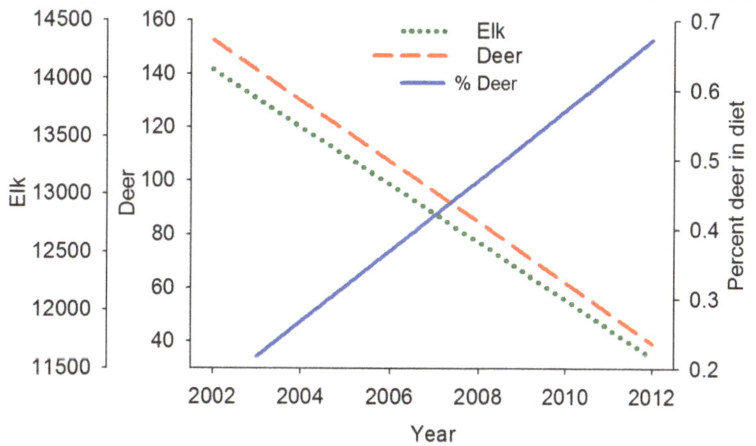

Fig. 2 Contrasting regression lines representing declining numbers of elk and deer, and the increasing proportions of deer in cougar summer diets over time

in summers and winters. We then repeated these analyses to test whether the proportion of deer killed by cougars in each season changed over time. Last, we tested whether the proportion of elk calves killed in both seasons changed over time. When we detected a change, we conducted a multiple regression analysis (GLM) to determine if variation in prey selection could be explained by prey availability, winter severity, the number of wolves in the system, or a combination of these parameters (JMP, SAS Institute Inc., Cary, NC, USA). We employed annual Wyoming Game and Fish Department elk counts and cow-calf ratios conducted via helicopter transects across our study area and modeled in POP II stochastic modeling software (Lubow and Smith 2004; Fossil Creek Software, Fort Collins, CO, USA) as an index for the local elk population (WGFD Job Completion Reports, http://wgfd.wyo.gov/web2011/wildlife-1000496.aspx), and annual Wyoming Game and Fish Department mule deer counts conducted during winter in the southern portion of our study area (Unit 150) as an index for the local deer population (WGFD Job Completion Reports, http://wgfd.wyo.gov/web2011/wildlife-1000496.aspx). We employed wolf numbers reported in annual wolf recovery reports for the study area (US Fish and Wildlife Service, Nez Perce Tribe, National Park Service, USDA Wildlife Services 2004, 2005, 2006, 2007, 2008, 2009, 2010, 2011, 2012).

For our actual analyses, we defined relative elk availability in terms of wolves (the number of elk per wolf) to account for the fact that wolves primarily eat elk in our study area (Stevenson et al. 2011), wolves are dominant over cougars, and competition over this resource should be most fierce (Ruth et al. 2015). We defined deer availability in terms of elk (the number of deer per elk) to emphasize the importance of changing prey ratios in determining prey availability (Garrott et al. 2007; Ruth et al. 2015). As a proxy for the potential influence of winter severity on predator and prey abundances, we used winter snow depths from November to April for each year (SNOTEL site, Gros Ventre Summit; Table 1); snow depths influence prey mobility and thus prey vulnerability and availability to predators in winter, as well as prey condition and vulnerability and availability in early summer (Parker et al. 1984). In our regression examining the proportion of elk calves killed over time, we included an additional variable, elk cow-calf ratios (number of calves per 100 cows, WGFD Job Completion Reports, http://wgfd.wyo.gov/web2011/wildlife-1000496.aspx). All generalized linear models were conducted with a Poisson distribution with a logit function identity.

Prior to analysis, we used pairwise coefficient correlations, and a correlation cutoff of 0.6, to test for significant correlation between independent variables. Wolf and elk numbers were highly correlated with elk availability, and deer and elk numbers were highly correlated with deer availability ($r > |0.80|$); thus, raw wolf and prey numbers were removed from the analyses. We then ran all possible combinations of the variables and calculated Akaike's information criterion adjusted for small sample size (AIC_c, ΔAIC_c, and Akaike weight w_i; Burnham and Anderson 2002) to assess model performance and determine variables that influenced prey selection from 2002 to 2012 (7 models for changing elk and deer selection, and 15 for changing elk). Models that had $\geq 2\ \Delta AIC_c$ were considered to have predictive power of significant difference from the next model (Burnham and Anderson 2002). In cases in which multiple top models were identified, we used model averaging to determine individual parameter weights to compare their relative influence on predictor variables (Burnham and Anderson 2002).

Table 1 Annual estimates of wolves reported by USFW, elk availability (elk per wolf), deer availability (deer per elk), cow elk to calf ratios, and mean snow depth in the Southern Yellowstone Ecosystem, WY, USA, followed by the percentage of elk and deer that constituted cougar diets, the percentage of the total elk killed that were calves, and the percentage of kill sites found in forested habitat for summer seasons

Year	Wolves	Elk/wolf	Deer/elk	Cow:calf	Snow depth (cm)	% elk	% deer	% calves	% forest
2002	15	997.07	0.012	20	72.29	n/a	n/a	n/a	n/a
2003	19	708.26	0.014	28	67.87	86.7	13.3	42	60
2004	15	915.33	0.006	26	77.51	63	18.8	60	78.1
2005	36	350.28	0.008	22	82.7	n/a	n/a	n/a	n/a
2006	30	428.50	0.007	25	69.68	26.3	68.8	40	68.6
2007	58	220.29	0.008	25	89.57	68.1	28.6	50	70
2008	81	155.33	0.004	27	78.22	9.9	63.4	60	82.2
2009	74	169.59	0.002	20	56.61	53.3	33.3	64	83.3
2010	91	128.47	0.010	21	92.93	16.3	63.8	50	95.4
2011	85	140.92	0.008	17	54.9	41.7	52.8	100	76.4
2012	83	144.36	0.002	25	60.11	27.3	69.8	83	79.1

Estimating kill rates and testing for increasing kill rates between 2005 and 2012

We estimated seasonal kill rates in animals killed per week for cougars wearing GPS collars between 2005 and 2012, when we had sufficient samples of cougars wearing GPS collars to conduct the research. We employed models that differentiated between GPS clusters with a high probability of being a kill versus GPS clusters with a low probability of being a kill. GPS collars were programmed to acquire location data at variable rates, ranging from 3- to 10-h intervals, with the majority of collars being programmed to acquire six to eight locations each day. To identify clusters, GPS data were analyzed with a Python script (Python Software Foundation Hampton, NH) developed by Knopff et al. (2009) to identify two or more locations within 100 m and ≤48 h of each other. We did not include cougars wearing VHF collars in our kill rate calculations.

First, we developed competing models to test four spatial and seven temporal attributes of clusters potentially predictive of kill sites (Table 1; Anderson and Lindzey 2003; Ruth et al. 2010; Miller et al. 2013). We tested these models against our database of a subset of GPS clusters we visited in the field, and for which we determined whether or not a kill was present. Additional sites where the presence of a kill could be confirmed were provided by visits based on VHF radio telemetry for clusters later identified with GPS data stored on collars. For cluster analysis, site searches conducted ≥4 months from the beginning date of cluster formation and site searches of questionable reliability (e.g., when snow fall may have obscured evidence) were omitted.

We employed logistic regression to assess the ability of univariate and multivariate models to predict at which clusters we would likely find a cougar kill. We employed residual plots of univariate models to assess assumptions of linearity and normality and transformed data accordingly. We also excluded models in which variables exhibited collinearity ≥0.6. We applied backwards and forwards model selection to select the best supported model, based on Akaike's information criterion, and obtain parameter estimates. We did not use weighted parameter estimates for predictor variables because of the large number of highly correlated variables among models. We assessed the fit of the top model from receiver operative characteristic (ROC) curves (Hosmer and Lemshow 2000).

We employed sensitivity and specificity curves to assess the efficiency of models in predicting kill sites and determining the optimum probability cut-point for clusters to identify kills (Hosmer and Lemshow 2000). High sensitivity may effectively classify clusters with kills successfully (reduce false negatives) but is likely to attribute kills to clusters without them (increase false positives). In contrast, high specificity may result in more false negatives but few false positives.

We applied our final model to GPS cluster data for all individuals that wore GPS collars to predict which clusters were likely to contain kills as follows:

$$\Pr(\text{kill}) = \exp^{(\beta_0 + \beta_1 * X_1 + \beta_2 * X_2 + \beta_n * X_n)} / 1 + \exp^{(\beta_0 + \beta_1 * X_1 + \beta_2 * X_2 + \beta_n * X_n)} \quad (1)$$

where β_0 is the intercept and β_n are coefficients for predictor variables X_n (Manly et al. 1993). We applied this model to identify clusters that likely contained kills in the entire GPS location data from all cats. We used these identified clusters in the full GPS dataset to estimate seasonal kill rates for each individual cougar. We conducted a test of whether there were differences in kill rates due to season or cougar sex with a two-way analysis of variance, where individual cougar was included as a random effect.

Due to the final model's low success rate in classifying clusters, we examined an alternative model for identifying kills. This model was based on previous research by Anderson and Lindzey (2003) and Ruth et al. (2010), which employed the number of nights a cougar visited a cluster as an efficient means to distinguish between kill and non-kill sites. In this approach, kills were assigned only to clusters that spanned more than one night. We reported kill rates determined using both the "best" and night > 1 models, because the benefit of the night > 1 model's ability to positively detect kills at sites may have outweighed the cost of over-attributing kills to clusters where none were found. As this method likely missed small prey that could be consumed in shorter time frames, we defined these kill rates as ungulate-only kill rates. Then, we tested for changes in kill rate over time using linear regression, in which we included individual cougar as a random effect.

Results

Cougar mortalities

We determined cause-specific mortality of 33 cougar kittens and 45 subadults and adults from 2002 to 2012. In order from greatest frequency to least, 10 cougar kittens died of starvation (38 %), 5 kittens were killed by wolves (19 %), 5 kittens died of unknown causes (19 %), 2 kittens died from an unidentified predator (8 %), 2 kittens were killed by other cougars (8 %), 1 kitten died from disease (4 %), and 1 was killed by wildlife managers (4 %). In order from greatest frequency to least, 16 subadult and adult cougars were legally harvested (36 %), 9 died from starvation (20 %), 5 were killed by other cougars (11 %), 5 died of disease (11 %), 5 died of undetermined cause (11 %), 3 were killed by wildlife managers (7 %), 1 was illegally killed (2 %), and 1 died of old age

(2 %). All kitten and adult starvations occurred after 2005, and there were zero cougars harvested from 2007 to 2012.

Seasonal prey selection and hunting habitats

Between January 2002 and October 1, 2012, we recorded 411 winter prey and 239 summer prey killed by 28 female and 10 male cougars, and an additional 37 prey items recorded for unmarked cougars (30 winter, 7 summer). Of these, only 29 prey were not ungulates. Deer composed 42.4 % of summer cougar diets but only 7.2 % of winter diets. Elk composed 38.3 % of summer cougar diets and 74.4 % of winter diets.

We detected different changes in the proportion of elk and deer killed by cougars in summer between 2002 and 2012. We detected a positive change in proportions of deer ($\chi^2_{1,7} = 5.89$, $p = 0.01$; Fig. 2) and a negative change in proportions of elk ($\chi^2_{1,7} = 4.47$, $p = 0.03$; Fig. 2) killed during the summer season. The top model explaining changing cougar prey selection of deer included only elk availability ($\beta = -0.003$; Table 2). In the case of cougar prey selection of elk, our analyses identified four top models (Table 3). Model averaging of top models identified elk availability ($\beta = -0.003$) as the predictive variable of greatest weight ($w = 1$), then deer availability ($\beta = -20.960$; $w = 0.49$), and then snow ($\beta = -0.013$; $w = 0.18$) (Table 1). In contrast, we did not detect a change in prey selection by cougars for either elk ($\chi^2_{1,8} = 0.28$, $p = 0.60$) or deer ($\chi^2_{1,8} = 0.29$, $p = 0.59$) during the winter season.

There was no change in the percentage of elk calves in cougar diets during winter seasons ($\chi^2_{1,7} = 2.63$, $p = 0.10$); however, there was a significant increase in the percentage of calves killed during the summer season ($\chi^2_{1,7} = 6.13$, $p = 0.01$). The results of our model comparisons identified three top models (within 2 ΔAIC; Table 4). Model averaging identified cow-calf ratios ($\beta = 0.240$; $w = 0.58$) as the strongest predictor of changing cougar selection for elk calves, followed by snow ($\beta = -0.050$; $w = 0.45$), then deer availability ($\beta = -137.345$; $w = 0.29$), and elk availability ($\beta = -0.001$; $w = 0.16$) (Table 1).

The percentage of kills found in forested habitats did not change over time in winter ($\chi^2_{1,9} = 0.60$, $p = 0.44$). During the summer season, however, there was a significant increase in the percentage of kills in forested habitats over the course of the study ($\chi^2_{1,7} = 4.35$, $p = 0.03$) (Table 1). Of all kills across years, 76 ± 14 % were found in forests in summer and 58 ± 14 % were found in forests in winter.

Cougar kill rates

GPS cluster model development. From 2005 to 2012, we collected location data adequate to identify 2729 clusters from 17 cougars wearing GPS collars. During this same time period, we visited 309 clusters from 14 individuals (average clusters visited/cat = 22; range 1–81 clusters/cat) to search for prey remains. Sites were visited between 0 and 98 days after initial cluster formation (average = 8 days). We found kills at 269 clusters and classified 40 clusters as non-kills, and with these data, we tested our kill rate models.

The final logistic regression model for determining if a cluster was likely a kill included the cluster spanning more than one night ($z = 3.042$; $p = 0.002$), the square root of the number of hours at night in cluster ($z = 2.084$; $p = 0.037$), fidelity ($z = 2.205$; $p = 0.027$), and distance to nearest consecutive cluster ($z = 2.284$; $p = 0.022$). The probability of finding a kill was positively associated with all four variables included in the final model. Model fit was moderate for the final model (AUC = 0.827). The probability cutoff of 0.887 maximized both sensitivity and selectivity in discriminating between clusters with and without kills; however, this cutoff resulted in only a 49.7 % success of correctly classifying our known kill and non-kill clusters, identified with field investigations.

The variable "night > 1" was found to be the strongest predictor of a cluster being a kill in both the multivariate models and among all univariate models (R^2 for night > 1 = 0.148; all other independent variables $R^2 < 0.057$). The night > 1 model was found to successfully identify clusters as having kills 86 % of instances but attributed

Table 2 Ranked models used to test explanatory variables (elk availability (elk/wolf), deer availability (deer/elk), and mean snow depth) for changes in proportions of deer in summer cougar diets over time, including AIC scores, ΔAIC, W_i, and likelihoods, in order of performance, best to worst

Model no.	No. of variables	Model	AIC	ΔAIC	W_i	Likelihood
1	1	{Elk/Wolf}	109.0864	0.00	0.82	1.00
2	2	{Deer/Elk + Elk/Wolf}	113.2963	4.21	0.10	0.12
3	2	{Elk/Wolf + Snow}	113.8864	4.80	0.07	0.09
4	2	{Deer/Elk + Elk/Wolf + Snow}	120.2914	11.21	0.00	0.00
5	1	{Deer/Elk}	139.6132	30.53	0.00	0.00
6	2	{Deer/Elk + Snow}	148.2827	39.20	0.00	0.00
7	1	{Snow}	153.6072	44.52	0.00	0.00

Table 3 Ranked models used to test explanatory variables (elk availability (elk/wolf), deer availability (deer/elk), and mean snow depth) for changes in proportions of elk in summer cougar diets over time, including AIC scores, ΔAIC, W_i, and likelihoods, in order of performance, best to worst

Model no.	No. of variables	Model	AIC	ΔAIC	W_i	Likelihood
1	1	{Elk/Wolf}	144.0788	0.00	0.33	1.00
2	2	{Deer/Elk + Elk/Wolf}	144.1373	0.06	0.32	0.97
3	2	{Elk/Wolf + Snow}	145.2729	1.19	0.18	0.55
4	3	{Deer/Elk + Elk/Wolf + Snow}	145.4837	1.40	0.16	0.50
5	2	{Deer/Elk + Snow}	165.3835	21.30	0.00	0.00
6	1	{Deer/Elk}	167.8292	23.75	0.00	0.00
7	1	{Snow}	183.3962	39.32	0.00	0.00

kills to 40 % of sites at which we did not find prey remains. This resulted in a 46.0 % classification success for the night > 1 model. The benefit of the night > 1 model's ability to positively detect kills at sites (86 % for night > 1 model vs. 75 % for full model) may have outweighed the cost of over-attributing kills to clusters where none were found (40 % for night > 1 model vs. 25 % for full model) (Table 4). Field crews often searched sites many days after the cluster had formed, and it was likely that they failed to find evidence of some kills that actually occurred. Given the potential bias in assigning non-kill clusters, it is reasonable to assume the model that over-attributed kills to clusters was more reliable.

We did not detect a change in cougar kill rates in animals killed per week between 2005 and 2012 in summer ($F_{1,25.3} = 0.090$, $p = 0.767$) or winter ($F_{1,31.1} = 0.073$, $p = 0.788$). Mean summer and winter kill rates were not significantly different (0.90 ± 0.23 animals/week and 0.99 ± 0.22 animals/week, respectively).

Discussion

Determining whether one species influences the realized niche of another is difficult, if not impossible in natural systems (Schoener 1983). We seized a rare opportunity to test whether wolves are currently competing with and influencing the realized niche of sympatric cougars, following wolf reintroductions in Yellowstone National Park, USA. In Lendrum et al. (2014), we reported that cougars exhibited spatial displacement from wolves when selecting home ranges. In this paper, we tested whether cougars changed their prey selection and kill rates between 2002 and 2012, as further evidence of ongoing competition between the two species.

Table 4 Ranked models used to test explanatory variables (elk availability (elk/wolf), deer availability (deer/elk), cow elk to calf ratios, and mean snow depth) for changes in proportions of elk calves in summer cougar diets over time, including AIC scores, ΔAIC, W_i, and likelihoods, in order of performance, best to worst

Model no.	No. of variables	Model	AIC	ΔAIC	W_i	Likelihood
1	2	{Deer/Elk + Cow Calf}	85.793	0.00	0.29	1.00
2	2	{Cow Calf + Snow}	85.8264	0.03	0.29	0.98
3	2	{Elk/Wolf + Snow}	87.0374	1.24	0.16	0.54
4	1	{Snow}	88.254	2.46	0.09	0.29
5	1	{Cow Calf}	88.8982	3.11	0.06	0.21
6	3	{Deer/Elk + Cow Calf + Snow}	89.8507	4.06	0.04	0.13
7	2	{Deer/Elk + Snow}	90.3895	4.60	0.03	0.10
8	3	{Elk/Wolf + Cow Calf + Snow}	91.6315	5.84	0.02	0.05
9	3	{Deer/Elk + Elk/Wolf + Cow Calf}	92.9873	7.19	0.01	0.03
10	2	{Elk/Wolf + Cow Calf}	93.1075	7.31	0.01	0.03
11	3	{Deer/Elk + Elk/Wolf + Snow}	93.6315	7.84	0.01	0.02
12	1	{Deer/Elk}	96.3017	10.51	0.00	0.01
13	2	{Deer/Elk + Elk/Wolf}	97.3428	11.55	0.00	0.00
14	1	{Elk/Wolf}	98.0101	12.22	0.00	0.00
15	4	{Deer/Elk + Elk/Wolf + Cow Calf + Snow}	101.6466	15.85	0.00	0.00

We found support for our first hypothesis that cougars would prey switch from elk to deer during summer as wolves increased in the system. Relative elk availability, defined as the number of elk per wolf, declined steeply over the course of the study. By 2012, elk availability was estimated to be 14 % of what it was in 2002. Relative deer availability, defined by their abundance relative to elk abundance, also decreased over the course of the study. As both elk and deer availability decreased over the course of the study, changes in cougar prey selection were unlikely driven by changing elk or deer availability. Instead, our data strongly suggest that competition with increasing wolves in the system was the likely driver of changes in cougar prey selection over the last 12 years.

Exploitation competition is indirect competition through a shared resource, where the exploitation of the resource by the dominant competitor negatively impacts the fecundity and survival of the subordinate competitor (Petren and Case 1996). If cougars prey switched from elk to deer because elk numbers had declined to levels that could no longer sustain them, then this would be evidence of exploitation competition (e.g., Kortello et al. 2007). However, elk numbers only dropped from 14,956 to 11,982 over the course of the study and far exceed the capacity to maintain the low density of cougars for the study area (4.2 resident adults/890 km^2 in 2011 in Elbroch et al. 2012). Instead, our results better support an alternate explanation for how competition may be playing out between wolves and cougars in the SYE—one that suggests that changing cougar prey selection was driven by changing cougar habitat use. If cougars have shifted to more complex habitats to mitigate direct interference competition with coursing wolves, as has been shown in Northern Yellowstone (Ruth et al. 2011, 2015), then this shift in habitat use likely led to increased encounter rates with more forest-dependent mule deer. In support of this hypothesis, we did in fact detect an increase in the percentage of kill sites found in forested habitats in summers over the course of the study.

Contrary to our prediction that cougars would decrease their predation on elk calves during wolf and grizzly bear recolonization (Griffin et al. 2011), cougars increased their predation of elk calves in summers over the course of the study, even while wolves, too, preferentially selected for elk calves (Stevenson et al. 2011). In contrast, cougars in Northern Yellowstone increased the percentage of adult elk in their diet as cow-calf ratios decreased after wolf reintroductions (Ruth et al. 2015). The fact that cow-calf ratios in the SYE remained consistent as wolves increased in the system over the course of the study is likely reflective of high elk abundance and high elk fecundity and survivorship supported by supplemental feeding (Smith 2007). Therefore, the fact that cougars increased the number and proportion of calves in their diet while wolves increased in the system may indicate that elk calves in our study system have yet to drop in abundance enough to become a limited resource for subordinate cougars.

Also contrary to our hypothesis, we did not detect any changes in kill rates for either summer or winter, as we would expect as evidence of interference competition. This was likely due to our methods rather than the ecological reality of the system, and a more intensive field effort may have yielded different results (Knopff et al. 2010; Elbroch and Wittmer 2013b). Because we employed the night > 1 model to quantify kill rates, we limited our ability to detect small prey or large prey when large competitors displaced cougars from their kills (Elbroch and Wittmer 2013a). This would influence both our comparison of selection for smaller ungulates and the predictive power of our kill rate models. The multiplicity of GPS fix rates employed over the life of the project also likely reduced the effectiveness of many explanatory variables in identifying clusters with and without kills. For these reasons, our methods may have failed to detect cougar prey switching to smaller prey that may have increased their kill rates.

We recorded 5 marked cougar kittens killed by wolves (23 % of known-cause mortalities) during the course of the study and 2 more that were killed by an undetermined predator which may or may not have been wolves. In addition, between 2005 and 2012, 9 adult cougars (20 % of known-cause mortalities) and 10 cougar kittens (45 % of known-cause mortalities) died of starvation, which may also be in part explained by competition with wolves (Kunkel et al. 1999; Ruth 2004b; Ruth et al. 2015). Direct mortalities provide strong evidence for interference competition between recolonizing wolves and resident cougars in the Southern Yellowstone Ecosystem.

Conclusions

The spatial displacement of cougars by wolves in the SYE reported in Lendrum et al. 2014, in combination with changes in cougar prey selection and interspecific competitive killing of resident cougars by recolonizing wolves (because kittens were not entirely consumed by wolves; Lourenço et al. 2013) that we report here, provides strong evidence for newly established and now ongoing competition between these two species. During the last 12 years, resident cougars in the SYE may have served as naïve intermediaries relearning to coexist with a dominant competitor absent since 1926, when the last wolf was killed in Yellowstone National Park (Haines 1996). Through competition, recolonizing wolves have impacted the realized niche of resident cougars in the SYE, and current resident cougars may now exhibit a realized niche more reflective of an era when these species were previously sympatric in the Yellowstone Ecosystem.

Competing interests
The authors declare that they have no competing interests.

Authors' contributions
ME, JN, and HQ designed the study. ME, JN, PL, HQ, and DT participated in fieldwork. ME, PL, and JN ran the analyses and wrote the manuscript. HQ and DT provided additional data integral to the study. All coauthors contributed to the editing of the manuscript. All authors read and approved the final manuscript.

Authors' information
ME is a Science Director for Panthera and the principal investigator for Panthera's Teton Cougar Project (PTCP) in northwest Wyoming. PL was a biologist for PTCP and is now a graduate researcher at Colorado State University. JN was a biologist for PTCP and now works for Montana Fish, Wildlife, and Parks. HQ initiated PTCP and is now Executive Director of Jaguar and Puma Programs for Panthera. DT is the Wyoming Game and Fish Department liaison to PTCP and an active collaborator on research.

Acknowledgements
We thank our cooperators, including D. Craighead (Craighead Beringia South), K. Murphy (Bridger-Teton NF), S. Cain (Grand Teton NP), T. Fuchs (WY Game and Fish Dept), and E. Cole (National Elk Refuge); our supportive funders, the Summerlee Foundation, The Richard King Mellon Foundation, The Charles Engelhard Foundation, The Laura Moore Cunningham Foundation, Tim and Karen Hixon Foundation, National Geographic Society, The Norcross Wildlife Foundation, Inc., Earth Friends Conservation Fund, the Cougar Fund, The Bay Foundation, Michael Cline Foundation, Eugene V. & Clare E. Thaw Charitable Trust, Connemara Fund, Hogan Films, National Fish and Wildlife Foundation, Community Foundation of Jackson Hole, The Oregon Zoo Foundation, Mr. and Mrs. G. Ordway, Mr. and Mrs. M. Manship, Mr. and Mrs. N. Jannotta, Mr. L. Westbrook, Mr. and Mrs. S. Robertson, Mr. R. Comegys; and several anonymous foundation and individual donors. Thanks also to T. Ruth for her generous review of our manuscript, which greatly enhanced its strength and clarity, and for the feedback of one more anonymous reviewer. No funders or funding organizations have any claim on the data collected as part of this research nor the products that were developed as part of this research, nor did any funders or funding organizations play a role in any decision-making in how the data was analyzed, presented, or published as part of this manuscript.

Author details
[1]Panthera, 8 West 40th Street, 18th Floor, New York, NY 10018, USA. [2]Craighead Beringia South, PO Box 147, Kelly, WY 83011, USA. [3]Wyoming Game and Fish, 260 Buena Vista, Lander, WY 82520, USA.

References
Anderson CR, Lindzey FG (2003) Estimating cougar predation rates from GPS location clusters. J Wildlife Manage 67:307–316
Atwood TC, Gese EM, Kunkel KE (2007) Comparative patterns of predation by cougars and recolonizing wolves in Montana's Madison Range. J Wildlife Manage 71:1098–1106
Bartnick TR, Van Deelen TR, Quigley HB, Craighead D (2013) Variation in cougar predation habits during wolf recovery in the southern Greater Yellowstone Ecosystem. Can J Zoolog 91:82–93
Birch LC (1957) The meanings of competition. Am Nat 91:5–18
Burnham KP, Anderson DR (2002) Model selection and multimodel inference: a practical information theoretic approach, 2nd edn. Springer, New York, USA
Connell JH (1983) On the prevalence and relative importance of interspecific competition: evidence from field experiments. Am Nat 122:661–696
Elbroch M (2003) Mammal track and signs: a guide to North American species. Stackpole Books, Mechanicsburg, USA
Elbroch M, Wittmer HU (2012) Puma spatial ecology of in open habitats with aggregate prey. Mamm Biol 77:377–384
Elbroch LM, Wittmer HU (2013a) Nuisance ecology: do scavenging condors exact foraging costs on pumas in Patagonia? PLoS One 8:e53595
Elbroch LM, Wittmer HU (2013b) The effects of puma prey selection and specialization on less abundant prey in Patagonia. J Mammal 94:259–268
Elbroch M, Quigley H, Craighead D (2012) Teton Cougar Project 2012 progress report. Craighead Beringia South, Kelly, WY, USA
Elbroch LM, Lendrum PE, Newby J, Quigley H, Craighead D (2013) Seasonal foraging ecology of non-migratory cougars in a system with migrating prey. PLoS One 8:e83375. doi:10.1371/journal.pone.0083375
Garrott RA, Bruggeman JE, Becker MS, Kalinowski ST, White PJ (2007) Evaluating prey switching in wolf-ungulate systems. Ecol Appl 17:1588–1597
Griffin KA, Hebblewhite M, Robinson HS, Zager P, Barber-Meyer SM, Christianson D, Creel S, Harris NC, Hurley MA, Jackson DH, Johnson BK, Myers WL, Raithel JD, Schlegel M, Smith BL, White C, White PJ (2011) Neonatal mortality of elk driven by climate, predator phenology and predator community composition. J Anim Ecol 80:1246–1257
Haines AL (1996) The Yellowstone story—a history of our first national park. University Press of Colorado, Niwot, CO, USA
Heffelfinger J (2010) Age criteria for Southwestern game animals. Special Report #19. Arizona Game and Fish Department, USA
Hosmer DW, Lemshow S (2000) Applied logistic regression. John Wiley and Sons, New York, USA
Husseman JS, Murray DL, Power G, Mack C, Wenger CR, Quigley H (2003) Assessing differential prey selection patterns between two sympatric large carnivores. Oikos 101:591–601
Hutchinson GE (1957) Concluding remarks. Cold Spring Harb Symp Quant Biol 22:415–427
Knight DH (1996) Mountains and plains: the ecology of Wyoming landscapes. Yale University Press, CT, USA
Knopff KH, Knopff AA, Warren MB, Boyce MS (2009) Evaluating global positioning system telemetry techniques for estimating cougar predation parameters. J Wildlife Manage 73:586–597
Knopff KH, Knopff AA, Kortello A, Boyce MS (2010) Cougar kill rate and prey composition in a multiprey system. J Wildlife Manage 74:1435–1447
Kortello AD, Hurd TE, Murray DL (2007) Interactions between cougars (*Puma concolor*) and gray wolves (*Canis lupus*) in Banff National Park, Alberta. Ecoscience 14:214–222
Kunkel KE, Ruth TK, Pletscher DH, Hornocker MG (1999) Winter prey selection by wolves and cougars in and near Glacier National Park, Montana. J Wildlife Manage 63:901–910
Lendrum PE, Elbroch LM, Quigley H, Thompson DJ, Jimenez M, Craighead D (2014) Home range characteristics of a subordinate predator: selection for refugia or prey availability? J Zool doi:10.1111/jzo.12153
Lourenço R, Penteriani V, Rabaça JE, Korpimäki E (2013) Lethal interactions among vertebrate top predators: a review of concepts, assumptions and terminology. Biol Rev doi: 10.1111/brv.12054
Lubow BC, Smith BL (2004) Population dynamics of the Jackson elk herd. J Wildlife Manage 68:810–829
Manly B, McDonald L, Thomas D (1993) Resource selection by animals. Chapman & Hall, New York, USA
Miller CS, Hebblewhite M, Miquelle DG, Seryodkin IV, Goodrich JM (2013) Estimating Amur tiger (*Panthera tigris altaica*) kill rates and consumption rates using GPS collars. J Mammal 94:845–855
Nelson AA, Kauffman MJ, Middleton AD, Jimenez M, McWhirter D, Barber J, Gerow K (2012) Elk migration patterns and human activity influence wolf habitat use in the Greater Yellowstone Ecosystem. Ecol Appl 22:2293–2307
Parker KL, Robbins CT, Hanley TA (1984) Energy expenditures for locomotion by mule deer and elk. J Wildl Manage 48:474–488
Peterson RO, Ciucci P, Mech LD, Boitani L (2003) The wolf as a carnivore. In: Wolves: behaviour, ecology and conservation. University of Chicago Press, Chicago, USA, pp 104–130
Petren K, Case TJ (1996) An experimental demonstration of exploitation competition in an ongoing invasion. Ecology 77:118–132
Quigley K (2000) Immobilization and biological sampling protocols. Hornocker Wildlife Institute/Wildlife Conservation Society, Bozeman, MT, USA
Riley SJ, Nesslage GM, Maurer BA (2004) Dynamics of early wolf and cougar eradication efforts in Montana: implications for conservation. Biol Conserv 119:575–579
Ruth TK (2004a) Ghost of the Rockies: the Yellowstone cougar project. Yellowstone Science 12:13–24
Ruth TK (2004b) Patterns of resource use among cougars and wolves in northwestern Montana and southeastern British Columbia. Dissertation, University of Idaho, Moscow, USA
Ruth TK, Murphy K (2010a) Competition with other carnivores for prey. In: Hornocker M, Negri S (eds) Cougar: ecology and conservation. University of Chicago Press, Chicago, USA, pp 163–172

Ruth TK, Murphy K (2010b) Cougar–prey relationships. In: Hornocker M, Negri S (eds) Cougar: ecology and conservation. University of Chicago Press, Chicago, USA, pp 138–162

Ruth TK, Buotte PC, Quigley HB (2010) Comparing ground telemetry and Global Positioning System methods to determine cougar kill rates. J Wild Manage 74:1122–1133

Ruth TK, Haroldson MA, Murphy KM, Buotte PC, Hornocker MG, Quigley HB (2011) Cougar survival and source-sink structure on Greater Yellowstone's Northern Range. J Wildlife Manage 75:1381–1398

Ruth TK, Buotte PC, Hornocker MG (2015) Yellowstone cougars: ecology before and during wolf reestablishment. University Press of Colorado, Boulder, CO, USA, in press

Sawyer H, Lindzey F, McWhirter D (2005) Mule deer and pronghorn migration in western Wyoming. Wildlife Soc B 33:1266–1273

Schoener TW (1983) Field experiments on interspecific competition. Am Nat 122:240–285

Seddon PJ, Armstrong DP, Maloney RF (2007) Developing the science of reintroduction biology. Conserv Biol 21:303–312

Sikes RS, Gannon WL, the Animal Care and Use Committee of the American Society of Mammalogists (2011) Guidelines of the American Society of Mammalogists for the use of wild mammals in research. J Mammal 92:235–253

Smith BL (2007) Migratory behavior of hunted elk. Northwest Sci 81:251–264

Stevenson J, Jimenez M, Dewey S, Cain S (2011) Wildlife research: wolf predation. In: Cain SL (ed) Wildlife conservation, management, and research. 2011 National Park Service, Grand Teton National Park, Moose, WY, USA, pp 55–56

US Fish and Wildlife Service, Idaho Department of Fish and Game, Montana Fish, Wildlife and Parks, Nez Perce Tribe, National Park Service, Blackfeet Nation, Confederated Salish and Kootenai Tribes, Wind River Tribes, Washington Department of Fish and Wildlife, Oregon Department of Fish and Wildlife, Utah Department of Natural Resources, USDA Wildlife Services (2012) Northern Rocky Mountain Wolf Recovery Program 2011 Interagency Annual Report. USFWS, Helena, MT, USA

US Fish and Wildlife Service, Montana Fish, Wildlife and Parks, Nez Perce Tribe, National Park Service, Blackfeet Nation, Confederated Salish and Kootenai Tribes, Wind River Tribes, Washington Department of Wildlife, Oregon Department of Wildlife, Utah Department of Natural Resources, USDA Wildlife Services (2011) Rocky Mountain Wolf Recovery 2010 Interagency Annual Report. USFWS, Helena, MT, USA

US Fish and Wildlife Service, Nez Perce Tribe, National Park Service, Montana Fish, Wildlife and Parks, Idaho Fish and Game & USDA Wildlife Services (2005) Rocky Mountain Wolf Recovery 2004 Annual Report. USFWS, Helena, MT, USA

US Fish and Wildlife Service, Nez Perce Tribe, National Park Service, Montana Fish, Wildlife and Parks, Idaho Fish and Game, USDA Wildlife Services (2006) Rocky Mountain Wolf Recovery 2005 Annual Report. USFWS, Helena, MT, USA

US Fish and Wildlife Service, Nez Perce Tribe, National Park Service, Montana Fish, Wildlife and Parks, Idaho Fish and Game, USDA Wildlife Services (2007) Rocky Mountain Wolf Recovery 2006 Annual Report. USFWS, Helena, MT, USA

US Fish and Wildlife Service, Nez Perce Tribe, National Park Service, Montana Fish, Wildlife & Parks, Blackfeet Nation, Confederated Salish and Kootenai Tribes, Idaho Fish and Game, USDA Wildlife Services (2008) Rocky Mountain Wolf Recovery 2007 Interagency Annual Report. USFWS, Helena, MT, USA

US Fish and Wildlife Service, Nez Perce Tribe, National Park Service, Montana Fish, Wildlife & Parks, Blackfeet Nation, Confederated Salish and Kootenai Tribes, Idaho Fish and Game, USDA Wildlife Services (2009) Rocky Mountain Wolf Recovery 2008 Interagency Annual Report. USFWS, Helena, MT, USA

US Fish and Wildlife Service, Nez Perce Tribe, National Park Service, Montana Fish, Wildlife & Parks, Blackfeet Nation, Confederated Salish and Kootenai Tribes, Idaho Fish and Game, USDA Wildlife Services (2010) Rocky Mountain Wolf Recovery 2009 Interagency Annual Report. USFWS, Helena, MT, USA

US Fish and Wildlife Service, Nez Perce Tribe, National Park Service, USDA Wildlife Services (2004) Rocky Mountain Wolf Recovery 2003 Annual Report. USFWS, Helena, MT, USA

20

Diagnosability and description of a new subspecies of Indo-Pacific humpback dolphin, *Sousa chinensis* (Osbeck, 1765), from the Taiwan Strait

John Y Wang[1,2,3*], Shih Chu Yang[4] and Samuel K Hung[5]

Abstract

Background: Subspecies recognition can affect how people (scientists and non-scientists alike) view organisms and thus has important implications for research on, as well as the conservation of, these entities. Recently, a small group of Indo-Pacific humpback dolphins was discovered inhabiting the waters off central western Taiwan. This geographically isolated population possesses pigmentation patterns that are subtly, but noticeably, different from their nearest conspecifics in the neighbouring waters of the Jiulong River Estuary and Pearl River Estuary of mainland China. Due to this population's low and declining numbers and the numerous threats it faces, it was assessed as critically endangered by the Red List of Threatened Species of the International Union for Conservation of Nature. The purpose of this study is to examine the degree of differentiation of the Taiwanese population to determine if subspecies recognition is warranted.

Results: Analysis of the degree of differentiation in pigmentation patterns revealed nearly non-overlapping distributions between dolphins from Taiwanese waters and those from the Jiulong River + Pearl River estuaries of mainland China (the nearest known populations). The Taiwanese dolphins were clearly diagnosable from those of the Jiulong River + Pearl River estuaries under the most commonly accepted '75% rule' for subspecies delimitation (with 94% of one group being separable from 99+% of the other). Evidence of geographical isolation and behavioural differences also provided additional support for the distinctiveness of the Taiwanese dolphins.

Conclusions: Together, the evidence strongly demonstrated that the Taiwanese humpback dolphin population is differentiated at the subspecies level and on an evolutionary trajectory that is independent from that of dolphins from adjacent waters of mainland China (i.e. Jiulong River + Pearl River estuaries). As a result, the taxonomy of *Sousa chinensis* was revised to include two subspecies: the Taiwanese humpback dolphin, *Sousa chinensis taiwanensis* subsp. nov., and the Chinese humpback dolphin, *Sousa chinensis chinensis* (the nominotypical subspecies). These subspecies are described, and the holotype and paratype specimens for *S. c. taiwanensis* are established.

Keywords: Indo-Pacific; Taiwanese; Humpback dolphin; New subspecies; *Sousa chinensis taiwanensis*; Diagnosability; '75% rule'

* Correspondence: pcrassidens@rogers.com
[1]CetAsia Research Group, Thornhill, Ontario L4J-7X1, Canada
[2]Department of Biology, Trent University, Peterborough, Ontario K9J-7B8, Canada
Full list of author information is available at the end of the article

Background

The value and importance of subspecies (versus other intraspecific designations, such as populations, evolutionarily significant units and distinct population segments) as a unit of organization in biology have been the subject of great debate (e.g. see Mayr 1982; Ryder 1986; Zink 2004; Phillimore and Owens 2006; Patten 2010; Remsen 2010). Some have argued that most of the criticism plaguing subspecies has been due to inconsistent and subjective delimitation of subspecies, most of which were established decades ago when analytical abilities were very limited and have not been re-examined since (see Patten and Unitt 2002; Remsen 2010). Although the subspecies debate is beyond the scope of this paper, we agree that when delimiting subspecies (or any other taxonomic unit), objective quantitative testing of clear hypotheses based on a specified conceptual framework is scientifically sensible and will result in less controversy. Furthermore, this will allow practicing taxonomists to determine if explicit criteria are satisfied without the need for personal philosophical views to be involved or to be forced into more esoteric debate about subspecies as a taxonomic entity.

The subspecies category is the only non-species taxonomic rank that is accompanied by formal taxonomic treatment, scientific (trinomial) naming and establishment of type specimens, and recognized and governed by the International Code of Zoological Nomenclature (ICZN 1999). In this regard, it is treated similarly to species. For conservation biologists and resource managers, subspecies designations can also be useful for drawing attention to or prioritizing resource allocation for protection of threatened unique biological entities (Haig et al. 2006). Several national and international lists of threatened wildlife recognize subspecies, and inclusion on such lists can have important legal and financial ramifications (Haig et al. 2006). Recognition of endemic taxa can often result in local citizens developing an increased sense of ownership of, pride towards, and also conservation responsibility for such wildlife. For example, in Taiwan, endemic subspecies such as the Formosan landlocked salmon *Oncorhynchus masou formosanus* (often referred to as the 'National Treasure Fish') and the Formosan black bear *Ursus thibetanus formosanus* are much celebrated and attract considerable attention from local citizens, news media, conservation groups, scientists and government agencies. Furthermore, subspecies are entrenched in some important wildlife conservation policy and thus have legal status (e.g. US Endangered Species Act, Wildlife Conservation Act of Taiwan, Canada's Species at Risk Act). Therefore, ignoring subspecies designations because of personal philosophy can be damaging to the conservation of some taxa because they may not receive the same recognition and attention as similarly distinct taxa that possess trinomial names.

Because cetaceans are charismatic, high-profile animals, it may be somewhat surprising that they are 'under-classified' with respect to the number of subspecies (i.e. many more subspecies likely exist than are recognized presently) (Reeves et al. 2004; BL Taylor, personal communications). The main reason for this apparent taxonomic deficit is that traditionally, evidence for recognizing cetacean subspecies has been primarily a combination of morphological differences and geographic separation. However, obtaining a sufficient series of specimens for many taxa is difficult or nearly impossible given their large size and often rarity. This shortcoming of the present taxonomy of cetaceans can have negative consequences for understanding evolutionary histories within this group and for their conservation because attention and resources may not be allocated optimally.

Definition of subspecies

Subspecies are generally considered to be a population or populations within a species that are found in different breeding locations (allopatry) and have been equated to 'geographic varieties'. A commonality among most, if not all, subspecies definitions is that subspecies represent groups that are diagnosably distinct rather than just exhibiting mean differences (e.g. see Mayr and Ashlock 1991; Patten and Unitt 2002). The most common operational definition of subspecies is based on an arbitrary '75% rule'. The origin of this rule is uncertain but was discussed and adopted by Amadon (1949) to mean that a population should only be recognized as a subspecies if the following conditions are met. For a given defining character or set of characters, 75% of one population is distinguishable from more or less all (99+%) members of another population (and vice versa). Another interpretation of the '75% rule', with 75% of one population's distribution lying outside 75% of the distribution of another population, was rejected by Amadon (1949) as being too easily satisfied.

During a workshop on the shortcomings of cetacean taxonomy in relation to the needs of conservation, the participants observed that subspecies can include two types of entities: those that may not differ enough to be considered species and those that should be species but insufficient data exist to justify species level distinction (Reeves et al. 2004). The following general descriptive guidelines for designating cetacean subspecies were also adopted, '[i]n addition to the use of morphology to define subspecies, the subspecies concept should be understood to embrace groups of organisms that appear to have been on independent evolutionary trajectories (with minor continuing gene flow), as demonstrated by morphological evidence or at least one line of appropriate genetic evidence. Geographical or behavioural differences can complement morphological and genetic evidence for establishing subspecies. As such, subspecies could be geographical forms or incipient species'.

Humpback dolphins, genus *Sousa*

There is uncertainty about the number of species within the genus *Sousa* (see review by Jefferson and Rosenbaum 2014). Until recently, the most widely accepted view was of two species: Indo-Pacific humpback dolphin, *Sousa chinensis*, and Atlantic humpback dolphin, *Sousa teuszii* (see Jefferson and Van Waerebeek 2004). However, mounting molecular and morphological evidence suggested that taxonomic revision was needed (see Frère et al. 2008, 2011; Mendez et al. 2013) and led to a new proposed arrangement that included four species within the genus (Jefferson and Rosenbaum 2014). *Sousa teuszii* remained unchanged while what previously comprised *Sousa chinensis* was divided into the Indo-Pacific humpback dolphin (*S. chinensis*), the Indian Ocean humpback dolphin (*S. plumbea*), and a newly described species, the Australian humpback dolphin (*S. sahulensis*). Due to concerns about poor sampling from important areas in these studies, this new four-species arrangement was accepted only conditionally by the Taxonomy Committee of the Society for Marine Mammalogy (2014). Although several subspecies were also suggested, none were formally described.

Humpback dolphins in Chinese waters

Regardless of the two- or four-species arrangements for *Sousa*, humpback dolphins found in Chinese waters remain firmly ensconced within *S. chinensis*. The type locality of *S. chinensis* is the Canton (=Pearl) River Estuary, where Osbeck (1765) observed this 'snow-white' dolphin swimming and named it *Delphinus chinensis*. However, no scientific specimen was available until 1867, when Robert Swinhoe collected a specimen from the waters of Quemoy (=Chinmen Island) in the Jiulong River Estuary that was later described in detail by Flower (1870) and named *Delphinus sinensis*. The nomenclature was later revised to the present name *S. chinensis* by Allen (1938). In Chinese waters, several populations of Indo-Pacific humpback dolphins have been suggested (Jefferson 2000; Jefferson and Hung 2004), but no studies have unambiguously confirmed the structure and boundaries of these provisional populations with the exception of the Taiwanese humpback dolphins (see below) and no subspecies have been described.

In 2002, a population of humpback dolphins, inhabiting the coastal nearshore waters off central western Taiwan (=eastern Taiwan Strait), was discovered (see Wang et al. 2004a) (Figure 1). The Taiwanese humpback dolphins are clearly closely related to others found in Chinese waters and exhibit the same typical general characteristics (i.e. lacking an obvious hump at the base of the dorsal fin and with older adults being pinkish white in colouration - see Jefferson and Karczmarski 2001; Jefferson and Van Waerebeek 2004; Jefferson and Rosenbaum 2014). A quantitative study of pigmentation differences revealed that the geographically isolated Taiwanese dolphins are phenotypically distinct from its nearest known neighbours, found in the Jiulong River and Pearl River estuaries (Wang et al. 2008a). The Taiwanese population appears to be a year-round resident (Wang and Yang 2011) with a highly restricted distribution (Wang et al. 2007) and comprises fewer than 100 individuals (Wang et al. 2007, 2012). Unfortunately, this small isolated population is facing numerous existing (and looming) threats to its continued existence (see

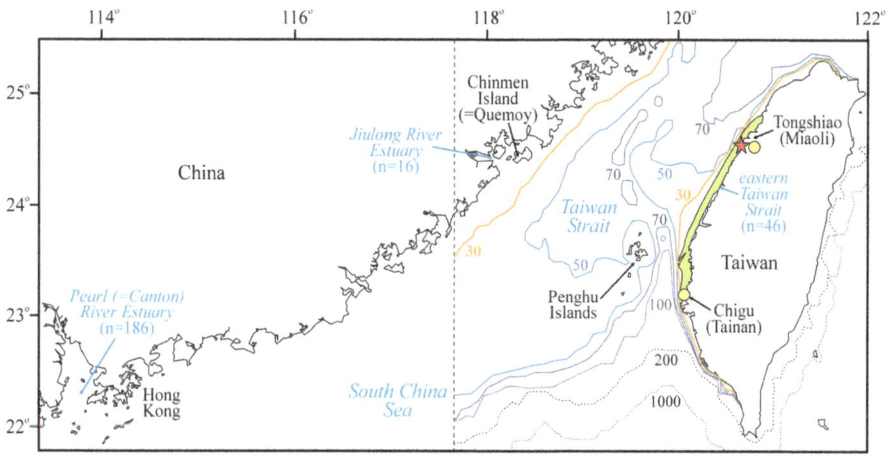

Figure 1 Map of the study area. Sampling locations of Indo-Pacific humpback dolphins analysed in this study. The green shaded area represents the known distribution of these dolphins, and the red star and yellow circles show the locations from where the holotype and paratype specimens of the Taiwanese humpback dolphin *Sousa chinensis taiwanensis* were collected, respectively. Some bathymetric contours in the Taiwan Strait are also shown.

Wang et al. 2004b; Ross et al. 2010; Dungan et al. 2011; Slooten et al. 2013), and an assessment of this population against the criteria of the IUCN Red List of Threatened Species resulted in a 'critically endangered' status (Reeves et al. 2008). The population's current very low potential biological removal level (Slooten et al. 2013) and precarious viability under various likely scenarios (Araújo et al. 2014; Huang et al. 2014) further confirm the direness of its Red List status.

In the original pigmentation study by Wang et al. (2008a), the degree of differentiation was not examined to determine if subspecies level recognition was warranted. The main purpose of this study was to rectify the oversight of the previous study by conducting a quantitative and objective analysis to determine if the Taiwanese population deserves subspecies recognition. We re-examine the pigmentation data of Wang et al. (2008a) (with additional new data) to test the null hypothesis that the Taiwanese humpback dolphins are not diagnosably distinct from those of the Jiulong River and Pearl River estuaries. We also compare osteological data from three Taiwanese specimens with published data on mainland Chinese humpback dolphin specimens and review available information on the species to better understand the degree of isolation that exists between these groups of dolphins.

Methods

Pigmentation

The raw scores of the spotting intensities on dolphin bodies and dorsal fins in Wang et al. (2008a) were reexamined. This dataset did not include completely grey individuals with little to no spotting (i.e. young calves and juveniles), which appeared to show no differences among the dolphins of the Pearl River Estuary, Jiulong River Estuary and eastern Taiwan Strait. Scores were obtained from nine independent observers who determined spotting intensity on a scale from 1 (least spotted) to 4 (most intensely spotted) for the bodies and dorsal fins of individual dolphins by examining photographs (for more details, see Wang et al. 2008a). To reduce the likelihood of various observer errors (e.g. mistakes due to data entry, fatigue, etc.), we excluded the high and low scores before calculating the means for each specimen. (However, including the high and low scores as in Wang et al. 2008a had no noticeable effect on the main results). Reexamination of the photographs used in Wang et al. (2008a) resulted in two eastern Taiwan Strait specimens being omitted from the present analyses. One specimen was excluded due to poor photographic quality, and the other specimen was eliminated because it represented the other side of an individual already in the dataset [note: since the study by Wang et al. (2008a), photographs of both sides of every individual in the Taiwanese population have been obtained by the authors so it was possible to confirm the double inclusion of this individual; none of the other specimens analysed presented this problem]. The total number of different individual dolphins examined in the present study was 248 with 186 from the Pearl River Estuary, 16 from the Jiulong River Estuary and 46 from the eastern Taiwan Strait (Figure 1). Of these, 6 and 17 new individuals from the Jiulong River Estuary and eastern Taiwan Strait, respectively, were photographed after Wang et al. (2008a) and added to the present study.

The photographs used in this study were primarily from individual identification catalogues. All photographs examined were of the two highest quality categories as described in Wang et al. (2012).

Data analyses

We repeated the statistical analyses of the previous study (Wang et al. 2008a) on the expanded dataset to determine if the main results remained unchanged. These analyses included multiple pairwise *t*-tests and one-way analysis of variance (ANOVAs) with *post hoc* comparisons (least significant difference, Scheffe's, Duncan's multiple range, Tukey's honestly significant difference and the Newman-Keuls methods). The level of significance for all tests was 0.05, but for the multiple pairwise *t*-tests, the critical threshold values were corrected using the sequential Bonferroni method (Rice 1989).

Because the main focus of this study was to examine the degree of differentiation between the Taiwanese humpback dolphins and those of their nearest known neighbours in mainland Chinese waters, the dolphins of the Jiulong River and Pearl River estuaries were pooled for the main analyses of this study. Given the small number of Jiulong River Estuary dolphins available in this study, keeping them in a separate group for comparison and diagnosability tests (see below) would not have been meaningful. Pooling was also supported by the findings of Wang et al. (2008a) and the analyses of the expanded dataset in this study (see 'Results'), which showed that Jiulong River Estuary dolphins were not significantly different or distinguishable from Pearl River Estuary dolphins in the pigmentation characters examined (Figure 2).

To better understand the degree of differentiation in the pigmentation of the Taiwanese and Jiulong River Estuary + Pearl River Estuary dolphins, we performed a discriminant analysis on the two variables (spotting intensities on the body and dorsal fin). Although the percentage of misclassifications can be used as a metric for how well specimens can be discriminated, merely reporting percent correct classifications from discriminant analyses was argued to be insufficient by Patten and Unitt (2002). Therefore, the discriminant scores of the specimens were analysed to determine if Taiwanese dolphins were diagnosably different

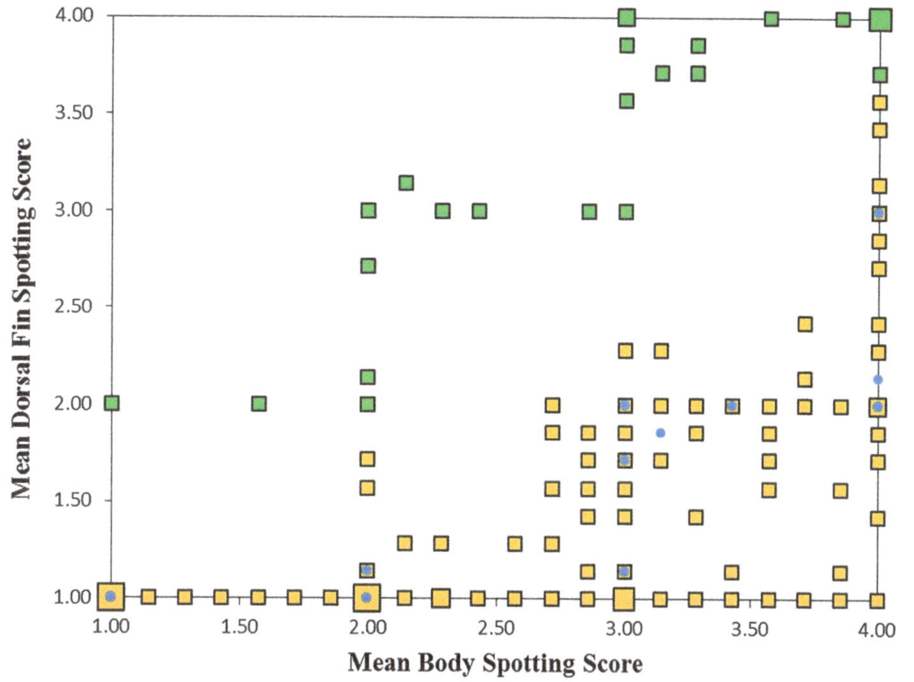

Figure 2 Scatterplot of spotting intensity on dorsal fins and bodies of Indo-Pacific humpback dolphins. Mean scores of the spotting intensities on the dorsal fins vs. bodies of Indo-Pacific humpback dolphins from the eastern Taiwan Strait (green squares), Pearl River Estuary (yellow squares) and the Jiulong River Estuary (blue circles). Small symbols = 1 to 5 individuals; medium symbols = 6 to 10 individuals; large symbols = >10 individuals.

from those of the Jiulong River Estuary + Pearl River Estuary (see diagnosability analysis below).

All statistical analyses were performed using STATISTICA version 12 (Statsoft, Inc. 2014, www.statsoft.com).

Diagnosability analysis for subspecies

With the possible exception of characters that are clearly and completely non-overlapping between two groups, diagnosability testing should be conducted to determine subspecies status in an objective and quantitative manner. To the best of our knowledge, Baker et al. (2002) was the first and only study that explicitly tested diagnosability under the operational definition of the '75% rule' for identifying a cetacean subspecies (in this case, Maui's dolphin, *Cephalorhynchus hectori maui*). To determine if the pigmentation exhibited by Taiwanese humpback dolphins was diagnosable from those of the Jiulong River Estuary + Pearl River Estuary, and hence supporting distinct subspecies, we also used the diagnosability index (*D*) of Patten and Unitt (2002) with the same interpretation of the operational definition ('75% rule') for subspecies as in Baker et al. (2002). Thus, for two groups to qualify as different subspecies, 75% of the distribution of one group must lie outside 99% or more of the other (and this must also be true when testing in the opposite direction - see Patten and Unitt 2002).

Osteology

For thoroughness, we also examined five new, complete (or nearly complete) skeletons of physically mature specimens from the eastern Taiwan Strait ($n = 3$) and Hong Kong's waters of the Pearl River Estuary ($n = 2$). Twenty-two of the cranial morphometric and meristic characters analysed in Jefferson and Rosenbaum (2014) were measured on the new specimens by JYW following Perrin (1975). Tooth counts were based on the largest number of alveoli counted in a single tooth row, and vertebral counts followed the sections described by Rommel (1990). Due to the small sample size, we only conducted simple comparisons with published data on specimens from mainland Chinese waters (Flower 1870; Huang et al. 1978; Wang and Sun 1982; Wang and Han 1996; Wang 1999; Zhou 2004; Jefferson and Van Waerebeek 2004; Jefferson and Rosenbaum 2014). Data on the proposed neotype for *S. chinensis* (Porter 2002) were excluded from this study due to serious concerns about the information presented as well as the specimen being immature (see Jefferson and Van Waerebeek 2004; Jefferson and Rosenbaum 2014).

Results

Pigmentation

Repeating the same analyses that were conducted by Wang et al. (2008a) but on the expanded dataset did not produce any real notable differences from previous results. However, the body spotting on the dolphins of the eastern Taiwan Strait and that on dolphins of the Pearl River Estuary were nearly significantly different (the *p* value was equal to the alpha, after sequential Bonferroni correction for multiple tests), whereas in the original study, differences in body spotting between dolphins from these two regions were not significant. The eastern Taiwan Strait dolphins had significantly greater dorsal fin spotting intensity and greater difference between the spotting on the dorsal fin and body than Jiulong River Estuary or Pearl River Estuary dolphins (all $p < 0.0001$), while neither of these characters were significantly different between dolphins of the Jiulong River Estuary and Pearl River Estuary (see Additional file 1 for detailed results of these analyses).

Not surprisingly, the discriminant analysis also showed very clear separation of the Taiwanese dolphins from the Jiulong River Estuary + Pearl River Estuary (pooled) dolphins (Wilk's lambda = 0.307, $F_{2,245} = 276.25$, $p < 0.0001$) with very little overlap (Figure 3) and thus provided more confirmation that the Taiwanese dolphins were differentiated. [Note: a discriminant analysis was also conducted on the original data, without the new specimens, which were reserved for cross-validation of the classification functions. The overall results were the same as above (Wilk's lambda = 0.376, $F_{2,222} = 183.90$, $p < 0.0001$), and the classification functions generated by the analysis correctly assigned 20 of the 21 (95.2%) new specimens to the region from which they originated].

Examination of the data (body spotting, dorsal fin spotting and discriminant scores) did not reveal any

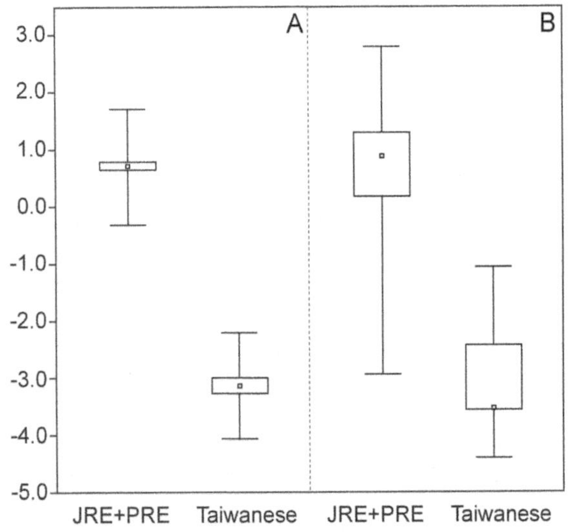

Figure 3 Box and whisker plots of canonical scores for Indo-Pacific humpback dolphins. The squares, boxes and whiskers represent means, standard errors and standard deviations, respectively, in **(A)** and medians, 25 and 75 quartiles and minima-maxima, respectively, in **(B)**. Indo-Pacific humpback dolphins from the Jiulong River Estuary + Pearl River Estuary (JRE + PRE) and Taiwanese waters.

significant deviations from normality or heteroscedasticity. Furthermore, there were no specimens with unusually large or small values in the original or discriminant scores to overly influence the means and variances, so correlations between these two parameters are unlikely.

The diagnosability indices provided clear evidence supporting subspecies recognition for the Taiwanese humpback dolphins following the '75% rule' ($D_{\text{eastern Taiwan Strait, Jiulong River Estuary + Pearl River Estuary}} = 0.91$ and $D_{\text{Jiulong River Estuary + Pearl River Estuary, eastern Taiwan Strait}} = 0.83$). In a mathematically similar manner as explained by Baker et al. (2002), we also estimated the percentage separation between the Taiwanese and Jiulong River Estuary + Pearl River Estuary dolphins using the formula of Patten and Unitt (2002). We found that the Taiwanese humpback dolphins were diagnosable from those of the Jiulong River Estuary + Pearl River Estuary at a very high level (94% from 99+%) (Figure 4).

Descriptive diagnosis

Differences in pigmentation between the Taiwanese and Jiulong River Estuary + Pearl River Estuary humpback dolphins can be best and most easily differentiated by dividing dolphins into two categories of spotting on their bodies. For dolphins with little to no spotting on the body, score = 1; all dolphins from the Jiulong River Estuary + Pearl River Estuary had less to no spotting on their dorsal fins (of 27 Jiulong River Estuary + Pearl River Estuary dolphins with a body spotting score of 1, 22 had bodies and dorsal fins that were both unspotted (Figure 5A), 5 had dorsal fins with noticeably less spotting, but none had a more spotted dorsal fin). There was only one eastern Taiwan Strait dolphin with a body score of 1 in the original dataset, and its dorsal fin was more spotted than its body. Supporting this pattern was another three eastern Taiwan Strait dolphins from the original dataset that have since lost enough spotting on their bodies to fit the category of little to no spotting (score = 1), and all have spotted dorsal fins (Figure 5B).

For dolphins with obvious spotting on the bodies, score = 2, 3 or 4; 96% and 98% of the specimens from the eastern Taiwan Strait and Jiulong River Estuary + Pearl River Estuary, respectively, would be correctly classified based on the spotting intensity on the dorsal fin being either 1) equal to or more than that on the body or 2) noticeably less than that on the body. The typical pattern for Jiulong River Estuary + Pearl River Estuary dolphins is that the dorsal fin is usually noticeably less spotted than the body (Figure 5C), whereas for eastern Taiwan Strait dolphins, the dorsal fin is equally or more intensely spotted than the body (Figure 5D). The observed frequencies of specimens divided into these categories of spotting intensity on the dorsal fin relative to

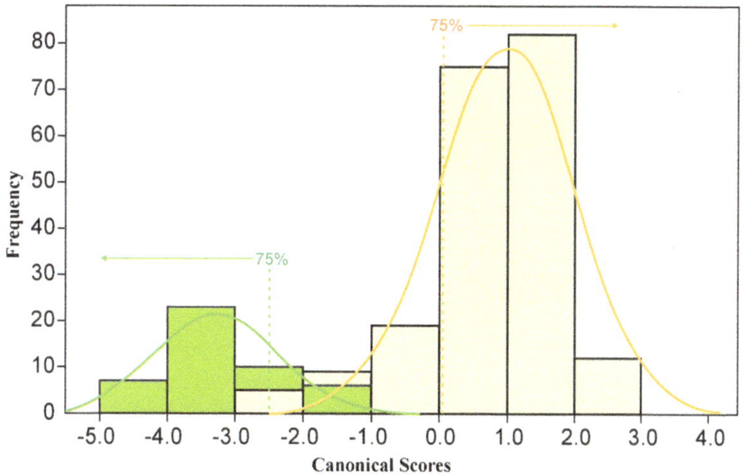

Figure 4 Frequency distributions of canonical scores of Indo-Pacific humpback dolphins. The distributions for Indo-Pacific humpback dolphins from Taiwanese waters and the Jiulong River Estuary + Pearl River Estuary are shown in green and yellow, respectively. The dotted lines indicate the locations where 75% of the distributions are found.

the body (Table 1) were significantly different ($p \ll 0.0001$, Fisher's exact test) between the eastern Taiwan Strait and Jiulong River Estuary + Pearl River Estuary. Furthermore, including the dolphins with little to no spotting on the bodies (score = 1) did not change these results ($p \ll 0.0001$, Fisher's exact test).

[Note: The lack of unspotted individuals in the eastern Taiwan Strait may be seen as reflecting a difference in age structure between the dolphins of Taiwanese waters and the Jiulong River Estuary + Pearl River Estuary (i.e. fewer Taiwanese dolphins are reaching an old enough age to become completely unspotted). However, we believe that becoming completely unspotted is not a characteristic shared by the Taiwanese population. All three stranded Taiwanese specimens had some spotting even though they were physically mature. Also, several of the least spotted adult Taiwanese dolphins (body scores = 2) have been monitored annually for more than 10 years, and minimal changes in their overall spotting intensity have been observed (JY Wang, unpublished observations). Regardless, even if there was a lack of the oldest individuals in the eastern Taiwan Strait population, the distinguishing pigmentation feature (i.e. the relationship between the spotting intensity on the dorsal fin and that on the body - best visualized in Figure 2), remains unaffected].

Osteology

Published data from a total of 38 specimens from mainland Chinese waters were included for comparison in

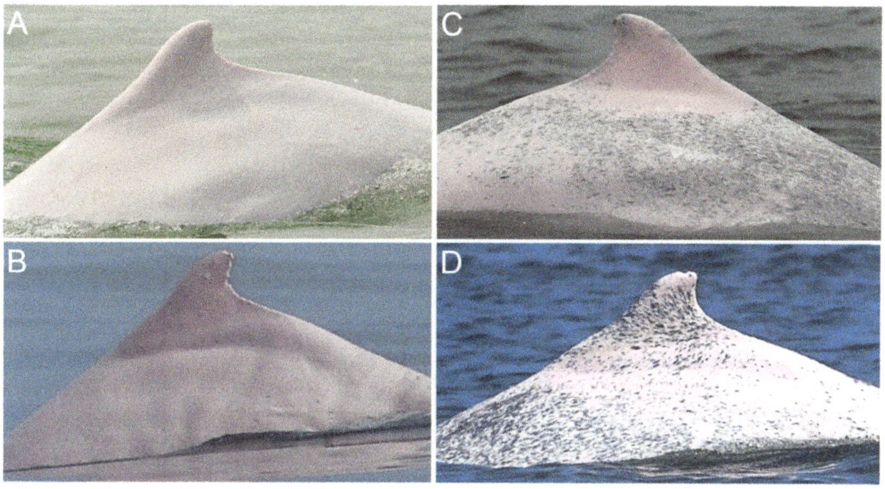

Figure 5 Typical Indo-Pacific humpback dolphins with unspotted and spotted bodies. Photographs of Indo-Pacific humpback dolphins from the Pearl River Estuary **(A, C)** and the eastern Taiwan Strait **(B, D)**. Photographs by J.Y. Wang/*FormosaCetus* Research and Conservation Group.

Table 1 Number of individual dolphins in categories based on relative spotting on dorsal fin versus body

	Dorsal fin spotting		Total
	≥ body	< body	
Eastern Taiwan Strait	43	2	45
Jiulong River Estuary + Pearl River Estuary	3	172	175

The dolphins in this table are individuals with obvious spotting on their bodies (score = 2, 3 or 4 in Wang et al. (2008a)). The dorsal fin spotting categories are as follows: spotting intensity on the dorsal fin is greater than or equal to that on the body, and spotting intensity on the dorsal fin is obviously less than that on the body. The observed frequencies were significantly different for the dolphins of the two regions ($p << 0.0001$, Fisher's exact test).

this study. Simple comparisons of data from the three Taiwanese specimens revealed no obvious differences in cranial characters (Additional file 2). For a few characters, absolute measurements of the Taiwanese specimens and Swinhoe's specimen described by Flower (1870) were slightly larger than the maximum published values. However, all proportions (relative to condylobasal length and the rostrum length to width ratio) were well within the minimum and maximum values reported. None of the post-cranial meristic values were beyond the reported data. Comparisons with the upper tooth counts versus length/breadth data in Figure three of Jefferson and Rosenbaum (2014) also showed no notable differences.

All three Taiwanese specimens had 52 vertebrae, whereas 17 specimens from mainland Chinese waters (including the two new Hong Kong and Swinhoe's specimens) had between 50 and 53 with a mode at 51 ($n = 9$). Only one specimen, from the Beibu Gulf, Beihai, Guangxi Province (Huang et al. 1978), had 53 vertebrae (note: Jefferson and Rosenbaum (2014) reported in error two specimens with 53 vertebrae - T.A. Jefferson, pers. comm.). This specimen was also reported to have 13 thoracic vertebrae, which is an unusual value for humpback dolphins from Chinese waters. Such outlying values suggest that this specimen should be re-examined. Although differences in the frequency distribution of vertebral counts may exist, more specimens (especially from the Taiwanese population) will be needed to test this hypothesis.

Other evidence

The preferred habitat of *Sousa* species is shallow (<30 m) coastal waters within a few kilometres of the shoreline, and at least in Chinese waters, humpback dolphins appear to have a close affinity to estuaries (see Ross et al. 1994; Zhou et al. 1995; Jefferson and Karczmarski 2001; Jefferson and Hung 2004; Jefferson and Rosenbaum 2014). With the exception of the narrow margins of inshore, estuarine waters along western Taiwan and mainland China, most of the Taiwan Strait (being deeper than 50 m and with considerable intrusions of oligotrophic oceanic waters from the Kuroshio and the South China Sea currents (see Jan et al. 2002)) does not possess the kind of habitat in which humpback dolphins are typically found. The shortest distance across the Taiwan Strait (which can be deeper than 70 m) between the Taiwanese population and nearest mainland Chinese humpback dolphins is about 180 km (Figure 1), whereas humpback dolphins tend to have minimal movement with linear distances of a few tens of kilometres (Hung 2000; Hung and Jefferson 2004). Although humpback dolphins in South African waters were reported to move across linear distances of up to 120 km (Durham 1994; Karczmarski 1996), these movements were always in shallow coastal waters along the shoreline and not over deep waters. Annual dedicated cetacean surveys in the eastern Taiwan Strait since 2002 have failed to record any humpback dolphins in waters deeper than about 30 m and beyond 3 km from the shoreline of western Taiwan (Wang et al. 2007; Ross et al. 2010; Dares et al. 2014). Even though there are many stranding, bycatch and sighting records of coastal small cetaceans (e.g. *Tursiops truncatus*, *T. aduncus*, *Neophocaena phocaenoides*, *Delphinus capensis*) from the Pescadores (=Penghu) archipelago, which is located between Taiwan and mainland China, there are no records of *Sousa* from these islands (Wang and Yang 2007). Furthermore, photographic identification of a large proportion of known Taiwanese individuals in the same waters year-round provided direct evidence that this population is almost certainly resident to the eastern Taiwan Strait (Wang and Yang 2011). Comparisons of the catalogues of humpback dolphins showed that no individual was common to both the eastern Taiwan Strait and the Jiulong River Estuary or the Pearl River Estuary (Wang et al. 2008a; Chou et al. 2013), which further supports the lack of exchange. The Taiwanese population has been well monitored annually by photo-identification methods since 2007. All non-calf individuals have been known since 2010, and there are no records of new individuals (with the exception of new calves), so there is no evidence of immigration from or emigration to adjacent populations. Finally, a comparative study on the social structure of humpback dolphins of the eastern Taiwan Strait and Pearl River Estuary showed that the Taiwanese dolphins also differed in social structure and organization (Dungan 2011; Dungan et al. 2012, 2015). When considered together, these many pieces of supplementary information provide strong supporting evidence for the lack of contemporary exchange of humpback dolphins across the Taiwan Strait.

Discussion

Colouration differences are important taxonomic characters for many species including cetaceans (e.g. Perrin 1972, 1975, 1990; Perrin et al. 1987; Heyning and Perrin 1994; Rosenbaum et al. 1995; Perrin 2009; Amano and Hayano 2007; Robineau et al. 2007). Complex colouration patterns (such as stripes and spotting) on individual cetaceans are not known to change with environmental conditions. None of the ephemeral or environmentally induced colour changes reviewed by Perrin (2009) for cetaceans apply to the pigmentation characters examined in this study, while there is some support for the genetic bases of colouration variation in cetaceans (e.g. Schaeff and Hamilton 1999; Ayoub et al. 2009; Polanowski et al. 2012). As such, the pigmentation differences in the humpback dolphins examined in this study are most likely to be phenotypic expressions of underlying genetic differences.

For *Sousa* species, Jefferson and Rosenbaum (2014) presented considerable descriptions of colouration differences between the species. However, they recognized that there was only one quantitative study of such differences (Wang et al. 2008a), which demonstrated subtle but clear differences between the Taiwanese and Jiulong River Estuary + Pearl River Estuary humpback dolphins. The present study fully corroborated the findings of this earlier pigmentation study (and with a larger dataset) but more importantly demonstrated that the Taiwanese population is diagnosably distinct from the dolphins of the Jiulong River Estuary + Pearl River Estuary and thus satisfied the '75% rule' for recognition as a subspecies. This was further supported by data on geographical separation, observations of the distribution of individuals and behavioural differences (i.e. in social structure - see Dungan 2011; Dungan et al. 2012, 2015). Moreover, the best available evidence shows that the Taiwanese humpback dolphin possesses the main features of subspecies (see Reeves et al. 2004; Patten 2010): they are restricted to the waters of western Taiwan, which means they are reproductively isolated (but note that reproductive isolation is not a prerequisite of subspecies, which by definition accepts or even expects some low level of continuing gene flow); the morphological differences that exist between the Taiwanese dolphins and their nearest neighbours are not clinal but are diagnosably distinct; the characters examined are not those that maybe environmentally induced (see Perrin 2009 for examples) but instead are likely a reflection of genetic and developmental differences; these features are consistent with the Taiwanese humpback dolphin being a lineage that is evolving independently from the dolphins of the Jiulong River Estuary + Pearl River Estuary. Although such lineages could also be considered distinct species under the unified species concept (de Queiroz 2007), we prefer to advocate subspecific designation until there is stronger evidence to elevate the status.

Zoogeography

Because of the coastal shallow water nature of the species, the biogeography of *S. chinensis* can be viewed more like that of a coast-dwelling terrestrial mammal than a typical pelagic marine species. It is most likely that humpback dolphins arrived at and were able to colonize the coastal waters of western Taiwan sometime during the last glacial maximum (about 17,000 to 18,000 years ago) when the sea level was much lower and the present-day Taiwan Strait was spanned by a land bridge (and its accompanying shallow waters) that connected the island of Taiwan to continental China and their associated coastal waters (e.g. see Voris 2000). It was likely that the shallow waters preferred by humpback dolphins appeared in the Taiwan Strait sometime just before the completion of the land bridge and allowed humpback dolphins to cross the 'deep' (relative to the preferred shallow water habitat of humpback dolphins) water barrier of the Taiwan Strait. Subsequent retreat of the ice resulted in increasing sea levels that first submerged the land bridge and then eliminated the shallow water conduit that permitted humpback dolphin movements across the Strait. The continuing deepening of the Taiwan Strait resulted in the effective isolation (geographically, behaviourally, reproductively and ultimately genetically) of the new colonizers of Taiwan. An interesting contrast with this biogeographical hypothesis, which resulted in the isolation of populations of *Sousa*, is that rising sea levels and the flooding of the Taiwan Strait during this period likely allowed secondary contact for newly formed finless porpoise (genus *Neophocaena*) species in this region after the land bridge barrier in the Taiwan Strait disappeared (Wang et al. 2008b).

Isolation of the Taiwanese humpback dolphins from those of mainland Chinese waters is likely to continue into the near future. With global climatic warming being observed and predicted into the future at least in the near and medium terms (hence further melting of polar ice and increasing sea levels), the Taiwan Strait will continue to deepen and further enhance this oceanographic barrier to movement of *Sousa* across the Taiwan Strait. Thus, it is reasonable to predict that the continued isolation of the Taiwanese humpback dolphins will further promote differentiation and, if sufficiently long, may even result in speciation.

Conclusions

Inconsistent descriptions, subjective conclusions, poor execution of delimitation criteria and lack of diagnosability testing have long plagued subspecies classifications. This has increased the amount of criticism levied against subspecies as being a useful level of biological organization. Although cetaceans may be under-classified with regard to the number of subspecies, this can actually be viewed as fortunate because much can be learned from the debates and experiences of researchers focused on other taxonomic groups that face the opposite problem (i.e. over-classification of subspecies). We believe that in order to build strong and objectively determined subspecies, future studies regarding cetacean subspecies should include diagnosability tests under clearly specified levels (or methods that are conceptually similar) as was used in Baker et al. (2002) and in the present study so that a common standard exists for determining subspecies across various taxonomic groups.

In the present study, clear diagnosability in pigmentation patterns between the humpback dolphins of Taiwanese and neighbouring waters was demonstrated in an objective and quantitative manner by using the method of Patten and Unitt (2002) for delimiting subspecies under the '75% rule'. Furthermore, additional supplementary lines of evidence also provided support for subspecies recognition. As such, we designate the geographically isolated and morphologically differentiated Taiwanese population of Indo-Pacific humpback dolphins as a new subspecies, *Sousa chinensis taiwanensis*. Consequently, all other humpback dolphins within *Sousa chinensis* are grouped into the nominotypical subspecies *Sousa chinensis chinensis*, until other subspecies can be demonstrated and described.

Subspecies descriptions

Order Cetartiodactyla Montgelard, Catzefils and Douzery 1997
Cetacea Brisson, 1762
Odontoceti Flower, 1867
Family Delphinidae Gray, 1821
Genus *Sousa* Gray, 1866

Sousa chinensis chinensis (Osbeck, 1765)

Holotype

The species was described by Osbeck (1765) from observing living animals, and because the practice of designating a holotype specimen was not in force at this point in history, no holotype exists; but the dolphins he observed would constitute syntypes (ICZN 1999). A near-complete skeleton of a physically mature specimen was acquired by Robert Swinhoe in 1867, thoroughly described and illustrated by Flower (1870), and would have been the prime candidate for designation as the neotype, if a need existed. Unfortunately, this specimen was destroyed during WWII (Pilleri 1979). Thus, no type specimen exists for the nominate subspecies. Although Porter (2002) proposed a neotype for *S. chinensis*, this endeavour failed to satisfy essential provisions in the International Code of Zoological Nomenclature (ICZN 1999) that govern neotype designation (Article 75, specifically 75.2, 75.3.1, 75.3.2 and 73.3.5). Given the detailed work presented by Flower (1870), there was (and still is) no controversy about the identity or type locality of this species and there was no 'exceptional need' (Article 75.3) for a neotype, as evidenced by the near absence of the reference to Porter's specimen in recent studies on *Sousa* taxonomy, including those co-authored by Porter (see Jefferson and Van Waerebeek 2004; Frère et al. 2008, 2011; Chen et al. 2008; Wang et al. 2008a; Chen et al. 2010; Lin et al. 2010, 2012; Mendez et al. 2013; Jefferson and Rosenbaum 2014). Not only was the proposed neotype not featured in these studies but also some authors specifically and directly recommended against comparisons to this specimen because of the unreliable information presented (see Jefferson and Van Waerebeek 2004; Jefferson and Rosenbaum 2014). Finally, with no examination of *Sousa* specimens from outside the type locality and by providing no new taxonomic information, the designation of the neotype by Porter (2002) can only be viewed as a curatorial exercise and hence is invalid (ICZN 1999) [a more detailed discussion about the invalidity of the proposed neotype and issues with the information presented in Porter (2002) is available in Additional file 3]. Any future need of reference specimens should consider the two complete adult topotype specimens from the Pearl River Estuary (SC03-08/09 and SC03-19/09) that were examined in this study as well as by Jefferson and Rosenbaum (2014). Presently, these topotypes and the detailed description of Swinhoe's specimen by Flower (1870) should be adequate to overcome the lack of a type specimen for this species.

Type locality

The species was first described from observing dolphins swimming in the Canton (=Pearl) River Estuary. Swinhoe's specimen that was described by Flower (1870) was collected from the waters of Quemoy (presently known as Chinmen Island) in the Jiulong River Estuary, which is roughly 500 km northeast of the Pearl River Estuary.

Etymology

The '*chinensis*' name refers to the general location from where the species was originally described. To maintain consistency with the scientific name of the subspecies, we suggest the following common names: Chinese humpback dolphin or the Chinese white dolphin (a direct translation of the most common local moniker, 'Zhonghua bai hai tun'). We recognize that before the present revision, Chinese white dolphin referred to the species. To avoid confusion, we recommend that Indo-Pacific humpback dolphin be retained as the common name for the species (following Jefferson and Rosenbaum 2014) while Chinese humpback dolphin or Chinese white dolphin is used in reference to the nominotypical subspecies *S. c. chinensis*. This nomenclatural arrangement should facilitate increased local acceptance and usage of the Indo-Pacific humpback dolphin name for the species because it does not negate the local common name for the local animals and would also allow easier (and logical) accommodations for any future subspecies that may be found within *S. c. chinensis* without the need to change the species' common name.

Diagnosis

This subspecies has complex pigmentation with dramatic developmental changes (see Jefferson et al. 2012; Jefferson and Rosenbaum 2014). It is medium to dark grey at birth and then transitions with age through various degrees of light and dark spotting while the base layer colour also changes from being primarily grey to white. In old individuals, spotting may completely disappear leaving only the white base colour, which may appear bright pink on living animals due to blood flushing to the surface of the skin. Young calves and juveniles (those possessing mostly grey pigmentation with little to no spotting) appear to be indistinguishable from the Taiwanese subspecies based on the characters examined in this study. For older individuals, the spotting intensity on the dorsal fin of this subspecies begins to diminish well before the spotting intensity on the area of the body adjacent to the dorsal fin. Thus, the colouration of the dorsal fin is less spotted (often strikingly so) and does not appear as a smooth continuation of the body pigmentation (Figure 5C). It is rare for dolphins with spotted bodies to have equal spotting intensity on the dorsal fin (only 3 of the 175 specimens examined), and none was observed to have a more intensely spotted dorsal fin. Very old dolphins from at least the Pearl River Estuary and Jiulong River Estuary can become completely spotless (both bodies and dorsal fin), and these appear to be mostly females, suggesting some sexual dimorphism in spotting loss (Figure 5A). It is important to note that dolphins in other parts of this subspecies range tend to exhibit much less white coloration, being primarily dark (see Jefferson and Rosenbaum 2014). It is likely that the present nominate subspecies comprises multiple undescribed subspecies (or possibly even species), but more research on pigmentation differences is needed to test this hypothesis.

Distribution

The nominate subspecies is found from the Yangtze River Estuary in central China south through Southeast Asia (east to at least Borneo) and extending west throughout the northern rim of the Indian Ocean to at least Orissa, India (Jefferson and Rosenbaum 2014). There appears to be at least eight major concentrations of this subspecies in the coastal waters of China (Jefferson 2000; Jefferson and Hung 2004), but others will almost certainly be found throughout the species' distribution. Further subspecies divisions are probable.

Sousa chinensis taiwanensis subsp. nov

Holotype

NMNS-14812 (=JYW-09-01). This was a physically mature (all vertebral epiphyses fused solidly to the centra) female about 250 cm in length and known to the authors since 2002 as 'TW-03' in the photographic catalogue of individuals of Taiwanese humpback dolphins (which was established in 2002 and maintained by the *FormosaCetus* Research and Conservation Group until 2013 when it was continued by the CetAsia Research Group). This was one of the least spotted and mostly pink individuals in the catalogue. Condylobasal length: 529.0 mm. The rostrum is curved slightly to the right. The lower jaw bones are unfused at the symphysis (but glued together) and also curve slightly to the right. Dental formula (alveolar counts): 34 or 35 (upper left), 33 or 34 (upper right), 31 or 32 (lower left), and 31 or 32 (lower right). Vertebral formula: Ce7 + T12 + L10 + Ca23 = 52 (the terminal caudal element counted as one vertebra). Fourteen chevrons were counted during examination of the flensed carcass, but only 13 chevrons were found when the cleaned skeleton was examined (the smallest and last chevron(s) was(were) missing). On each side, there were five double-headed and seven single-headed vertebral ribs (no floating ribs were observed, but loss of small floating ribs could not be precluded; this uncertainty would affect the thoracic and lumbar counts but not the total number of vertebrae) and seven sternal ribs. Three mesosternal elements are fused solidly into a single unit, and the basihyal and thyrohyals are also fused. A small diaphragmatic bone (measuring 40.9 mm long and 5.5 mm wide at about the midpoint of the bone) was also present and found along the midline of the diaphragm and just ventral to the vertebral column (likely an ossification in the central tendon). On vertebrae Ca3 to Ca5, there are osteopathological growths with an especially large (about 5.9 cm long and 3.2 cm wide) mass on the right dorsal posterior surface of Ca4. The seventh right vertebral rib was broken and partially healed with signs of infection, which may have resulted in the bone not healing properly. The skeleton specimen is nearly complete, but the scapulae and teeth were not available when we examined the specimen; the tympanic bones are glued to the skull (likely to the wrong sides). The skeleton is maintained in the collection of the National Museum of Natural Science (NMNS), Taichung City, Taiwan. The dorsal view of the calvaria of the holotype is shown in Figure 6A; other views of its calvaria and bones are available in Additional file 4.

Type locality

The specimen was found stranded on a beach of Tongshiao, Miaoli County (Taiwan) on 25 September 2009. The specimen's condition was fresh (code 2 of Geraci and Lounsbury 2005) and highly emaciated and showed clear signs of net entanglement (around the tail stock, flukes, rostrum, flippers and possibly the dorsal fin) that likely resulted in her death.

Paratypes

NMNS-6366 (=JYW-00-08): An adult of 247.0 cm (total body length) that was found stranded on a beach of

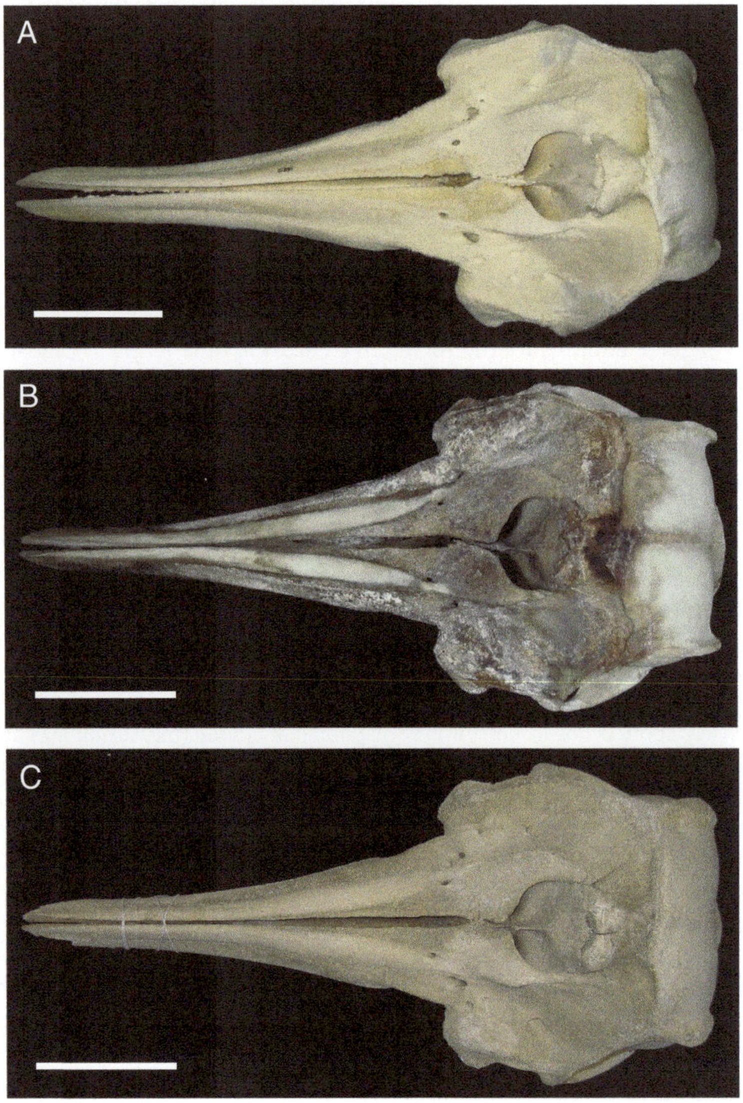

Figure 6 Dorsal views of the calvariae of *Sousa chinensis taiwanensis* specimens. **(A)** holotype (NMNS-14812 (=JYW-09-01)) and paratypes **(B)** NMNS-6366 (=JYW-00-08) and **(C)** TN-2005-35 (=JYW-05-12). The white thick bars represent 10 cm in each photograph. Photographs by J.Y. Wang/*FormosaCetus* Research and Conservation Group.

Tongshiao, Miaoli County (Taiwan) on 8 August 2000. Physically mature. Condylobasal length: 504.5 mm. Vertebral formula: Ce7 + T12 + L9 + Ca24 = 52. This skeleton is also in the collection of the National Museum of Natural Science. The dorsal view of the specimen's calvaria is shown in Figure 6B.

TN-2005-35 (=JYW-05-12): An adult measuring approximately 250 cm (total body length) that was found stranded on a beach near Chigu, Tainan County (Taiwan) on 24 April 2005. Physically mature. Condylobasal length: 496.0 mm. Vertebral formula: Ce7 + T12 + L + Ca33 = 52. The specimen is maintained at the interpretation centre of the Si-Cao Protected Area, Tainan County, Taiwan. The dorsal view of the calvaria is shown in Figure 6C.

More details about these paratype specimens are available in Additional file 5.

Etymology

The proposed subspecies name reflects the endemic distribution of the subspecies in the waters of western Taiwan and in the same flavour as other endemic subspecies of Taiwan. Similar to the common names for *S. c. chinensis*, we suggest Taiwanese humpback dolphin or Taiwanese white dolphin, which is the direct translation of a local common name, 'Taiwan bai hai tun'.

Diagnosis

The general overall pigmentation and the complex developmental changes that occur are similar to those of

S. c. chinensis. Young calves and juveniles (those possessing mostly grey pigmentation with little to no spotting) appear to be indistinguishable from those of *S. c. chinensis*. For older individuals of this subspecies, clear but subtle differences exist. Unlike those of the Pearl River Estuary and Jiulong River Estuary, spotting intensity on the dorsal fin of the Taiwanese subspecies is generally equal to or greater than that on the area of the body adjacent to the dorsal fin (Figure 5D). Thus, the overall colouration of the dorsal fin usually appears to be a smooth continuation of the body or is slightly darker in appearance (of 46 Taiwanese specimens examined, only 2 had dorsal fins that were less spotted than their bodies and then only slightly so). This is the opposite pattern found on *S. c. chinensis* from the Pearl River Estuary and Jiulong River Estuary. However, individuals that are heavily scarred by conspecific aggression (e.g. biting) may exhibit dorsal fins that appear less spotted due to depigmented scarred areas. It is also very rare for Taiwanese dolphins to have little to no spotting on the body, and none documented had an unspotted dorsal fin (Figure 5B).

Distribution

This subspecies is endemic to the shallow coastal waters of central western Taiwan (=eastern Taiwan Strait) including the waters of Miaoli, Taichung, Chunghua, Yunlin, Chiayi and Tainan counties. Dolphins have been reported from as far north as the waters influenced by the Zhonggang and Houlong rivers and south to the waters influenced by the Tzengwen River (near Chigu, Tainan County). With the exception of the intertidal waters inshore of large sandbars of Changhua County, they are generally found less than 3 km from the shore, in waters <30 m deep, and tend to be more commonly observed in and near estuaries (Dares et al. 2014). The known area occupied by this subspecies is about 600 km^2, but the full distribution is likely larger, extending potentially further north to the influences of the Danshuei River Estuary, where suitable habitat appears to be present (Wang et al. 2007a; Ross et al. 2010).

Competing interests

The authors declare that they have no competing interests.

Authors' contributions

JYW developed the design of the study, conducted the analyses, drafted the manuscript and obtained funding for and undertook the field work that resulted in the data analysed in this study. SCY and SKH also contributed to data collection in the field, initial processing and analyses of the photographs, and scoring of the photographs. They also helped with the logistics, obtained funding for the field work and provided intellectual discussions that helped in the study. All authors read and approved the final manuscript.

Acknowledgements

This work and the new subspecies name have been registered with ZooBank under urn:lsid:zoobank.org:pub:65944560-EE0D-4664-BDC6-00ABD1139149. We would like to thank Yen-Jyun Chen (collection manager) and her staff at the National Museum of Natural Science for facilitating the authors' access to specimens in the NMNS collection; Jian-Ping Wang for allowing examination of the Chigu specimen; and the Agriculture, Fisheries and Conservation Department of Hong Kong for access to the two Pearl River Estuary specimens examined in this study. We would like to thank Claryana Costa Araújo, Lauren E. Dares, Sarah Z. Dungan, Timothy R. Frasier, Jordan M. Hoffman, Shiva Javdan, Thomas A. Jefferson, Michelle N. Klein, Christine MacFarquhar, William F. Perrin, Dale W. Rice, Kimberly N. Riehl, Eduardo R. Secchi, Elizabeth Slooten, Bradley N. White, Robin Winkler and Bernd Würsig for intellectual discussions that resulted in this study and/or comments on earlier drafts of this work. This paper was also improved by the comments from Dr. Patricia E. Rosel and an anonymous reviewer. Chin-Lien Chen also helped JYW with translations of publications in Chinese. The research that resulted in the data analysed in the present study was funded primarily by the Hong Kong Cetacean Research Project, Hong Kong Dolphin Conservation Society, *FormosaCetus* Research and Conservation Group, Ocean Park Conservation Foundation of Hong Kong, Matsu Fish Conservation Union, Wild At Heart Legal Defense Association, and Small Cetacean Fund of the International Whaling Commission. Supplementary support was also provided by the American Cetacean Society, the Professional Development Fund jointly provided by the Canadian Union of Professional Employees (CUPE) 3908-1 and Trent University, and H. Yang.

Author details

[1]CetAsia Research Group, Thornhill, Ontario L4J-7X1, Canada. [2]Department of Biology, Trent University, Peterborough, Ontario K9J-7B8, Canada. [3]National Museum of Marine Biology and Aquarium, Pingtung 944, Taiwan. [4]*FormosaCetus* Company Ltd, Hualien 970, Taiwan. [5]Hong Kong Cetacean Research Project, Lam Tin, Kowloon, Hong Kong, China.

References

Allen GM (1938) The mammals of China and Mongolia. American Museum of Natural History, New York

Amadon D (1949) The seventy-five percent rule for subspecies. Condor 51:251–258

Amano M, Hayano A (2007) Intermingling of *dalli*-type Dall's porpoises into a wintering *truei*-type population off Japan: implications from color patterns. Mar Mamm Sci 23:1–14

Araújo CC, Wang JY, Hung SK, White BN, Brito D (2014) Viability of the critically endangered eastern Taiwan Strait population of Indo-Pacific humpback dolphins *Sousa chinensis*. Endang Species Res 24:263–271

Ayoub NA, McGowen MR, Clark C, Springer MS, Gatesy J (2009) Evolution and phylogenetic utility of the melanocortin-1 receptor gene (MC1R) in Cetartiodactyla. Mol Phylogen Evol 52:550–557

Baker AN, Smith ANH, Pichler FB (2002) Geographical variation in Hector's dolphin: recognition of new subspecies of *Cephalorhynchus hectori*. J Royal Soc NZ 32:713–727

Chen H, Zhai K, Chen J, Chen Y, Wen H, Chen S, Wu Y (2008) A preliminary investigation on genetic diversity of *Sousa chinensis* in the Pearl River Estuary and Xiamen of Chinese waters. J Genet Genom 35:491–497
Chen L, Caballero S, Zhou K, Yang G (2010) Molecular phylogenetics and population structure of *Sousa chinensis* in Chinese waters inferred from mitochondrial control region sequences. Biochem Systemat Ecol 38:897–905
Chou L-S, Chang W-L, Wu Y-J, Yu H-Y (2013) Indo-Pacific humpback dolphins in the waters of Kinman: distribution, habitat usage and population structure. J Nat Park (Taiwan) 23:1–10
Dares LE, Hoffman JM, Yang SC, Wang JY (2014) Habitat characteristics of the critically endangered Taiwanese humpback dolphin (*Sousa chinensis*) population of the eastern Taiwan Strait. Aquat Mamm 40:368–374
de Queiroz K (2007) Species concepts and species delimitation. Syst Biol 56:879–886
Dungan SZ (2011) Comparing the social structures of Indo-Pacific humpback dolphins (*Sousa chinensis*) from the Pearl River Estuary and Eastern Taiwan Strait. Trent University, Dissertation
Dungan SZ, Riehl KN, Wee A, Wang JY (2011) A review of the impacts of anthropogenic activities on the critically endangered eastern Taiwan Strait Indo-Pacific humpback dolphins. J Mar Anim Ecol 4(2):3–9
Dungan SZ, Hung SK, Wang JY, White BN (2012) Two social communities in the Pearl River Estuary population of Indo-Pacific humpback dolphins (*Sousa chinensis*). Can J Zool 90:1031–1043
Dungan SZ, Wang JY, Araújo CC, Yang SC, White BN (2015) Social structure in a critically endangered Indo-Pacific humpback dolphin (*Sousa chinensis*) population. Aquat Conserv Mar Freshw Ecosyst (in press)
Durham B (1994) The distribution and abundance of the humpback dolphin (*Sousa chinensis*) along the Natal coast. Dissertation, University of Natal, South Africa
Flower WH (1870) Description of the skeleton of the Chinese white dolphin (*Delphinus sinensis*, Osbeck). Trans Zool Soc London 7:151–160
Frère CH, Hale PT, Porter L, Cockcroft VG, Dalebout ML (2008) Phylogenetic analysis of mtDNA sequences suggests revision of humpback dolphin (*Sousa* spp.) taxonomy is needed. Mar Freshw Res 59:259–268
Frère CH, Seddon J, Palmer C, Porter L, Parra G (2011) Multiple lines of evidence for an Australasian geographic boundary in the Indo-Pacific humpback dolphin (*Sousa chinensis*): population of species divergence? Conserv Genet 12:1633–1638
Geraci JR, Lounsbury VJ (2005) Marine mammals ashore: a field guide for strandings, 2nd edn. National Aquarium in Baltimore, Baltimore
Haig SM, Beever EA, Chambers SM, Draheim HM, Duggar BD, Dunham S, Elliott-Smith E, Fontaine JB, Keisler DC, Knaus BJ, Lopes IF, Loschl P, Mullins TD, Sheffield LM (2006) Taxonomic considerations in listing subspecies under the U.S. Endangered Species Act. Conserv Biol 20:1584–1594
Heyning JE, Perrin WF (1994) Evidence for two species of common dolphins (genus *Delphinus*) from the eastern North Pacific. Contrib Sci (Los Angeles) 442:1–35
Huang W, Wen Y, Tang Z (1978) Preliminary studies on *Sousa chinensis*. Fudan J (Natural Sciences) 1:105–110
Huang S-L, Chang WL, Karczamarski L (2014) Population trends and vulnerability of humpback dolphins *Sousa chinensis* off the west coast of Taiwan. Endang Species Res 26:147–159
Hung SK (2000) Ranging patterns of Indo-Pacific humpback dolphins (*Sousa chinensis*) in the Pearl River Estuary. University of San Diego, People's Republic of China, Dissertation
Hung SK, Jefferson TA (2004) Ranging patterns of Indo-Pacific humpback dolphins (*Sousa chinensis*) in the Pearl River Estuary, People's Republic of China. Aquat Mamm 30:159–174
International Commission on Zoological Nomenclature (ICZN) (1999) International code of zoological nomenclature, 4th edn. The International Trust for Zoological Nomenclature, London
Jan S, Wang J, Chern C-S, Chao S-Y (2002) Seasonal variation of the circulation in the Taiwan Strait. J Mar Systems 35:249–268
Jefferson TA (2000) Population biology of the Indo-Pacific hump-backed dolphin in Hong Kong waters. Wildl Monogr 144:1–65
Jefferson TA, Hung SK (2004) A review of the status of the Indo-Pacific humpback dolphin (*Sousa chinensis*) in Chinese waters. Aquat Mamm 30:149–158
Jefferson TA, Karczmarski L (2001) *Sousa chinensis*. Mamm Species 655:1–9
Jefferson TA, Rosenbaum HC (2014) Taxonomic revision of the humpback dolphins (*Sousa* spp.), and description of a new species from Australia. Mar Mamm Sci 30:1494–1541
Jefferson TA, Van Waerebeek KV (2004) Geographic variation in skull morphology of humpback dolphins (*Sousa* spp.). Aquat Mamm 30:3–17
Jefferson TA, Hung SK, Robertson KM, Archer FI (2012) Life history of the Indo-Pacific humpback dolphin in the Pearl River Estuary, southern China. Mar Mamm Sci 28:84–104
Karczmarski L (1996) Ecological studies of humpback dolphins *Sousa chinensis* in the Algoa Bay region, Eastern Cape. Dissertation, University of Port Elizabeth, South Africa
Lin W, Zhou R, Porter L, Chen J, Wu Y (2010) Evolution of *Sousa chinensis*: a scenario based on mitochondrial DNA study. Molec Phylogenet Evol 57:907–911
Lin W, Chang L, Frère CH, Zhou R, Chen J, Chen X, Wu Y (2012) Differentiated or not? An assessment of current knowledge of genetic structure of *Sousa chinensis* in China. J Exp Mar Biol Ecol 416–417:17–20
Mayr E (1982) Of what use are subspecies? Auk 99:593–595
Mayr E, Ashlock PD (1991) Principles of systematic zoology, 2nd edn. McGraw-Hill, New York
Mendez M, Jefferson TA, Kolokotronis S-O, Krützen M, Parra GJ, Colline T, Minton G, Baldwin R, Berggren P, Särnblad A, Amir OA, Peddemors VM, Karczmarski L, Guissamulo A, Smith B, Sutaria D, Amato G, Rosenbaum HC (2013) Integrating multiple lines of evidence to better understand the evolutionary divergence of humpback dolphins along their entire distribution range: a new dolphin species in Australian waters? Mol Ecol 22:5936–5948
Osbeck P (1765) Reise nach Ostindien und China (English translation 1771). Koppe, Rostok
Patten MA (2010) Null expectations in subspecies diagnosis. Ornithol Monogr 67:35–41
Patten MA, Unitt P (2002) Diagnosability versus mean differences of sage sparrow subspecies. Auk 119:26–35
Perrin WF (1972) Variation and taxonomy of spotted and spinner porpoises (genus *Stenella*) of the eastern tropical Pacific and Hawaii. University of California, Dissertation
Perrin WF (1975) Variation of spotted and spinner porpoise (genus *Stenella*) in the eastern Pacific and Hawaii. Bull Scripps Inst Oceanogr 21:1–206
Perrin WF (1990) Subspecies of *Stenella longirostris* (Mammalia: Cetacea: Delphinidae). Proc Biol Soc Washington 103:453–463
Perrin WF (2009) Coloration. In: Perrin WF, Würsig B, Thewissen JCM (eds) Encyclopedia of marine mammals. Academic Press, San Diego, pp 236–244
Perrin WF, Mitchell ED, Mead JG, Caldwell DK, Caldwell MC, van Bree PJH, Dawbin WH (1987) Revision of the spotted dolphins. Mar Mamm Sci 3:99–170
Phillimore AB, Owens IPF (2006) Are subspecies useful in evolutionary and conservation biology? Proc Royal Soc London B 273:1049–1053
Pilleri G (1979) Pehr Osbeck (1723–1805) and the discovery of the Chinese white dolphin (*Sousa chinensis* Osbeck, 1765). Investig Cetacea 10:333–334
Polanowski AM, Robinson-Laverick SM, Paton D, Jarmin SN (2012) Variation in the tyrosinase gene associated with a white humpback whale (*Megaptera novaeangliae*). J Heredity 103:130–133
Porter LJ (2002) A redescription of *Sousa chinensis* (Osbeck, 1765) (Mammalia, Delphinidae) and designation of a neotype. Bull Nat Hist Mus Zool (London) 68:27–37
Reeves RR, Perrin WF, Taylor BL, Baker CS, Mesnick SL (2004) Report of the workshop on shortcomings of cetacean taxonomy in relation to the needs of conservation and management, April 30 – May 2, 2004 La Jolla, California. NOAA-TM-NMFS-SWFSC-363, National Oceanic and Atmospheric Administration, La Jolla, California
Reeves RR, Dalebout ML, Jefferson TA, Karczmarski L, Laidre K, O'Corry-Crowe G, Rojas-Bracho L, Secchi ER, Slooten E, Smith BD, Wang JY, Zhou K (2008) *Sousa chinensis* (eastern Taiwan Strait subpopulation). In: IUCN 2014. IUCN Red List of Threatened Species. Version 2014.3 Available via IUCN. www.iucnredlist.org. Accessed 12 March 2015
Remsen JV Jr (2010) Subspecies as a meaningful taxonomic rank in avian classification. Ornithol Monogr 67:62–78
Rice WR (1989) Analyzing tables of statistical tests. Evolution 43:223–225
Robineau D, Goodall RNP, Pichler F, Baker CS (2007) Description of a new subspecies of Commerson's dolphin, *Cephalorhynchus commersonii* (Lacépède, 1804), inhabiting the coastal waters of the Kerguelen Islands. Mammalia 71:172–180
Rommel S (1990) Osteology of the bottlenose dolphin. In: Leatherwood S, Reeves RR (eds) The bottlenose dolphin. Academic Press, San Diego, pp 29–49
Rosenbaum HC, Clapham PJ, Allen J, Nicole-Jenner M, Jenner C, Florez-González L, Urbán RJ, Ladrón GP, Mori K, Yamaguchi M, Baker CS (1995) Geographic variation in central fluke pigmentation of humpback whale *Megaptera novaeangliae* populations worldwide. Mar Ecol Prog Series 124:1–7
Ross GJB, Heinsohn GE, Cockcroft VG (1994) Humpback dolphins, *Sousa chinensis* (Osbeck, 1765), *Sousa plumbea* (G. Cuvier, 1829) and *Sousa teuszii*

(Kukenthal, 1892). In: Ridgway SH, Harrison R (eds) The first book of dolphins, vol 5, Handbook on marine mammals, vol. Academic Press, San Diego, pp 23–42
Ross PS, Dungan SZ, Hung SK, Jefferson TA, MacFarquhar C, Perrin WF, Riehl KN, Slooten E, Tsai J, Wang JY, White BN, Würsig B, Yang SC, Reeves RR (2010) Averting the baiji syndrome: conserving habitat for critically endangered dolphins in Eastern Taiwan Strait. Aquat Conserv Mar Freshw Ecosyst 20:685–694
Ryder OA (1986) Species conservation and systematics: the dilemma of subspecies. Trends Ecol Evol 1:9–10
Schaeff CM, Hamilton PK (1999) Genetic basis and evolutionary significance of ventral skin color markings in North Atlantic right whales (*Eubalaena glacialis*). Mar Mamm Sci 15:701–711
Slooten E, Wang JY, Dungan SZ, Forney KA, Hung SK, Jefferson TA, Riehl KN, Rojas-Bracho L, Ross PS, Wee A, Winker R, Yang S-C, Chen CA (2013) Impacts of fisheries on the critically endangered humpback dolphin *Sousa chinensis* population in the eastern Taiwan Strait. Endang Species Res 22:99–114
Taxonomy Committee of the Society for Marine Mammalogy (2014) List of marine mammal species and subspecies. http://www.marinemammalscience.org/species-information/list-of-marine-mammal-species-subspecies/. Accessed 12 March 2015
Voris HK (2000) Maps of Pleistocene sea levels in Southeast Asia: shorelines, river systems, and time durations. J Biogeogr 27:1153–1167
Wang P (1999) Chinese cetaceans. Ocean Enterprises Limited, Hong Kong Special Administrative Region
Wang P, Han J (1996) On the Chinese white dolphin found in the Xijiang River and the mouth of the Yangtze River and its status of resource distribution. Aquat Products Sci 15:3–8
Wang P, Sun J (1982) Studies on the Zhonghua white dolphin, *Sousa chinensis*, from the South China Sea. Trans Liaoning Zool Soc 3:67–76
Wang JY, Yang SC (2007) An identification guide to the dolphins and other small cetaceans of Taiwan. Jen Jen Publishing Company and National Museum of Marine Biology and Aquarium, Taiwan
Wang JY, Yang SC (2011) Evidence for year-round occurrence of the eastern Taiwan Strait Indo-Pacific humpback dolphins (*Sousa chinensis*) in the waters off western Taiwan. Mar Mamm Sci 27:625–658
Wang JY, Hung SK, Yang S-C (2004a) Records of Indo-Pacific humpback dolphins, *Sousa chinensis* (Osbeck, 1765), from the waters of western Taiwan. Aquat Mamm 30:187–194
Wang JY, Yang SC, Reeves RR (eds) (2004b) Report of the first workshop on conservation and research needs of Indo Pacific humpback dolphins, *Sousa chinensis*, in the waters of Taiwan, February 2004. National Museum of Marine Biology and Aquarium, Checheng
Wang JY, Yang SC, Hung SK, Jefferson TA (2007) Distribution, abundance and conservation status of the eastern Taiwan Strait population of Indo-Pacific humpback dolphins, *Sousa chinensis*. Mammalia 71:157–165
Wang JY, Hung SK, Yang SC, Jefferson TA, Secchi ER (2008a) Population differences in the pigmentation of Indo-Pacific humpback dolphins, *Sousa chinensis*, in Chinese waters. Mammalia 72:302–308
Wang JY, Frasier TR, Yang SC, White BN (2008b) Detecting recent speciation events: the case of the finless porpoise (genus *Neophocaena*). Heredity 101:145–155
Wang JY, Yang SC, Fruet PF, Secchi ER (2012) Mark-recapture analyses of the critically endangered eastern Taiwan Strait population of Indo-Pacific humpback dolphins (*Sousa chinensis*): implications for conservation. Bull Mar Sci 88:885–902
Zhou K (2004) Fauna sinica mammals volume 9 Cetacea, Phocoidea, Sirenia. Science Press, Beijing, Carnivora
Zhou K, Leatherwood S, Jefferson TA (1995) Records of small cetaceans in Chinese waters: a review. Asian Mar Biol 12:119–139
Zink RM (2004) The role of subspecies in obscuring avian biological diversity and misleading conservation policy. Proc Royal Soc London B 271:561–564

21

Alona kaingang (Crustacea, Cladocera, Aloninae): a new species of the *pulchella*-group, with identification key to Neotropical species

Francisco Diogo Rocha Sousa[1,2*], Lourdes Maria Abdu Elmoor-Loureiro[2] and Sandro Santos[1]

Abstract

Background: The subfamily Aloninae has been the focus of extensive studies on the Chydoridae because it has a higher diversity of species and due to the necessity of a detailed redescription of many taxa, especially using characters related to thoracic appendages. The polyphyletic genus *Alona* was redefined recently, and many species complexes were translocated to natural groups. The *pulchella*-group is a candidate for removal from *Alona* because it forms a well-defined complex of species, besides possessing higher species diversity. At present, 50 % of the known species occur in the Neotropics.

Results: In this study, we described *Alona kaingang*, a new species of *pulchella*-group distributed in southern South America. The new species shares with other species similarities in general features of limbs but, for instance, differs from South American *Alona glabra* because it has a short postabdomen. *Alona kaingang* sp. nov. differs from Andean *Alona altiplana* because it has thick denticles on the postabdomen and longitudinal lines on the carapace are absent. Differences in the limbs were also observed.

Conclusions: *Alona kaingang* sp. nov. is part of the group that contains species with complete connection between the main head pores. Like other species of the *pulchella*-group, *A. kaingang* sp. nov. also has similarities with the genus *Ovalona*. The relationship between these two species groups still needs to be tested.

Keywords: Chydoridae; *Alona anamariae*; *Ovalona*; Head pores; Limb; Postabdomen

Background

Chydoridae Dybowski & Grochowski, 1894 *emend.* Frey, 1967 is the most speciose family of the superorder Cladocera (Crustacea: Branchiopoda). The majority of its species inhabit the littoral zones of inland waters, usually associated with aquatic vegetation or fine sediments (Smirnov 1971; Kotov 2006). The subfamily Aloninae Dybowski & Grochowski, 1894 *emend.* Frey, 1967 has been extensively studied for two main reasons: (1) the high diversity of species and (2) the need for detailed redescriptions of taxa, especially using characters related to thoracic appendages (Smirnov 1998; Kotov 2000a, b; Sinev and Kotov 2000; Sinev 2001a; Sinev and Kotov 2001; Kotov and Elías-Gutiérrez 2002; Kotov 2003; Van Damme et al. 2003; Kotov and Sanoamuang 2004; Sinev and Elmoor-Loureiro 2010; Van Damme et al. 2010; Sinev and Atroschenko 2011).

Many of the recent morphological investigations and redescriptions of species within this subfamily were related to the genus *Alona* Baird, 1843 and resulted in the description of new genera or species translocations to correlated groups, such as *Karualona* Dumont & Silva-Briano, 2000 (Dumont and Silva-Briano 2000; Sinev and Hollwedel 2005), *Nicsmirnovius* Chiambeng & Dumont, 1999 (Kotov 2003; Van Damme et al. 2003), *Parvalona* Van Damme, Kotov & Dumont, 2005 (Van Damme et al. 2005), *Armatalona* Sinev, 2004 (Sinev 2004a), *Matralona*

* Correspondence: sousa_bio@yahoo.com.br
[1]Núcleo de Estudos em Biodiversidade Aquática, Programa de Pós-graduação em Biodiversidade Animal, Universidade Federal de Santa Maria—UFSM, Av. Roraima 1000, Camobi, Santa Maria CEP 97105-900, RS, Brazil
[2]Laboratório de Biodiversidade Aquática, Universidade Católica de Brasília—UCB, QS7 lote 1, Bloco M, sala 204, Taguatinga CEP 71966-700, DF, Brazil

Van Damme & Dumont, 2009 (Van Damme and Dumont 2009), *Maraura* Sinev & Shiel, 2008 (Sinev and Shiel 2008), *Miralona* Sinev, 2004 (Sinev 2004b), and *Leberis* Smirnov, 1989 (Sinev et al. 2005). Together, these studies highlighted the polyphyletic nature of *Alona*, which has been supported by phylogenetic analysis (e.g., Elmoor-Loureiro 2004; Sacherová and Hebert 2003).

Recently, Van Damme and Dumont (2008a) redescribed *Alona quadrangularis* (O.F. Müller 1776) using specimens from the Palearctic region, where it was originally described. Based on the morphology of the postabdomen, terminal claws, and limb setae, these authors defined consistent diagnostic characters for the genus *Alona*. Subsequently, some species complexes that had previously been attributed to *Alona* sensu lato were studied and allocated to new genera, such as *Phreatalona*, which corresponds to the *protzi*-complex (Van Damme et al. 2009); *Coronatella*, which corresponds to the *rectangula*-complex (Van Damme and Dumont 2008b); *Brancelia* (Van Damme and Sinev, 2011) which corresponds to the *hercegovinae*-complex; and *Anthalona* Van Damme, Sinev & Dumont, 2011, which includes species of the *verrucosa*-complex (Van Damme et al. 2011). However, several groups still require removal from *Alona* sensu lato, such as the *pulchella*-group (Van Damme et al. 2010).

The *pulchella*-group is a well-defined complex of species of *Alona* sensu lato characterized by postabdomen with marginal denticles merged and moderately developed and inner distal lobe (IDL) with three setae, seven setae on the exopodite of the third limb, reduced filter comb on the fifth limb, and two lateral aesthetascs on male antennules (for more details, see Sinev et al. 2012). The *pulchella*-group has at least 16 valid species and six species waiting for revision, which have been reported for all continents, except Antarctica (Sinev 2001b, c; Sinev 2002a; Sinev 2009; Kotov et al. 2010; Van Damme et al. 2010; Sinev et al. 2012; Sinev and Silva-Briano 2012).

In the Neotropics, the *pulchella*-group is highly diverse presenting species with the entire morphological range known for the group, including the specialized species *Alona bromelicola* Smirnov, 1988 (Sinev 2002a). Specifically for South America, there are reports of three species (Sinev 2001b; Kotov et al. 2010); however, the knowledge of the *pulchella*-group remains incipient. This manuscript is part of an extensive review concerning populations of *Alona* sensu lato in Brazil, aims to evaluate the morphology, and describes a new species of the *pulchella*-group.

Methods

Selected specimens were transferred to slides containing glycerin and dissected under a stereomicroscope. The morphology of appendages and other structures was studied using a phase-contrast microscope. Some animals were dehydrated in an acetone series (50, 70, 90 and 100 %) and critical point-dried or air-dried, mounted on aluminum stubs, coated with gold, and examined under a JEOL-JSM 7001F scanning electron microscope. Enumeration of limb setae and other structures proceeded from the epipodite to the gnathobase, without relation to homology, according to the recent literature (Van Damme and Dumont 2007; Sinev and Kobayashi 2012). Drawings were prepared using a camera lucida.

Results

Systematics

Class Branchiopoda Latreille, 1817

Order Anomopoda Sars, 1865

Family Chydoridae Dybowski & Grochowski, 1894 *emend.* Frey, 1967

Subfamily Aloninae Dybowski & Grochowski, 1894 *emend.* Frey, 1967

Genus *Alona* Baird, 1843

Alona kaingang Sousa, Elmoor-Loureiro & Santos (Figs. 1, 2, 3, and 4)

Type material

Holotype. Undissected, adult parthenogenetic female in a tube with 90 % ethanol deposited at the Museum of Zoology of the University of São Paulo under access number MZUSP 29593. The label of the holotype is "*Alona kaingang* n. sp., 1 parth. ♀ from Mostardas—Lagoa de São Simão, RS, Brazil, Holotype."

Paratypes. Two undissected adult parthenogenetic females in tubes with 90 % ethanol, deposited at the Museum of Zoology of the University of São Paulo under access number MZUSP 29594. One adult parthenogenetic female and one adult male, undissected in tubes with 90 % ethanol, deposited at the National Museum of Rio de Janeiro under access numbers MNRJ 23739 and MNRJ 23740. Eight specimens selected as paratypes (access number EL02477), and several unstudied individuals (access numbers EL02165 and EL02166) are deposited at the Laboratório de Biodiversidade Aquática, Universidade Católica de Brasília.

Type locality

Pond in São Simão, Mostardas, Rio Grande do Sul state, Brazil (30° 57′ 06″ S, 50° 42′ 29″ W).

Etymology

The name *kaingang* refers to indigenous people who inhabited the southern portion of South America before the European colonization. The *kaingang* people were distributed in Southeastern and Southern Brazil, where the new species occurs.

Material examined

Twenty-three adult parthenogenetic females and five males from the pond in São Simão, Mostardas, Rio Grande do Sul state, Brazil (30° 57′ 06″ S, 50° 42′ 29″ W). Material collected by Lourdes M. A. Elmoor-Loureiro on 01.i.2002. Four adult parthenogenetic females from the Limnology

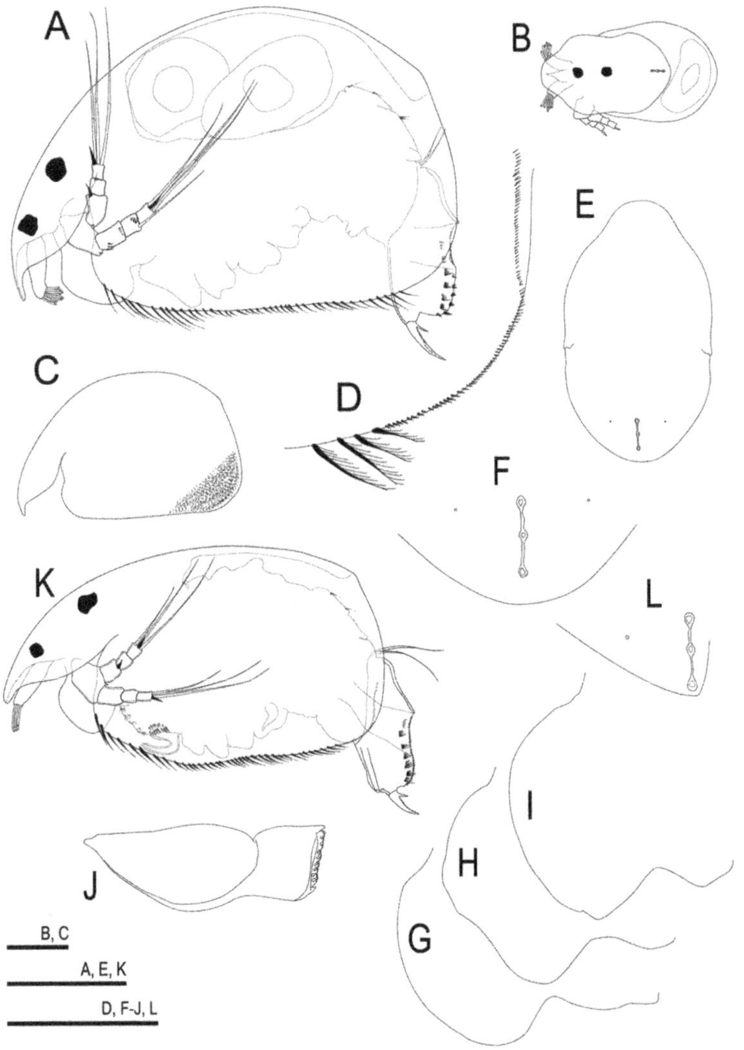

Fig. 1 *Alona kaingang* sp. nov., (***a–j***) adult parthenogenetic females from São Simão pond, Mostardas, Rio Grande do Sul, Brazil. (***a***) Habitus, lateral view; (***b***) dorsal view; (***c***) ornamentation on the carapace; (***d***) posteroventral valve corner; (***e***) head shield; (***f***) head pores; (***g–i***) labral keel; (***j***) mandible; (***k–l***) adult male from São Simão pond, Mostardas, Rio Grande do Sul, Brazil; (***k***) habitus, lateral view; (***l***) head pores. *Scale bar* = 50 μm

Laboratory cultivation facility, Instituto de Ciências da Natureza, Universidade Federal de Alfenas; specimens collected at Furnas Reservoir, Minas Gerais state, Brazil (21° 03′ 24″ S, 46° 00′ 18″ W), leg. M. J. Santos-Wisniewski (EL02167). Four adult parthenogenetic females from a marginal pond in São Simão Reservoir, Minas Gerais state, Brazil (18° 48′ 50″ S, 50° 22′ 58″ W); material collected by José Roberto Debastiani-Júnior (EL02173).

Diagnosis

Female. Of small size, length 0.33–0.40 mm. Body ovoid with dorsal margin strongly arched. In dorsal view, it shows weak lateral compression, dorsal keel absent. *Head.* Head shield with posterior margin rounded, rostrum short, three main head pores subequal in size, narrowly connected, postpore distance (PP) less than 0.2 interpore distance (IP). Lateral pores tiny, inserted midway between the main pores and the head shield margin. Labral keel wide, apex obtuse, without setae or denticles. *Carapace.* No striations or longitudinal lines. Ventral margin with 40–50 setae arranged in three groups. Posteroventral corner without denticles, with numerous spinules. *Antennules.* About 2–2.5 times as long as wide, four rows of setules on antennular body, distalmost row shorter than the others. Antennular sensory seta slender and long, about 1.5 times smaller than antennular body. Nine terminal aesthetascs of different sizes. *Antenna.* Antennal formula: spines 001/101, setae 113/003. Spine on first endopodite segment not exceeding the length of the second segment. Apical setae of endopodite and exopodite with lateral spinules. *Postabdomen.* Of moderate

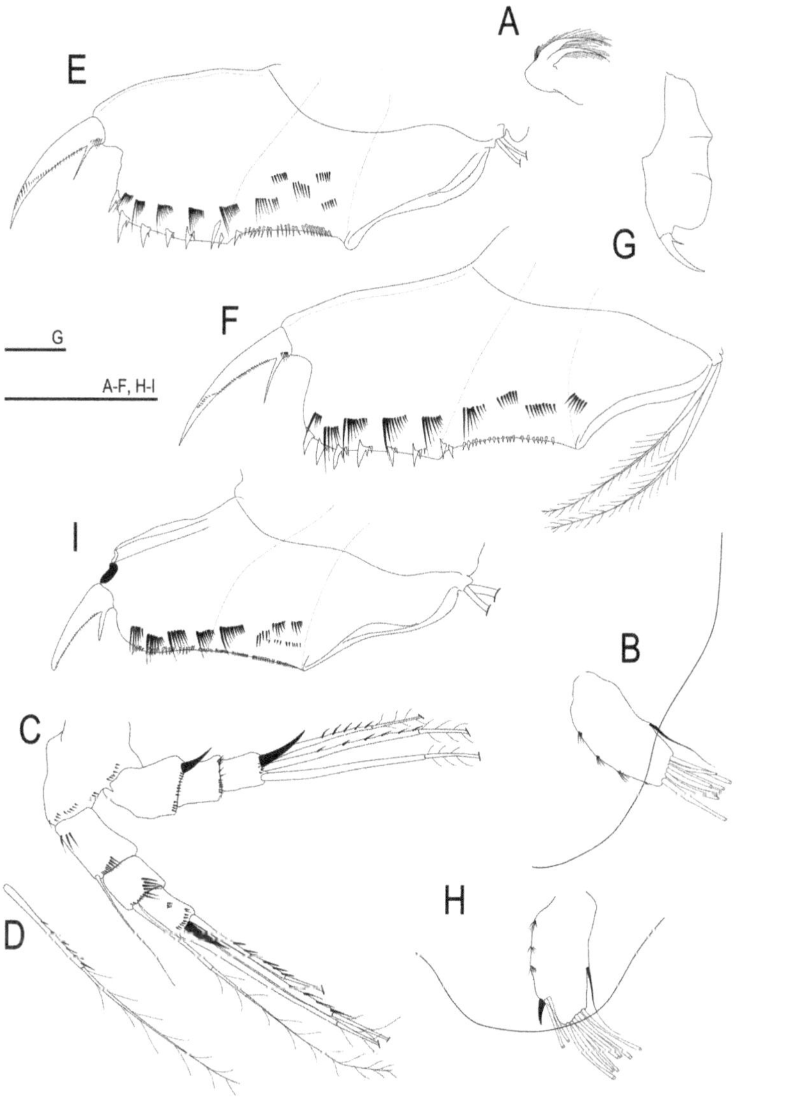

Fig. 2 *Alona kaingang* sp. nov., (***a–g***) adult parthenogenetic female. (***a***) first maxilla; (***b***) antennule; (***c***) antenna; (***d***) antenna terminal setae; (***e***) postabdomen (Lagoa de São Simão, Mostardas, Rio Grande do Sul); (***f***) postabdomen (São Simão Reservoir, Minas Gerais, Brazil); (***g***) idem; (***h–i***) adult male from São Simão pond, Mostardas, Rio Grande do Sul, Brazil; (***h***) antennule; (***i***) postabdomen. *Scale bar* = 50 μm

size (about 2–2.4 times as long as wide). Pre-anal margin longer than anal margin and shorter than postanal margin. Anal margin inserted in a depression, with distinct angles. Lateral fascicles arranged in five to seven groups, first spines of each group thick and exceeding margin of postabdomen. Six to seven clusters of marginal denticles. Postabdominal setae with proximal portion naked, distal portion with short setules armed bilaterally. *Terminal claw.* Longer than anal margin, with one row of spinules on its base. *Basal spine.* Long, without spinules or setules. *Limb I.* First endite with two marginal setae, IDL with three setae; setae 2–3 bisegmented, long and with strong proximal spines. *Limb II.* Exopodite elongated, with setae present; scrapers 3, 5, 6, 7, and 8 with robust denticulation. *Limb III.* Exopodite with seven marginal setae; first seta longer than the second, fifth, and seventh setae less than half length of sixth seta, fourth setae geniculated. *Limb IV.* Exopodite with six marginal plumose setae; fourth seta thick and more than half length of fifth seta; sixth seta half as long as fifth seta. *Limb V.* Exopodite with slight depression in middle portion, with four plumose setae; seta 3 smaller than setae 1 and 2; Gnathobase without elements or setae. *Male.* Body elongated, smaller than female (length 0.27–0.30 mm). Ventral margin armed with 40 setae followed by fine spinules. Posteroventral corner without denticles. Head with short rostrum. Main pores nearly connected. Lateral pore inserted between distal and median pores. *Antennules.* With three rows of short setules on the antennular body. Nine (juvenile) to 11 (adult) aesthetascs present: seven to nine terminal and

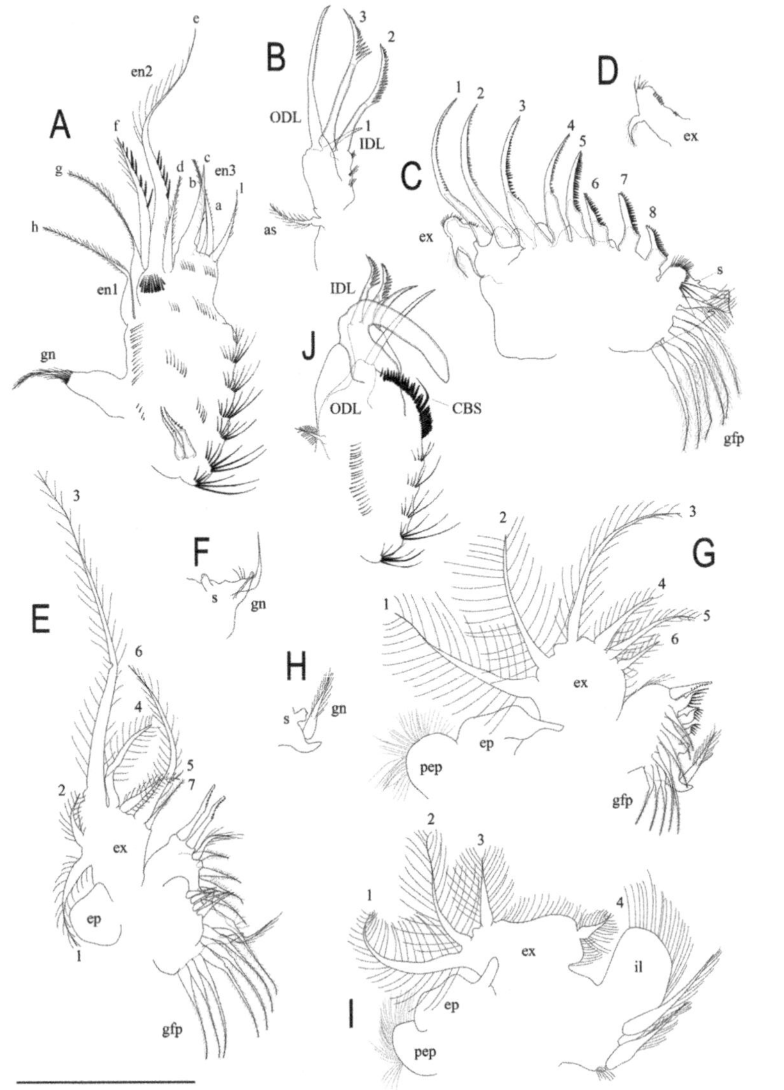

Fig. 3 *Alona kaingang* sp. nov., adult parthenogenetic females from São Simão pond, Mostardas, Rio Grande do Sul, Brazil. (***a***) limb I; (***b***) idem; (***c***) limb II; (***d***) idem; (***e***) limb III; (***f***) idem, sensillum and gnathobasic elements; (***g***) limb IV; (***h***) idem, gnathobase; (***i***) limb V; (***j***) limb I (male). *Scale bar* = 50 μm

two lateral. Male seta short, thick, about 4.6 times shorter than length of antennular body. *Postabdomen.* Narrowing toward distal portion, anal angle not defined. Marginal clusters of spinules present, at least five postanal fascicles. Gonopores opening ventrally, subapically to terminal claw. *Terminal claw.* Smaller than that of female, tip blunt, with slender spinules implanted at base. Basal spine about one-third length of terminal claw. *Limb I.* With copulatory hook U-shaped; copulatory brush seta slender; IDL with three setae. Seta 1 long, thick, subequal in size to setae 2 and 3.

Description of adult parthenogenetic female

Habitus (Fig. 1(a–c)). Compared to other species of the genus, it is a small-sized animal (0.33–0.40 mm), colorless, and transparent. Carapace ovoid in lateral view, about 1.5 times as long as high, dorsal margin strongly arched. In dorsal view, it shows weak lateral compression. Dorsal keel absent.

Head (Fig. 1(a, e, f)). Ocellus as large as eye. Head shield about 1.7 times as long as wide, with posterior margin rounded. Rostrum short, blunt, projected toward ventral margin. Three main head pores of subequal size, narrowly connected (Fig. 4(a)), PP less than 0.2 IP. Lateral pores tiny, inserted midway between main pores and head shield margin. *Labrum* (Fig. 1(g–i)). Labral keel wide in lateral view, anterior margin slightly convex, apex obtuse, lateral projections not observed, without setae or denticles. In some specimens, the margin is subquadrangular (Fig. 1(h)) or with a small depression near

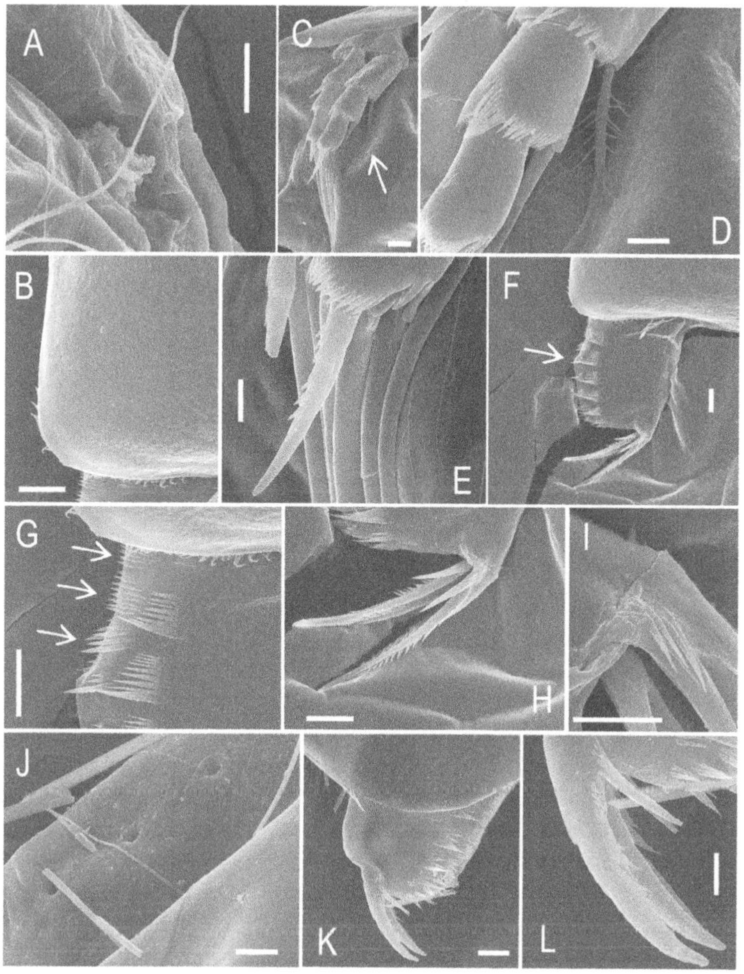

Fig. 4 *Alona kaingang* sp. nov., (***a–i***) adult parthenogenetic females from São Simão pond, Mostardas, Rio Grande do Sul, Brazil. (***a***) main head pores; (***b***) valves. (***c***) antenna—*arrow* showing seta on the first exopod segment; (***d***) idem. (***e***) antenna—apical spine on the first endopod segment; (***f***) postabdomen—*arrow* showing spinules on the anal margin; (***h***) terminal claw; (***i***) idem—base with spinules; (***j–l***) adult male from Lagoa de São Simão, Mostardas, Rio Grande do Sul, Brazil. (***j***) main head pores. (***k***) valves and postabdomen; (***l***) terminal claw. *Scale bar* indicates 5 μm to figures (***d***), (***e***), (***i***), (***j***), (***l***); 10 μm to figures (***a***), (***b***), (***c***), (***f***), (***g***), (***h***), (***k***)

the apex (Fig. 1(i)). *Mandible* (Fig. 1(j)). Well developed in relation to body size (Fig. 1(a)). *First maxilla* (Fig. 2(a)). Relatively well developed, with two long setulose setae.

Carapace (Figs. 1(c, d) and 4(b)). No striations or longitudinal lines (Fig. 4(b)), punctuated ornamentation observed in optical microscopy. Ventral margin straight, with 40–50 setae arranged in three groups, proximal group long, middle group short, posterior group increasing in size toward posteroventral corner. These setae are followed by spinules not arranged in groups, proximalmost spinule projecting beyond margin.

Antennules (Fig. 2(b)). Reaching the tip of the rostrum, setae sometimes exceeding it, about 2–2.5 times longer than wide, four rows of setules on antennular body, distalmost row shorter than the others. Antennular sensory seta slender and long, about 1.5 times smaller than antennular body, inserted at two thirds of antennule length, counting from the base. Nine terminal aesthetascs of different sizes. None of these aesthetascs exceed the length of the antennules.

Antenna (Figs. 2(c, d) and 4(e)). Coxal setae not studied. Basipodite thick, with many spinules and one long spine. First exopodite segment with long thick spinules near the base (3–4) and terminal portion. Second exopodite segment with long thick spinules on the terminal portion. Endopodite segments with short spinules. Antennal formula: spines 001/101, setae 113/003. Seta on first exopodite segment thick and plumose (Fig. 4(c, d)), not reaching mid length of terminal setae. Seta on second exopodite bisegmented, long, with many setules. Spine on first endopodite segment without denticles, reaching but not exceeding distal end of second segment. Apical spines of endopodite and exopodite of different sizes, with denticles (Fig. 4(e)). Two apical setae of endopodite and exopodite with spinules, which are larger on the first segment.

Abdomen. About two times shorter than thorax. Two or three abdominal setae (Fig. 1(a)).

Postabdomen (Fig. 2(e–g)). Moderate size, relatively wide, about 2–2.4 times as long as wide, ventral margin slightly rounded, with at least three rows of spinules. Preanal margin longer than anal margin and shorter than postanal margin. Anal margin concave, clearly inserted in a depression and with two distinct angles (Fig. 4(f–g)), with many spinules forming groups (Fig. 4(g)). Postanal margin rounded, not projected. Lateral fascicles arranged in five to seven groups, first spines of each group thick and exceeding margin of postabdomen. Six to seven clusters of marginal denticles, distal ones with two to three marginal denticles (apparently fused when observed in optical microscopy but superimposed in layers when observed in SEM (Fig. 4(f)); proximal denticles may be separated; length of the marginal denticles on the postanal margin of the postabdomen about two times longer than the width of the base of the denticles. *Postabdominal seta* (Fig. 2(e)). Approximately half as long as postabdomen, proximal portion naked. Short setules armed bilaterally, inserted toward distal portion. *Terminal claw* (Fig. 2(e–g)). Implanted at projected basis from the postabdomen, longer than anal margin, uniformly curved, with one row of spinules on the base (Fig. 4(i)); pecten armed with rows of internal and external spinules; inner row composed of robust spinules; outer row with spinules increasing in size toward distal region (Fig. 4(h)). Basal spine long, about two times as long as width of claw at its base, without spinules or setules.

Five pairs of limbs

Limb I (Fig. 3(a, b)). Epipodite not studied. Outer distal lobe (ODL) with thin seta, serrulate in distal part, about same size as the longest IDL seta. Accessory seta relatively small, setulated, and implanted near base of ODL. IDL with three groups of spinules on its face, three setae present; seta 3 smaller than the others, naked. Setae 2–3 bisegmented, long, and with strong proximal spines and narrow distal portion with short denticles. Third endite with four setae, one of them thinner than the others (1); inner seta densely setulated, longer than the other two (a, b). Second endite with one row of spinules; three setae of different sizes, seta 4 (e) about 2.7 times longer than seta 3 (f) and about 4.3 times longer than seta 5 (d); setae 3–4 (d–f) with thick spinules on the lateral face, seta 4 (e) with spinules not exceeding median portion; seta 5 (d) about half length of seta 3 (f). First endite with two marginal setae (g, h), bisegmented, similar in size and slightly setulated. No specialized elements on endites. Ejector hooks similar in size. Ventral face of limb with seven rows of setules organized in clusters, decreasing toward distal portion. Gnathobase thick, corresponding to a densely setulated seta.

Limb II (Fig. 3(c, d)). Exopodite elongated, with tiny proximal spinules, and distally setulated. Seta on exopodite present, setulated, about two to three times smaller than the exopodite itself. Endite armed with eight scrapers gradually decreasing in size toward the gnathobase, sixth and eighth scrapers smaller than seventh. Scrapers 1, 2, and 4 armed with fine denticulation, other scrapers with robust denticulation. Proximal portion of gnathobase wide, apex armed with fine spinules followed by long setules (in the populations from Minas Gerais state, these appear shorter, thinner, and more numerous), distal portion armed with sensillum and three elements, first element geniculated and setulated. Filter comb with seven setae, first seta short and densely setulated; other setae long, with setules and implanted from median portion.

Limb III (Fig. 3(e, f)). Epipodite round, without evident projections (in the populations from Minas Gerais state, long projections were observed). Exopodite subquadrangular, with seven marginal setae arranged in 2 + 5. First seta longer than second; third seta about 1.8 times longer than sixth seta; fourth seta geniculated; fifth and seventh setae less than half length of sixth seta; all setae are plumose, except the seventh. Distal endite with three setae, two scraper-like of different sizes, third seta geniculated and armed with many setules implanted bilaterally; four plumose setae, increasing in size towards the gnathobase. Basal endite with four soft setae, increasing in size towards the base. Gnathobase armed with four elements, first a cylindrical sensillum, second a strong geniculated seta, setulated on the base, third and fourth elements with acute tip and without setules. Filter comb with seven long setae.

Limb IV (Fig. 3(g, h)). Pre-epipodite round, densely setulated, epipodite oval with short projections. Exopodite round with six marginal plumose setae; setae 1 and 2 subequal in size, third seta longer than all others, fourth seta thick exceeding the half length of fifth seta; sixth seta half as long as fifth seta. Distal endite with four setae, one scraper-like and three flaming-torch-like, decreasing in size toward the base. Basal endite with three soft setae, increasing in size toward the base. Gnathobase armed with one globular sensillum and a setulated seta implanted on robust base. Filter plate with five slender setae.

Limb V (Fig. 3(i)). Pre-epipodite round and densely setulated, epipodite with short projections. Exopodite rectangular, not divided in lobes, about twice as long as wide, slight depression in the middle portion; four plumose setae, decreasing in size toward internal lobe; setae 1 and 2 subequal in size, seta 3 markedly smaller than setae 1 and 2; seta 4 about half length of seta 3. Internal lobe wide, oval and with long setules; two setulated setae of similar size on the inner face of the lobe, no elements between these setae. Gnathobase reduced, without elements or setae.

Ephippial female: Unknown.

Description of male

Habitus (Fig. 1(k)). Smaller than female, length 0.27–0.30 mm, about 1.6 times as long as high. Body elongated, weak lateral compression, body less arched than in female. *Head* with short rostrum, ocellus smaller than eye. Main pores close to each other, nearly connected, lateral pores tiny and arranged in different positions than in females (between distal and median pores; Fig. 1(l)). *Carapace* without punctuated ornamentations or longitudinal lines (Fig. 4(k)), ventral margin armed with 40 setae followed by fine spinules, the most proximal exceeding the line of posterior margin.

Antennules (Fig. 2(h)). Slightly exceeding the tip of the rostrum, about two times as long as wide, with three rows of short setules on antennular body; seven (juvenile) to nine (adult) aesthetascs of different sizes (none of these aesthetascs exceeds the length of antennules). Two lateral aesthetascs of different sizes, about half as long as antennular body; sensory seta long, about 1.4 times shorter than antennular body. Male seta short, thick, about 4.6 times shorter than length of antennular body, inserted at three thirds of antennular length, counting from base.

Antenna as described for females.

Postabdomen (Fig. 2(i)). Smaller than in female, narrowing distally; anal margin longer than postanal margin, angles not defined; marginal clusters of spinules present (Fig. 4(k)); at least five postanal fascicles, distalmost fascicle exceeding the margin line; anal margin with spinules differentiated in three groups; gonopores opening ventrally, subapically to terminal claw. *Terminal claw* smaller and thicker than in female, tip not acute (Fig. 4(l)). Slender spinules implanted at the base of terminal claw (as in the female). Basal spine about one-third length of terminal claw.

Limb I (Fig. 3(j)). Smaller than in female, copulatory hook U-shaped. Copulatory brush present, copulatory brush seta slender; IDL with three setae, smaller than corresponding setae of female, seta 1 long, thick, subequal in size to the other two setae, setae 2 and 3 bisegmented and with strong proximal spines (as in female).

Differential diagnosis

Alona kaingang sp. nov. is a member of the *pulchella*-group because its posteroventral corner of the carapace is armed with many short and thin spinules not arranged in groups, three main head pores, marginal denticles of postabdomen merged and developed, lateral fascicles developed, IDL with three setae, seven setae on the exopodite of limb III, fourth setae of limb III well-developed, gnathobase of limb V reduced, and limb VI absent. *Alona kaingang* sp. nov. also has the fourth seta on the exopodite of limb III geniculated, like many species of the *pulchella*-group (except in *Alona archeri*). As an alternative to identification, the *pulchella*-group may be separated into two groups based on the morphology of the main head pores: one group contains species with interrupted connection (including those species with main head pore isolated) and the other contains species with complete connection. Thus, *A. kaingang* sp. nov., having a complete connection between the main head pores, could be differentiated from species belonging to do the first group: *Alona capensis* (Rüher, 1914), *Alona setulosa* (Megard, 1967), *Alona cambouei* (Guerne & Richard, 1893), *Alona aguascalientensis* Sinev & Silva-Briano, 2013, *Alona nigra* Smirnov, 1996, *Alona setuloides* Smirnov & Timms, 1983, *Alona azorica* Frenzel & Alonso 1989, *Alona anastasia* Sinev Alonso, Miracle & Suhaquillo, 2012, and *Alona nuragica* Margaritora, 1971. Regarding species with complete connection between main head pores, *A. kaingang* sp. nov. is distinguished by the presence of spinules on apical setae of antenna and proximal spines on IDL setae. Besides, *A. kaingang* sp. nov. has single morphological traits such as the slight depression in the middle portion of the exopodite of the limb V.

Ecology and distribution

Alona kaingang sp. nov. has populations occurring in water bodies of the subtropical region of South America. Until now, *A. kaingang* sp. nov. has been found only in Minas Gerais state (São Simão and Furnas Reservoirs) and Rio Grande do Sul state (São Simão pond), Brazil. However, we expect that its geographical distribution is broader in the subtropical zone, because *A. kaingang* sp. nov. might be confused with *Coronatella rectangula* (Sars, 1861). This latter species likely does not occur in the Neotropics (Van Damme and Dumont 2008b), but there are many records in different Brazilian hydrographic regions (see Green 1972; Smirnov and Santos-Silva 1995; Eskinazi-Sant'Anna et al. 2005; Sterza and Fernandes 2006; Rocha et al. 2011), including the Furnas Reservoir (Santos et al. 1994; Viti et al. 2013). The review of *Coronatella* in Brazil confirmed that *C. rectangula* does not occur in Brazil, and probably not in the Neotropics (Sousa et al. 2015).

São Simão pond is located on the coastal plain of Rio Grande do Sul state and is part of the coastal wetland complex between the Atlantic Ocean and Patos Lagoon. São Simão pond does not connect with the ocean. The abiotic parameters of the pond water were as follows: pH 6.8–7.1, electrical conductivity 97.3–110 $\mu S.cm^{-1}$, total phosphorus 0.104–0.139 $mg.L^{-1}$, ammoniacal nitrogen 0.051–0.11 $mg.L^{-1}$, dissolved oxygen 7.2–9.1 $mg.L^{-1}$, and temperature 17–17.5 °C (data available from State Foundation of Environmental Protection Henrique Luiz Roessler—FEPAM, referring to the year when the type material was collected).

Furnas Reservoir, located in Minas Gerais state, is the largest reservoir in southeastern Brazil, with a total area of approximately 1440 km^2. Recent data indicate that water temperature ranges between 24 and 28 °C, pH is approximately 7, electrical conductivity 19–22 $\mu S.cm^{-1}$, total nitrogen 210–318 $\mu g.L^{-1}$, total phosphorus 13–29 $\mu g.L^{-1}$, and dissolved oxygen 8–9 $mg.L^{-1}$ (Santos et al. 2010).

São Simão Reservoir is located on the boundary between Goiás and Minas Gerais states. Currently, it is being impacted by the uncontrolled increase in urban areas and farmland in its surroundings. This has led to eutrophication of the reservoir (Fonseca 2010).

The biology of *A. kaingang* sp. nov. is not well known. It inhabits the littoral zones of water bodies, probably among aquatic vegetation. Based on the present records, it should be found in different kinds of ecosystems: reservoirs, wetlands, and coastal lagoons. We believe that this species may occur in habitats with varying degrees of salinity. *Alona kaingang* sp. nov. is presumed to be an active scraper of substrates, based on the strong scrapers found in the limbs I and II (Fryer 1968; Van Damme et al. 2011).

Identification key for species from the *pulchella*-group with occurrence in the Neotropics (adult parthenogenetic females)

1. Basal spines very short, less 0.1 of the length claw, aesthetascs length about 2.6 times shorter than antennular body................*A. bromelicola*-Central America (Sinev 2002b)

– Basal spine longer than 0.1 of the length of the claw, aesthetascs length about 1.6–2 times shorter than antennular body... 2

2. Labral keel with two rows of short setules...........3

– Labral keel naked.. 4

3. Postabdomen narrowing distally, basal segment of the apical setae of antenna without spines, one seta markedly longer than the others on the anterior group on the ventral margin of the carapace,....................*Alona anamariae*—Endemic from the Central Plateau of Mexico (Sinev and Silva-Briano 2012)

– Postabdomen wide and truncated, basal segment of the apical setae of antenna with spines, setae of the anterior group on the ventral margin of the carapace not differentiated in length................................*A. setulosa*—From North Canada to North Mexico (Sinev, 2009; Sinev and Silva-Briano 2012)

4. Abdomen armed with two to three rows of setules..5

– Abdomen armed with five to six rows of long setules.........*A. nigra*—Bolivia and North Chile on the Andes (Kotov et al. 2010)

5. Main head pores connected..............................6

– Main head pores disconnected......................... *A. aguascalientensis*—Endemic of the Central Plateau of Mexico (Sinev and Silva-Briano 2012)

6. Antenna, spine on the first segment of the endopodite shorter than second segment..7

– Antenna, spine on the first segment of the endopodite of the same length of the second segment.................*A. kaingang* sp. nov—Southern South America

7. Carapace ornamented with longitudinal lines, length of the postabdomen about 2.5 height, postanal margin armed with 5–6 denticles.........*Alona altiplana*—Endemic from Andes (Kotov et al. 2010)

– Carapace without longitudinal lines, length of the postabdomen about 2.8–3 height, postanal margin armed with seven to eight denticles.......................... *Alona glabra*—Common in the Neotropics (Sinev 2001b; Sinev and Silva-Briano 2012)

Discussion

The presence of main head pores with complete connection may be taken as the main morphological trait uniting *A. kaingang* sp. nov. with valid species *Alona pulchella* (Herrick, 1884), *A. archeri* (Sars, 1888), *A. altiplana* Kotov, Sinev & Berrios, 2010, *A. glabra* (Sars, 1901), *A. bromelicola* (Smirnov, 1988), *A. karelica* Stenroos, 1897, and *A. anamariae* Sinev & Silva-Briano, 2012. Among these species, the postabdomen of *A. kaingang* sp. nov. is quite similar to other species that present short postabdomen: the Chilean *A. altiplana* and the Mexican *A. anamariae*. All other species aforementioned have an elongated postabdomen (Sinev and Silva-Briano 2012).

Alona kaingang sp. nov. (length 0.33–0.40 mm) is smaller than *A. altiplana* (0.35–0.50 mm). Besides, *A. kaingang* sp. nov. differs from *A. altiplana* because its carapace does not have longitudinal lines (Fig. 4(b)). In

A. kaingang sp. nov., the length of the marginal denticles on the postanal margin of the postabdomen is about two times longer than the width of the base of the denticles (Fig. 2(e–f)) while in *A. altiplana*, this proportion is about 2.5–3 times longer (Kotov et al. 2010). The following morphological traits of the limbs are observed in *A. kaingang* sp. nov. but not in *A. altiplana*: (1) seta e on the endite 2 of the limb I about two times longer than seta f, (2) setae 2–3 of the IDL markedly different in length, (3) setae 2–3 of the IDL armed with proximal spines, (4) scrapers 5–8 armed with strong denticles, (5) absence of specialized structure on the endite of the limb III, (6) seta 5 on the exopodite of the limb IV longer than setae 5–6, and (7) seta 3 on the exopodite of the limb V shorter than setae 1–2.

Regarding the size, *A. kaingang* sp. nov. (length 0.33–0.40 mm) is longer than *A. anamariae* (0.28–0.30 mm) (Sinev and Silva-Briano 2012). In *A. anamariae*, the labral keel has two rows of setules, while in *A. kaingang* sp. nov., the labral keel is naked (Fig. 1(g–i)). In *A. kaingang* sp. nov., PP is less than 0.2 IP while in *A. anamariae*, PP is about 0.2–0.3 IP. Another important morphological trait observed in *A. anamariae* and that is absent in *A. kaingang* sp. nov. is the longest setae on the anterior group of margin of the carapace (Sinev and Silva-Briano 2012). The postabdomen of *A. anamariae* does not have postanal angle as evident as in *A. kaingang* sp. nov (Fig. 2(e–f)). On the limbs, *A. kaingang* sp. nov. differs from *A. anamariae* because it has (1) setae 2–3 of the IDL markedly of different length, (2) setae 2–3 of the IDL armed with proximal spines, (3) scrapers 5–8 armed with strong denticles, (4) absence of a specialized structure on the endite of the limb III, (5) first flaming torch relatively robust, and (6) seta 5 on the exopodite of the limb III shorter than seta 7.

Recently, Sinev et al. (2012) described the Mediterranean *A. anastasia*, which has a distinct short postabdomen; however, this species may be easily separated from *A. kaingang* sp. nov. by the interrupted connection of the main head pores and absence of seta on the exopodite of the limb II. *Alona setulosa* is another species of the *pulchella*-group with a short postabdomen (Sinev 2009), and the shape of this structure is quite similar to that observed in *A. kaingang* sp. nov. Besides that, *A. setulosa* and *A. kaingang* sp. nov. share similar armature of the apical setae of the antenna and denticulation of scrapers 6–8. The armature of the apical setae of the antenna observed in *A. setulosa* and *A. kaingang* sp. nov. was also found in phylogenetically distant groups such as *Armatalona* (Sinev 2004a) and *Acanthalona* (Sinev and Kobayashi, 2012), which suggests that this morphological similarity may be the result of convergence. *Alona setulosa* differs from *A. kaingang* sp. nov. because it has a labral keel with two rows of setules and incomplete connection between main head pores. They differ also in fine features of the limbs (see Sinev 2009).

The differences between the other species that possess main head pores with complete connection and *A. kaingang* sp. nov. go beyond the shape of the postabdomen. The South American *A. glabra* has the third seta on the exopodite of the limb V longer than the second seta, and the first flaming torch on the limb IV is relatively thinner (Sinev 2001b). Besides that, the geographical distribution of *A. glabra* extends from Argentina to the extreme North of the Neotropics, while *A. kaingang* sp. nov. is restricted to the southern regions. *Alona archeri* has a PP less than 0.5 IP, well-developed distal denticles on the postabdomen transformed in setules toward the anal margin and seta 4 on exopodite of the limb III relatively straight and markedly long (Sinev 2002b). The paleotropical *A. pulchella* has thin setules on the scrapers of the limb II and relatively long second setae on the exopodite of the limb III (Sinev 2001c). The specialized *A. bromelicola* has a markedly short seta on the exopodite of the limb II, setae 6–4 on the exopodite of the limb IV of similar length, seta 3 is of similar length to setae 1–2 and seta 5 is markedly long, and the basal spine is very short (Sinev 2002a). Some other species that are presumed to be members of the *pulchella*-group still await revision (Van Damme et al. 2010), and their affinities and differences with *A. kaingang* sp. nov. cannot be addressed yet.

Although there are several morphological differences between species in the *pulchella*-group and those related to species groups, Van Damme et al. (2013) indicated the possibility of identification mistakes. For example, some species of the *pulchella*-group has a postabdomen of a similar shape to the one observed in the *Alona gutatta*-group (e.g., *A. archeri*, *A. glabra*, and *A. bromelicola*). However, the *Alona gutatta*-group is a member of the *Hexalona*-branch, and the observation of the limbs is necessary (or at least the use of high magnification) for correct identification. Under low magnification, the postabdomen of *A. kaingang* sp. nov. resembles *C. rectangula* (= *Alona rectangula*), and many records of this species in Brazil (see Green 1972; Santos et al. 1994; Smirnov and Santos-Silva 1995; Eskinazi-Sant'Anna et al. 2005; Sterza and Fernandes 2006; Rocha et al. 2011; Viti et al. 2013) perhaps may be attributed to new species of the *pulchella*-group described here; this, however, needs to be checked.

From the evolutionary point of view, the *pulchella*-group presents an interesting duality because its species do not have many specializations on the limbs (their morphology is almost constant) but differ in habitus and, mainly, in the morphology of the postabdomen. An extreme case was observed in *A. capensis*, which is similar in its limb morphology to other species of the group, but has an elongated body, a bulge on the rostrum, and a very peculiar postabdomen (Van Damme et al. 2013). Besides, the variability of shapes of the postabdomen in other species of *pulchella*-group (elongated, short and narrowing, short and massive, short and rounded; marginal denticles thick or thin; lateral fascicles short or long) suggests speciation mediated by evolutionary pressure on the movement, once postabdomen is the main morphological trait involved in such purpose, being also related to food collecting and shelter searching. The phylogenetic interpretation of the diversity of shapes of postabdomen in the *pulchella*-group still remains unclear, but elongated and short forms in different species may be arisen from adaptive convergence associated to similar environmental pressure.

As in other species group (e.g., *Anthalona*, *Nicsmirnovius*, *Euryalona*, *Monospilus*), the morphology of head pores may be important phylogenetically for the *pulchella*-group because of the absence of reductions or specializations in the limbs, that make difficult to do suggestions of affinities between species. However, in species with both morphologies of main head pores, a considerable level of affinities related to general shape of the postabdomen and fine structure on the limbs are observed: *A. cambouei*, a species with broken connection, has armatures of postabdomen and seta 5 of the limb III similar to species with complete connection, such as, *A. pulchella* and *A. glabra* (Sinev 2001b, c). When compared with the broken connection species *A. nigra*, *A. aguascalientensis*, and *A. setulosa*, *A. cambouei* presents differences in the armature of antenna, postabdomen, and proportions of setae on limbs I and III (Sinev 2001c, 2009; Sinev and Silva-Briano 2012). Therefore, morphology of main head pores could also be the result of convergence and to test relationship in *pulchella*-group may be useful to elucidate these issues.

The morphological boundaries of the *pulchella*-group indicate that it does not belong to the *Alona* sensu stricto (Van Damme and Dumont 2008a). It is probable that in the future, a new genus will be created and that will include *A. kaingang* sp. nov. (Van Damme et al. 2010). Moreover, the *pulchella*-group has many affinities with *Ovalona* Van Damme & Dumont, 2008, a group recently removed from *Alona*. These two groups present affinities in habitus and limbs (mainly with *Ovalona weinecki*), and the only feature that is truly different is the presence of the merged (*pulchella*-group) and not-merged (*Ovalona*) denticles on the postanal margin of the postabdomen (Van Damme and Dumont 2008b). The use of molecular tools can be valuable to define the relationship between the *pulchella*-group and *Ovalona*.

Conclusions

The *pulchella*-group is now composed of at least 15 valid species. The Neotropical fauna of the *pulchella*-group is extremely diverse and contains 50 % of the known species. *Alona kaingang* sp. nov. has connection between main head pores complete; therefore, it is separated from *A. setulosa*, *A. cambouei*, *A. aguascalientensis*, *A. nigra*, *A. setuloides*, *A. azorica*, *A. nuragica*, *A. capensis*, and *A. anastasia*. *Alona kaingang* sp. nov. resembles *A. altiplana* but differs from it mainly in the morphology of the marginal denticles of the postabdomen and fine features on the limbs. The relationship between *pulchella*-group and *Ovalana* needs to be better studied.

Abbreviations

As: accessory seta; CBS: copulatory brush seta; EL: Elmoor-Loureiro collection, at Universidade Católica de Brasília, Brazil; en: endite; ep: epipodite; ex: exopodite; fc: filtercomb; gfp: gnathobasic filter plate; gn: gnathobase; IDL: inner distal lobe; il: inner lobe; IP: interpore distance (distance between anterior and posterior major head pores); MNRJ: National Museum of Rio de Janeiro, Brazil; MZUSP: Museum of Zoology of the University of São Paulo, Brazil; ODL: outer distal lobe; pep: pre-epipodite; PP: postpore distance (distance between posterior major head pore and posterior border of head shield); s: sensillum; SEM: scanning electron microscopy.

Competing interests

The authors declare that they have no competing interests.

Authors' contributions

FDRS, LMAEL, and SS wrote the manuscript. FDRS and LMAEL produced the drawings, photographs on the SEM, and taxonomic discussion. SS discussed the taxonomic parts, made the intellectual contributions, and reviewed the manuscript. All authors read and approved the final manuscript.

Acknowledgements

We thank Prof. Dr. Maria José dos Santos Wisniewski and Dr. José Roberto Debastiani Júnior for providing the material from Furnas and São Simão reservoirs, respectively. Thanks to Prof. Dr. Ciro Joko for the suggestion about the digital scientific illustrations. We also thank Dr. Sônia N. Bao, Ingrid C. M. da Silva, and Márcia C. O. da Rocha (Electron Microscopy Laboratory of the University of Brasília), who made the SEM micrographs possible. We also thank MSc. Diego M. Magalhães and Ms. Susan Casement for revising the English version of this manuscript. We are grateful to Dr. Gilmar Perbiche Neves, Dr. Ricardo Lourenço Pinto, Carla Bender Kotizian, and Dr. Alexey Kotov for the suggestions on the initial draft of this manuscript. We are grateful to the anonymous reviewers for their valuable comments and suggestions on the manuscript. FDRS was awarded a grant from the Coordenação de Aperfeiçoamento de Pessoal de Nível Superior (CAPES). We are also grateful to the CNPq (Conselho Nacional de Desenvolvimento Científico e Tecnológico) for the productivity grant to SS (308598/2011-3).

References

Dumont JH, Silva-Briano M (2000) *Karualona* n. gen. (Anomopoda: Chydoridae), with a description of two new species, and a key to all known species. Hydrobiologia 435:61–78

Elmoor-Loureiro LMA (2004) Phylogenetic relationships among families of the order Anomopoda (Crustacea, Branchiopoda, Cladocera). Zootaxa 760:1–26

Eskinazi-Sant'Anna EM, Barbosa PMM, Rietzler AC (2005) Zooplankton biodiversity of Minas Gerais State: a preliminary synthesis of present knowledge. Acta Limnol Bras 17:199–218

Fonseca GAB (2010) Contribuição antrópica na poluição de reservatórios hidrelétricos: o caso da usina hidrelétrica de São Simão – GO/MG. Universidade Federal do Rio de Janeiro, Dissertation

Fryer G (1968) Evolution and adaptive radiation in the Chydoridae (Crustacea: Cladocera): a study in comparative functional morphology and ecology. Philos Trans R Soc Lond B Biol Sci 254:221–384

Green J (1972) Freshwater ecology in the Mato Grosso, Central Brazil. II. Associations of Cladocera in meander lakes of the Rio Suiá Missú. J Nat Hist 6:215–227
Kotov AA (2000a) Redescription and assignment of the chydorid *Indialona ganapati* Petkovski, 1966 (Branchiopoda: Anomopoda: Aloninae) to Indialonini, new tribus. Hydrobiologia 439:161–178
Kotov AA (2000b) Analysis of *Kozhowia* Vasiljeva & Smirnov, 1969 (Chydoridae, Anomopoda, Branchiopoda), with a description of *Parakozhowia* n. gen. Hydrobiologia 437:17–56
Kotov AA (2003) Notes on Aloninae & Grochowski, 1894 *emend.* Frey, 1967 (Cladocera: Anomopoda: Chydoridae): 1. Translocation of *Alona incredibilis* Smirnov 1984 to the genus *Nicsmirnovius* Chiambeng & Dumont, 1999. Arthropoda Sel 12:167–170
Kotov AA (2006) Adaptations of Anomopoda crustaceans (Cladocera) to the benthic mode of life. Entomol Rev 8:210–225
Kotov AA, Elíaz-Gutiérrez M (2002) Analysis of the morphology of Spinalona anophtalma Ciros-Pérez & Elíaz-Gutiérrez, 1997 (Aloninae, Anomopoda, Cladocera). Hydrobiologia 486:185–192
Kotov AA, Sanoamuang L (2004) Comments on the morphology of *Nicsmirnovius eximius* (Kiser, 1948) (Aloninae, Anomopoda, Cladocera) in Thailand, with description of its male. Hydrobiologia 519:117–125
Kotov AA, Sinev AY, Berrios VL (2010) The Cladocera (Crustacea: Branchiopoda) of six high altitude water bodies in the North Chilean Andes, with discussion of Andean endemism. Zootaxa 2430:1–66
Rocha O, Santos-Wisniewski MJ, Matsumura-Tundisi T (2011) Checklist de Cladocera de água doce do estado de São Paulo. Biota Neotrop 11:571–592
Sacherová V, Hebert PDN (2003) The evolutionary history of the Chydoridae (Crustacea: Cladocera). Biol J Linn Soc 79:629–643
Santos GB, Maia-Barbosa PM, Vieira F, López CM (1994) Fish and zooplankton community structure in reservoirs of southeastern Brazil: effects of the introduction of exotic predatory fish. In: Pinto-Coelho RM, Giani A, Sperling EV (eds) Ecology and human impact on lakes and reservoirs in Minas Geraia. SEGRAC Press, Belo Horizonte, pp 115–130
Santos RM, Negreiro NF, Silva LC, Rocha O, Santos-Wisniewski MJ (2010) Biomass and production of Cladocera in Furnas Reservoir, Minas Gerais, Brazil. Braz J Biol 70:879–887
Sinev AY (2001a) Redescription of *Alona iheringi* Sars, 1901 (Chydoridae, Anomopoda, Branchiopoda), a South American species related to *A. rustica* Scott, 1895. Hydrobiologia 464:113–119
Sinev AY (2001b) Redescription of *Alona glabra* Sars, 1901, a South American species of the *pulchella*-group (Chydoridae, Anomopoda, Branchiopoda). Arthropoda Sel 10:273–280
Sinev AY (2001c) Separation of *Alona cambouei* Guerne & Richard, 1893 from *Alona pulchella* King, 1853 (Branchiopoda, Anomopoda, Chydoridae). Arthropoda Sel 10:5–18
Sinev AY (2002a) Place of a Central American bromeliad-inhabiting cladoceran *Alona bromelicola* Smirnov, 1988 within the genus (Branchiopoda: Anomopoda: Chydoridae). Arthropoda Sel 11:109–116
Sinev AY (2002b) Redescription of Australian cladocera *Alona archeri* Sars, 1888 (Branchiopoda: Anomopoda: Chydoridae). Arthropoda Sel 11:247–254
Sinev AY (2004a) *Armatalona* gen. n.—a new genus of subfamily Aloninae (Anomopoda, Chydoridae), separated from genus *Alona* Baird, 1840. Hydrobiologia 520:29–47
Sinev AY (2004b) *Miralona* gen. n.—a new genus of the subfamily Aloninae (Anomopoda, Chydoridae) from Australia. Hydrobiologia 526:3–14
Sinev AY (2009) Notes on morphology and taxonomic status of some North American species of the genus *Alona* Baird, 1843 (Cladocera: Anomopoda: Chydoridae). Fundam Appl Limnol 175:59–77
Sinev AY, Atroschenko MM (2011) Revision of the genus *Alonopsis* Sars, 1862 and its position within Aloninae (Cladocera: Anomopoda: Chydoridae). Zootaxa 2800:1–17
Sinev AY, Elmoor-Loureiro LMA (2010) Three new species of chydorid cladocerans of subfamily Aloninae (Branchiopoda: Anomopoda: Chydoridae) from Brazil. Zootaxa 2390:1–25
Sinev AY, Hollwedel W (2005) Translocation of *Alona muelleri* Richard, 1897 into the genus *Karualona* Dumont & Silva-Briano, 2000 (Branchiopoda: Anomopoda: Chydoridae). Arthropoda Sel 14:93–101
Sinev AY, Kobayashi T (2012) Redescription of the endemic Australian cladoceran *Alona willisi* (Smirnov, 1989) and its assignment to *Acanthalona* gen. nov. (Cladocera: Anomopoda: Chydoridae). Zootaxa 3390:43–45
Sinev AY, Kotov AA (2000) Redescription of *Alona labrosa* Vasiljeva& Smirnov, 1969 (Chydoridae, Anomopoda, Branchiopoda). Hydrobiologia 439:125–139
Sinev AY, Kotov AA (2001) Redescription of *Alona setosocaudata* Vasiljeva & Smirnov, 1969 (Anomopoda, Branchiopoda). Hydrobiologia 452:47–58
Sinev AY, Shiel RJ (2008) Redescription of *Alona macracantha* Smirnov and Timms, 1983 and its assignment to *Maraura* gen. nov. (Cladocera: Anomopoda: Chydoridae). J Nat Hist 42:2809–2824
Sinev AY, Silva-Briano M (2012) Cladocerans of genus *Alona* Baird, 1843 (Cladocera: Anomopoda: Chydoridae) and related genera from Aguascalientes State, Mexico. Zootaxa 3569:1–24
Sinev AY, Van Damme K, Kotov AA (2005) Redescription of tropical-temperate cladocerans *Alona diaphana* King, 1853 and *Alona davidi* Richard, 1895 and their translocation to *Leberis* Smirnov, 1989 (Branchiopoda: Anomopoda: Chydoridae). Arthropoda Sel 14:183–205
Sinev AY, Alonso M, Miracle MR, Sahuquillo M (2012) The West Mediterranean *Alona azorica* Frenzel & Alonso, 1988 (Cladocera: Anomopoda: Chydoridae) is composed of two species. Zootaxa 3276:51–68
Smirnov NN (1971) *Chydoridae fauny mira. Fauna USSR.* Rakoobraznie, Leningrad. (English translation: Chydoridae of the world. Israel Program for Scientific Translations, Jerusalem, 1974
Smirnov NN, Santos-Silva EN (1995) Some littoral anomopods (Crustacea) from Central Amazonia. Hydrobiologia 315:227–230
Smirnov NN (1998) A revision of the genus Camptocercus (Anomopoda, Chydoridae, Aloninae). Hydrobiologia 386:63–83
Sousa FDR, Elmoor-Loureiro LMA, Santos S (2015) Redescription of the *Coronatella poppei* (Richard, 1897) (Crustacea, Branchiopoda, Chydoridae) and a revision of the genus in Brazil, with descriptions of new taxa. Zootaxa 3955:211–244
Sterza JM, Fernandes LL (2006) Distribution and abundance of Cladocera (Branchiopoda) in the Paraíba do Sul River estuary, Rio de Janeiro, Brazil. Braz J Oceanogr 54:193–204
Van Damme K, Dumont HJ (2007) Limb morphology of the carnivorous anomopods *Anchistropus emarginatus* Sars, 1862 and *Pseudochydorus globosus* (Baird, 1843) (Crustacea: Branchiopoda: Anomopoda). Ann Limnol – Int J Limnol 43:271–284
Van Damme K, Dumont JH (2008a) The 'True' genus *Alona* Baird, 1843 (Crustacea: Cladocera: Anomopoda): position of the *A. qudrangularis*-group and description of a new species from Democratic Republic of Congo. Zootaxa 1943:1–25
Van Damme K, Dumont JH (2008b) Further division of *Alona* Baird, 1843: separation and position of *Coronatella* Dybowski & Grochowski and *Ovalona* gen.n. (Crustacea: Cladocera). Zootaxa 1960:1–44
Van Damme K, Dumont JH (2009) Notes on chydorid endemism in continental Africa: *Matralona* gen. n., a monotypic Alonine from the Fouta Djalon Plateau (Guinea, West Africa) (Crustacea: Cladocera: Anomopoda). Zootaxa 2051:26–40
Van Damme K, Sinev AY (2011) A new genus of cave dwelling microcrustaceans from the Dinaric Region (south-east Europe): adaptations of true stygobitic Cladocera (Crustacea: Branchiopoda). Zool J Linn Soc 161:31–52
Van Damme K, Brancej A, Dumont HJ (2009) Adaptations to the hyporheic in Aloninae (Crustacea: Cladocera): allocation of *Alona protzi* Hartwig, 1900 and related species to *Phreatalona* gen. nov. Hydrobiologia 618:1–34
Van Damme K, Chiambeng G, Maiphe S, Dumont JH (2003) New species in the rheophilous genus *Nicsmirnovius* Chiambeng & Dumont, 1999 (Branchiopoda: Anomopoda: Chydoridae) and reassignment of *Alona eximia* Kiser, 1948 and *Alonella fitzpatricki* Chien, 1970. Hydrobiologia 499:25–49
Van Damme K, Kotov AA, Dumont HJ (2005) Redescription of *Leydigia parva* Daday, 1905 and assignment to *Parvalona* gen. nov. (Cladocera: Anomopoda: Chydoridae). J Nat Hist 39:2125–2136
Van Damme K, Kotov AA, Dumont HJ (2010) A checklist of names in *Alona* Baird 1843 (Crustacea: Cladocera: Chydoridae) and their current status: an analysis of the taxonomy of a lump genus. Zootaxa 2330:1–63
Van Damme K, Sinev AY, Dumont HJ (2011) Separation of *Anthalona* gen.n. from *Alona* Baird, 1843 (Branchiopoda: Cladocera: Anomopoda): morphology and evolution of scraping stenothermic Alonine. Zootaxa 2875:1–64
Damme V, Bekker E, Kotov AA (2013) Endemism in the Cladocera (Crustacea: Branchiopoda) of Southern Africa. J Limnol 72:440–463
Viti T, Wisniewski C, Orlando TC, Santos-Wisniewski MJ (2013) Life history, biomass and production of *Coronatella rectangula* (Branchiopoda, Anomopoda, Chydoridae) from Minas Gerais. Iheringia, Sér Zool 103:110–117

22

Morphological variability of the cranium of *Lontra longicaudis* (Carnivora: Mustelidae): a morphometric and geographic analysis

Pablo César Hernández-Romero[1*], José Antonio Guerrero[2] and Carolina Valdespino[1]

Abstract

Background: Studies of morphometric variation make it possible to delimit species and geographic intraspecific variation, mainly in species with wide distribution ranges. In the Neotropical otter, *Lontra longicaudis*, variation in the shape of the rhinarium of three potential subspecies has been described but it is not known whether there is a pattern to the morphometric variation in the skull throughout the distribution of this species. We analyzed morphological variation in the cranium (ventral view) and the mandible (lateral view) of the Neotropical otter, comparing male and female specimens and evaluating the differences between specified geographic units utilizing methods from geometric morphometrics. Specimens from the entire distribution of the species were analyzed. Between sexes, variability in the shape was determined by calculating the Procrustes distances and using Goodall's *F* test. Geographic variation was analyzed using a discriminant analysis, an analysis of variance (ANOVA) on a matrix of partial warp scores, and a cluster analysis with Mahalanobis distances, allowing for similarities in shape to be identified between different geographic units. Variation in the size of the two structures was calculated based on the values for centroid size using a one-way ANOVA with a Bonferroni correction and a 95 % confidence interval.

Results: There was sexual dimorphism in shape for both views, with males the largest. In general, there was geographic variation in the shape and size of both the cranium and the mandible in the Neotropical otter, exhibiting a pattern that resembled Bergmann's rule. Variation in shape between geographic units could result from the presence of geographic barriers, the spatial configuration of hydrological regions, and/or the large distances between populations throughout this species' distribution.

Conclusions: The Neotropical otter exhibits dimorphism in the size, but not in the shape of the skull. There is geographic variation between geographic units, and our results suggest that *L. longicaudis* could be a group of species. An integrative study using molecular and morphological data could elucidate its taxonomy.

Keywords: Geographic variation; Geometric morphometrics; *Lontra longicaudis*; Sexual dimorphism

Background

The Neotropical otter (*Lontra longicaudis*) is an aquatic species with a wide distribution that is limited to the terrestrial-aquatic ecotone that corresponds to lagoon and riparian systems (Gallo-Reynoso et al. 2008). This species is Neotropical in origin and has a continuous distribution from Argentina in South America to the Trans-Mexican Volcanic Belt (Fig. 1). The latter is a mountain range that divides the country in two, follows the outline of both the Atlantic and Pacific slopes, and ultimately reaches its northern extreme in the state of Sonora on the Pacific slope with its other extreme in Tamaulipas on the Gulf of Mexico slope (Gallo-Reynoso 1997). Sexual dimorphism has been reported for *L. longicaudis*, with males larger than females (Kruuk 2006). Along its distribution, this otter occupies habitats such as coastal lagoons, rivers, and lakes that are located in distinct geographic and climatic regions, from arid zones to wetlands and from tropical to temperate forests (Larivière 1999). The composition of its diet is also variable between regions

* Correspondence: pabloecologico@yahoo.com.mx
[1]Instituto de Ecología A.C., Red de Biología y Conservación de Vertebrados, CP 91070, Carretera antigua a Coatepec 351, Xalapa, Mexico
Full list of author information is available at the end of the article

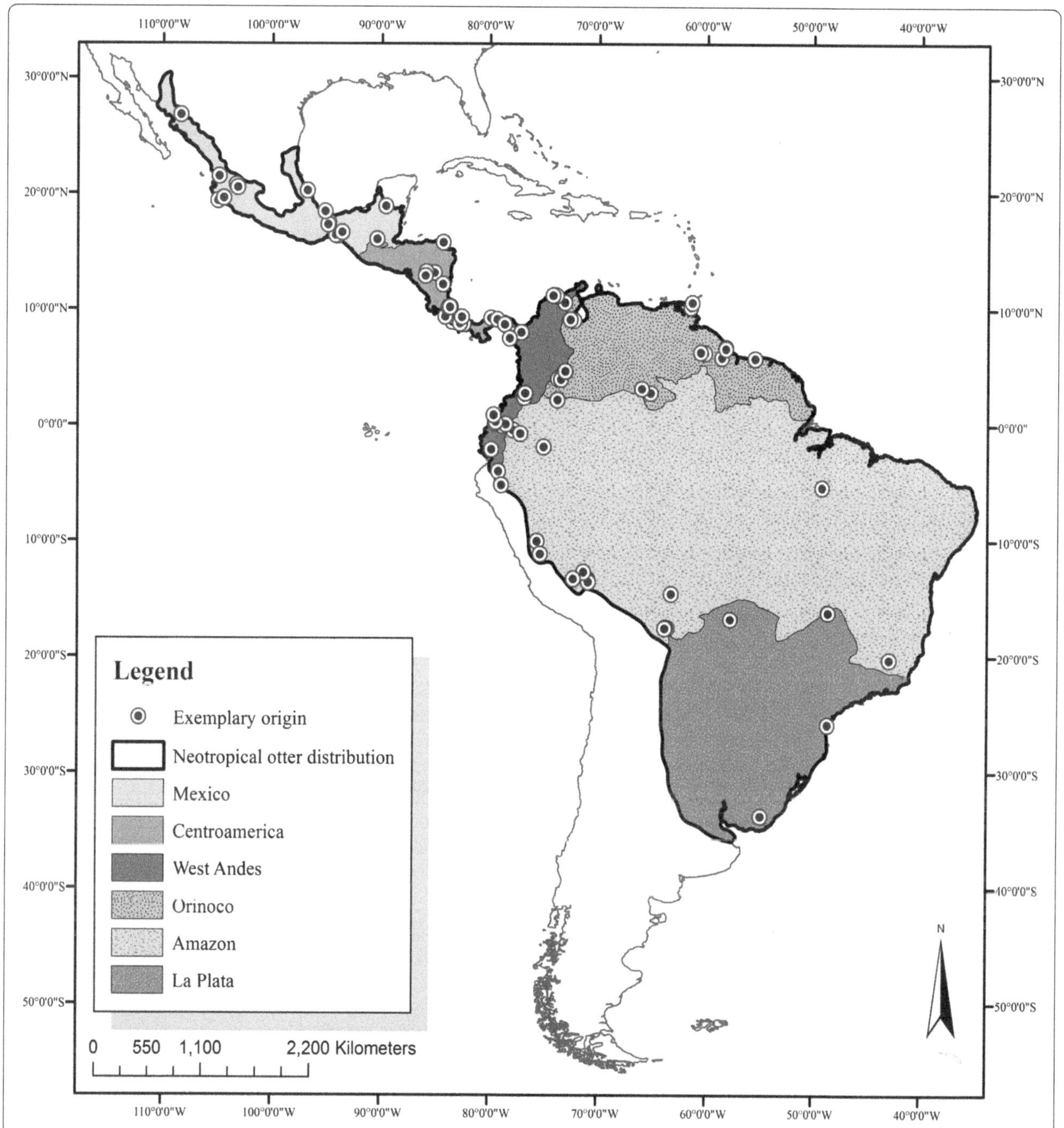

Fig. 1 Geographic distribution of *Lontra longicaudis*, indicating the origin of the specimens used, and the six geographic units: Mexico (MEX), Central America (CEAM), Western Andes (AND), Orinoco (ORI), Amazon (AMA), and La Plata (PLA)

and seasons, though different studies have shown its diet is largely made up of fish and crustaceans and to a lesser extent amphibians, mammals, birds, and turtles (Gallo-Reynoso et al. 2008; Carvalho-Junior et al. 2010;Chemes et al. 2010; Platt and Rainwater 2011). Owing to these factors and the variation in its habitat in terms of vegetation, climate, and food availability throughout its extensive distribution, potential changes in the cranial morphology of *L. longicaudis* may be interpreted as local adaptations to environmental variables such as different types of prey.

The wide geographic distribution of *L. longicaudis* and the variation in its aquatic habitats give rise to two hypotheses of morphological variation. The first hypothesis is that there should be a differentiation in the shape and size of the cranium and mandible between the sexes, and the second is that these variables differ between populations of the different geographic regions throughout its distribution.

In order to test these hypotheses, we used techniques based on geometric morphometrics (Kendall 1984; Bookstein 1991; Lawing and Polly 2010) which describe and interpret changes in the shape of different anatomical structures of organisms, based on comparisons of homologous structures (Adams et al. 2004). This analysis of anatomical shape allows for a better understanding of morphological evolution and adaptive effects (Zelditch et al. 2004) since variation in shape is influenced by environmental factors (Rohlf and Marcus 1993).

Morphometric analysis of the cranium of several species has indicated that patterns of morphological variation in species with large distribution areas might be adaptations to the range of environmental conditions occurring throughout their distribution (Zelditch et al. 2004; Cardini et al. 2007). This pattern has been observed in mammals such as the puma (*Puma concolor*), the tiger (*Panthera tigris*), and the brown bear (*Ursus arctos*), for which variation in the size and configuration of the cranium are correlated with the different geographic regions throughout which their populations are distributed (Gay and Best 1996; Baryshnikov and Puzachenko 2011; Mazák 2011). These patterns of morphological variation are associated with habitat variables such as temperature, precipitation, and latitude (Platz et al. 2011). Patterns of change in the mandible associated with contrasting habitats have been observed in several species of rodents (Duarte et al. 2000). This also occurs for species like the titus monkey, genus *Leontopithecus* (Freitas et al. 1999), and for bats of the genus *Monophyllus* (Mancina et al. 2010). And for the green monkey, *Cercopithecus aethiops*, there is a close relationship between rainfall pattern and the length of its face (clinal variation) (Cardini et al. 2007).

Patterns of morphological change associated with geographic regions are even more notable in species with a high level of habitat specialization, such as aquatic species (Langerhans et al. 2003; Jonsson and Jonsson 2001). For the cichlid *Biotodoma wavrini* and the characid *Bryconops caudomaculatus*, distinct morphological adaptations have been found for their two contrasting habitats, river channels and lagoons, where morphological variation increases with increasing geographic distance between populations and with variation in other environmental factors (Langerhans et al. 2003).

Morphological variation in the Eurasian otter (*Lutra lutra*) has only been assessed using linear morphometric analysis. For example, Ruiz-Olmo et al. (1998) reported that the age of *L. lutra* may be determined on the basis of its cranial morphology. And variation in cranium size has also been found over the range of this species, with more variability in animals in the tropics (Baryshnikov and Puzachenko 2012). This technique demonstrated that there is a significant difference in cranial morphology between healthy populations of otters and populations considered at risk due to decreased population size and habitat quality (Pertoldi et al. 2000). In addition, these studies have made it possible to explore any effects of contaminants on the morphology and ontogenesis of the cranium of *L. lutra* (Pertoldi et al. 1998) and other species such as *Ursus maritimus* (Pertoldi et al. 2012).

Several taxonomic studies of *L. longicaudis* have postulated that it is potentially composed of a number of species based primarily on the differentiation of its rhinarium (Pohle 1920; Cabrera 1957; Harris 1968). The most recent studies have suggested that populations with distinct rhinarium morphology may be conspecific and that these populations may be subdivided geographically into three subspecies: (I) *L. longicaudis annectens*, found from Mexico to Colombia and Ecuador; (II) *L. longicaudis enudris*, distributed in French Guiana, Suriname, Peru, and Trinidad and Tobago; and (III) *L. longicaudis longicaudis*, distributed throughout most of South America from Brazil to Uruguay and Argentina (Van Zyll 1972; Larivière 1999). However, this classification has also been questioned recently by Trinca et al. (2012) as a result of new insights into the phylogenetic structure of the South American populations based on mitochondrial sequence analysis. These authors mentioned that an intraspecific subdivision continues to be controversial due to genetic differentiation within the different geographic regions, and indicated that an integrative study would be necessary to examine morphological variation in the species, including an analysis of cranial morphometrics.

The present study has the objectives of evaluating changes in the morphology of the cranium and mandible of *L. longicaudis* using geometric morphometric techniques and examining the variation in the configuration (shape and size) of the cranium and the mandible between sexes and among the different geographic regions of its distribution. Morphological variation associated with distinct geographic regions throughout which the otter is distributed would support previously reported genetic differences.

Methods

Sampling

The crania of 151 specimens of *L. longicaudis* from localities throughout its distribution were examined (Fig. 1). The specimens were housed in the National Collection of Mammals of the National Autonomous University of Mexico, the Mammal Collection of the Autonomous University of Campeche, the Vertebrate Collection of the Biological Station of the National Autonomous University of Mexico in Los Tuxtlas, the Mammal Collection of the University of Kansas, the American Museum of Natural

History, the Mammal Collection of the Field Museum, the National Museum of Natural History of the Smithsonian Institute, the Carnegie Museum, and the Michigan University Museum (see list of specimens in Additional file 1). Since the Neotropical otter is a threatened species and protected in the majority of the countries where it is found, the analysis was limited to cranium samples from the aforementioned collections.

Each specimen was assigned to one of the three age categories using the criteria of Van Zyll (1972): juvenile (presence of deciduous teeth), subadult (a full set of fully erupted teeth, presence of nasal sutures, and occipital-parietal sutures), and adult (absence of maxillary-nasal and occipital-parietal sutures). For the morphometric analysis, only adult samples were used to avoid the errors associated with the modifications that the cranium undergoes when the animal is growing (Caumul and Polly 2005).

The samples were grouped a priori into six geographic units (GUs): Mexico, Central America, Western Andes, Orinoco, Amazon, and La Plata (Fig. 1) based on the main hydrological delimitations throughout this otter's distribution and on patterns of morphological and genetic structure observed among the different populations of other aquatic species in South America (Banguera-Hinestroza et al. 2002; Hubert and Renno 2006; Arzamendia and Giraudo 2009), including the Neotropical otter (Trinca et al. 2012).

Digitization of images, recording landmarks, and semi-landmarks

The analysis of variation in cranial morphology was based on digital images of the cranium in ventral view ($n = 64$) and the mandible in lateral view ($n = 87$). All photographs were taken by the same person to avoid any variation associated with possible different points of view of observers. Specimen placement was standardized, and the crania were photographed on the right side of the face, from a distance of 40 cm from the camera lens and oriented at an angle of 90°. All images were captured using a Sony Alpha 200 reflex camera with a resolution of 3872 × 2592 pixels. A ruler was included in each photograph as a reference scale.

The use of only the right side of the cranium and the mandible to locate the landmarks assumes bilateral symmetry in these structures. The configuration of the cranium and the mandible was recorded using landmarks (LMs), which are natural points located on homologous sites and that can be easily located on any specimen (Bookstein 1997), and semi-landmarks (SLMs), which are located along the homologous outlines of the structures and are based on templates drawn between the LM points (Bookstein 1991; Gunz and Mitteroecker 2013). Outlines were drawn using MakeFan6 software (Sheets 2003) to facilitate the location of the SLMs and were based on two LMs (numbers 1 and 8 for the mandible and 1 and 16 for the cranium, Fig. 2). LMs and SLMs were digitized for each specimen using the TpsDig 2.15 software (Rohlf 2001), and configuration matrices obtained for each view. For these matrices, a generalized Procrustes superposition adjustment (Rohlf and Slice 1990) was applied using the software CoorGen6f (Sheets 2005a), where variation associated with the effects of positioning, orientation, and scale is minimized by calculating the sum of the squared differences between corresponding landmarks (Bookstein 1991; Zelditch et al. 2004). The placement of the outlines for the positioning of the SLMs results in a shape distortion during the Procrustes adjustment (Bookstein 1996; Gunz and Mitteroecker 2013), so a second adjustment was made to eliminate variation by sliding the SLMs based on the alignment distance criteria of the Semiland software (Sheets 2002).

For each configuration of LMs and SLMs (2 LMs and 21 SLMs on the ventral cranium view; 2 LMs and 16 SMLs on the lateral mandible view), an eigen analysis was done following the methodology of Bookstein (1991) and Rohlf (1996). This analysis allows a tangent space to be converted to a Euclidian space, making it possible to use conventional multivariate statistical analyses (Rohlf 1996). The average general shape was used as a reference, and each sample was described in terms of its deviation from the average form for all specimens. The resulting matrices describe change in shape (partial warp scores, PWS) and size as measured from the centroid size (CS). These variables were used in subsequent analyses to describe the change in shape and size of the skull and mandible.

Statistical analysis

A two-way analysis of variance (ANOVA) for centroid size and a two-way multivariate analysis of variance (MANOVA) for shape were performed to test a priori the differences between sexes and among GUs. Specimens from La Plata were excluded from both analyses of the ventral view because only females were available for this GU. These analyses were run using Statistica 8.0 software (StatSoft 2007).

Sexual dimorphism in shape and size

Because sexual dimorphism has been reported for several species of otter (Kruuk 2006), we analyzed shape and size differences between sexes, based on 70 mandible images (36 male and 34 female) and 56 cranium images (29 male and 27 female). To this end, a principal component analysis (PCA) and two-way MANOVA were run to test for sexual dimorphism in the shape of the mandible and the cranium. Subsequently, a

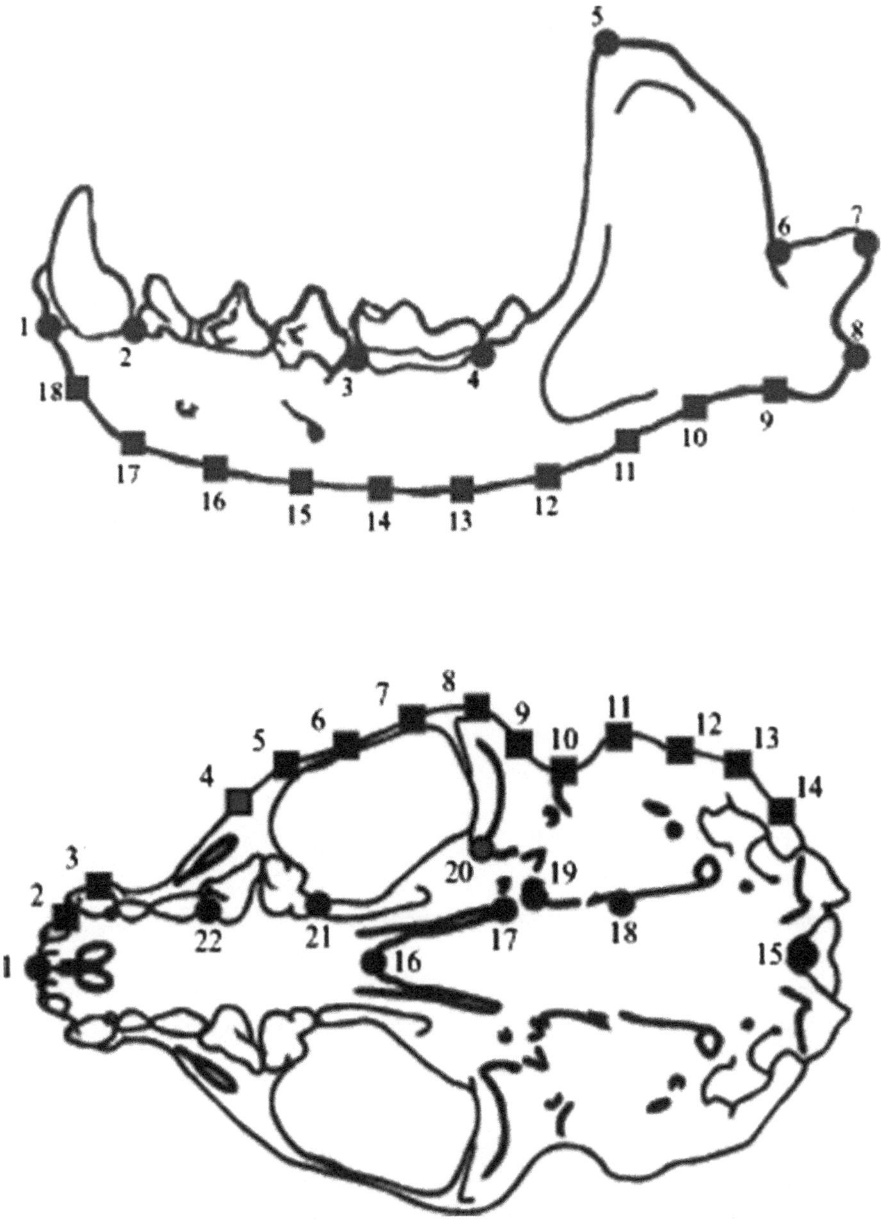

Fig. 2 Landmarks used to capture the configuration of the mandible (LM = 8 and SLM = 10) and the ventral view of the cranium (LM = 9 and SLM = 14) of *Lontra longicaudis*. *Circles* are LM and *squares* SLM

permutation test was performed using Goodall's *F* test, designed to compare Procrustes coordinates (Goodall 1991, Zelditch et al. 2004), as implemented in the software TwoGroup6 (IMP series; Sheets and Zeldich 2001). The significance of the differences was tested with 2500 bootstrap permutations. The variation in size between the sexes was analyzed using an ANOVA with Bonferroni correction and a 95 % confidence interval and run in Statistica 8.0 (StatSoft 2007). These analyses were run to determine whether the data for males and females could be pooled for subsequent tests.

Geographic variation in shape and size

Due to the low number of specimens used for the analysis of geographic variation and the fact that no significant sexual dimorphism was found for either structures (see "Results"), the data for both sexes were combined to analyze for geographic variation.

Canonical variables were extracted in a multidimensional space in order to explain the majority of the variation between groups for which the partial warp scores matrix was used. To assess shape variation between GUs in both views, a canonical variate analysis (CVA) was run in CVAGen6 software (Sheets 2005b). We obtained thin-plate spline plots (deformation grids describing shape changes) by regressing the PWS onto the first and second canonical variables. Subsequently, a discriminant function analysis (DFA) using Mahalanobis distances was calculated to assess the validity of the a priori classification of the specimens into six GUs.

Applying a MANOVA, a significant difference in shape was detected between the distinct GUs. To describe morphological similarity between GUs, a cluster analysis with the Unweighted Pair Group Method with Arithmetic Mean (UPGMA) was used, using the Mahalanobis distance between GUs. The DFA and UPGMA were run in Statistica 8.0 (StatSoft 2007).

Variation in cranium and mandible size between the six GUs of *L. longicaudis* was analyzed using a one-way ANOVA, with log-natural-transformed values of centroid size. Post hoc comparisons among GUs were run using a Bonferroni correction and a 95 % confidence interval in Statistica 8.0 (StatSoft 2007).

Results

Sexual dimorphism in shape and size

The PCA of the skull and the mandible indicate a small proportion of the variance was explained by the first two principal components, 63.6 and 42.9 %, respectively. Plots of these two components for each structure showed continuous variation between male and females, indicating the absence of sexual dimorphism (Fig. 3). This was confirmed by Goodall's *F* tests which showed that there were no significant differences in the shape of either structure between sexes (mandible: Goodall's $F_{(32)} = 0.75$, $p = 0.84224$; skull: Goodall's $F_{(42)} = 1.62$, $p = 0.1271$). Hence, we pooled female and male samples in subsequent analyses. The results of the ANOVA for centroid size indicated that males are significantly different from females in both skull size ($F_{(1,54)} = 6.2030$, $p = 0.0158$) and mandible size ($F_{(1,68)} = 9.8040$, $p = 0.0025$), with males larger (median CS mandible = 103.76, median CS skull = 169.76) than females (median CS mandible = 95.76, median CS skull = 161.53) for both views.

Geographic variation in the shape of the mandible and the cranium

The MANOVA detected significant differences in skull (Wilks' $\lambda = 0.014$, $F = 1.362$, $p < 0.05$) and mandible (Wilks' $\lambda = 0.013$, $F = 1.606$, $p < 0.001$) shapes between GUs. The CVA plots depict the patterns of change between the GUs defined in this study for both structures (Fig. 4). For both views, the samples from La Plata differed the most from the samples from the other GUs. The shape of the cranium had the most marked differences between GUs. For the change in the deformation axes of the mandible between GUs, the VC1 values indicate a change in the height of the coronoid process, which includes the emergence of the first and second molars, and in the width of the set of teeth. For the second canonical correlation (VC2), a change in shape was found between landmarks 1 and 8, which corresponds to the length of the mandible.

The deformation grid of the cranium between the GUs along VC1 (Fig. 5) shows a change in the shape of the posterior neurocranium (LMs 12–14) and the foramen

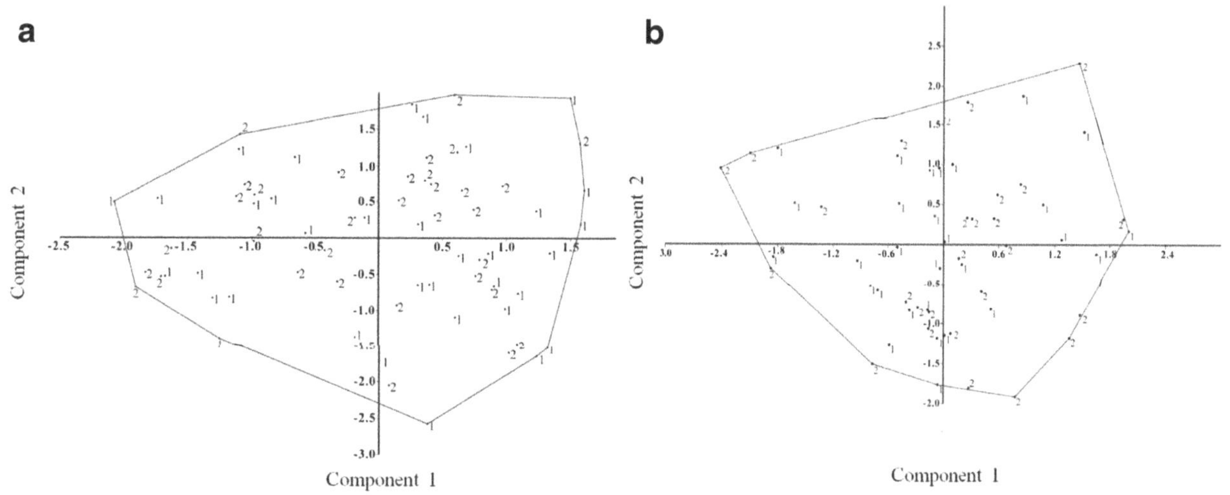

Fig. 3 Principal component analysis (PCA) of the variation in shape between sexes, where 1 = males and 2 = females. **a** PCA of the mandible with 49.9 % of the variation explained by the first axis and 12.7 % by the second. **b** PCA of the ventral view of the cranium with 30.3 % of the variation explained by the first axis and 12.6 % by the second

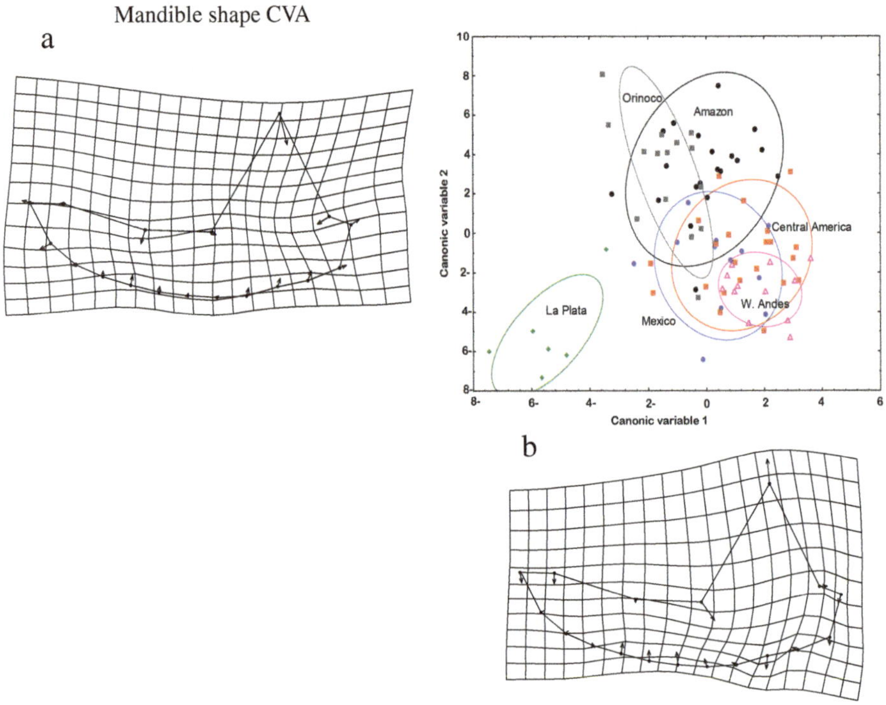

Fig. 4 a, b Analysis of the canonical variables and planes of deformation for the shape of the mandible in lateral view between geographic units. The *arrows* indicate the direction and magnitude of change

magnum (LM 15), in addition to a decrease in the width of the zygomatic arch. For VC2, there is an increase in the length of the mastoid bone (LM 11), along with an increase in the distance between the pterygoid muscles (LM 17) and a decrease in the distance between the rows of molars and premolars (LMs 21 and 22). The DFA of the mandible demonstrated that the a priori grouping of the specimens in six GUs is appropriate, with the majority of mandible samples correctly assigned (89.6 %) (Fig. 6). The analysis of the mandible for

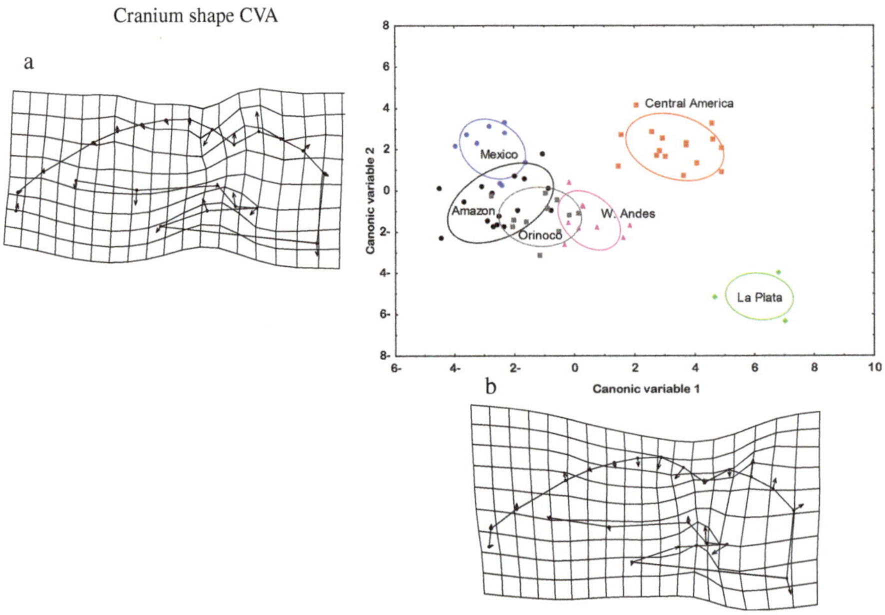

Fig. 5 a, b Analysis of the canonical variables and planes of deformation for the shape of the cranium in ventral view between geographic units. The *arrows* indicate the direction and magnitude of change

specimens from the six GUs using the Mahalanobis distances resulted in three groups. The first group is formed by the Amazon region, the Western Andes, Central America, and Mexico, Mexico being the region with the most differences within this grouping. The second group has the specimens from the Orinoco, and the third group is formed by the specimens from La Plata, the region with the lowest level of similarity with respect to the other GUs (Fig. 6).

The deformation grid of the cranium between the GUs along VC1 (Fig. 5) shows a change in the shape of the posterior neurocranium (LMs 12–14) and the foramen magnum (LM 15), in addition to a decrease in the width of the zygomatic arch. For VC2, there is an increase in the length of the mastoid bone (LM 11), along with an increase in the distance between the pterygoid muscles (LM 17) and a decrease in the distance between the rows of molars and premolars (LMs 21 and 22).

The DFA of the cranium samples indicated that 98.43 % were correctly assigned to the GUs by the discriminant analysis (Fig. 6). The analysis of the grouping of the six GUs using the Mahalanobis distances shows that the variation in shape of the mandible of *L. longicaudis* allowed the specimens to be classified into three groups. The first group includes specimens from the Amazon, Orinoco, and Mexico, Mexico being the unit with the most differences in shape within this larger grouping. The second most different group is that of the Western Andes and Central America. The third group has the samples from La Plata, the region with the lowest similarity with respect to the other GUs (Fig. 6).

Geographic variation in size

Differences were found ($F_{(5,81)} = 3.9603$, $p = 0.0029$) in the size of the mandible between the specimens from distinct GUs. The multiple comparison test with a Bonferroni correction classified the GUs into three groups. The first group included Western Andes, the Amazon, and Central America (average centroid size 107.27); the second Orinoco and La Plata (average centroid size 112.28), and the third Mexico (average centroid size 122.59). There were marginally significant differences in size of the cranium ($F_{(5,58)} = 2.3123$, $p = 0.0553$) between the GUs.

Discussion

Sexual dimorphism in shape and size

This is the first study to use geometric morphometric techniques to describe the sexual and geographic variations in the cranium and mandible of *L. longicaudis* throughout its distribution. The results indicate that there is no sexual dimorphism in the shape of the mandible or the skull of the Neotropical otter, though both structures do differ in size. This difference corresponds to a pattern of sexual dimorphism common in the Mustelidae family where males are generally larger than females (Moors 1980). The Eurasian otter *L. lutra* (Lynch and O'Sullivan 1993; Lynch et al. 1996; Baryshnikov and Puzachenko 2012) and the river otter *Lontra canadensis* (Pertoldi et al. 2000; Kruuk 2006) also have this pattern. Given that the difference in the size of the cranium and the mandible of the otters is correlated with body size (Pertoldi et al. 1998), the difference in size between sexes of *L. longicaudis* could be associated with its characteristically solitary habit and territorial behavior. The males of *L. lutra* must defend their territory, and a larger body ensures a higher position in the hierarchy and potentially better territory (Pertoldi et al. 1997; Kruuk 2006). For *L. canadensis*, social behavior has been reported and there are even cases of overlapping territories between males (Blundell et al. 2002; Melquist et al. 2003), though the territories of the males are larger

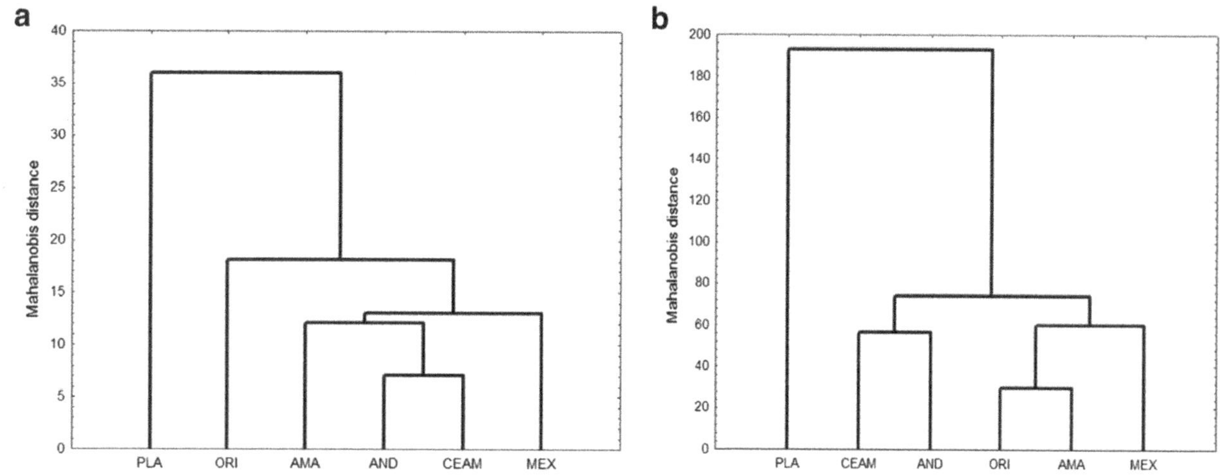

Fig. 6 UPGMA dendrogram of geographic samples based on the Mahalanobis distances matrix for the **a** mandible and **b** cranium. *MEX* Mexico, *CEAM* Central America, *PLA* La Plata, *AND* Western Andes, *AMA* Amazon, and *ORI* Orinoco

in size than those of the females possibly due to the larger energetic requirement of having a larger body (Sandell 1996; Thom et al. 2004). On the other hand, females do not engage in territorial disputes and are smaller, with lower energy requirements for daily maintenance. It is thought that the excess of energy is allocated to raising the litters (Moors 1980; Pertoldi et al. 1997; Gorman et al. 2006).

Geographic variation in the size and shape of the skull and mandible

The size of the mandible differs between GUs: animals from Mexico and La Plata have larger mandibles relative to those from the other GUs. These two GUs are the most northern (Mexico) and southern (La Plata) distributions of *L. longicaudis.* This variation is congruent with the size patterns reported for other species, like the American puma, *P. concolor,* which in addition to having a large distribution, like the Neotropical otter, also has a larger cranium size in temperate regions and with increasing latitudinal distance from the Equator (Iriarte et al. 1990; Gay and Best 1996). These patterns of change correspond to Bergmann's rule, which dictates that populations located furthest from the Equator have a relatively larger body. It is assumed that this pattern is the result of an adaptation that allows the animal to maintain its body temperature and confers greater endurance during aestivation (Blackburn and Hawkins 2004).

Cranium size does not follow the pattern observed for the mandible: the specimens from Mexico had a larger centroid size, while for the specimens from La Plata, it was smaller. In this study, only four samples were used from the La Plata region, so it is necessary to increase the sample size to determine whether this pattern would still be observed.

Our results indicated that significant variation in the shape of the cranium and the mandible of the Neotropical otter is associated with geography. In other species of Mustelidae such as weasels, European minks, and European badgers (Lee and Mill 2004), it is assumed that this type of variation responds to the variation in environmental conditions throughout the distribution of the species (Kruuk 2006). The morphology of a species responds to select factors in the environment, in addition to limitations in the response of the phenotype to such factors. This has led to the proposal that the ecological and geographic distribution of populations is a reflection of the ecological range of a phenotype (Ricklefs and Miles 1994). In contrast, it has been reported that geographic differences do not always lead to a homologous effect on the shape and size of an anatomical structure, in this case the cranium, due to the high levels of plasticity and adaptation in size that have been observed in other mammals (Cardini et al. 2007). Our data indicates that for *L. longicaudis,* there are changes in shape as well as size between the GUs studied, suggesting the existence of different selective pressures to which the shape and size of the cranium and the mandible respond and different means of adaptation.

In general, variation in shape corresponded to an increase in distance between the GUs. For example, the shortest Mahalanobis distances were between the Orinoco and the Amazon for the cranium and between Central America and the Western Andes for the mandible. In both cases, the GUs are adjacent hydrological regions and the proximity of the regions could allow for regional migration. The similarity between neighboring hydrological regions is exemplified by the genetic distance between the giant otter populations of *Pteronura brasiliensis,* located in the Amazon Basin, the Orinoco, and Guyana, suggesting that these populations originated from an ancestral population that inhabited the central region of the Amazon. Several populations of giant otter subsequently diverged and were partially isolated, although similarity may have been maintained between them by gene flow (Pickles et al. 2011). Likewise, for those GUs that are geographically most distant, such as Mexico and La Plata, greater Mahalanobis distances were also found for both structures. To elaborate, it was expected that these two regions would have the greatest differences in shape for the two views: not only due to geographic distance but also because these GUs are located at the limits of the Neotropical otter's distribution. Organisms that live at the limits of their distribution generally exhibit the greatest morphological change, as they are generally subject to suboptimal conditions that may challenge their limits of tolerance (Gay and Best 1996).

Morphological variation is not always associated with geographic distance, as in some cases it may be explained by the presence of different evolutionary units that are delimited by geographic barriers that successively give origin to vicariance events (Patton and Smith 1994; De Queiroz and Good 1997; Smith et al. 1997). This pattern is observed in the GUs of the Western Andes, for which there was less variation in shape compared with Central America than with the adjacent regions of the Orinoco and the Amazon. The Andes mountain range is a potential geographic barrier that may give rise to vicariance events.

The hypothesis of vicariance events caused by the Andes supports the genetic distances reported by Trinca et al. (2012), who found large genetic distances between the phylogenetic groups of South America and those of the Western Andes, where genetic distances are greater between populations of *L. longicaudis* than with its sister species, *Lontra provocax.* This separation also corresponds to a subspecies division: *L. longicaudis annectens* is distributed throughout the Western Andes and *L. longicaudis enudris* is distributed throughout the Eastern Andes (Larivière 1999). The effect of population separation by the Andes mountain range has also been reported for other vertebrate species including some South American fish (Hubert et al. 2007) and bats belonging to the genus *Artibeus* (Larsen et al. 2007), where the elevation of the Andes prevents gene flow and has resulted in different species on each side of the mountain range. The results of the present study offer morphological information that concurs with the molecular data from Trinca et al. (2012) on the separation of populations of the Eastern and Western Andes. This indicates that sufficient time has passed for both the morphological and the genetic divergence of the populations of *L. longicaudis* on both sides of the mountain range.

Another important pattern that reinforces the possibility of divergence events for the Neotropical otter was the notable differentiation in the shape of the mandible and cranium of the animals from La Plata with respect to those from the other five GUs. This reinforces the possibility of divergence events for the Neotropical otter. This dissimilarity for La Plata corresponds to the geographic group labeled Eastern South America and reported by Trinca et al. (2012), which was formed by a clade that does not share a single haplotype with the other clades that were reported on by the authors. This pattern of differentiation corresponds to the genetic variation between the populations of Pantanal, which belongs to the hydrological region of La Plata, and the other populations of the giant otter *P. brasiliensis* (Pickles et al. 2011).

In addition, this pattern of differentiation in shape may be influenced by the two biogeographically distinct regions in the La Plata Basin—tropical in the north and subtropical-temperate in the south—where climatic zoning has promoted changes in distribution and diversification patterns in both aquatic and terrestrial (Arzamendia and Giraudo 2009) species, and has possibly had the same effect on the Neotropical otter. Added to this effect is the presence of prairies that may be functioning as a barrier between the Amazon and La Plata during the dry season when bodies of water become scarce, preventing the Neotropical otter from migrating from one hydrological region to another.

Finally, it is important to address the pattern found for the Mahalanobis distance values and for the UPGMA, and the variation in the shape of the mandible and the cranium for specimens found in Mexico. These specimens showed no association with any of the other GUs, thereby indicating that the otters of this region could be under selective forces that are allowing for this differentiation. A similar pattern occurred in the morphology of the cranium of *Panthera onca* (Larson 1997), confirmed by molecular analysis (Eizirik et al. 2001), for which there is also a pattern of geographic division for populations in Mexico and Central America in comparison to the populations of South America. This pattern has also been found for the ocelot, *Leopardus pardalis*, and the margay, *Leopardus wiedii*, using molecular data (Eizirik et al. 1998), species which both have distributions similar to that of the Neotropical otter.

This supports the idea that the otters in Mexico and Central America have been separated from the more southern populations and are subject to a distinct set of selective pressures, resulting in important morphological variations that are reflected in the shape and size of the cranium and the mandible. On the other hand, since the Pacific and Gulf of Mexico slopes where *L. longicaudis* is distributed are separated by the Central Mexican Plateau, it is possible that the shape and size of the cranium also vary between the two coastlines. We recommend that a morphometric study be undertaken to compare populations from the two regions separated by the Trans-Mexican Volcanic Belt that bifurcates and creates the Pacific and the Atlantic ridge. This would allow for a better description of any vicariance event that has happened between these hydrological regions of Mexico.

The hypothesis that *L. longicaudis* is a species complex is supported by the geographic distances, the presence of geographic barriers, and the morphological adaptations to distinct environmental conditions that exist throughout its distribution, as well as by the congruence between the morphological differences we report here and the genetic information reported in Trinca et al. (2012). This hypothesis should be further tested using an integrative approach that includes phylogeographic information and geometric morphometrics with an appropriate number of sequences and representative sampling for the entire distribution of the Neotropical otter, especially in Mexico and the La Plata region. This would elucidate the current status of *L. longicaudis* as a possible group of species, as proposed by other authors based on the form of the rhinarium (Pohle 1920; Cabrera 1957; Harris 1968). The results of such a study would have important consequences for the knowledge and conservation of this species given that *L. longicaudis* is on the Red List of Threatened Species of the International Union for Conservation of Nature (IUCN) and is currently categorized as data deficient, due to uncertainty about the rate decline of its populations and debate over its taxonomic status (IUCN 2015).

Conclusions

The present results add new information about morphometric skull variation in *L. longicaudis*. There is significant variation in the size of the cranium between male and female Neotropical otters, with the males' crania the biggest. In contrast, the variation in shape is not significant between sexes.

The cranial and mandible shape analysis indicated the morphometric distance was greatest between the La Plata and the other GUs, whereas the smallest distance was found between the Western Andes and Central America for the cranium and Amazonia and Orinoco for the mandible.

For size variation, both views indicated that specimens from Mexico are the biggest relative to the other GUs, in concurrence with Bergmann's rule. The presence of geographic barriers and the large distances between the GUs could have given rise to the morphological variation observed in the Neotropical otter. The results of this study and previous research on genetic variability suggest that *L. longicaudis* could actually be a group of species and therefore requires an integrative study to clarify its taxonomy.

Competing interests

The authors declare that they have no competing interests.

Authors' contributions

PCHR conceived this study and acquired the data. PCHR and JAG coordinated this study and performed data analysis. PCHR, JAG, and CV interpreted the data and drafted and revised the manuscript. All authors read and approved the final manuscript.

Acknowledgements

This study was started with a Seed Fund Grant awarded by the State University of New York (SUNY) and funding from the Instituto de Ecología A.C. (INECOL). PCHR received a doctoral studies scholarship from CONACyT. Carla Gutierrez and Ella Vázquez contributed to this article with ideas and comments. We thank all the curators for their kindness and permission to use their *Lontra longicaudis* samples and for access to the following collections: National Collection of Mammals of the National Autonomous University of Mexico, the Mammal Collection of the Autonomous University of Campeche, the Vertebrate Collection of the Biological Station of the National Autonomous University of Mexico in Los Tuxtlas, the Mammal Collection of the University of Kansas, the American Museum of Natural History, the Mammal Collection of the Field Museum, the National Museum of Natural History of the Smithsonian Institute, the Carnegie Museum, and the Michigan University Museum. Bianca Delfosse revised the English.

Author details

[1]Instituto de Ecología A.C., Red de Biología y Conservación de Vertebrados, CP 91070, Carretera antigua a Coatepec 351, Xalapa, Mexico. [2]Facultad de Ciencias Biológicas, Universidad Autónoma del Estado de Morelos, C. P. 62210 Cuernavaca, Morelos, Mexico.

References

Adams DC, Rohlf FJ, Slice DE (2004) Geometric morphometrics: ten years of progress following the "revolution". Italian J Zool 71:5–16

Arzamendia V, Giraudo AR (2009) Influence of large South American rivers of the Plata Basin on distributional patterns of tropical snakes: a panbiogeographical analysis. J Biogeogr 36:1739–1749

Banguera-Hinestroza E, Cárdenas H, Ruiz-García M, Marmontel M, Gaitán E, Vázquez R, García-Vallejo F (2002) Molecular identification of evolutionarily significant units in the Amazon River dolphin *Inia* sp. (Cetacea: Iniidae). Journal of Heredity 93:312–322

Baryshnikov GF, Puzachenko AY (2011) Craniometrical variability in the cave bears (Carnivora, Ursidae): multivariate comparative analysis. Quat Int 245:350–368

Baryshnikov GF, Puzachenko AY (2012) Craniometrical variability of the Eurasian otter (*Lutra lutra*: Carnivora: Mustelidae) from the Northern Eurasia. Zoological Institute Russian Academy of Sciences 316(3):203–222

Blackburn TM, Hawkins BA (2004) Bergmann's rule and the mammal fauna of northern North America. Ecography 27:715–724

Blundell GM, Ben-David M, Bowyer RT (2002) Sociality in river otters: cooperative foraging or reproductive strategies? Behav Ecol 13:134–141

Bookstein EL (1991) Morphometric tools for landmark data: geometry and biology. Cambridge University Press, London

Bookstein EL (1996) Biometrics, biomathematics and the morphometric synthesis. Bull Math Biol 58:313–365

Bookstein EL (1997) Landmark methods for forms without landmarks: morphometrics of group differences in outline shape. Med Image Anal 1:225–243

Cabrera A (1957) Catálogo de los mamíferos de América del Sur. I (Metatheria-Unguiculata-Carnivora). Revista del Museo Argentino de Ciencias Naturales "Bernardino Rivadavia" e Instituto Nacional de Investigación de las Ciencias Naturales. Ciencias Zoológicas 4:1–307

Cardini A, Jansson AU, Elton S (2007) A geometric morphometric approach to the study of ecogeographical and clinal variation in vervet monkeys. J Biogeogr 34:1663–1678

Carvalho-Junior O, Macedo-Soares L, Birolo AB (2010) Annual and interannual food habits variability of a Neotropical otter (*Lontra longicaudis*) population in Conceicao Lagoon, south of Brazil. IUCN Otter Special Group Bulletin 27(1):1–15

Caumul R, Polly PD (2005) Phylogenetic and environmental components of morphological variation: skull, mandible, and molar shape in marmots (Marmota, Rodentia). Evolution 59:2460–2472

Chemes BS, Graudo RA, Gil G (2010) Dieta de *Lontra longicaudis* (Carnivora: Mustelidae) en el Parque Nacional El Rey (Salta, Argentina) y su comparación con otras poblaciones de la Cuenca del Paraná. Mastozoología Neotropical 17(1):19–29

De Queiroz K, Good DA (1997) Phenetic clustering in biology: a critique. Q Rev Biol 72:3–30

Duarte LC, Rabello ML, Von Zuben FJ, Furtado DS (2000) Variation in mandible shape in *Thrichomys apereoides* (Mammalia: Rodentia): geometric analysis of a complex morphological structure. Syst Biol 49(3):563–578

Eizirik E, Bonato SL, Johnson P, Crawshaw PG Jr, Vié JC, Brousset DM et al (1998) Phylogeographic patterns and evolution of the mitochondrial DNA control region in two Neotropical cats (Mammalia, Felidae). J Mol Evol 47:613–624

Eizirik E, Kim J, Menotti-Raymond M, Grawshaw PG Jr, O'Brien SJ, Johnson WE (2001) Phylogeography, population history and conservation genetics of jaguars (*Panthera onca*, Mammalia, Felidae). Mol Ecol 10(1):67–79

Freitas BCH, M-d L, Pissinatti A (1999) Cranial and mandibular morphometry in *Leontopithecus* Lesson, 1840 (Callitrichidae, Primates). Am J Primatol 48:185–196

Gallo-Reynoso JP (1997) Situación y distribución de las nutrias en México, en énfasis en *Lontra longicaudis annectens* MAJOR, 1897. Revista Mexicana de Mastozoologia 2:10–32

Gallo-Reynoso JP, Ramos-Rosas N, Rangel-Aguilar O (2008) Depredación de aves acuáticas por la nutria neotropical (*Lontra longicaudis annectens*), en el Río Yaqui, Sonora, México. Revista Mexicana de la Biodiversidad 79:275–279

Gay SW, Best TL (1996) Relationships between abiotic variables and geographic variation in skulls of pumas (*Puma concolor*: Mammalia, Felidae) in North and South America. Zool J Linn Soc 117:259–282

Goodall CR (1991) Procrustes methods in the statistical analysis of shape (with discussion and rejoinder). J R Stat Soc, Ser B 53:285–339

Gorman TA, Erb JD, McMillan BR, Martin DJ (2006) Space use and sociality of river otters (*Lontra canadensis*) in Minnesota. J Mammal 87(4):740–747

Gunz P, Mitteroecker P (2013) Semilandmarks: a method for quantitative curves and surfaces. Hystrix 24(1):103–109

Harris CJ (1968) Otters: a study of the recent Lutrinae. Weinfield and Nicholson, London
Hubert N, Renno JF (2006) Historical biogeography of South American freshwater fishes. J Biogeogr 33:1414–1436
Hubert JM, Duponchelle F, Nuñez J, Garcia-Davila C, Paugy D, Renno JF (2007) Phylogeography of the piranha genera *Serrasalmus* and *Pygocentrus*: implications for the diversification of the Neotropical ichthyofauna. Mol Ecol 16:2115–2136
Iriarte JA, Franklin WL, Johnson WE, Redford KH (1990) Biogeographic variation of food habits and body size of the America puma. Oecologia 85:185–190
IUCN (2015) The IUCN Red List of Threatened Species. Version 2015.1. <www.iucnredlist.org>.
Jonsson B, Jonsson N (2001) Polymorphism and speciation in Arctic charr. J Fish Biol 58:605–638
Kendall DG (1984) Shape-manifolds, procrustean metrics, and complex projective spaces. Bull Lond Math Soc 16:81–121
Kruuk H (2006) Otters: ecology, behaviour and conservation. Oxford University Press, Great Britain
Langerhans BR, Layman CA, Langerhans AK, Dewitt TJ (2003) Habitat-associated morphological divergence in two Neotropical fish species. Biol J Linn Soc 80:689–698
Larivière S (1999) *Lontra longicaudis*, Olfers 1818. Mamm Species 609:1–5
Larsen PA, Hoofer SR, Bozeman MC, Pedersen SC, Genoways HH, Phillips CJ, Pumo DE, Bajer RJ (2007) Phylogenetics and Phylogeography of the *Artibeus jamaicensis* complex based on Cytochrome-b DNA sequences. J Mammal 88(3):712–727
Larson SE (1997) Taxonomic re-evaluation of the jaguar. Zoo Biol 16:107–120
Lawing AM, Polly PD (2010) Geometric morphometrics: recent applications to the study of evolution and development. J Zool (Lond) 280:1–7
Lee S, Mill PJ (2004) Cranial variation in British mustelids. J Morphol 260:57–64
Lynch JM, O'Sullivan WM (1993) Cranial form and sexual dimorphism in the Irish otter *Lutra lutra*. Biology and Environment: Proceedings of the Royal Irish Academy 93B:97–105
Lynch JM, Conroy JWH, Kitchener AC, Jefferies DJ, Hayden TJ (1996) Variation in cranial form and sexual dimorphism among five European populations of the otter *Lutra lutra*. Zoology 238:81–96
Mancina CA, Balsero F (2010) Variación en la forma de la mandíbula en *Monophyllus redmani* (Chiroptera: Phyllostomidae): análisis geométrico de la variación sexual y geográfica. Mastozoología Neotropical 17(1):87–95
Mazák JH (2011) Craniometric variation in the tiger (*Panthera tigris*): implications for patterns of diversity, taxonomy and conservation. Mamm biol 75:45–68
Melquist WE, Polechla P Jr, Toweill D (2003) River otter. In: Feldhamer GA, Thompson BC, Chapman JA (eds) Wild mammals of North America: biology, management and conservation, 2nd edn. Johns Hopkins University Press, Baltimore, pp 708–734
Moors PJ (1980) Sexual dimorphism in the body size of Mustelids (Carnivora): the roles of food habits and breeding systems. Oikos 34:147 158
Patton JL, Smith MF (1994) Paraphyly, polyphyly, and the nature of species boundaries in pocket gophers (genus *Thomomys*). Syst Biol 43:11–26
Pertoldi C, Loeschcke V, Bo Madsen A, Randi E (1997) Developmental stability in the Eurasian otter (*Lutra lutra*) in Denmark. Ann Zool Fenn 34:187–196
Pertoldi C, Bo Madsen A, Randi E, Braun A, Loeschcke V (1998) Variation of skull morphometry of Eurasian otters (*Lutra lutra*) in Denmark and Germany. Ann Zool Fenn 35:87–94
Pertoldi C, Loeschcke V, Braun A, Bo Madsen A, Randi E (2000) Craniometrical variability and developmental stability. Two useful tools for assessing the population viability of Eurasian otter (*Lutra lutra*) population in Europe. Biol J Linn Soc 70:309–323
Pertoldi C, Sonne C, Wiig Ø, Baagøe HJ, Loeschcke V, Bechshøft TØ (2012) East Greenland and Barents Sea polar bears (*Ursus maritimus*): adaptive variation between two populations using skull morphometrics as an indicator of environmental and genetic differences. Hereditas 149:99–107
Pickles RSA, Groombridge JJ, Zambrana-Rojas VD, Van Damme P, Gotelli D, Kundu S et al (2011) Evolutionary history and identification of conservation units in the giant otter *Pteronura brasiliensis*. Mol Phylogenet Evol 61:616–627
Platt GS, Rainwater RT (2011) Predation by Neotropical otter (*Lontra longicaudis*) on turtles in Belize. IUCN Otter Specialist Group Bulletin 28(1):4–10
Platz S, Hertwig ST, Jetschke G, Krüger M, Fischer M (2011) Comparative morphometric study of the Slovakian wildcat population (*Felis silvestris silvestris*): evidence for a low rate of introgression? Mamm Biol 76:222–233
Pohle H (1920) Die Unterfamilie der Lutrinae. Eine systematisch tiergeographische studie an dem Material der Berliner Messen Archiv für Naturgeschichte 85:1–247
Ricklefs RE, Miles DB (1994) Ecological and evolutionary inferences from morphology: An ecological perspective. In: Wainwright PC, Reilly SM (eds) Ecological morphology: Integrative organismal biology. University of Chicago Press, USA
Rohlf FJ (1996) Morphometric spaces, shape components, and the effects of linear transformations. In: Marcus LE, Corti M, Loy A, Naylor G, Slice DE (eds) Advances in morphometrics. Plenum Publishing Corporation, New York, pp 117–129
Rohlf FJ (2001) TPSdig, version 1.23. State University at Stony Brook, New York, software
Rohlf FJ, Marcus LF (1993) A revolution in morphometrics. Trends Ecol Evol 8:129–132
Rohlf FJ, Slice D (1990) Extensions of the Procrustes method for the optimal superimposition of landmarks. Syst Zool 39:40–59
Ruiz-Olmo J, Delibes M, Zapata SC (1998) External morphometry, demography and mortality of the Otter *Lutra lutra* (Linneo, 1758) in the Iberian Peninsula. Galemys 10:239–251
Sandell M (1996) The mating tactics and spacing patterns of solitary carnivores. In: Gittleman JL (ed) Carnivore behavior, ecology, and evolution. Cornell University Press, Ithaca, New York, pp 164–182
Sheets HD (2002) Semiland6. A tool for processing semi-landmarks. Physics Department, Canisius College, Buffalo, New York, http://www.canisius.edu/~sheets/morphsoft.html
Sheets HD (2003). MakeFan6 [Online]. Available from http://www3.canisius.edu/~sheets/.
Sheets H D (2005a). CoordGen6, Coordinate Generation program for calculating shape coordinates. http://www3.canisius.edu/~sheets/morphsoft.html.
Sheets H D (2005b). CVAGen6, Canonical Variates Analysis program for the analysis of shape, based on partial warp scores. http://www3.canisius.edu/~sheets/morphsoft.html.
Sheets HD, Zeldich ML (2001) Tmorphgen6traditional morphometrics variables generation utility part of IMPIntegrated Morphometrics Package. Physics Department, Canisius College, Buffalo, New York, http://www.canisius.edu/~sheets/morphsoft.html
Smith HM, Chiszar D, Montanucci RR (1997) Subspecies and classification. Herpetological Review 28:13–16
StatSoft (2007) STATISTICA (data analysis software system). StatSoft, Inc, Tulsa, Oklahoma
Thom MD, Harrington LA, MacDonald DW (2004) Why are American mink sexually dimorphic? A role for niche separation. Oikos 105:522–535
Trinca CS, Thoisy B, Rosas F, Waldemarin HF, Koepfli KP, Vianna J et al (2012) Phylogeography and demographic history of the Neotropical otter (*Lontra longicaudis*). J Hered 103(4):479–492
Van Zyll CG (1972) A systematic review of the Nearctic and Neotropical river otters (Genus *Lutra*, Mustelidae, Carnivora). Royal Ontario Museum Life Sciences Contributions 80:112
Zelditch ML, Swidwerski DL, Sheets HD (2004) Geometric morphometrics for biologists: a primer. Academic Press of Elsevier, Second edition

Permissions

All chapters in this book were first published in ZS, by Springer; hereby published with permission under the Creative Commons Attribution License or equivalent. Every chapter published in this book has been scrutinized by our experts. Their significance has been extensively debated. The topics covered herein carry significant findings which will fuel the growth of the discipline. They may even be implemented as practical applications or may be referred to as a beginning point for another development.

The contributors of this book come from diverse backgrounds, making this book a truly international effort. This book will bring forth new frontiers with its revolutionizing research information and detailed analysis of the nascent developments around the world.

We would like to thank all the contributing authors for lending their expertise to make the book truly unique. They have played a crucial role in the development of this book. Without their invaluable contributions this book wouldn't have been possible. They have made vital efforts to compile up to date information on the varied aspects of this subject to make this book a valuable addition to the collection of many professionals and students.

This book was conceptualized with the vision of imparting up-to-date information and advanced data in this field. To ensure the same, a matchless editorial board was set up. Every individual on the board went through rigorous rounds of assessment to prove their worth. After which they invested a large part of their time researching and compiling the most relevant data for our readers.

The editorial board has been involved in producing this book since its inception. They have spent rigorous hours researching and exploring the diverse topics which have resulted in the successful publishing of this book. They have passed on their knowledge of decades through this book. To expedite this challenging task, the publisher supported the team at every step. A small team of assistant editors was also appointed to further simplify the editing procedure and attain best results for the readers.

Apart from the editorial board, the designing team has also invested a significant amount of their time in understanding the subject and creating the most relevant covers. They scrutinized every image to scout for the most suitable representation of the subject and create an appropriate cover for the book.

The publishing team has been an ardent support to the editorial, designing and production team. Their endless efforts to recruit the best for this project, has resulted in the accomplishment of this book. They are a veteran in the field of academics and their pool of knowledge is as vast as their experience in printing. Their expertise and guidance has proved useful at every step. Their uncompromising quality standards have made this book an exceptional effort. Their encouragement from time to time has been an inspiration for everyone.

The publisher and the editorial board hope that this book will prove to be a valuable piece of knowledge for researchers, students, practitioners and scholars across the globe.

List of Contributors

Cecilia Yanina Di Prinzio, Guillermo Omad, María Laura Miserendino and Ricardo Casaux
Consejo Nacional de Investigación en Ciencia y Tecnología (CONICET), Rivadavia 1917, 1033 Buenos Aires, Argentina
Centro de Investigaciones Esquel de Montaña y Estepa Patagónica (CIEMEP), Sarmiento 890, 9200

Esquel, Argentina
Laboratorio de Investigación en Ecología y Sistemática Animal (LIESA), Universidad Nacional de la Patagonia, Sarmiento 890, 9200 Esquel, Argentina

Ricardo Casaux
Instituto Antártico Argentino, Cerrito 1248, 1010 Buenos Aires, Argentina

Sameer Kumar Pati
Zoological Survey of India, Western Regional Centre, Vidyanagar, Sector 29, P.C.N.T. Post, Akurdi, Pune 411 044, Maharashtra, India

Arath Raghavan Sudha Devi
Department of Zoology and Research Centre, Mary Matha Arts and Science College, Vemom P.O., Mananthavady 670 645, Kerala, India

Aneesh Panakkool Thamban, Sudha Kappalli and Helna Ameri Kottarathil
Post Graduate Department of Zoology and Research Centre, Sree Narayana College, Kannur 670 007, India

Anilkumar Gopinathan
School of Biosciences and Technology, VIT University, Vellore 632014, Tamil Nadu, India

Trilles Jean Paul
UMR 5119 (CNRS-UM2-IRD-UM1-IFREMER), Equipe Adaptation Ecophysiologique et Ontogenèse, University of Montpellier 2, CC. 092, 34095 Montpellier, Cedex 05, France

Xianyuan Zeng, Junhua Ji, Yujiang Hao and Ding Wang
Institute of Hydrobiology, Chinese Academy of Sciences, 7 South Donghu Road, Wuchang District, Wuhan 430072, China

Xianyuan Zeng and Junhua Ji
University of Chinese Academy of Sciences, 19 Yuquan Road, Shijingshan District, Beijing 100049, China

Simone da Silva Ximenez
Instituto de Ciências Biológicas, Universidade Federal do Rio Grande—FURG, Av. Itália km 8, Rio Grande RS 96203-900, Brazil

Alexandro Marques Tozetti
Laboratório de Ecologia de Vertebrados Terrestres, Universidade do Vale do Rio dos Sinos—UNISINOS, São Leopoldo Av. Unisinos, 950, Cristo Rei, São Leopoldo RS 93022-000, Brazil

Zi-Ying Fu, Xing-Yue Dai, Na Xu, Qing Shi, Gao-Jing Li, Bin Li, Juan Li, Jie Li, Jia Tang and Qi-Cai Chen
School of Life Sciences and Hubei Key Lab of Genetic Regulation and Integrative Biology, Central China Normal University, Wuhan 430079Hubei, China

Philip Hung-Sun Jen
College of Special Education, Binzhou Medical College, Yantai, Shandong, China
Department of Life Science, National Taiwan Normal University, Taipei, China
Division of Biological Sciences, University of Missouri-Columbia, Columbia, MO 65211, USA

Chia-Hao Chang and Yeong-Shin Lin
Department of Biological Science and Technology, National Chiao Tung University, Hsinchu, Taiwan

Chia-Hao Chang
Department of Biology, St. Louis University, St. Louis, MO, USA

Yi Ta Shao, Wen-Chung Fu and Hong Young Yan
Sensory Physiology Laboratory, Marine Research Station, Academia Sinica, I-Lan, Taiwan

Yi Ta Shao
Present Address: Institute of Marine Biology, National Taiwan Ocean University, Keelung, Taiwan

Kazuhiko Anraku
Faculty of Fisheries, Kagoshima University, Kagoshima, Japan

Yeong-Shin Lin
Institute of Bioinformatics and Systems Biology, National Chiao Tung University, Hsinchu, Taiwan

Hong Young Yan
Hanse-Wissenschaftskolleg Institute of Advanced Study, Delmenhorst, Germany

Sofya Zvonareva, Yuri Kantor and Temir Britayev
Laboratory of Ecology and Morphology of Marine Invertebrates, A.N. Severtsov Institute of Ecology and Evolution, 33 Leninskij prosp, Moscow 119071, Russia

Xinzheng Li
Institute of Oceanology, Chinese Academy of Sciences, 7 Nanhai Road, Qingdao, 266071, China

J Delton Hanson and Stephen B Cox
RTLGenomics, 4321 Marsha Sharp Fwy Door 2, Lubbock, TX 79407, USA

Guillermo D'Elía
Instituto de Ciencias Ambientales y Evolutivas, Universidad Austral de Chile, Valdivia, Chile

Sheri B Ayers
True Health Diagnostics, Frisco, TX, USA

Santiago F Burneo
Museo de Zoología Pontificia Universidad Católica del Ecuador, Quito, Ecuador

Thomas E Lee Jr
Abilene Christian University, Abilene, TX, USA

Shanfeng Jiang, Yunfang Gao, Yangmei Zhang, Kun Liu and Huiping Wang
Key Laboratory of Resource Biology and Biotechnology in Western China (Northwest University), Ministry of Education, Xi'an 71069, China

Nandu Goswami
Institute of Physiology, Center of Physiological Medicine, Medical University Graz, Graz, Austria

Osvaldo Villarreal M and Adriano B Kury
Departamento de Invertebrados, Museu Nacional/UFRJ, Quinta da Boa Vista, São Cristóvão, 20.940-040 Rio de Janeiro, RJ, Brazil

Ricardo Pinto-da-Rocha
Departamento de Zoologia, Instituto de Biociências, Universidade de São Paulo, Caixa Postal 11461, 05422-970 São Paulo, SP, Brazil

Ram Kumar
Ecosystem Research Laboratory, Department of Zoology, Acharya Narendra Dev College (University of Delhi), Govindpuri, Kalkaji, New Delhi 110 019, India
Centre for Environmental Sciences, School of Earth Biological and Environmental Sciences, Central University of Bihar, Patna, India

Jiang-Shiou Hwang and Ram Kumar
Institute of Marine Biology, National Taiwan Ocean University, Pei Ning Road, Keelung 202, Taiwan

Priyanesh Muhid
Australian Rivers Institute, Griffith University, Nathan, QLD 4111, Australia

Hans-Uwe Dahms
Department of Biomedical Science and Environmental Biology, KMU - Kaohsiung Medical University, No. 100, Shin-Chuan 1st Road, Kaohsiung 80708, Taiwan

Jaigopal Sharma
Department of Biotechnology, Delhi Technological University, Delhi 110042, India

R Jayabarathi, G Padmavati and I Anandavelu
Department of Ocean Studies and Marine Biology, Pondicherry University, Brookshabad Campus, P. Bag. No. 01, Chakkargoan, Port Blair 744112, Andaman and Nicobar Islands, India

Bin Huo, Cong Xin Xie, You Jian Duan, Xue Feng Yang and Hai Ping Huang
College of Fisheries, Huazhong Agricultural University, Wuhan, Hubei 430070, China

Bao Shan Ma
Yangtze River Fisheries Research Institute, Chinese Academy of Fishery Sciences, Wuhan, Hubei 430223, China

G. Vineetha, N. V. Madhu, K. K. Kusum and P. M. Sooria
CSIR-National Institute of Oceanography, Regional Centre, Dr. Salim Ali Road, Kochi, India

Koji Mochida and Mamoru Toda
Tropical Biosphere Research Center, University of Ryukyus, Nishihara, Okinawa 903-0213, Japan

Wan-Yu Zhang
Graduate School of Engineering and Science, University of Ryukyus, Nishihara, Okinawa 903-0213, Japan

Wahiba Zaabar, Rym Zakhama-Sraieb, Faouzia Charfi-Cheikhrouha and Mohamed Sghaïer Achouri
S11UR11 Bio-Ecologie et Systématique Evolutive, Département de Biologie Animale, Faculté des Sciences de Tunis, Université Tunis El Manar, 2092 Manar II, Tunis, Tunisia

Prakash Kumar Paudel, Martin Hais and Pavel Kindlmann
Department of Biodiversity Research, Global Change Research Centre AS CR, Bělidla 4a, 603 00 Brno, Czech Republic

Prakash Kumar Paudel and Martin Hais
Faculty of Sciences, University of South Bohemia, Branišovská 31, 370 05 České Budějovice, Czech Republic

Pavel Kindlmann
Institute for Environmental Studies, Faculty of Natural Sciences, Charles University, Benátská 2, Prague, Czech Republic

Prakash Kumar Paudel
Current address: Center for Conservation Biology, Kathmandu Institute of Applied Sciences, PO Box 23002, Kathmandu, Nepal

L Mark Elbroch, Patrick E Lendrum and Howard Quigley
Panthera, 8 West 40th Street, 18th Floor, New York, NY 10018, USA

Jesse Newby
Craighead Beringia South, PO Box 147, Kelly, WY 83011, USA

Daniel J Thompson
Wyoming Game and Fish, 260 Buena Vista, Lander, WY 82520, USA

John Y Wang
CetAsia Research Group, Thornhill, Ontario L4J-7X1, Canada
Department of Biology, Trent University, Peterborough, Ontario K9J-7B8, Canada
National Museum of Marine Biology and Aquarium, Pingtung 944, Taiwan

Shih Chu Yang
FormosaCetus Company Ltd, Hualien 970, Taiwan

Samuel K Hung
Hong Kong Cetacean Research Project, Lam Tin, Kowloon, Hong Kong, China

Francisco Diogo Rocha Sousa and Sandro Santos
Núcleo de Estudos em Biodiversidade Aquática, Programa de Pós-graduação em Biodiversidade Animal Universidade Federal de Santa Maria – UFSM, Av. Roraima 1000, Camobi, Santa Maria CEP 97105-900, RS, Brazil

Francisco Diogo Rocha Sousa and Lourdes Maria Abdu Elmoor-Loureiro
Laboratório de Biodiversidade Aquática, Universidade Católica de Brasília – UCB, QS7 lote 1, Bloco M, sala 204, Taguatinga CEP 71966-700, DF, Brazil

Pablo César Hernández-Romero and Carolina Valdespino
Instituto de Ecología A.C., Red de Biología y Conservación de Vertebrados, CP 91070, Carretera antigua a Coatepec 351, Xalapa, Mexico

José Antonio Guerrero
Facultad de Ciencias Biológicas, Universidad Autónoma del Estado de Morelos, C. P. 62210 Cuernavaca, Morelos, Mexico